Alluvial Archaeology in Britain

Proceeding of a Conference sponsored by the RMC Group plc
3–5 January 1991, British Museum

Edited by
Stuart Needham and Mark G. Macklin

Published by
Oxbow Books, Park End Place, Oxford OX1 1HN

ISBN 0 946897 52 2

This book is available direct from
Oxbow Books, Park End Place, Oxford, OX1 1HN
(Phone: 0–865–241249; Fax: 0–865–794449)

and

The David Brown Book Company
PO Box 5605, Bloomington, IN 47407, USA
(Phone: 812–331-0266; fax: 812–331-0277)

Printed in Great Britain by
The Short Run Press, Exeter

Contents

Section 4: Archaeology and alluvial landscapes in the English Midlands: lowland river systems

Section 5: Archaeology and alluvial landscapes in southern and eastern Britain: lowland river systems

Concluding remarks

List of Figures

List of Tables

List of Contributors

I.K. Bailiff, Luminescence Dating Laboratory, Archaeology II, University of Durham, South Road, Durham DH1 3LE

Martin Bell, Department of Archaeology, St David's University College, University of Wales, Lampeter, Dyfed SA48 7ED

A.G. Brown, Department of Geography & The School of Archaeological Studies, The University of Leicester, LE1 7RH

Paul J. Burrin, Goldsmiths College, University of London, London SE14 6NW

A.J. Clark, 19 The Crossways, Onslow Village, Guildford, Surrey GU2 5QG

Patrick Clay, Leicestershire Archaeological Unit, Museums Annexe, 16 Humberside Drive, Leicester LE5 0RD

Bryony Coles, Department of History & Archaeology, University of Exeter, The Queen's Drive, Exeter EX4 4QH

P. Davies, School of History & Archaeology, University of Wales College of Cardiff, P O Box 909, Cardiff CF1 3XU

James Dinn, Archaeology Section, Hereford & Worcester County Council, Tetbury Drive, Warndon, Worcester WR4 9LS

J.G. Evans, School of History & Archaeology, University of Wales College of Cardiff, P O Box 909, Cardiff CF1 3XU

C.A.I. French, Department of Archaeology, Downing Street, Cambridge CB2 3DZ

Graham D. Keevill, Oxford Archaeological Unit, 46 Hythe Bridge Street, Oxford, OX1 2EP

M.K. Keough, Department of Geography, The University of Leicester, Leicester LE1 7RH

George Lambrick, Oxford Archaeological Unit, 46 Hythe Bridge Street, Oxford OX1 2EP

John Lewin, Institute of Earth Studies, University of Wales, Aberwystwyth SY23 3DB

John S.C. Lewis, Museum of London Archaeological Services, Museum of London, London Wall, London EC2Y 5HN

Susan Limbrey, Department of Ancient History and Archaeology, University of Birmingham, Edgbaston, Birmingham B15 2TT

Mark G. Macklin, Department of Geography, University of Newcastle upon Tyne, Newcastle upon Tyne, NE1 7RU

Richard Macphail, Institute of Archaeology, University College London, 31-4 Gordon Square, London WC1H 0PY

Nick Merriman, Museum of London, London Wall, London, EC2Y 5HN

R. Mount, School of History & Archaeology, University of Wales College of Cardiff, P O Box 909, Cardiff, CF1 3XU

Stuart Needham, Department of Prehistoric and Romano-British Antiquities, British Museum, London WC1B 3DG.

David G. Passmore, Department of Geography, University of Newcastle upon Tyne, Newcastle upon Tyne NE1 7RU

Mark Robinson, The University Museum, Parks Road, Oxford, OX1 3PW

Rebecca Roseff, Department of Archaeology, University of Birmingham P O Box 363, Edgbaston, Birmingham B15 2TT

Barbara T. Rumsby, Department of Geography, University of Newcastle upon Tyne, Newcastle upon Tyne NE1 7RU

C.R. Salisbury, 165 Tollerton Lane, Tollerton, Nottingham NG12 4FT

Robert G. Scaife, Department of Geography, University of Southampton, Southampton SO9 5NH

Richard Tipping, Nether Kidston Cottage, Nether Kidston Farm, By Peebles, Borders Region, EH45 8PJ

D. Williams, School of History & Archaeology, University of Wales College of Cardiff, P O Box 909, Cardiff CF1 3XU

Patricia E.J. Wiltshire, Institute of Archaeology, University College London, 31-4 Gordon Square, London WC1H 0PY

J.J. Wymer, The Vines, The Street, Great Cressingham, Thetford, Norfolk IP25 6NL

Preface

This volume represents in large measure the proceedings of a conference held at the British Museum from 3–5 January 1991, entitled "Archaeology under Alluvium: archaeology and the river environment in Britain". The conference arose out of a feeling that, whilst valley bottom deposits were increasingly being studied from a number of viewpoints, no forum had offered the prospect of reviewing the inferences being drawn, or the questions asked. Although they should not be divorced from broader landscape studies, valley bottoms have special features which affect landscape and archaeological record alike in a unique way and deserve special attention. From an archaeological perspective attention has already been drawn to various interrelated ecological zones and themes: on coastal change (Thompson 1980), the exploitation of wetlands (Murphy and French 1988), or on wetlands generally (e.g. Coles and Coles 1989; Coles 1992; Purdy 1988). Meanwhile fluvial geomorphologists have been investigating valley deposits in great depth and detail in order to model and explain change. A series of volumes are testimony to this (Gregory 1983; Gregory, Lewin and Thornes 1987) and we believe this volume will interdigitate with those cited and others.

The coverage of this volume is, as we have already indicated, an artificially circumscribed subject area, considered to have suffered relative neglect at large debating forums. The subject is conveniently called *alluvial archaeology* and shares the concerns and approaches of the geoarchaeology movement (Davidson and Shackley 1976; Butzer 1982). Whilst the study of colluvial archaeology (e.g. Bell 1983) is not within our remit, the volume contains reminders of the importance of colluvium as a source of sediment in river systems. Also to be excluded from detailed consideration are marine-laid deposits on coast-lines and in estuaries, since they have received specific treatment in the recent past (e.g. Eddison and Green 1988; Thompson 1980; Wilkinson 1989). Again, one cannot deny their interactive role in alluvial archaeology owing to the direct effect of sea-level changes on lower river gradients, and their consequent effect on the survival of floodplain topographies and the entrapped archaeology. Similarly, extensive fen zones such as those of East Anglia, the Somerset Levels or the Vale of Pickering have been played down, having received much attention (e.g. Coles and Coles 1986; *Somerset Levels Papers*; Hall *et al* 1987; Cloutman 1988a, 1988b), although there is no neat cut-off; such basins generally contain complex interleavings of fen deposits and alluvial material.

The conference was largely concerned with Holocene rather than earlier river systems. Partly this is a reflection of the burgeoning of recent work on Holocene alluvial archaeology, but we are also implicitly acknowledging the fact that there has been a long-standing geoarchaeological dialogue in Pleistocene studies (e.g. Wymer 1988; Schick 1986), a subject for which a mutual information base has been essential in formulating both natural and cultural sequences.

Inevitably perhaps there have been a few changes between the papers delivered orally and those committed to print here. Papers presented at the conference, but not present in this volume, are listed below. The papers by Wymer and Lambrick have been added.

Although many of the papers have thematic emphases, this does not provide a convenient basis for dividing the volume. Similarly, chronological scope was all too often very wide ranging and we were left with geography and physiography as the most obvious means of creating useful sections. Several contributions are however more strictly thematic and are covered in section 1, on dating and prospection, and section 2, on characterisation in river environments, which includes broadly based soil and biological studies.

The site-specific and regional contributions are split three ways for convenience, section 3 covering upland and piedmont rivers of northern and western Britain, whilst sections 4 and 5 deal respectively with lowland river systems in the Midlands and southern and eastern England. At this point we should acknowledge the tremendous regional imbalance in the contents of the volume (as in the conference itself). The upland river systems of northern and western Britain are very poorly represented. Although their geoarchaeological potential might generally be less (but see Macklin *et al* 1991), we do not believe this is the full story since relic alluvial landscapes frequently survive as terraces from many periods of the Holocene. We believe the imbalance may reflect in large measure the low density of fieldwork specifically directed to the issues with which we concern ourselves here.

The collection of papers we have been able to assemble is, nevertheless, testimony to a relative upsurge of interest in alluvial archaeology. A high proportion involves new fieldwork, very little of which has yet reached publication, but we also wanted to include some fresh overviews of continuing and prominent pieces of valley research, for which the integration of larger quantities of more diverse evidence allows greater interpretive scope. The conference was not intended merely to receive description and interpretation of the physical record of valley deposits in its various facets, but was also open to thoughts on priorities in the evaluation, exploitation and preservation of the landscapes in question. These issues come through in several of the papers.

Mark G. Macklin
Stuart Needham
April 1992

References

Bell, M. 1983. Valley sediments as evidence of prehistoric land-use on the South Downs. *Proceedings of the Prehistoric Society*, 49, 119–50.

Butzer, K. 1982. *Archaeology as human ecology*. Cambridge: University Press.

Cloutman, E.W. 1988a. Palaeoenvironments in the Vale of Pickering. Part 1: stratigraphy and palaeogeography of Seamer Carr and Flixton Carr. *Proceedings of the Prehistoric Society*, 54, 1–20.

Cloutman, E.W. 1988b. Palaeoenvironments in the Vale of Pickering. Part 2: environmental history at Seamer Carr. *Proceedings of the Prehistoric Society*, 54, 21–36.

Coles, B.J. (ed.) 1992. *The wetland revolution in prehistory*. Exeter: Department of History and Archaeology, University of Exeter; Prehistoric Society and Wetland Archaeological Research Project.

Coles, B.J. and Coles, J.M. 1986. *Sweet track to Glastonbury: the Somerset Levels in prehistory*. London: Thames and Hudson.

Coles, J.M. and Coles, B.J. (eds.) 1989. *The archaeology of rural wetlands in England*. Exeter: Wetland Archaeological Research Project.

Davidson, D.A. and Shackley, M.L. (eds.) 1976. *Geoarchaeology: earth science and the past*. London: Duckworth.

Eddison, J. and Green, C. (eds.) 1988. *Romney Marsh: evolution, occupation, reclamation*. Oxford: University Committee for Archaeology, Monograph no 24.

Gregory, K.J.(ed.) 1983. *Background to Palaeohydrology: a perspective*. Chichester: John Wiley.

Gregory, K.J., Lewin, J. and Thornes, J.B. (eds.) 1987. *Palaeohydrology in practice*. Chichester: John Wiley.

Hall, D., Evans, C., Hodder, I. and Pryor, F. 1987. The Fenlands of East Anglia, England: survey and excavation. In J.M. Coles and A.J. Lawson (eds.), *European wetlands in prehistory*, 169–201. Oxford: Clarendon Press.

Macklin, M.G., Passmore, D.G., Stevenson, A.C., Cowley, D.C., Edwards, D.N. and O'Brien, C.F. 1991. Holocene alluviation and land-use change on Callaly Moor, Northumberland, U.K. *Journal of Quarternary Science*, 6(3), 225–32.

Murphy, P. and French, C. (eds.) 1988. *The exploitation of wetlands*. Oxford: British Archaeological Reports, British series, 186.

Purdy, B.A. (ed.) 1988. *Wet site archaeology*. New Jersey: Telford Press, Caldwell.

Somerset Levels Papers 1975–89, vols. 1–15. Somerset Levels Project, Department of History and Archaeology, University of Exeter.

Schick, K.D. 1986. *Stone Age sites in the making: experiments in the formation and transformation of archaeological occurrences*. Oxford: British Archaeological Reports, International Series, 319.

Thompson, F.H. (ed.) 1980. *Archaeology and coastal change*. London: Society of Antiquaries, Occasional Paper (new series) 1.

Wilkinson, T.J. 1989. The archaeological survey of coastal and estuarine wetlands. *In* J.M. Coles and B.J. Coles (eds.) 1989, 23–26.

Wymer, J. 1988. Palaeolithic archaeology and the British Quaternary Sequence. *Quaternary Science Reviews*, 7, 79–98.

Acknowledgements

First and foremost we wish to thank the RMC Group Ltd, whose financial support considerably eased the burdens of floating the conference. In particular we would like to acknowledge the sustained interest of Richard Fox and Tony Fage, and to thank John Cooper for opening the conference. The smooth running of the conference and, beyond, the preparation of the publication volume owe much to the assistance of administrative and secretarial staff in the Department of Prehistoric and Romano-British Antiquities; we are conscious of their significant input. We also thank James Tullett of the computing section, Department of Scientific Research, for assistance with translation of magnetic media.

Several participants at the conference kindly did us the service of chairing sessions, conducting this all too difficult task with consummate skill and judgement. To Professors Leendert Louwe Kooijmans and Jean-Paul Bravard we offer especial thanks for their contributions, which injected a welcome sense of breadth to our otherwise insular preoccupations at the conference. Finally, we thank deeply all the unfortunates who undertook to help in the peer-refereeing process for the written manuscripts; their reward is in the assurance we can give that they have had a considerable collective impact on the contents of these proceedings.

Conference papers not represented in this volume

Lithostratigraphical and sedimentological models applicable to waterlain deposits, London.
Tony Barham and John Dillon

Airborne remote sensing of alluvial deposits in the Trent valley for the evaluation of past landscapes and potential archaeological sites.
J.M. Allsop and D. Greenbaum

Archaeological assessment and the role of developers in alluvial archaeology.
R.C. Turner

High Lodge — pre-Anglian drainage and human occupation.
J. Rose, S.G. Lewis, J.Cook and N. Ashton

Anslows Cottages, Berkshire: investigation of a palaeochannel and riverbank activity on the Kennet floodplain.
Sue Lobb

The fluvial site of Lugdunum (Lyon, France) in the context of Holocene changes of the river Rhône.
Jean-Paul Bravard

Archaeology under alluvium — the Dutch experience.
L.P. Louwe Kooijmans

1

Introduction to the volume

Stuart Needham and Mark G. Macklin

Two dating techniques were covered in the conference owing to their pioneering application to alluvial sedimentary environments in recent years. Although both luminescence and magnetic dating have limitations at the present, there is optimism from current progress that progressive refinements will be made. When the characteristics of deposits are favourable, these techniques are already proving to be useful in indicating broad periods of deposition, where other dating sources fail.

In chapter 3 Bailiff leads the reader through a labyrinthine complex of potential problems in obtaining ages from luminescence measurements on sediments. The problems relate generally to the physics of stored luminescence energy in different minerals under different conditions of sediment exposure, deposition and subsequent burial. Most have to be resolved at the laboratory stage, but specific sampling on site will obviate or mitigate some risks, as is well illustrated by Bailiff's reported work on coastal sequences which shows that it is necessary to sample sediments away from any interfaces with peat. He stresses that the understanding of the dynamics of deposition and subsequent burial processes is an essential prerequisite for successful dating.

Underpinning current directions in luminescence dating is the realisation that, while thermoluminescence (TL) is usefully applied to sediments (including those waterlain) beyond 15 ka, over shorter timespans its utility is seriously impaired by the uncertainties over residual dose which give a proportionately greater inaccuracy and imprecision. The more recently applied optically stimulated luminescence techniques (OSL) are however stepping into the breach for this younger time-scale, because of substantially improved zeroing characteristics which mean severely lessened residual doses.

In the choice of precise techniques to be employed, Bailiff considers the relative merits of feldspars against quartz, of different grain sizes, and of signal/response features at different points of the light spectrum. Of crucial relevance to our present subject area is the sizable effect water has on the transfer of background radiation into the storage traps (which is the source for the measurement). If the water content can not be calculated and allowed for, results can be significantly discrepant from the true age of deposition. Basically, to get valid results the hydrological history of the deposits needs to be known. Consequently it is likely that permanently waterlogged sediments will be more conducive to good dating than those intermittently waterlogged.

In the field of magnetic dating, Clark (chapter 4) is working with a better background of data. A calibration, or reference, curve exists for Britain (curves are regionally dependent) back to 10,000 cal BP, thus spanning effectively the Holocene. However, most of Clark's paper is devoted to evaluating apparent discrepancies between the existing curves based on lake sediment cores and various archaeomagnetic dating evidence. Some of the discrepancies seem to originate in deficiencies in the lake core data, others are explained as probably due to depositional environments. Primary in the latter category is the effect of particle shape on average inclination values; when elongate, particles will tend to settle out more horizontally than their magnetic axes would otherwise dictate. In the light of this, Clark finds it difficult to foresee the prospect of matching absolute values for inclinations, whereas probable errors in the declination values of the reference curves may well be steadily remedied by further archaeomagnetic data. It is already possible to propose a correction of 5 %W for the lake core reference curves to make them compatible with new land-based measurements.

Single-level dating can be achieved by multiple sampling, but becomes more prone to match-ambiguity the earlier the date. At present it is the *relative* changes of direction through a sequence that more often allow confident matching with the reference curves; this is already well shown for alluvial deposits amongst others. Even so, matching is not always easy, especially if sequences are discontinuous or experienced very variable rates of accumulation. The problem is compounded by the fact that the British reference curves lack the varve-chronometer of, for example, the North American, and suffer instead the vagaries and imprecisions of detrital radiocarbon measurement. Clark looks forward to the development of computer programs to objectivise the matching process and perhaps quantify the degree of fit.

A somewhat tangential point made in this paper, which is nevertheless acutely relevant to wider issues in alluvial archaeology, is that remnant magnetism in conjunction with magnetic intensity and susceptibility measurements can offer information about the history and nature of the sedimentation process.

Although not treated individually here, radiocarbon remains the mainstay of valley chronologies, as witnessed for example by its abundant use throughout this volume. But there is continuing concern, voiced for example by Macklin and Needham in chapter 2, and again by Bell in his concluding thoughts, over the dangers of residuality, and indeed contamination through water mobility, of the organic matter often relied on for dating purposes. This does mean that in many cases radiocarbon-dated sequences can only be regarded as providing approximate chronologies. Archaeologists and geomorphologists in Britain are perhaps lagging behind their North American and Australian colleagues in the use of small-sample AMS radiocarbon assay for dating Holocene alluvial sequences. This technique has considerable potential especially for dating the small organic inclusions within fine-grained flood sediments where the likelihood of residuality is very small (e.g. Prosser 1991).

Dendrochronology sees rather less frequent mention in this volume, reflecting the relative rarity of sites yielding wood capable of dendrochronological treatment. Bell (chapter 25), however, reminds us of the breadth of information that can on occasion come from dendrochronological studies, as well illustrated by the Alpine lake evidence.

The question of the recognition of the existence of archaeological sites in alluvial landscapes is central to a number of the papers, and some go further to touch on the explicit management plans being developed to deal with such intractable terrains when faced with destruction. Two crucial messages emerge from the volume: firstly, that the alluviated parts of valley bottoms contain a similar feature density to that well known on flanking terraces; secondly, that in the absence of guarantees that techniques of prospection will find all sites, those areas that are apparently archaeologically blank must be investigated to be sure of identifying the important hidden sites in adequate time and in a cost-effective manner.

The question of discovery is directly confronted by Clark's paper on archaeogeophysical prospection in alluvial terrains (chapter 5). He deals comprehensively with ground-based techniques which lend themselves best to mapping the underlying topography, identifying broad zones of significant activity, or giving advance indications of features and intra-site arrangements. Throughout Clark advises on the relative facility with which techniques may be applied and he concludes with some practical suggestions on the course to follow in given situations.

If there is pessimism about the potential of now conventional prospection techniques to probe substantial thicknesses of alluvium, it would appear to be not entirely well founded. Clark presents calculated values for hypothetical features at selected depths to show that features can be detected beneath surprising thicknesses of alluvium by gradiometers. Naturally, the deeper one is dealing with, the more the various survey techniques will discriminate against small and modest size features and show up only large cut features or sizable burnt structures. This does mean that we should be wary of a possible increasing selectivity in site type with depth amongst those brought to light by prospection techniques alone. Nevertheless, the prospects are encouraging for the detection of much buried archaeology in these often seemingly blank areas and the techniques outlined by Clark will, in appropriate combinations and backed up by site watching briefs and broader aerial reconnaissance (e.g. remote sensing as covered by Allsop and Greenbaum at the conference; see also Donnoghue and Shennan 1988), are surely indispensible first steps in any evaluation of alluvial tracts.

The loss of beavers through much of the British landscape somewhat before the 12th century AD has meant that they have been relatively neglected here as a potential agent of environmental change as compared with parts of Europe (where the animal survives in places to the present day). Coles (chapter 9), in drawing attention to the accumulated archaeological evidence for beaver, is able to argue that they were a persistent aspect of our valley fauna for the greater part of the Holocene. The existence of the animal, being witnessed increasingly by the distinctive teeth marks they have left on wood as much as by their bones, does not, however, answer the all-important questions surrounding their specific behaviour, which will determine their aggregate effect on valley conditions. Coles reviews the as yet tantalising evidence for specific beaver structures, hints of possible dams, lodges or canals, but looks towards future better directed research to confirm their existence.

Her example of contrasting patterns of fundamental

behaviour between neighbouring communities of beaver in the Narev river system in Poland is a warning that no conclusion for a given site, however firm, will necessarily be generally applicable. In addition to the search for hard physical evidence, Coles discusses the possibilities for perceiving beaver clearings from the palynological record.

Although in certain situations (e.g. the modern Polish one again) farmers and beavers can coexist in relative harmony (the farmers do take steps to limit disadvantageous effects on water levels etc), it is thought that more normally competition for land would have led in the past ultimately to the beaver being squeezed out. Attention is drawn to the benefits to farmers of ready cleared land and fertile soil provided by beaver meadows and silted, dried out beaver ponds. One might add mention here of the growing evidence that early agriculture was often being practised right down to the valley bottoms before metamorphosis led to more constraints on land use (see below).

The paper by Evans, Davies, Mount and Williams (chapter 7) is concerned with fauna that is a passive reflector, rather than a modifier, of habitats. The authors take a fresh look at the treatment of molluscan assemblages with the object of indexing immediate environmental conditions. To be sure of dealing with assemblages that are meaningful, the authors make use of a context-selective approach, thereby steering away from the uniform presentation of data as histograms which can imply that all is of equal weight or significance regardless of context, quantity or condition. This philosophy is undeniably important for the floodplain situation, where the allochthonous contribution is prone to be widespread and persistent. It does not reduce problems of interpretation, however, especially where no good modern analogues exist. However, Evans *et al* contend that the recurrence of associations under specified conditions will have empirical significance, and that other, non-molluscan, associated data may give vital clues about that significance.

Evans *et al* concentrate entirely on the fairly stable ground surfaces which suffered minimal input from flooding. This focus has the virtue of dealing with the surfaces that would have been occupied and farmed in the past and, if the various taxocenes formulated here stand the test of time (for the chalk-rich environment specified), the scheme promises subtle distinctions in the habitats borne by these low-lying surfaces. Such information is doubly important above the water-table where the insect evidence gives out and, furthermore, it bears on a tighter locality than either insects or pollen.

Scaife and Burrin (chapter 8) set out to address the issue of the significance of pollen extracted from inorganic alluvial sediments. One school of thought, that such material is unreliable as an indicator of contemporary vegetational compositions and changes, is questioned by comparing equivalent records from peat sequences. While their evaluation suggests that alluvial pollen is not necessarily the less reliable, they identify special additional taphonomic factors to be considered when interpreting such assemblages. Their further discourse on the ecological significance of organic versus inorganic accumulations in valley bottoms should serve to caution against any simplistic equations between these responses and their origins in vegetational history, precipitation, run-off, or base-level changes, particularly since their study valleys from a small part of Britain do not show uniform patterns. A similar point also comes through strongly from Robinson's comparisons across different Midland valleys (chapter 19). The small southern wealden valleys investigated by Scaife and Burrin have surprised many on account of the depths of Postglacial sediment present, much of it dating to later prehistory, for which period evidence for intensive land use is at present lacking. There is, however, a growing appreciation of significant settlement (e.g. Gardiner 1990) and this may account for the redeposition of material which appears to be largely loessic in ultimate origin.

Scaife and Burrin are also able to draw some deductions from their projection of long valley profiles. For example, the trend for older fills to lie downstream and later fills to reach progressively further up river. This may well be applicable to other southern valleys, but as a trend-model, it should not be allowed to obscure sight of local variations due to vagaries in topography or land use.

Although Limbrey's paper (chapter 6) draws almost exclusively on her detailed soil and sediment studies in a Wiltshire chalk valley, it illustrates well the wide range of inferences that such studies offer. She makes it plain that, revolutionary though micromorphological studies may be, they are merely the apex of a tier of methods.

The evidence that can come through for cultivation is fundamental, particularly since such evidence can be wholly exiguous where fields were unbounded, or ploughmarks never left discernible fills. Limbrey found cultivation evidence associated with two flood-free phases (Neolithic and Romano-British to Medieval). Soil studies may also suggest by implication the extent of surface stripping caused by flooding, or what soil processes took place prior to hydrological changes, or the rapidity of overbank accumulation (as judged by the absence of decalcification), all details imperative to our understanding of the survival, distribution and nature of archaeological sites.

With regard to the water meadows clayey alluvium and the classic agricultural economy known to be associated, Limbrey perceives a conundrum in the balance between water contributions in the winter from flooding as opposed to adjacent-slope run-off. The integrated land-use system outlined depended on the relative warmth of the floodwaters to encourage early growth of the grass, yet the available evidence suggests a substantial input from run-off, and therefore colder water.

One major preoccupation with the calcareous valley deposits is the origin of the frequently high calcium carbonate component. This will have a bearing on many other things, such as the rate of erosion of source soils, or transfer mechanisms for calcium within the catchment. Limbrey reviews a surprising array of biomineralisation processes involved in the production or reprocessing of calcium. She is inclined to the view that some of these made a substantial contribution to calcareous floodplain deposits, a conclusion that could have drastic implications for the extent of up-valley erosion or our ability to gauge the fluvial conditions of deposition by means of particle size analysis.

Few though the contributions on upland rivers may be, they illustrate a range of very detailed work being undertaken in those environments. Lewin sets the stage (chapter 10) by reviewing the aspects of upland Holocene sequences that have drawn most attention, and stating those that have been neglected. His main thesis, however, is to emphasise that we should not be looking for norms in the past behaviour of British rivers, even over comparatively small distances. Much weight is certainly given to this by his investigations in the Vyrnwy valley, where four very divergent patterns can be demonstrated over a 10 km reach of a single valley. He explains how these will give rise to correspondingly divergent geoarchaeological potentials.

Lewin's introductory passage gives an outline of sedimentary processes and the resulting characteristics, with special reference to upland rivers. There follow certain insights into, *inter alia*, the hydrological conditions which lead to the deposition of available sediment. Crucial to recognise is that competence-limitation is especially important in the case of coarser bed sediments, whilst supply-limitation is generally the more important with finer materials. Nevertheless, recognising reduced inter-dependence does not make judgement on origins and causation the easier, as Lewin is keen to stress with regard to discerning climatic and anthropogenic effects from the sedimentary record, especially when the two are interacting.

Tipping expresses concern in chapter 11 about the ways in which geomorphic changes in the valley context, particularly aggradation, are linked to potentially causative environmental change. He urges the consideration, wherever possible, of proxy palaeoenvironmental records which document the *preconditions* to the aggradations under study. His study material is drawn from the high-energy river systems draining the Cheviot massif where he focuses on alluvial gravel deposition which has been shown to mark dramatic changes in the fluvial sedimentary record at surprisingly late dates in the Holocene. The dating evidence permits the suggestion of two major episodes of high-energy fluvial activity (ca. 3500 and 2000 BP), both of which appear to be triggered by major phases of woodland clearance witnessed in several local pollen sequences. Consideration of the problems of time-lag involved in the formation process leads to differing conclusions on the relative chronology of the coarse aggradations and the immediately proximate pollen records. Well shown is the potential for fuller understanding of developments by looking at the different manifestations of environmental change (vegetation, clearance, sedimentary instability etc) independently in several localities within a region. Attention is also paid to possible correlations with the settlement record in that zone, but the challenge here is now for archaeology to individualise the settlement histories of the various catchments to test alluvial models.

Immediately to the south, another study (chapter 12) is concerned with the Tyne drainage system, which also rises in part in the Cheviots. Macklin, Passmore and Rumsby tackle the problem of causes of alluviation and, equally important, erosion throughout this grand catchment, from upper to lower reaches, giving interesting contrasts. These authors set out a range of geomorphological, environmental and cultural elements that need to be elucidated in order to enable the respective contributions of anthropogenic and climatic factors to be adduced. In the Tyne basin these preconditions have already been met to a reasonable degree, thus allowing Macklin, Passmore and Rumsby to construct a complex interaction of possible causes and effects for their Holocene sequence. While some of the patterns they highlight are similar to those elsewhere in Britain, others are more surprising, for example, the lateness of the radical change in the alluvial regime found in the upland reaches of the Tyne. There is reason to belive that many British uplands, including the Cheviots and the Pennines, saw significant clearance during the second millennium BC, which poses the question of what circumstances prevented a perceptible increase in run-off during that period, but allowed it on a very noticeable scale later. Explanation might come from the scale and location of clearance, the use of the land, or climatic change.

Macklin *et al* have also been able to model certain hydrological changes on the basis of the behaviour of the Tyne in the past century. Two periods which experienced floods of greater than normal magnitude and frequency are known to have caused significant incision.

Dinn and Roseff's case study at Wellington Quarry on the river Lugg (chapter 13) illustrates a good combination of prior prospection and characterisation followed by selective excavation and successful negotiation for partial preservation. The dating evidence so far points to the thick alluvial deposit of three major units representing accretion over relatively restricted timespans when seen against the duration of the Holocene. One resulting conclusion is that one part of the floodplain was stable for several thousand years. Just such a situation has been recognised at other sites and it will doubtless come to be a familiar phenomenon, although the scale of survival will have depended on later floodplain history.

Dinn and Roseff outline in their conclusion a staged evaluation scheme now operative in Herefordshire, one similar to those set up previously in Cambridgeshire and Berkshire. The impact of the new focus on valleys in their region is perhaps given the more emphasis because it is not an area that has enjoyed a high level of fieldwork in the past; nevertheless, the fact that 20% of valley enclosures currently identified are situated on alluvium is truly impressive, a statistic which surely mirrors that for sites generally in the major river valleys of eastern England (see for example Lambrick; Robinson; French, Macklin and Passmore; Merriman).

Salisbury and Clay are concerned in their respective papers (chapters 14 and 15) with different aspects of the Trent valley around Hemmington. Nominally well inside 'lowland' Britain, this river is clearly atypically powerful and active. This seems to result, in the reaches studied in detail at any rate, in the regular movement of coarse sediments including gravel. This gives rise to particularly difficult juxtapositions of gravel fills against gravel banks, with concomitant lowering of prospects for mapping information by means of prospection prior to destruction. Conditions are not uniform, however, even in the Hemington area, where there is a difference in the degree of reworking of the gravels in the meander belt just above and just below the Derwent/Trent confluence.

Salisbury shows how systematic observation and documentation over many years is the only way of building up a comprehensive picture of a set of valley deposits faced with wholesale destruction. He goes on to set out the kinds of sources, such as cartographic and historical documents, which are indispensible in comprehending the later history of our valleys. His example of an land exchange deed of 1402, which promised also the fixing of the river by piling and planting, leaves us in little doubt about the direct influence wrought on the riverscape by man from early times. Indeed similar influences are suggested by Needham at Runnymede in late prehistoric times (chapter 23). Salisbury's work also documents the physical evidence of this interference, in the form of many piled structures, doubtless of varied origins including river bank revetment.

This theme of interference also underlies Clay's case-study (chapter 15), an inevitability given the context of mills and their dams. His citation of the remit of a Royal Commission in the late 14th century gives us a vivid impression of the riverscaping and extent of water control that must have been enforced by the 'many weirs, mill dams, pales and kiddles fixed or rased in the water of the Trent impeding the passage of ships.' This illustrates rather clearly the regular conflict of interests between economic benefit directly to the immediate hinterland (ready source of power; fish supply) and that to the region and state (trade) before the imposition of gate-locks partially reconciled these needs.

The large rivers of the south and east Midlands are the common thread running through five papers. Geographically there is much overlap, with both the upper Thames of Lambrick and the Nene of Brown and Keough covered also by Robinson in his overview, which takes in the Ouse as well. However, the contents of the respective papers show quite divergent approaches and themes.

Robinson's comprehensive overview (chapter 19) deals with a range of issues across the three study valleys (Upper Thames, Nene, Ouse) with the object of comparing combined histories of hydrology, sedimentation and settlement. Using a range of archaeological and biological evidence, he is able to show non-synchronous occurrence of rising groundwater levels (documented by levels of organic preservation in deep features of different periods), consequent flooding and the (not synonymous) major phases of alluviation from valley to valley. He concludes that different conjunctions of topography, environment and human land use explain the catchmentised scale of these factors.

Within his summary of the archaeological sites of these Midland valleys, Robinson deduces variation in their functions. Although there was patently considerable anthropogenic *effect* on these systems as a consequence of settlement and land use, there was probably little overt *control* until Roman times. Similarly, the reclamation associated with suburbs spreading onto floodplains at Oxford, Bedford and Northampton, (redolent of many Roman and Medieval waterfront reclamations and encroachments into channels), epitomises a generally late trend: human alteration of the valley landscape in the face of more dramatic changes to channels and persistent flooding, in this region at least.

The Upper Thames area avowedly saw the first emergence of a concerted look at a British floodplain landscape with archaeological eyes. Lambrick's paper (chapter 20) builds on much that has gone before, summarising current perception of stages of development in the history of the alluviated zone and culminating in some forward-looking issues to be addressed. Throughout he argues for a predominantly anthropogenic causation for the changing patterns identified. Amongst other things it is suggested that the character of the alluvium may have altered in response to Roman encroachment onto the clay lands.

Using the Medieval evidence from the Upper Thames valley, Lambrick picks up on the potential of floodplain archaeology to investigate social circumstances rarely suspected otherwise; for example, the contrast in diet evident between Blackfriars and the general urban populace, or again, the suggestion that utilisation of the floodplain in the later Iron Age was geared to a rather specific socio-economic structure, in which horse-raising played an important part.

He is also concerned with the significance of the newly won archaeological experience in valley bottoms and cites briefly the results of a combined TWA/OAU survey to evaluate the discovery potential of floodplain and terrace. There is a plea for the further monitoring of dewatering effects on adjacent sites that are nominally 'preserved' amidst otherwise archaeology-emptied landscapes. On the question of ancient water levels there is also a surprise finding that the tributary Windrush had an unexpectedly high water-table in the Early to Mid Holocene when compared to the Thames, yet another cautionary to attempting wide-scale hydrological generalisations.

Studies on the Nene and its tributary the Soar (chapter 18) are set by Brown and Keough within a broader conceptual framework for the geoarchaeological potential of valleys during the Holocene. They see this potential as being very uneven in response to the frequency of channel avulsion, lateral channel migration and floodplain groundwater conditions. In the catchments under study in this paper, for example, this entails less potential during the late Early Holocene.

Brown and Keough talk of a metamorphosis of floodplains and channels during the Holocene and believe that at times in later prehistory some changes must have taken place on human timescales. The anastomosing channel patterns thought to be a dominant form in lowland systems at earlier stages of the Holocene were, they suggest, partly inherited from Lateglacial braided systems. The fact that changing configurations were largely brought about by avulsions rather than significant migration was probably due to the increasing load of fine silt carried by the system. Not only did this alter the immediate environmental conditions of the water itself and the banks, but it also led to the plugging of some channels and thus a trend towards reduction in the number of concurrent channels. Accompanying palynological studies are also summarised and include an extensive programme of radiocarbon dating of selected wood species from their sedimentary sequences. The dominance of alder in the Middle Holocene, and the later rise in scrubby species are becoming familiar themes in valley studies.

The other Midland contributions, by Keevill (chapter 17) and French, Macklin and Passmore (chapter 16), give insights into the wealth and diversity of archaeological remains that just small patches of these fertile valleys can yield when systematically targeted for archaeological attention. Both in the Stanwick/Raunds area on the Nene (Keevill) and at Etton/Maxey on the Welland system (French *et al*) the prehistoric record is dominated by concentrations of Neolithic to Early Bronze Age monuments, ranging from causewayed enclosure, henge and cursus to frequent round barrows. These are supplemented by more elusive categories of site such as the midden dumps at Etton or the brushwood dump at West Cotton, reminiscent of that at Runnymede. One suspects that with more intensive fieldwork in valley bottoms more such exiguous but vital counter-balancing evidence will materialise to help give a more rounded picture of the organisation of valley activities. One poignant message from these distributions is that, although there may be evidence from the precise siting of monuments of some acknowledgement of the subtle pre-alluvial topographies, they convey a landscape which was not highly differentiated between flood zone and drier terraces, a point driven home by Robinson in his paper. It would be wrong to extrapolate too widely from these instances, but nevertheless there are prospects for the existence of many similar land surfaces dipping under later Holocene alluvium, just as also known beneath marine silts in certain coastal zones (Wilkinson 1989).

Keevill goes on to summarise the impressive later sequence at Redlands Farm, Stanwick, spanning Roman and Early Medieval: the transition from mill to villa agricultural complex in Roman times, and then later the contraction and decline of villa building, accompanied by new architectural elements, the sunken huts. All this can be set against the changing tide of the alluvial regime which led to increasing marginality by the late phase of settlement. From about 600 AD the deposition of a thorough covering of alluvium put the stone-rich villa remains out of the sight of those involved in the resurgence of building in stone, another facies of the benevolent preservation of these environments.

One of the main preoccupations for French, Macklin and Passmore is the study of the local palaeochannel systems; much can be elucidated from air photographs alone in areas like East Anglia. Different channel morphologies in juxtaposition in a very restricted reach of the river system must imply very different flow conditions and rates, and in turn a significant time lapse. On the strength of these data the authors are able to advance a scheme of fluvial evolution and can even posit the interrelationships of the neighbouring archaeological sites, providing models to test by excavation.

Wymer's review of the yield of archaeological information from alluvial contexts of the Palaeolithic (chapter 21) is perforce focused on the southern counties of England where terraces of various ages have escaped wholesale glacial destruction. Ironically, alluvial deposits, which are so prone to the vagaries of erosion, are one of few situations to yield consistently stratified Palaeolithic archaeology in this country. Wymer, however, examines in some depth the significance of the best known stratified groups. The taphonomy of the lithic artifacts and, where it survives, of the bone as well, is shown to be a central ingredient of such studies, but is not a universal solution to problems of interpretation. It is now thought, partly on the strength of experimental work, that artifacts in a more rolled condition do not necessarily imply a greater age than those which are relatively fresh. Wymer takes us carefully through all of the key sites, assessing each on its individual merits. The assemblages most confidently regarded as *in situ* tend to lie in fine-grained alluvium, as might be predicted, but this should not become dogma. Often a given site contains differently characterised assemblages which have undergone different transformations, sometimes attributable to humans, more usually to natural agencies. The deductions on the formation processes at Swanscombe led to the conclusion that the cultural material probably derived from a nearby bank, having been transported by perhaps just one flood. In his discussion on the Furze Platt complex near Maidenhead, Wymer again emphasises the potential importance of material in secondary context, particularly when it transpires that transportation has been relatively limited. Understanding of alluvial sequence has always been acknowledged to be central to Quaternary studies, and it is clear from the review presented here that further refinement is necessary not only to reconstruct accurately sites and localities, but also to get assemblages into correct temporal order.

Three papers deal with different parts of the lower Thames basin based on specific site studies (Lewis, Wiltshire and Macphail; Needham; Merriman). Far from presenting a coherent picture, they yet again show how just one basin (albeit large by British standards) can experience different alluvial histories and thereby focus on quite distinct archaeological-environmental records. That the Runnymede (Needham, chapter 23) and Three Ways Wharf, Uxbridge (Lewis, Wiltshire and Macphail, chapter 22) sequences have most in common is only to be expected from their relative proximity. Both have sizeable alluvial bodies of Early to Mid Holocene, and both show rich *surface-intact* archaeological horizons. While different emphases on the time periods represented could perhaps be dismissed as a result of the particular areas of investigation within each locality, different responses to particular situations, such as the Boreal rapid sea-level rise, require explanation. However, the organic response at Three Ways Wharf as opposed to the inorganic one at Runnymede are not readily explained by the obvious characteristics of their respective catchments and may be due to topography on a rather local scale.

The explorations at Uxbridge have provided the first outstanding opportunity in modern times to build on Lacaille's earlier work on the Colne valley deposits and their contained archaeology. Detailed palynological and soil micromorphological studies are the mainstay of this paper, and the palaeoenvironmental reconstructions drawn are interrelated with the evidence for Early Holocene settlement in the Colne valley. The possibility of a shift in the settlement pattern is discussed; so too is the problematic occurrence of horizons rich in comminuted charcoal, a phenomenon well known elsewhere in this time bracket.

The complex alluvial sequence at Runnymede Bridge (chapter 23) is formulated in terms of alluvial parcels. Much of the paper is devoted to outlining and explaining this sequence of channel movements and variable overbank deposition in the context of episodes of human activity and environmental change in the catchment. Special attention is given to a coarse overbank aggradation unit (gravel), dating to ca.4000 BP. Although such deposits are not totally unknown in lowland Britain, (gravel spreads also being reported in alluvial sequences in Wiltshire valleys by Limbrey -this volume), they constitute evidence of severe flooding.

Merriman's selected sites in the central London area (chapter 24) inevitably show dominant influences in the later stages of the Holocene from sea-level changes in the form of both fresh-water and then saline depositional environments. Good correlations seem to be possible across this zone, although care is being taken to date independently each major exposure before confirming wider conclusions. Despite the limited extent of exploration under the alluvium, a variety of archaeological contexts is in evidence: marshy carr zones, typically prone to spasmodic human activity, foreshore situations, and low-altitude banks, which were nevertheless dry and able to support occupation and even arable agriculture. Given the diversity and evident quality of evidence in the London sites, Merriman argues that their location cannot be left entirely to chance and advocates the collation of the existing borehole information and systematic trial trenching to give a topographic baseline which is central to any planning.

The ardmarks at Phoenix Wharf, along with other such well preserved remains of ploughing in the upper Thames at Hamel and Drayton (Lambrick), or in the Nene valley at Warren Villas (Robinson) — here remarkably with associated waterlogged seeds — deserve comment. To have chanced upon such instances at this relatively early stage of serious alluvial archaeology surely indicates a massive potential for the recovery of *in situ* early agricultural evidence in these valley bottom settings. These locations often have the joint benefit of having been basically dry land surfaces at some stage during the Holocene and yet having been protected from the depredations of more recent interference by coverings of alluvium.

Bell draws the volume to a close (chapter 25) with a review of many issues which the conference raised (both overtly and tacitly) salted with his own thoughts on future research directions in alluvial archaeology. He is concerned that the alluvial record should neither be divorced from consideration of the broader landscape, nor be taken necessarily as representative of it. Particular taphonomic factors characterise the alluvial record, but the depth of their effects, he suggests, will depend on catchment size. Bell explores the often perceived opposition between anthropogenic and climatic roles in shaping the alluvial landscape, placing the debate furthermore in the context of contemporary environmental concerns with associated funding implications. This is a salutary warning against the subject becoming prey to political designs, however indirectly.

On the question of anthropogenic inputs, he argues for further evaluation of rates of soil erosion, its pre-conditions and character. While woodland clearance and agricultural disturbance dominate discussion by most workers, the effects of other human activities, and indeed of fauna such as beaver, aurochsen and red deer, should not be overlooked. Bell is not optimistic on the subject of pin-pointing causative factors, especially when fine chronological control is lacking. He believes that where fluctuations in the controlling variables are not extreme, significant changes may only register geomorphologically under specific land-use regimes. He argues that different thresholds for registration may well apply between different parts of Europe, and he also illustrates the value of inter-continental comparisons, which may help in isolating the more influential variables.

The benevolent situation where dendrochronology-yielding tree remains survive on a site is held up as a beacon. By virtue of their contexts and contained ring sequences, such trees can offer several interlocking viewpoints on valley histories.

Despite the undeniably catastrophic effect climate can on occassion have on river regimes and human populations, Bell bemoans the dearth of detailed documentation for individual British floods (historical timescale). In the section on hazard response Bell introduces us to some important concepts, such as coping mechanisms, which are the means by which society buffers itself against adverse conditions. These are as variable as society itself and caution us against uniform projections of human response to environmental changes or 'events'.

Finally and fittingly Bell joins others in his concern for the very survival of the resource base for alluvial archaeology. The extent to which future work can build on these formative offerings will depend absolutely on the success with which the resource is managed by all interested parties.

References

Donnoghue, D. and Shennan, I. 1988. The application of multispectral remote sensing techniques to wetland archaeology. In P. Murphy and C. French (eds.), *The exploitation of wetlands*, 47–59. Oxford: British Archaeological Reports, British Series, 186.

Prosser, I.P. 1991. A comparison of past and present episodes of gully erosion at Wangrah Creek, southern Tablelands, New South Wales. *Australian Geographical Studies*, 29(1), 139–54.

Wilkinson, T.J. 1989. The archaeological survey of coastal and estuarine wetlands. In J.M. Coles and B.J. Coles (eds.), *The archaeology of rural wetlands in England*, 23–26. Exeter: Wetland Archaeological Research Project.

2

Studies in British alluvial archaeology: potential and prospect

Mark G. Macklin and Stuart Needham

The strong interest that both geomorphologists and archaeologists have in the alluvial landscape is not new. Neither is collaboration between specialists of very disparate backgrounds. Yet we feel that the last ten years or so have seen a surge in the exploration of alluvial environments accompanied by more regular collaboration across the relevant disciplines. Prior to that such ventures were rather sporadic and took the form of isolated local studies, though we would not wish to diminish their pioneering importance. Nor should we forget developments outside Britain, notably in North America (e.g. Bryan 1941; Hack 1942; Waters 1988) and the Mediterranean (e.g. Bailey *et al* 1990; Bintliff 1975; Higgs and Vita-Finzi 1966; Lewin *et al* 1991; Pope and Van Andel 1984: Vita-Finzi 1969, 1975).

We cannot overstress that there is no separate subject of *alluvial archaeology*, as commented in the preface. Consideration of this very specific environmental circumstance is aimed at exploring ways of cross-fertilising our studies and resolving methodological and practical issues. The ultimate goal should be to interlink floodplain activity, both human and geomorphological, with that in the hinterland — on terraces and slopes — thereby taking account of the catchment or territory as a whole. Despite the fact that we are aiming for a composite view, we find that current preoccupations in alluvial archaeology are still most conveniently dealt with according to whether they have a primarily cultural or primarily natural environmental orientation; hence the major sections below.

Comprehension of soils, sediments and topography is fundamental to all archaeology, so what, if anything, is different or unusual about alluvial archaeology? The special circumstances are perhaps to be found in:

i) the potential for large-scale natural inputs into the record on a variety of time-scales,
ii) the potential for the high mobility and sometimes great displacement of those material inputs,
iii) and the potential for excellent preservation of organics below water- tables.

Of course there is also general acknowledgement that river valleys, like certain other topographic zones, often give rise to deeply stratified sequences of sediments which may contain archaeological material. It is clear now that alluvial archaeology can include whole sites and fragments of early landscapes, and that, where these are well sealed by later deposits thereby escaping certain destructive agencies such as the plough, they may be in good condition. The rewards anticipated from this degree of preservation lie in rich combinations of artifacts, ecofacts and structures, which offer prospects for meaningful associations and sequences based on clear stratification of contexts.

The potential high mobility of sediments and objects in the floodplain environment is legendary and needs no amplification. This might at first seem to work against the positive benefits of multiple associations within stratified sequences. This is a valid concern and it forces us more than ever to consider the origins of the various recovered organisms and materials. It is an especially acute concern perhaps for the biological data. However, once ways have been found of disentangling the autochtonous from the allochtonous components — using such techniques as differential condition, biases between sexes, age groups or different anatomical parts, and the depositional history of the enveloping sediment — then there is actually a prospect of reaping a double reward: we gain an understanding of the landscape in the catchment, as well as in the immediate site environs. Obviously 'catchment' itself will be something variable according to the nature of the particular branch of evidence being considered in conjunction with the geomorphological processes in train.

Assessment of the movement of artifacts under fluvial conditions is equally vital (e.g. in Butzer 1982), though not necessarily easily predicted, as has become clear from experimental work, such as that by Schick (1986) or Harding *et al* (1987). It emerges that artifacts need to be regarded as rather specialised particles, and each weight/shape category may have its own pattern of displacement. Not only does this have implications for gross 'dispersal' (as defined by Butzer 1982: 100), but also for micro-movement such as the 'sinking' of certain objects into a river bed, pitching, or the closing up of small clusters of material under certain conditions (Schick 1986: 43–59). These kinds of observations will become crucial if we are to understand properly the detailed positions of artifacts on excavated sites. Schick also documents closely the kind of assemblage modifications that can take place for Quaternary assemblages (Schick 1986: 60–89). More such experimental work on a wider variety of materials and artifact types needs to be carried out, both in the the laboratory (flume experiments) and in the field.

The geomorphological history of an archaeological 'site' is a positive asset. The causes of natural aggradation, truncation, incision will tend to lie away from the site. The fact that there are likely to be some, or many, geomorphological changes to a valley floor within the human time-scales that concern us means that we can interrelate the effects of natural and anthropogenic activities across a broad terrain with those at a specific location. One accepts that there may be some time-lag to be allowed for. Nevertheless, by investigating the site we are also probing the hinterland.

CULTURAL ORIENTED PERSPECTIVES

Something needs to be said about the contrast between rural and urban environments in alluvial archaeology. Urban archaeology inevitably has a close association with major rivers and the great amount of work undertaken during the 1970's and 1980's has yielded much information on the land/water interfaces (Milne and Hobley 1981; Herteig 1985; Good *et al* 1991). The rewards may be exemplified perhaps by London, where many excavations on both the City and the Southwark shores have contributed to reconstructions – particularly for the Roman period – of water margin topographies, attendant water levels and estimation of tidal reach (e.g. Brigham 1990; Milne *et al* 1983; Yule 1988). These things, added to the intrinsic interest of the archaeological structures themselves, are vital background information for a better understanding of the siting of *Londinium*, its communication routes, or its suitability as a port.

Whilst urban archaeology does have a lot to offer our specific subject, its deep stratified sequences can have a significant contribution from make up of direct human origin, even where they directly border rivers. For activity sites away from urbanisation, however, the balance shifts markedly towards the predominance of natural origins, and consequently fluvial geomorphology here has a more overt part to play in interpretation. There are inevitably exceptions to this urban/rural dichotomy, particularly where large-scale industries in rural settings have generated large bodies of material and deposited them in valley bottoms. Such is the case at Cwmystwyth, mid Wales, where thick beds of rock waste from 18th and 19th century lead mining were dumped in the valley and have since in part been reworked by the river (Lewin *et al*, 1977); they cannot have failed to alter the hydrology locally.

Clearly the analytical and interpretative processes associated with alluvial archaeology are fraught with difficulties. However, the overall results can be far reaching, and they may be further strengthened by the frequent survival of a diverse range of environmental and structural evidence; amongst this diversity one line of evidence can act to check and qualify the interpretations drawn from others. The benefits of this interplay between different kinds of information will lead to multi-dimensional understanding of valley morphology and attendant environment. Since these factors are likely, in turn, to be major factors in governing site location and type, then this brings us full circle in having direct ramifications on cultural interpretations.

A framework for formation and survival in alluvial archaeology

Prediction of the archaeological potential of river valley deposits is hazardous since we are often dealing with well-masked topographies and rapid variations in geomorphological process. Obviously it is vital to take into account all available information from previously studied exposures in conjunction with any evidence of broad formations from prospection, whether aerial or ground-based (many of the latter are dealt with by Clark, chapter 5 of this volume). A broad framework for temporal and spatial change might thus be gleaned, thereby allowing hypotheses to be developed on the period sequences likely to be present in the given location. Any previously established archaeological horizons in the vicinity might well add refinement.

Such direct evidence, however, is almost invariably going to be patchy and we therefore need to consider on a more theoretical level what governs *formation* and *survival*, and what characteristics of the cultural archaeological record are likely to be encountered in given situations. It is useful to try to treat formation and survival separately, although at times this is not easy as they are intertwined parts of a particular hydrological regime.

On formation, the character of the alluvial archaeological record is determined by both cultural and geomorphological operators working in concert. It is useful to think in terms of different topographic elements of valley floor and floodplain which broadly govern human activities and much else besides. For convenience a tripartite division is shown here to schematically represent the key elements that might be present in an alluvial topography (at altitudes below any raised terraces):

i) 'dry' banks, i.e. normally a dry land surface, but not beyond the reach of floods,
ii) water margins, this to include backwater swamps and floodplain wetlands, and
iii) channels.

Figure 2.1 illustrates how this zonation might affect the character of both structural and artifactual material. When we are considering the kind of activities which often depend on the driest bank – particularly those to do with occupation – then we are likely to see minimal natural accretion around the resulting residues during the activity. This means that conflation of phases will be a normal feature of such material, except where there was a significant man-made contribution in the accumulation of deposits, as for example can happen with large-scale refuse dumping. Otherwise, some limited temporal sequencing might only be reflected under very specific conditions associated with biologically-caused downward migration. There are a number of problems here: only a relatively short-term sequence would actually be spread through the worm-sorted zone, any earlier material generally having collected around the lower boundary; for part of the sequence to be preserved at all the faunal turbation responsible would have to be curtailed at or very soon after the close of human activity; finally, there is a likelihood of differential downward movement according to size etc

'DRY' BANKS	WATER MARGINS AND FLOODPLAIN WETLANDS	ACTIVE CHANNELS
***In situ* structures (ie. buildings and dug features):**		
Permanent settlement (if flood risk considered acceptable) Seasonal encampments and specialised structures (eg. religious)	Waterfrontages - revetments, wharves, barriers, etc. For beaching craft For water-dependent pursuits - eg. milling, saltings, large-scale cooking Consolidation for occupation or activity	Land-ancillary - eg. jetties, bridges For specialised subsistence activities - eg. fish weirs For water control - eg. weirs, constrictions, locks Non-functional - eg. water cult Suspended settlement structures
Artifacts - *in situ*:		
Relating to any type or intensity of activity	'Primary' refuse from certain types of activity and deliberate dumping of 'secondary' refuse - survive *in situ* if favorable aggradational conditions	Rather limited materials, tending to be heavy items - eg. net weights, punt irons, weapon/tool losses and ritual deposits
← *wide ranging possibilities*		*limited possibilities* →
Artifacts - reworked:		
Generally limited and erratic contribution dependent on hydrological conditions, topography and artifact morphology/density Potentially derived from wide area	Beach-stranded and lag-stranded river-worked material; bank collapse material	Very varied material and histories before final incorporation in stable deposits
← *low proportion of bank assemblages*		*high proportion of channel assemblages* →
Aggradation rates and consequent chronometric control:		
Usually rather slow; more rarely rapid bursts Generally limited vertical separation of phases Nevertheless, sequence potentially representative of lengthy period	Variable and episodic; sometimes rapid Often sequences with very fine chronological resolution However, surviving sequences likely of relatively short duration	Variable and episodic; often rapid Often sequences with very fine chronological resolution However, surviving sequences likely of short duration

***Fig. 2.1.** The main archaeological features of different zones within alluvial topographies.*

of the material, and therefore the danger of false associations at specific horizons within the full artifact-bearing zone.

Obviously during periods when regular increments of flood silt were sufficient to heighten the floodplain surface on human timescales, then there would be far more scope for the preservation of true sequences. Such hydrological circumstances are likely, however, to have affected the way in which floodplains were used, in particular limiting permanent occupation. One drastic solution could have been to raise the floors of buildings (cf. Coles and Coles 1992), but this in itself would have implications for the nature of the archaeological record formed, since many activities would no longer be represented by material *in situ* in the ground contemporary ground surface.

'Channel margins' pose more varied formation conditions. Formation potential in a river channel is good when its border is aggrading. The nature of aggradation itself will clearly have profound implications for contained artifacts or ecofacts. Times of high flow (not necessarily beyond bank-full stage) could result in severe disturbance of material recently dropped on the surface. In this way there could be introductions of extraneous material and subtractions of indigenous material, as well as the redistribution of the remaining debris.

Gentler flow under standard flow conditions would be more conducive to the preservation of material in its primary, culturally-determined resting place, although one should not rule out a degree of 'winnowing' and consequent deflation of the cultural elements. Persistent gentle aggradation would then help by giving undistorted sequences, free from major and differential natural transformations. Prolonged aggradation of this kind is most likely to occur in subsidiary or cutoff channels, or on lake shores, in all cases with the precondition that there was a continuing sedimentary input. It might also, however, occur on the convex banks of slowly migrating main channels where seasonal flow was never very turbulent during the given depositional sequence, or in more artifically-protected margins, such as behind the waterfront at Runnymede (Needham 1991: 58–67, 114).

River margins thus offer good prospects generally for creating sequences featuring chononometric sensitivity, especially when aggradation is significant on human time-scales (e.g. > 2 mm per annum on average). However, understanding such sequences will not necessarily be easy; they may contain alternations of accretion, stasis and even erosion, or have the effects of significant faunal turbation superimposed on the broader formation trends (for instance by burrowing ostracods).

Active channels away from margins will be deprived of many of the advantages of the bank-side zone. Aggradation, whether on river bed or channel bars, will tend to be transient until such time as that branch of the river becomes slack causing the formations to fossilise. Insofar as these formations survive, they may have very similar potential to that of the margins for chronometrically fine sequences; however, their potential for the entrapment of primary archaeological material is rather different and would appear logically to focus on non-land based activities. Except in the rare circumstance of settlement structures suspended above the open water, one would normally expect culturally-ejected refuse from settlements to have arrived in channel contexts through fluvial reworking. To recognise the secondary deposition of eroded material does not necessarily rob it of archaeological significance; it might answer important questions, particularly when interlinked with other local evidence. For example, diagnostic artifacts could give a valuable *terminus post quem* for a phase of erosion from a site nearby. On the other hand, the channels will have recieved primary residues from certain activities (notably sunken boats and their cargoes, ballast, fishing weights etc) and similarly specialised structural remains.

There is then a degree of symbiosis between the zones we have defined, not only in terms of the original human utilisation, but also in terms of the depth and character of the archaeological record created. It follows that the best combined alluvial archaeological results will stem from studies where it has been possible to locate a range of topographic elements representing a particular period river system and associated floodplain; but this of course depends on their having survived.

On the question of survival, it is self-evident that various aspects of rivers and their attendant hydrological regimes will severely condition the nature and indeed the very survival of useful archaeological data (as well documented in this volume – e.g. Lewin, chapter 10). The chains of cause and effect are complex. Much of the serious destruction that floodplain relicts and their entrapped archaeology have experienced will have resulted from progressive lateral erosion. Also significant will have been the surface erosion caused by the first floods to succeed the particular activity on the floodplain, echoing Butzer's comment for archaeological sites more generally: " The greatest dispersal hazard generally coincides in time with site burial and tends to be a function of the net energy crossing the site surface."(1982: 100). Crucial factors here will be: the time-lapse between the activity and the first floods and whether this has allowed material to become incorporated into a soil profile; the severity of the floods and how much scouring they caused; and the success with which they deposited sufficient load to protect the archaeological deposits from later scouring. According to these factors, it is likely that some or all of a cultural assemblage on a floodplain will suffer reworking. Occasionally everything might remain essentially *in situ* (excluding micro-dispersals); this would be most likely where a lengthy period of nil or minimal disturbance had allowed the cultural remains to become well incorporated in the soil profile, and where succeeding flooding was not accompanied by any significant surface stripping. The corrollary of these 'ideal' conditions for *in situ* survival on the floodplain, is however that the ground surface in question was untroubled by ravaging floods for some while and in consequence that surface would be highly suitable for renewed dry-land cultural processes, either those destructive of *in situ* situations (e.g. ploughing) , or those causing cultural additions (e.g. settlements) and leading potentially to a degree of conflation between materials of different phases. There is therefore perhaps something of a trade-off between the benefits from sites totally undisturbed fluvially and those rapidly sealed, albeit with the risk of some reworking. On a rather different aspect of survival, concerning the submerged parts of sites, changes in water-tables are of paramount importance, as these will critically affect the level of organic preservation and therefore the quality of information retrievable (Coles 1986; French and Taylor 1985).

Channels and their margins will obviously be very vulnerable to periodic scouring due to either seasonal fluctuations or changes in river regime. Even relict channels seem often to have been reactivated at a later stage so that former fills may suffer wholesale or large-scale erosion. This suggests that any useful sequences formed in river depositional contexts may be more prone to reworking than those on dry banks.

Another aspect of survival is that in general terms there will be a higher potential for the preservation of archaeological surfaces and structures on the downstream side of islands or meander lobes. Upstream sides, on the other hand, would generally suffer the brunt of erosional forces. On those occasions when, due to sudden hydrological changes, the upstream banks aggraded, then there would be a tendency to have marked discontinuities preserved in the sequence. At Runnymede, for example, a sequence of Post-medieval channel silts are directly juxtaposed aqainst a deep prehistoric sequence of Mesolithic and Neolithic silt accumulation. Meanwhile there is a better prospect of 'downstream' sequences containing a number of temporally close juxtapositions as point-bars or similar lateral accretionary bodies.

Degrees of association

From the foregoing it should be clear that we need to refine as far as possible our characterisation of *in situ* versus reworked artifacts and of autochthonous versus allochthonous organisms, and, furthermore, refine the chronologies of the sequences we are dealing with, both with respect to periods of formation and those of stasis. Indeed these things are interlinked, for the *in situ* status of material, whether biological or cultural, is crucial to judgement on whether sequences of data have a significant temporal dimension. At present the vast majority of radiocarbon determinations and artifact associations for alluvial sequences provide nothing more than a *terminus post quem* for the formation of the layer they were in (for an example of how discrepant results can be in a lake-bed sequence, see Edwards and Rowntree 1980: 215 fig.15.5). In the early stages of an alluvial archaeological project such dating may be adequate, but as research proceeds the interpretative interplay between natural and anthropogenic causes and effects will demand better and better controls over relative and absolute dating, because successful interpretation will depend on the minimisation of anachronisms. Hopefully, means of direct scientific dating of the sediments themselves will become suitably refined and confident. In the meantime it is necessary to pay close attention to the contextual integrity of the material being used for dating.

Aspects of chronology are often closely tied to preservation; this is especially so at stratigraphic discontinuities. Obviously it is desirable, if optimistic, to be able to identify and quantify the duration of a hiatus. For understanding the buried archaeology, however, it is of crucial importance when such a hiatus occurs immediately above an *in situ* horizon. All too often the initial impression may be of a well preserved site simply because it is well sealed under alluvial sediments. In the respect that this might have prevented damage from modern deep ploughing, or root systems, such sites are indeed subject to good preservation. However, such land surfaces may frequently have suffered earlier agriculture, before being sealed and protected (as at Phoenix Wharf -Merriman, this volume). Similarly, intensive researches on the Essex shore have now shown that Neolithic sites within that land surface were exposed to natural post-depositional transformations for two thousand years before inundation began (Wilkinson and Murphy 1988: 134). In all situations there is a necessity to look for signs of these processes through rigorous study of stratigraphy, structures and artifact condition and mixing. This is axiomatic in all good archaeological practice today, but if ignored in alluvial archaeology the dangers of misinterpretation are prodigious.

NATURAL ENVIRONMENT ORIENTED PERSPECTIVES

In the second part of this paper we want to briefly review some of the major themes which have been recurrent in alluvial archaeological studies in Britain amongst fluvial geomorphologists over the last decade or so, and then go on to critically evaluate opportunities and obstacles emerging in this field. Five themes are chosen for special consideration: (i) Holocene alluvial histories; (ii) dating Holocene alluvial sequences; (iii) Postglacial channel and floodplain typologies; (iv) palaeoflood analysis and (v) provenancing alluvial sediments. All of these focus on the Holocene period for the reasons outlined at the beginning of the preface. Sections (i), (ii) and (iii) provide a general overview of Postglacial river histories and environments in Britain, whilst sections (iv) and (v) consider two relatively recent methodological innovations in alluvial archaeology.

British Holocene river erosion and sedimentation histories

There is an ongoing and hotly contested debate, which was highlighted by a number of speakers, as to the roles and relative importance of human activity as opposed to climate in Holocene river development in Britain. Such controversy, however, is not restricted to Britain, as shown by the climate and landuse protagonists of Holocene alluviation in the Mediterranean (e.g. Vita Finzi 1975; Van Andel and Zangger 1990).

Efforts to resolve this apparent conundrum have been given new impetus by anticipated future climate change associated with global warming. With Britain's long history of ecosystem and landscape disturbance by humans, and the smaller magnitude of climate change in the Holocene compared to earlier periods of the Quaternary, it has become conventional wisdom amongst most archaeologists and many earth scientists that British rivers have played what one might term a "passive" role in Postglacial alluviation. For example, a large proportion of extant river basin landforms and sediments are directly inherited from the Pleistocene or earlier periods, and many British rivers appear to have been comparatively ineffective in re-fashioning their valley floors during the Holocene. Thus the timing and magnitude of river sedimentation, particularly over the Late Holocene in southern and eastern Britain, has been considered to reflect directly the scale, type and history of catchment disturbance, (Brown and Barber 1985; Burrin and Scaife 1985; Robinson and Lambrick 1984). The apparent non-synchronous nature and regional pattern of alluviation has been taken to exclude an underlying climatic control (Bell 1982; Burrin and Scaife 1988).

Macklin and Lewin (in press) have recently explored this theme in a paper that examined the Holocene river alluviation record in Britain. In this they attempted to systematically review all relevant papers (more than a hundred) published in both the archaeological and geomorphological literature up to 1989. Although this was an interim survey, it was possible with some confidence to: (i) identify significant episodes of Postglacial alluviation; (ii) establish regional and national alluviation events and their degree of synchrony and (iii) evaluate links between floodplain sedimentation, Holocene climate, vegetation and landuse change. In Figure 2.2 alluvial units (dated by radiocarbon, archaeological material, pollen analysis or soil development) are plotted on a latitudinal basis from north to south, together with their thickness and lithofacies type. 5000 BP appears to mark a significant change in Holocene river activity in Britain with sedimentation recorded for the first time since the Late-glacial on many valley floors, as far afield as northern Scotland (river Etive) and the south coast of England (rivers Cuckmere and Ouse). Alluviation in most valleys began around 4800 BP but had slowed or ceased by 4200 BP. Between ca. 3800 and 3300 BP significant gravel deposition occurred in a number of British rivers, most notably in southern England within the tributaries of the Great Ouse and Lower Thames. Towards the end of the period (ca. 3400 BP) alluviation is recorded on the Avon, Ouse and Cuckmere floodplains. Furthermore, it is also evident that in catchments south of the Severn fine-grained alluvial units, considered to be associated with human catchment disturbance, generally pre-date similar post-settlement alluvium (in more northern river basins), which appear to have been deposited after 3000 BP.

Major alluviation occurred in southern Britain between 2800 and 2400 BP; more limited valley floor sedimentation has also been documented at this time in Northumberland and Cumbria. The periods 2000 to 1600 BP and 1200–800 BP represent probably the most important phases of Holocene river activity in Britain. In upland catchments of north west England (Harvey and Renwick 1987) and northern Scotland (Brazier and Ballantyne 1989) these periods were characterised by extensive fan gravel deposition. In the last 800 years, however, there is a conspicuous regional dichotomy in river activity: alluviation dated to between 800–400 BP is on the whole confined to river valleys south and south-east of the Trent, whilst in a number of northern British rivers significant gravel depositions are documented for the last 400 years.

Since October 1989 subsequent investigations in the Severn basin (Brown 1988; 1990), river Dane, Cheshire (Hooke *et al* 1990) and river Aln, north-east England (Macklin *et al* 1991) have confirmed that in the periods 4800–4200, 3800–3300, 2700–2400, 1900–1600, 800–600 and 400–250 BP significant alluviation or accelerated river channel change occurred in many British river valleys. Gaillard (1985), Karlen (1991), Starkel (1991), and Macklin *et al* in this volume (chapter 12), have shown that these episodes coincide with marked hydroclimatological change in Britain, Scandinavia and northern Europe. It is important to emphasise, however, that the record of Postglacial alluviation compiled by Macklin and Lewin (in press) for Britain only highlights accretionary events, and it is much more difficult to date phases of increased lateral channel movement, floodplain incision or erosion that are represented by hiatuses in the alluvial record. The timing, causes and, in particular, rates of channel and valley floor degradation are especially poorly researched in Britain, although recent investigations of river behaviour over the last 300 years in the Northern Pennines (Macklin, Rumsby and Heap 1992) have shown channel incision and enlargement to be associated with an increase in flood magnitude and frequency, controlled primarily by non-random decadal-scale fluctuation in rainfall. This suggests that major episodes of channel cutting and filling during the Holocene, particularly those in gravel-bed rivers of the British uplands, are most likely to reflect significant changes in catchment runoff.

It still remains difficult to draw firm conclusions concerning the relationship between Holocene fluvial activity in Britain and climatic or anthropogenic controls. Precisely dated evidence from key sites is still lacking; questions of preservation and bias, dating difficulties (discussed below), and the necessarily generalized comparisons between the occurrence of dated alluvial episodes and environmental controls do urge some caution in interpreting British Postglacial alluvial sequences. Nevertheless, this new overall view of the evidence does suggest a shift in emphasis and the two following conclusions are ventured.

(a) The widespread synchrony of Holocene fluvial episodes in Britain, particularly over the last 5000 years, suggests that climate may have more directly controlled river sedimentation and erosion than has hitherto been generally believed. The relative degree and nature of fluvial response, however, varied considerably between river basins reflecting catchment physiography, vegetation and land-use histories, and availability and calibre of sediment. But broad parallels in Postglacial river activity evident in Britain, northern Europe, Scandinavia and especially the USA (Knox 1983), would appear to underline the importance of large-scale, sub-hemispheric climate change during the Holocene.

(b) Differences in the response of British river systems to climatic fluctuations in the Early Holocene compared to those in the Middle and

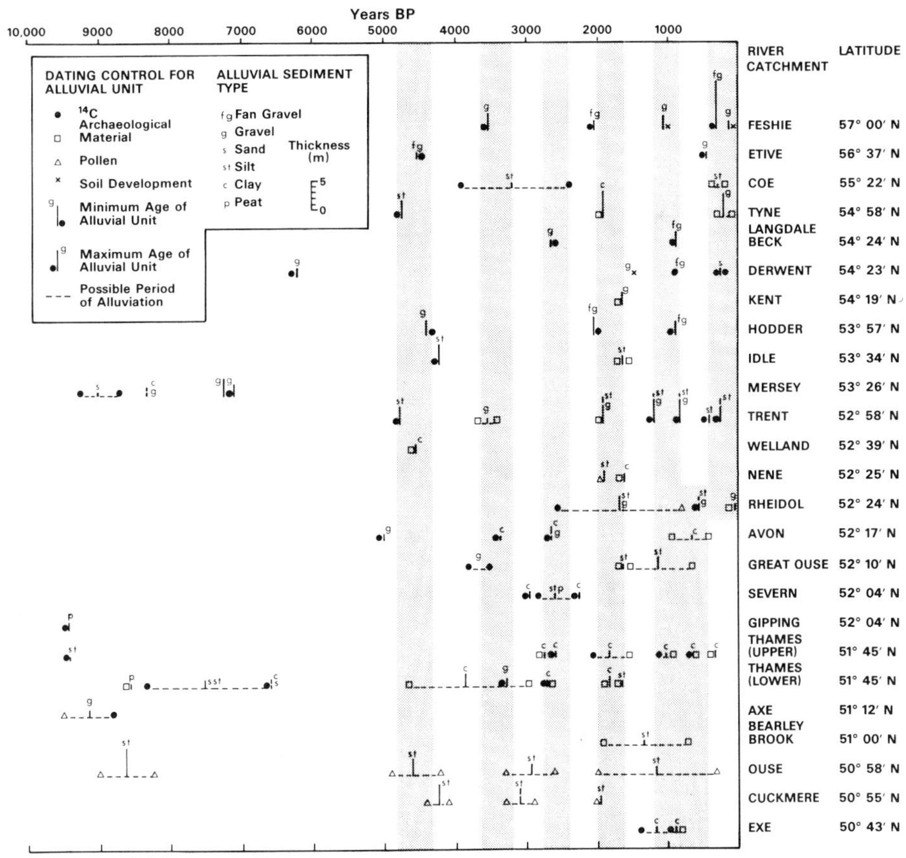

Fig. 2.2. Plot of Holocene alluvial units in Britain showing thicknesses of alluvium and lithofacies type. The frequency maxima of dated alluvial units deposited since 5000 years BP are shown by diagonal shading (after Macklin and Lewin, in press).

Late Holocene are considered to reflect changes in sediment supply and channel-slope coupling. In particular, increased rates of soil erosion and delivery of fine sediment to valley floors following the expansion and intensification of cultivation that began in the Neolithic. Although forest clearance and agricultural practices were undoubtedly important for initiating erosion, significant redistribution of this material and widespread alluviation would appear to have occurred during relatively brief periods (10–100 years?) of abrupt climate shift that were characterised by major changes in flood frequency and magnitude. Thus, Macklin and Lewin (in press) conclude that the British Holocene fluvial record can be considered to be an alluvial sequence that has been climatically driven but culturally blurred.

Dating Holocene alluvial sequences

Despite the comparatively large number of papers which have examined Postglacial river development in Britain, Macklin and Lewin (in press) estimated that up to 1989 only around 123 alluvial units have been dated within 25 major river catchments (Fig. 2.3). In upland Britain, although there are fewer dated alluvial units, nearly 71% are time-bracketed by ^{14}C dates. Whereas in lowland river valleys only a third of all alluvial units are ^{14}C dated, and the ages of most alluvial episodes, especially over the last two millennia, have been determined by reference to archaeological material. Even when alluvial deposits appear to be tightly constrained by datable material, it is often very difficult to determine precisely when alluviation began or ceased. Consequently, the length of time over which sedimentation is believed to have taken place has almost certainly been overestimated in many cases.

Sediment-based dating techniques such as OSL (Bailiff, this volume) and palaeomagnetism (Clark, this volume) together with small sample AMS ^{14}C, however, could potentially revolutionise the geochronological framework of Holocene river development in Britain. Dating sediments themselves will enable alluvial fills to be dated at selected key horizons rather than relying on chance finds of organic material or artefacts, which invariably are not recovered in the most appropriate stratigraphic context for elucidating river histories. In this respect dating Holocene alluvium may be as valuable as remote sensing of alluvial-covered landscapes for predicting patterns of archaeological survival on valley floors. Zoning Holocene alluvial bottom-lands on the basis of age could be a very useful starting point for identifying where sites of a particular period are likely to be upstanding, buried or absent.

Postglacial channel and floodplain typologies

Contemporary channel activity, floodplain forms and Quaternary valley fills, when examined in the context of sediment supply and streampower provides the basis for a broad but convenient classification of British drainage basins into "upland" and "lowland" types (Macklin and Lewin 1986). This separates upland valley systems of northern and western Britain which have been glaciated several times in the Pleistocene, and generally have steeper gradients and higher unit discharges, from lowland catchments of southern and eastern Britain, with lower relief and unit discharges, that lay beyond the limits of glaciation, or experienced major ice incursions only in their headwaters. Generally speaking both streampower and rates of floodplain reworking reach a peak in the middle courses of a river (Graf 1982; Lewin 1982); streampower is less upstream, where discharges are very small and may be less downstream, or in lowland rivers where gradients become minimal.

In northern and western Britain the most active channels are often found in piedmont locations where rivers leave the uplands (Newson 1981). Here low-sinuosity, "wandering gravel" rivers are common (Hooke and Redmond 1989; Smith 1989) that have channel and floodplain morphologies intermediate to upland torrents and sinuous rivers more typical of the lowlands. The Holocene erosion and sedimentation history of river basins in Britain affected by Devensian glaciation have generally followed a paraglacial sequence of valley floor development (cf. Church and Ryder 1972). Thus upland (Harvey and Renwick 1987; Richards *et al* 1987; Brazier *et al* 1988) and piedmont (Macklin and Lewin 1986; Hooke *et al* 1990) rivers in response to declining sediment supply since deglaciation, and isostatic adjustment, have progressively incised their valley floors during the Holocene resulting in locally well developed river terraces and entrenched tributary valley alluvial fans. Incision, however, was episodic and partial refilling of valley floors followed land-use change and agricultural development (Harvey *et al* 1981; Brazier and Ballantyne 1989; Macklin *et al* 1991) and metal mining in some catchments (Lewin *et al* 1983; Macklin and Lewin 1989). In contrast, the vertical tendency of many rivers in southern and eastern lowland England over the last 10,000 years has been one of progressive alluviation, as a result of glacioeustatic sea-level rise, particularly in the Early Holocene (Waller, Burrin and Marlow 1988), and accelerated catchment erosion in the Late Holocene as a consequence of human activity (Shotton 1978; Brown 1983).

It is worth pointing out, however, that categorizing a river on a reach scale either as lowland or upland may be misleading. Valley floors display similar channel planform and sedimentation styles as a consequence of similar stream powers, rates of sediment supply and transport, and calibre of alluvial material all of which to a greater or lesser extent have Pleistocene and/or anthropogenic antecedence. Vertical accretion of fine-grained alluvium, for example, generally considered to be the predominant mode of Holocene floodplain construction in lowland southern and eastern Britain, is also important in the lower reaches of many rivers in western and northern Britain (e.g. Samuels and Buckland 1978; Passmore *et al* 1992), and in former glacial lake basins within upland areas (e.g. Leeks *et al* 1988). Conversely, although gravel transport and deposition in British rivers is at present generally restricted to the north and west of the country, Early-Middle Holocene palaeochannels in south and east England developed in gravel (now buried beneath fine-grained alluvium) testify to significant coarse sediment movement during this period.

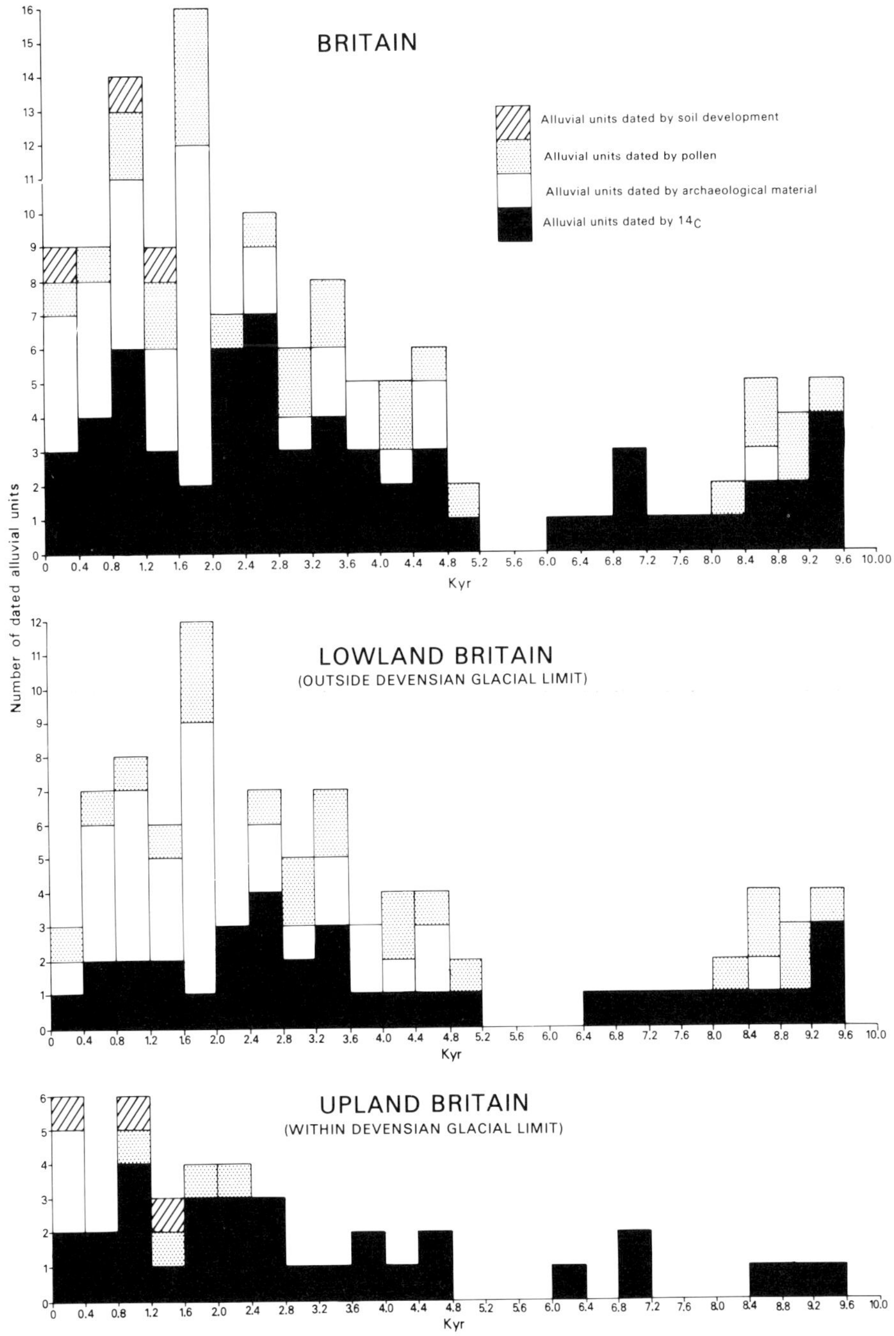

Fig. 2.3. Histograms of dated Holocene alluvial units in lowland and upland Britain (after Macklin and Lewin, in press).

Although contemporary channel morphologies and floodplain sedimentary sequences provide a useful template for interpreting earlier alluvial fills, there are a number of river types present in the Holocene alluvial record which are rare in Britain today. The most notable are divided channel systems which largely as the result of river regulation and land drainage are either moribund or have been totally transformed. For example, anastomosed river systems (cf. Smith 1983), consisting of multiple, interconnected channels bordered by wetlands and floodbasins, appear to have been the equilibrium channel forms in many low gradient valleys in southern and eastern England (e.g. Seale 1979; French *et al*, this volume). Similarly, fairly extensive active braiding is presently restricted to one major British river, the Spey (Lewin and Weir 1977) and a few smaller rivers draining catchments of 50–500 km^2 (e.g. River Feshie, Werritty and Ferguson 1980) in central Wales, Scottish Highlands, Southern Uplands, Pennines and Lake District (Ferguson 1981). Historic maps of the eighteenth and nineteenth centuries, however, show extensive braided reaches in the River Tyne catchment, north east England (Macklin 1986; Macklin and Lewin 1989), on the Afon Ystwyth, mid-Wales (Lewin *et al* 1983) and in the River Tay, Scotland (Gilvear, in press), where at present single thread channels are evident. The reduction this century in the degree of braiding in the Rivers South Tyne and Ystwyth is certainly partly linked with the cessation of metal mining, while in the Tay bank protection works appear to have been an important factor. Ongoing work in the Tyne (Macklin, Passmore and Rumsby, unpublished), however, suggests that braiding may have been a response to climate change, coincident with the Little Ice Age (Grove 1988), and a greater frequency of large floods especially those arising from snowmelt. A similar pattern has been demonstrated by Bravard (1989) in the river systems of the French Alps and related to climatic deterioration in the fourteenth and eighteenth centuries. There is also evidence of channel metamorphosis and braided river development in earlier neoglacial phases elsewhere in northern Europe and Scandinavia (Starkel 1991).

What is becoming clear is that the Holocene was characterised by a greater diversity of channel types and floodplain sedimentation styles than currently are found in Britain. We therefore must move away from interpretation of alluvial sequences based solely on recent river history or catchment physiography towards classification and interpretation of river sediments using process-based facies models. Those recently developed by Allen (1983) and Miall (1985), for example, encourage identification of "fluvial architectural elements" (eg lateral accretion deposits, overbank fines) or "facies associations" which then enable depositional environments and river types to be identified. Investigations of floodplain sedimentation processes, river behaviour (vertical and lateral), and the resultant alluvial sequences are considered not only to hold the key in assessing the utilisation of fluvial environments by past communities, but also for quantifying the spatial and temporal pattern of archaeological survival in river valleys.

Holocene palaeoflood analysis

For more than a decade North American geomorphologists have successfully been employing alluvial stratigraphic analysis in historic and Holocene flood reconstruction, often in conjunction with excavation of riverside archaeological sites (e.g. Patton and Dibble 1982). Yet in continental Europe and Britain, although documentary sources have been used to extend flood records beyond the period of gauging (e.g. Probst 1989), with the exception of Macklin *et al*'s (1992a, 1992b) and Rumsby's (1991) recent studies in the Tyne catchment the alluvial record has generally not been used to reconstruct Holocene flood histories in British river basins on an event by event basis. Palaeoflood analysis has been previously hindered in this country by difficulties in both identifying and especially dating flood units. Although the deposits of major flood events in high relief upland catchments (typically forming extensive boulder bars, lobes and sheets) are relatively easy to recognise and date, for example using lichenometric techniques (e.g. Harvey *et al* 1984; Macklin *et al* 1992a), distinguishing individual flood units in fine-grained floodplain sequences is more problematic.

Recent investigations in the Lower Tyne valley, north-east England have identified major floods in fine-grained vertically accreted alluvium (deposited in the last 300 years) by textural discontinuities and prominent sand units that reverse the overall fining upward sequence (Macklin *et al* 1992b; Rumsby 1991). Flood units in this instance were dated using heavy metal chemistry (cf. Macklin 1985) which can be related to documented periods of lead and zinc mining in the Tyne catchment. Earlier fine-grained alluvial sequences have been successfully dated using palaeomagnetic techniques (Macklin *et al*, in this volume; Macklin, Passmore and Ridgway, unpublished) and a pilot study is currently in progress to assess the suitability of OSL techniques for dating single flood units. British river catchments developed on Carboniferous lithologies (such as the Tyne) that have a significant sand load and relatively rapid rates of floodplain accretion would therefore appear to be highly suitable for detailed Holocene flood frequency and magnitude reconstruction. It may be more difficult, however, to discern individual flood events in floodplains comprised of silts and clays such as those found in the English Midlands, Thames valley and East

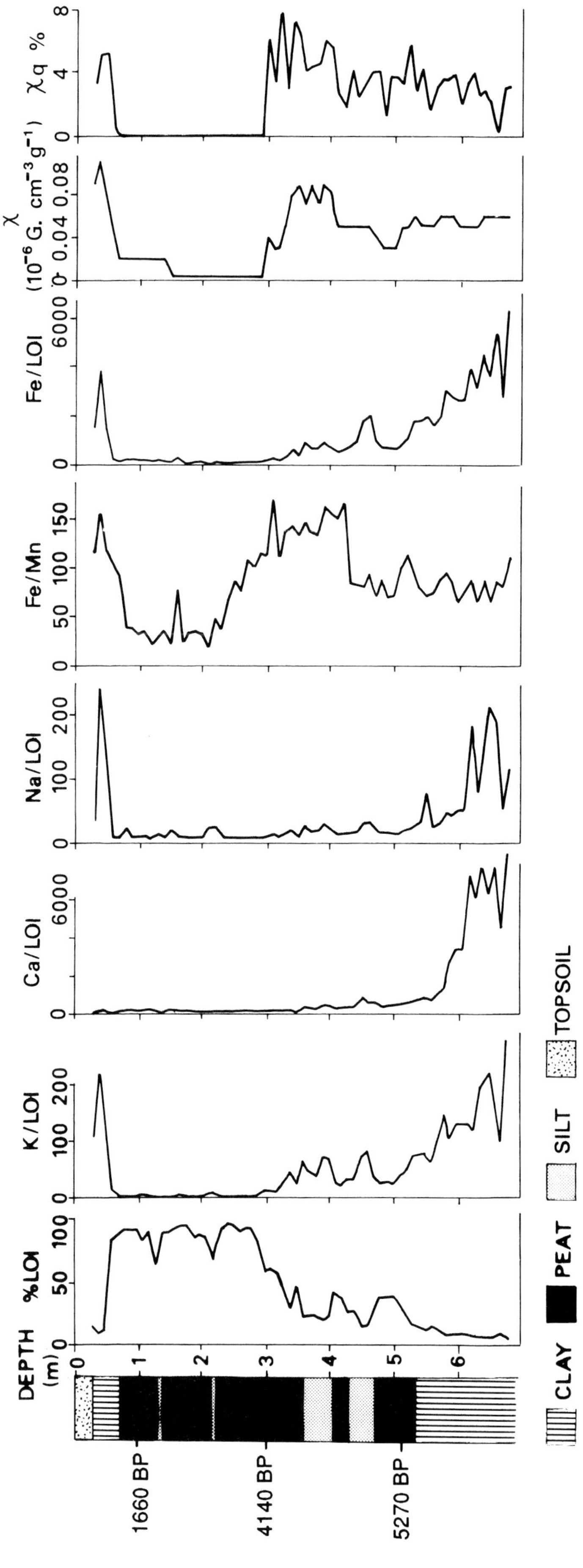

Fig. 2.4. Geochemical and mineral magnetic profiles through Holocene alluvium and peat deposits in the Fornaught Strand basin, south-east Ireland.

Anglia. Holocene river systems in these areas are generally characterised by low accretion rates, and original sedimentary structures (diagnostic of hydraulic regime) are liable to disturbance and destruction by bioturbation and soil processes. In this context, micromorphology or X-ray analysis could be particularly useful both for assessing the integrity of floodplain sediments for palaeoflow analysis, as well as for improving stratigraphic resolution.

There is clearly considerable potential, though at present largely unrealised, for Holocene palaeoflood studies in British rivers especially those with sandy floodplain sequences which can be dated by palaeomagnetic or luminescence techniques. Palaeoflood analysis could reveal longer-term fluctuations in hydroclimates (complimentary to, though independent of, other proxy climate records), establish flood hazard at a site or determine rates of floodplain accretion which may influence fertility or contamination of alluvial soils.

Provenancing alluvial sediment

Palaeolimnologists routinely analyse chemical, physical and mineralogical properties of lacustrine sediments to evaluate the nature and causes of environmental change within lake catchments, particularly those associated with human disturbance (e.g. Flower *et al* 1989). Similar opportunities also exist in river basins where past and present agricultural, mining and industrial activity have introduced soil, sediment and pollutants into channel systems. Surprisingly, with the exception of catchments affected by coal (Macklin *et al* 1991) or metal mining (Davies and Lewin 1974; Macklin 1985), fine-grained alluvium has not usually been analysed for provenancing purposes by earth scientists working either in "on" or "off-site" archaeological contexts. Analysis of the physical (e.g. charcoal concentrations, grain size (Burrin 1981), grain morphology (Gordon, in Macklin and Lewin 1986), heavy mineral assemblages (Limbrey and Robinson 1988) mineral magnetics (Hooke *et al* 1990)) and chemical properties (Macklin *et al* 1992b; Macklin *et al*, this volume) of sediment can reveal the parent material from which alluvium was derived and where in a drainage basin erosion was taking place at a particular time. For example, soil erosion in Fornaught Strand, south-east Ireland (Macklin, Passmore and Heap, unpublished), associated with agricultural activity in the Middle Neolithic, resulted in minerogenic sedimentation on the valley floor which can be identified by distinctive mineral magnetic assemblages in alluvium of this period (Fig. 2.4). In Macklin *et al*'s (this volume) study of the Tyne, north-east England, phases of Holocene alluviation in the lower reaches of the basin were related to episodic erosion within two major tributaries (rivers North and South Tyne) by virtue of their contrasting geologies. These types of analysis could be extended to many other British river basins and could considerably enhance our understanding of the relative importance and role of land-use change in promoting erosion and accelerating floodplain sedimentation. In the future, it should be possible, for example, to identify the first significant influx of eroded soil onto valley floors and establish whether or not it was synchronous with clearance and/or climate change indicated by archaeological, palynological and proxy-climate records.

CONSERVATION OF THE ALLUVIAL ARCHAEOLOGICAL RESOURCE

Although the planning and legislative frameworks for the preservation of archaeological sites is now well established in Britain, the notion of geological conservation, and protecting landforms and stratigraphic sequences, is a very recent innovation in this country. The growing appreciation of the importance of "off-site" environmental studies in archaeology needs to be reflected in greater parity between geological and archaeological conservation, particularly in alluvial contexts where these records are so intimately entwined. Both are increasingly under threat from mineral extraction, road construction and land drainage, though this has to be balanced by the fact that many important archaeological and geological discoveries would not have come to light without valley floor development (e.g. Needham 1991). Nevertheless, when-ever opportunities arise to study alluvial sequences, every effort should be made to assess their archaeological and environmental significance in both a regional and national context and, if considered appropriate, make provision for their protection. The Nature Conservancy Council in their recent geological conservation review (NCC 1990), and RIGS (Regionally Important Geological/Geomorphological Sites) initiative launched in December 1990, have gone a significant way in heightening the public's awareness of our geomorphological and geological heritage, but a great deal more needs to be done.

Holocene alluvial sequences are arguably unique in the way that they integrate and record environmental change (natural and anthropogenic) over a wide range of spatial and temporal scales. They therefore constitute an invaluable "library" which can be read using appropriate geological, archaeological and palaeoecological techniques. This important environmental and cultural resource should be protected for the benefit of future generations of archaeologists and earth scientists.

References

Allen, J.R.L. 1983. Studies in fluviatile sedimentation: bars, bar-complexes and sandstone sheets (low sinuosity braided streams) in the Brownstones (L Devonian), Welsh Borders. *Sedimentary Geology*, 33, 237–293.

van Andel, T.H. and Zangger, E. 1990. Landscape stability and destabilization in the prehistory of Greece. In G. Bottema, Entsesnieborg and van Zeist (eds.), *Man's role in shaping of the Eastern Mediterranean Landscape*, 139–157. Rotterdam: Balkema.

Bailey, G.N., Lewin, J., Macklin, M.G. and Woodward, J.C. 1990. The "older fill" of the Voidomatis valley, north-west Greece, and its relationship to the Palaeolithic archaeology and glacial history of the region. *Journal of Archaeological Science*, 17, 145–50.

Bell, M. 1982. The effects of land-use and climate on valley sedimentation. In A.F. Harding (ed.), *Climatic Change in the late Prehistory*, 127–142. Edinburgh: University Press.

Bintliff, J.L. 1975. Mediterranean alluviation: new evidence from archaeology. *Proceedings of the Prehistoric Society*, 41, 78–84.

Bravard, J.P. 1989. La Metamorphose des rivières des alpes Françaises a la fin du moyen-age et à l'époque moderne. *Bulletin de la Société Geographique de Liège*, 25, 145–157.

Brazier, V. and Ballantyne, C.K. 1989. Late Holocene cone evolution in Glen Feshie, Western Cairngorm Mountains, Scotland. *Transactions of the Royal Society of Edinburgh: Earth Sciences*, 80, 17–24.

Brazier, V., Whittington, G. and Ballantyne, C.K. 1988. Holocene debris cone evolution in Glen Etive, Western Grampian Highlands. *Earth Surface Processes and Landforms*, 13, 525–531.

Brigham, T. 1990. The Late Roman waterfront in London. *Britannia*, 20, 99–183.

Brown, A.G. 1983. Floodplain deposits and accelerated sedimentation in the lower Severn basin. In K.J. Gregory (ed.), *Background to Palaeohydrology*, 413–431. Chichester: John Wiley & Sons.

Brown, A.G. 1988. The palaeoecology of *Alnus* (alder) and the Postglacial history of floodplain vegetation. Pollen percentage and influx data from the West Midlands, United Kingdom. *New Phytologist*, 110, 425–436.

Brown, A.G. 1990. Holocene floodplain diachronism and inherited downstream variations in fluvial processes: a study of the river Perry, Shropshire, England. *Journal of Quaternary Science*, 5(1), 39–51.

Brown, A.G. and Barber, K.E. 1985. Late Holocene palaeoecology and sedimentary history of a small lowland catchment in central England. *Quaternary Research*, 24, 87–102.

Bryan, K. 1941. Pre-Columbian agriculture in the south-west as conditioned by periods of alluviation. *Annals of the Association of American Geographers*, 31, 219–42.

Burrin, P.J. 1981. Loess in the Weald. *Proceedings of the Geologists Association*, 92, 87–92.

Burrin, P.J. and Scaife, R.G. 1984. Aspects of Holocene valley sedimentation and floodplain development in southern England. *Proceedings of the Geologists Association*, 95(1), 81–96.

Burrin, P.J. and Scaife, R.G. 1988. Environmental thresholds, catastrophe theory and landscape sensitivity: their relevance to the impact of Man on valley alluviation. In J.L. Bintliff, D.A. Davidson and E.G. Grant (eds.), *Conceptual Issues in Environmental Archaeology*, 211–232. Edinburgh: University Press.

Butzer, K. 1982. *Archaeology as human ecology*. Cambridge: University Press.

Church, M. and Ryder, J.M. 1972. Paraglacial sedimentation: a consideration of fluvial processes conditioned by glaciation. *Bulletin of the Geological Society of America*, 83, 3059–3072.

Coles, J.M. 1986. Precision, purpose and priorities in wetland archaeology. *Antiquaries Journal*, 66, 227–47.

Coles, J. and Coles, B. 1992. The wetland revolution: a natural event. In B. Coles (ed.), *The wetland revolution in prehistory*, 147–53. University of Exeter: Wetland Archaeological Research Project, Occasional Paper 6.

Davies, B.E. and Lewin, J. 1974. Chronosequences in alluvial soils with special reference to historic lead pollution in Cardiganshire, Wales. *Environmental Pollution*, 6, 49–57.

Edwards, K.J. and Rowntree, K.M. 1980. Radiocarbon and palaeoenvironmental evidence for changing rates of erosion at a Flandrian stage site in Scotland. In R.A. Cullingford, D.A. Davidson and J. Lewin (eds.), *Timescales in Geomorphology*, 207–24. Chichester: Wiley.

Ferguson, R.I. 1981. Channel form and channel changes. In J. Lewin (ed.), *British Rivers*, 90–125. London: Allen and Unwin.

Flower, R.J., Stevenson, A.C., Dearing, J.A., Foster, I.D.L., Airey, A., Rippey, B., Wilson, J.P.F. and Appleby, P.G. 1989. Catchment disturbance inferred from palaeolimnological studies of three contrasted sub-humid environments in Morocco. *Journal of Palaeolimnology*, 1, 293–322.

French, C. and Taylor, M. 1985. Dessication and destruction: the immediate effects of dewatering at Etton, Cambridgeshire. *Oxford Journal of Archaeology*, 4, 139–58.

Gaillard, M.J. 1985. Postglacial palaeoclimatic changes in Scandinavia and central Europe. A tentative correlation based on studies of lake level fluctuations. *Ecologia Mediterranea*, 11(1), 159–175.

Gilvear, D.J. (in press). Natural and man-induced river planform changes on the regulated River Tay, Scotland: the use of historical analysis to flood hazard management. *Regulated Rivers*.

Good, G.L., Jones, R.H. and Ponsford, M.W. (eds.) 1991. *Waterfront archaeology: proceedings of the third international conference, Bristol, 1988*. London: Council for British Archaeology, Research Report no 74.

Graf, W.L. 1982. Spatial variation of fluvial processes in semi-arid lands. In C.E. Thorn (ed.), *Space and Time in Geomorphology*, 193–217. London: George Allen and Unwin.

Grove, J.M. 1988. *The Little Ice Age*. London: Methuen.

Hack, J. 1942. The changing physical environment of the Hopi Indians of Arizona. *Harvard University Peabody Museum Archaeology and Ethnology Papers*, 35, 1–85.

Harding, P., Gibbard, P.L., Lewin, J., Macklin, M.G. and Moss, E.H. 1987. The transport and abrasion of flint handaxes in a gravel-bed river. In G.de G. Sieveking and M.H. Newcomer (eds.) *The human uses of flint and chert*, 115–26. Cambridge: University Press.

Harvey, A.M., Alexander, W.and James, P.A. 1984. Lichens, soil development and the age of Holocene valley floor landforms: Howgill Fells, Cumbria. *Geografiska annaler*, 66A(4), 353–366.

Harvey, A.M., Oldfield, F., Baron, A.F. and Pearson, G.W. 1981. Dating of Post-glacial landforms in the Central Howgills. *Earth Surface Processes and Landforms*, 6, 401–412.

Harvey, A.M. and Renwick, W.H. 1987. Holocene alluvial fan and terrace formation in the Bowland Fells, north west England. *Earth Surface Processes and Landforms*, 12, 249–257.

Herteig, A.E. (ed.) 1985. *Conference on waterfront archaeology in north European towns, no 2, Bergen, 1983.*

Higgs, E. and Vita-Finzi, C. 1966. The climate, environment and industries of Stone-age Greece. *Proceedings of the Prehistoric Society*, 32, 1–29.

Hooke, J.M., Harvey, A.M., Miller, S.Y. and Redmond, C.E. 1990. The chronology and stratigraphy of the alluvial terraces of the River Dane valley, Cheshire, N W England. *Earth Surface Processes and Landforms*, 15, 717–737.

Hooke, J.M. and Redmond, C.E. 1989. River channel changes in England and Wales. *Journal of the Institute of Water and Environmental Management*, 3, 328–335.

Karlen, W. 1991. Glacier fluctuations in Scandinavia during the last 9000 years. In L. Starkel, K.J. Gregory and J.B. Thornes (eds.), *Temperate Palaeohydrology*, 395–412. Chichester: John Wiley & Sons.

Knox, J.C. 1983. Responses of river systems to Holocene climates. In H.E. Wright Jnr (ed.), *Late-Quaternary Environments of the United States, The Holocene, vol.2*, 26–41. Minneapolis: University of Minnesota Press.

Leeks, G.J., Lewin, J. and Newson, M.D. 1988. Channel change, fluvial geomorphology and river engineering: the case of the Afon Trannon, mid-Wales. *Earth Surface Processes and Landforms*, 13, 207–223.

Lewin, J. 1982. British floodplains. In B.H. Adlam, C.R. Fenn and L. Morris (eds.), *Papers in Earth Studies*, 21–36, Norwich: Geobooks.

Lewin, J., Bradley, S.B. and Macklin, M.G. 1983. Historical valley alluvation in mid-Wales. *Geological Journal*, 18, 331–350.

Lewin, J., Davies, B.E. and Wolfenden, P.J. 1977. Interactions between channel change and historic mining sediments. In K.J.Gregory (ed.), *River channel changes*, 353–67. Chichester: Wiley.

Lewin, J., Macklin, M.G. and Woodward, J.C. 1991. Late Quaternary fluvial sedimentation in the Voidomatis basin, Epirus, Northwest Greece. *Quaternary Research*, 35, 103–15.

Lewin, J. and Weir, M.J.C. 1977. Morphology and recent history of the lower Spey. *Scottish Geographical Magazine*, 93, 45–51.

Limbrey, S. and Robinson, S. 1988. Dry land to Wet land: Soil resources in the Upper Thames Valley. In P. Murphy and C.A.I. French (eds.), *The Exploitation of Wetlands*, 129–144. Oxford: British Archaeological Reports, British Series, 186.

Macklin, M.G. 1985. Floodplain sedimentation in the upper Axe valley, Mendip, England. *Transactions of the Institute of British Geographers* (NS), 10, 235–244.

Macklin, M.G. 1986. Channel and floodplain metamorphosis in the River Nent, Cumberland. In M.G. Macklin and J. Rose (eds.), *Quaternary River Landforms and Sediments in the Northern Pennines*, 19–33. BGRG/QRA Field Guide.

Macklin, M.G. and Lewin, J. 1986. Terraced fills of Pleistocene and Holocene age in the Rheidol Valley, Wales. *Journal of Quaternary Science*, 1(1), 21–34.

Macklin, M.G. and Lewin, J. 1989. Sediment transfer and transformation of an alluvial valley floor: the River South Tyne, Northumbria, UK. *Earth Surface Processes and Landforms*, 14, 233–246.

Macklin, M.G. and Lewin, J. (in press). Holocene river alluviation in Britain. In I. Douglas and J. Hagedorn (eds.), *Geomorphology and Geoecology, Fluvial Geomorphology*. Zeitschrift fur Geomorphologie (supplement) 85.

Macklin, M.G., Passmore, D.G., Stevenson, A.C., Cowley, D., Edwards, D.N. and O'Brien, C.F. 1991. Holocene alluviation and land-use change on Callaly Moor, Northumberland, UK. *Journal of Quaternary Science*, 6(3), 225–232.

Macklin, M.G., Rumsby, B.T. and Heap, T. 1992a. Flood alluviation and entrenchment: Holocene valley floor development and transformation in the British uplands. *Geological Society of America Bulletin* 104, 631–643.

Macklin, M.G., Rumsby, B.T. and Newson, M.D. 1992b. Historical floods and vertical accretion of fine-grained alluvium in the Lower Tyne valley, northeast England. In P. Billi, R.D. Hey, C.R. Thorne and P. Tacconi (eds.), *Dynamics of Gravel-bed rivers*, 573–89. Chichester: Wiley.

Miall, A.D. 1985. Architectural-element analysis: a new method of facies analysis applied to fluvial deposits. *Earth Science Reviews*, 22, 261–308.

Milne, G., Battarbee, R., Straker, V. and Yule, B. 1983. The River Thames in London in the mid 1st century AD. *Transactions of the London and Middlesex Archaeological Society*, 34, 19–30.

Milne, G. and Hobley, B. (eds.) 1981. *Waterfront archaeology in Britain and northern Europe*. London: Council for British Archaeology, Research Report no 41.

Nature Conservancy Council 1990. *Earth Science Conservation in Great Britain.* A strategy.

Needham, S.P. 1991. *Excavation and Salvage at Runnymede Bridge 1978: The Late Bronze Age Waterfront Site.* London: British Museum Press.

Newson, M.D. 1981. Mountain Streams. In J. Lewin (ed.), *British Rivers*, 59–89. London: Allen and Unwin.

Passmore, D.G., Macklin, M.G., Stevenson, A.C., O'Brien, C.F. and Davies, B.A.S. 1992. A Holocene alluvial sequence in the lower Tyne valley, northern Britain: a record of river response to environmental change. *The Holocen* 2(2), 138–147.

Patton, P.C. and Dibble, D.S. 1982. Archaeologic and geomorphic evidence for the palaeohydrologic record of the Pecos River in West Texas. *American Journal of Science*, 282, 97–121.

Pope, K.O. and van Andel, T.H. 1984. Late Quaternary alluviation and soil formation in the southern Argolid: its history, causes and archaeological implications. *Journal of Archaeological Science*, 11, 281–306.

Probst, J.L. 1989. Hydroclimatic fluctuations of some European rivers since 1800. In G.E. Petts (ed.), *Historical change of large alluvial rivers: Western Europe*, 41–55. Chichester: John Wiley and Sons.

Richards, K.S., Peters, N.R., Robertson-Rintoul, M.S.E. and Switzur, V.R. 1987. Recent valley sediments in the North York Moors: evidence and interpretation. In V. Gardiner (ed.), *International Geomorphology 1986 Part 1*, 869–883. Chichester: John Wiley & Sons.

Robinson, M.A. and Lambrick, G.H. 1984. Holocene alluviation and hydrology in the upper Thames basin. *Nature*, 308, 809–814.

Rumsby, B.T. 1991. Flood Frequency and Magnitude Estimates Based on Valley Floor Morphology and Floodplain Sedimentary Sequences: The Tyne Basin, N E England. Unpublished PhD Thesis, University of Newcastle upon Tyne, 205 pp.

Samuels, J. and Buckland, P.C. 1978. A Romano-British settlement at Sandtoft, South Humberside. *The Yorkshire Archaeological Journal*, 50, 65–75.

Schick, K.D. 1986. *Stone Age sites in the making*. Oxford: British Archaeological Reports, International Series, 319.

Seale, R.S. 1979. Ancient courses of the Great and Little Ouse in Fenland. *Proceedings of the Cambridgeshire Antiquarian Society*, 69, 1–19.

Shotton, F.W. 1978. Archaeological inferences from the study of alluvium in the lower Severn-Avon valleys. In S. Limbrey and J.G. Evans (eds.), *Man's Effect on the Landscape: The Lowland Zone*, 27–32. London: Council for British Archaeology, Research Report No 21.

Smith, D.G. 1983. Anastomosed fluvial deposits: modern examples from Western Canada. In J.D. Collinson and J. Lewin (eds.), *Modern and Ancient Fluvial Systems*, 155–168. Special Publication No 6, International Association of Sedimentologists. Oxford: Blackwell Scientific Publications.

Smith, S.A. 1989. Sedimentation in a meandering gravel-bed river: the River Tywi, South Wales. *Geological Journal*, 24, 193–204.

Starkel, L. 1991. The Vistula River Valley: A case study for central Europe. In L. Starkel, K.J. Gregory and J.B. Thornes (eds.), *Temperate Palaeohydrology*, 171–88. Chichester: John Wiley & Sons.

Vita-Finzi, C. 1969. *The Mediterranean valleys: geological changes in historical times*. Cambridge: University Press.

Vita-Finzi, C. 1975. Chronology and implications of Holocene alluvial history of the Mediterranean basin. *Biuletyn Geologiczny*, 19, 137–147.

Waller, M., Burrin, P.J. and Marlow, A. 1988. Flandrian sedimentation and Palaeoenvironments in Pett Level, the Brede and Lower Rother valleys and Walland Marsh. In J. Eddison and C. Green (eds.), *Romney Marsh: Evolution, Occupation, Reclamation*, 3–29. Oxford University Committee for Archaeology, Monograph No 24.

Waters, M.R. 1988. Holocene alluvial geology and geoarchaeology of the San Xavier reach of the Santa Crux River, Arizona. *Geological Society of America Bulletin*, 100, 479–91.

Werritty, A. and Ferguson, R.I. 1980. Pattern changes in a Scottish braided river over 1, 30 and 200 years. In R.A. Cullingford, D.A. Davidson and J. Lewin (eds.), *Timescales in geomorphology*, 53–68. Chichester: John Wiley & Sons.

Wilkinson, T.J. and Murphy, P. 1988. The Hullbridge Basin Survey, 1987: interim report no.8. Essex County Council, Planning Department, Archaeology Section (circulation report).

Yule, B. 1988. Natural topography of north Southwark. In *Excavations in Southwark, 1973–76, Lambeth, 1973–79*, 13–18. London: Joint Publication no.3 of the London and Middlesex Archaeological Society and the Surrey Archaeological Society.

SECTION 1

Dating and geophysical prospection on alluvial sediments

3

Luminescence dating of alluvial deposits

I. K. Bailiff

Summary: *The introduction of thermoluminescence (TL) techniques for dating unheated sediments and their subsequent widespread use represented a significant advance for absolute dating methods in the Quaternary period (Wintle and Huntley 1980; 1982). They have provided the means to obtain absolute dates directly associated with depositional events for a wide range of sedimentary environments. A substantial number of TL ages have been produced for loess, buried soils and water-lain silts, ranging across much of the Quaternary period. The use of TL techniques for dating alluvial sediments with an archaeological context, particularly deposits of the last 15,000 years, has been restricted by a number of technical problems, the most important of which are concerned with the zeroing mechanism of the 'TL clock'. However, the development of new ways of stimulating luminescence has substantially improved dating prospects in this field. The new techniques employ light rather than heat to release the stored energy in dating minerals, producing what is generally termed optically stimulated luminescence (OSL). In this paper the basis of the new techniques and factors which are expected to affect the prospects for dating alluvial deposits from archaeological sites are discussed.*

INTRODUCTION: THE LUMINESCENCE CHRONOMETER

Luminescence techniques for dating both heated and unheated materials are based on a common property of luminescent minerals. The property is the acquisition of stored energy with time and occurs in minerals found in sediments such as quartz and feldspar. The source of energy is ionizing radiation released from naturally occurring radioactive elements within the surrounding medium. There are three major types of ionizing radiation which comprise alpha, beta and gamma rays. Each type penetrates matter to differing degrees; ranging from only tens of microns (1 micron = 1 millionth of a meter) for alpha rays to tens of centimeters for gamma rays. As ionizing radiation passes through matter, energy is transferred and absorbed. In this manner, a radiation dose is delivered to the medium concerned (the physical definition of radiation dose is in terms of quantification of the energy absorbed, and both dose and energy absorbed will be used synonymously in this paper). The imperfect structure of most crystalline materials provides a means of storing a fraction of the dissipated energy. When luminescent minerals which have received a radiation dose are subsequently stimulated (by heating, producing thermoluminescence (TL) or, as discussed at a later stage in this paper, by exposure to light of particular wavelengths), the stored energy is released in the form of light (luminescence), which can be measured using modern photomultiplier detectors. Stimulation depletes the stored energy until it is exhausted. The result of a TL measurement is shown in Figure 3.1 (curve 1). The process is repeatable and if a dose of the same size is again administered to the crystals, the same quantity (in principle) of luminescence will be released during further stimulation. Thus, a dose delivered to crystals may be quantified by comparing the luminescence emitted during the first stimulation and that obtained after the administration of a known 'calibrating' dose. In this manner, detected luminescence intensities can be expressed in terms of absorbed energy (or dose) for a given measurement.

The way in which the property of registration of dose due to ionizing radiation is used to form the basis of a dating method is illustrated schematically in Figure 3.2. For sediments to be suitable for dating they must have been 'zeroed' before burial so that any extant stored energy was reduced to a negligible level (see figure). After burial, the stored energy accrues with time at a rate (referred to as the dose-rate) which depends on the radioactivity of the surrounding medium and other factors such as the water content of the burial medium. For most situations, a constant dose-rate is assumed, and it can be evaluated after determination of the radioactivity of the burial medium, the water content and other physical parameters associated with the interaction of ionizing radiation with matter.

Thus, for a sediment dating sample, there are two quantities to be determined experimentally: a) the energy accrued in luminescent minerals since burial and b) the energy deposited in them per year. With these two quantities an expression for the time elapsed since zeroing and burial is given by the equation:

$$\text{Luminescence Age (years)} = \frac{\{\text{energy accrued since burial}\}}{\{\text{energy deposited per year}\}}$$

ZEROING OF THERMOLUMINESCENCE

When minerals such as quartz and feldspar (which have received a radiation dose either during burial or in the laboratory) are exposed to sunlight before measurement, a proportion of the TL signal is erased. This effect, referred to as optical bleaching, is illustrated in Figure 3.1. The realization that zeroing of the 'TL clock' could be achieved by optical bleaching under sunlight illumination was the crucial element which provided the basis of Wintle and Huntley's proposal that TL techniques could be used for dating sediments (1980; 1982; for recent review see Berger 1988). Further investigation of the bleaching characteristics of these minerals indicated that the TL of feldspars bleached more rapidly than quartz and that, in general, exposure to direct sunlight of several hours was found to be required to reduce the TL signal to a negligible level. Indeed, in some cases a substantial TL residual signal remained after prolonged exposure to light, depending on the type of sample tested and the spectrum of light used. Since the level of residual signal forms a baseline in calculating the TL age, uncertainty in its value will be reflected in the error associated with the TL age. A potentially significant problem arises for young (i.e. Holocene) deposits if, immediately after burial, the residual energy stored in the dating minerals was a significant fraction of that accrued during burial (referred to as the palaeodose) and could have been reduced by further exposure to light.

For water-lain sediments the exposure period for zeroing is expected to be substantially longer than, for example, aeolian sediments since the spectrum of sunlight reaching TL minerals is more strongly

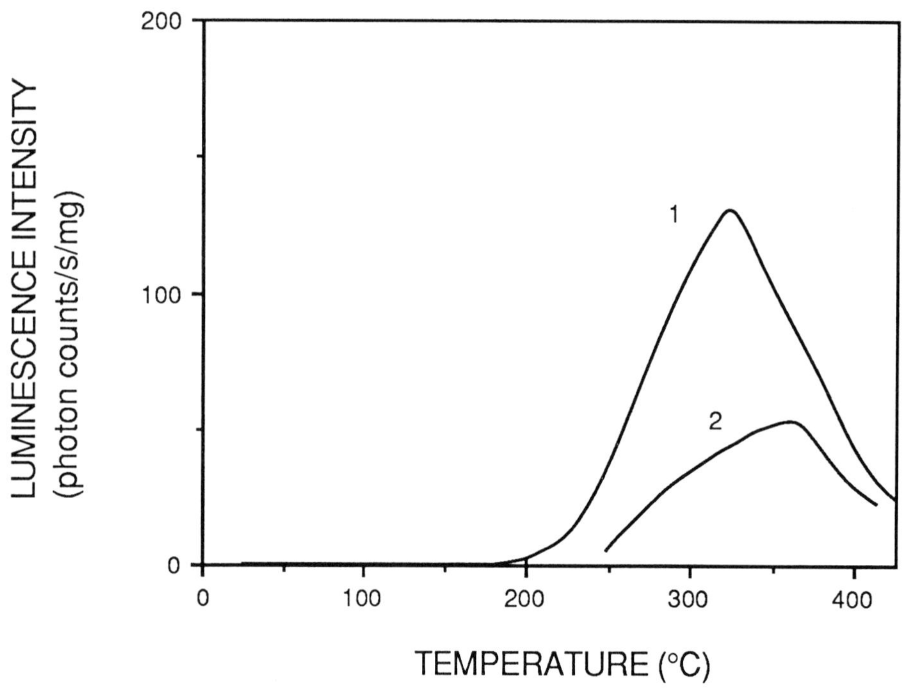

Fig. 3.1 Thermoluminescence (TL) glow curves obtained with a sample containing quartz grains extracted from a Holocene alluvial sediment. The curves shown in the figure are referred to as TL 'glow curves' which represent TL intensity (detected using a photomultiplier tube) recorded as sample aliquots are heated rapidly (usually 10 °/sec) to high temperatures. For an aliquot (curve 1) of sample, without any additional treatment, heated to about 450 °C at a constant rate, the intensity of luminescence increases sharply above 200 °C, rising to a maximum at ca. 325 °C and then decreasing rapidly at higher temperatures. Curve 2 shows the TL glow curve obtained with another aliquot which was exposed to sunlight for several hours before measurement.

attenuated after passing through water and suspended sediment in blue and infra-red wavelength regions relative to green (Berger 1990); surface coatings on grains may also reduce the intensity of light reaching the inner volume. Although the characteristics of signal reduction with exposure to light (in terms of wavelength and intensity) can be determined in the laboratory, reconstructing the exposure history prior to burial is fraught with difficulty. Within any randomly selected group of grains, the variability of conditions of transport and dispersal within the carrying medium and of duration of light exposure is potentially enormous. As the TL techniques were applied to different types of sedimentary deposit it became clear that in testing alluvial sediments complete zeroing before burial should not be assumed. This had been recognized by Wintle and Huntley (1982; also Huntley 1984) and although an experimental procedure known as the partial bleach TL technique was proposed (Wintle and Huntley 1980; 1982) to deal with samples which have been incompletely zeroed, it has since been argued (Bailiff and Poolton 1989) that its success in circumventing the problem is not expected on the basis of the underlying assumptions of the technique. Wintle and Huntley also advised that tests on modern sediments from similar depositional environments should be performed to check whether 'zero' TL ages could be obtained. Using fine grains Berger (1990) determined the TL ages for 'zero age' deposits for a number of types of sediment. He obtained TL ages of circa 200 years (with uncertainties comparable with the age) in the most favourable cases (tidal flat and delta front muds). For a fluvial sand, zeroing appears to have been less effective and an age of circa 1000 years was obtained.

Clearly a very careful examination of the circumstances of deposition is necessary when attempting to obtain TL dates for Holocene alluvial sediments and where there is doubt concerning the exposure of sediments before burial.

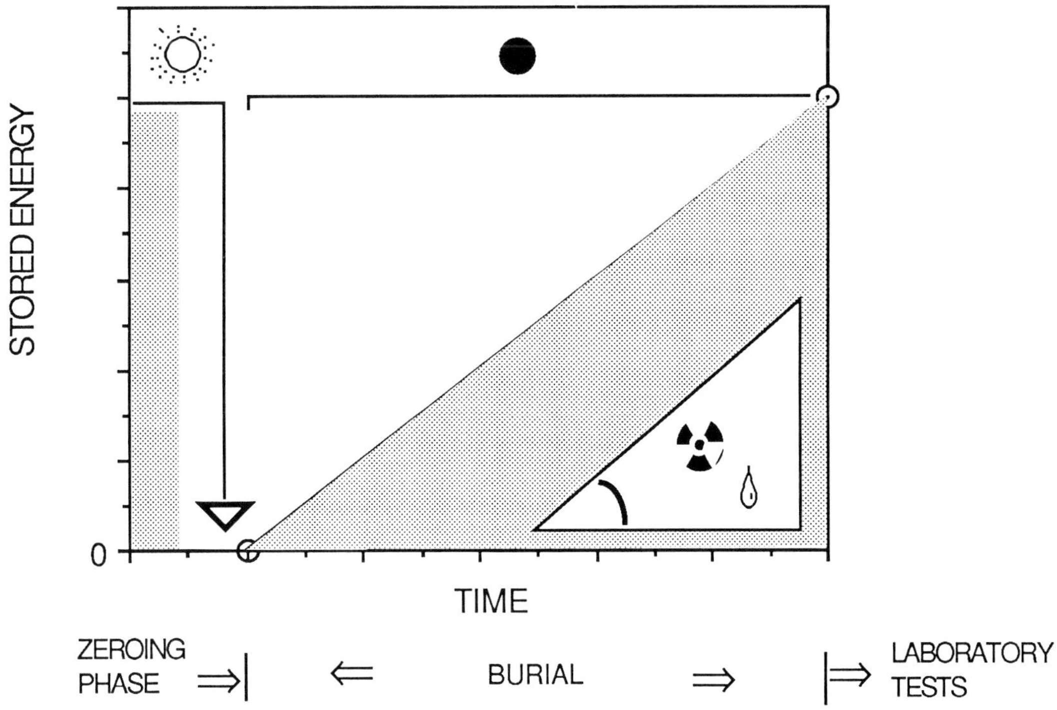

Fig. 3.2. Graphical representation of the luminescence chronometer. Before burial of luminescent minerals within a sediment, a period of exposure to sunlight (zeroing phase) reduces any residual stored energy to a negligible level or to a level which cannot be reduced by further exposure to light. After burial, during which the sediments must remain under dark conditions, energy is deposited in the burial medium by ionizing radiation from naturally occurring radioactive elements within the sediment; in luminescent crystals a fraction of the available energy is stored and accrues with time. The gradient of the build-up curve (as illustrated by inset) is determined by the quantity of energy deposited in the luminescent minerals per unit time (the dose-rate). For a medium with a fixed quantity of radioactivity, which is reasonable to assume in many cases, the dose-rate is affected by the quantity of water within the pore structure, causing it to be reduced with increased water content. This is because some of the energy that would otherwise have been deposited within the luminescent crystals is absorbed by the water. Hence steady-state conditions are to be preferred (e.g. dry or waterlogged).

Symbols: *= exposure to sunlight;* *= darkness;* *= moisture;* *= natural radioactivity*

OPTICALLY STIMULATED LUMINESCENCE

It is in this milieu that a means by which a reduction in the uncertainties associated with zeroing was sought. The introduction of optically stimulated luminescence (OSL) (Huntley *et al* 1985) promised the necessary advance in technique because its zeroing characteristics are substantially better than those of TL (although they are related processes). The first experiments with optically stimulated luminescence employed highly monochromatic green light from a laser (of 514 nanometers, nm, wavelength). Under constant illumination from the laser light, blue/ultra-violet(uv) luminescence was detected which fell rapidly to a negligible level within a number of seconds, producing a characteristic decay curve (see below). Both separated quartz and polymineral (containing quartz and feldspar) sand and silt samples (including aeolian deposits of Holocene age) were found to produce strong optically stimulated luminescence signals, the intensity of which increased with the ionizing radiation dose received by the sample before measurement – as is the case for TL and as required for dating. The significance of this development for dating alluvial sediments was that, compared with TL, zeroing of the OSL signal under solar illumination could be more fully and more rapidly achieved.

Studies performed with separated quartz and feldspar samples have provided further detail of their characteristics under optical stimulation, the most important being the differences in luminescence intensity with wavelength of stimulating light. The data for quartz are limited, but those obtained so far are sufficient (from Godfrey-Smith *et al* 1988) to indicate that the efficiency of stimulation is very low for red to infra-red wavelengths, rising strongly between green (circa 550 nm) and blue (circa 450 nm) wavelengths. Greater resolution of the wavelength response has been obtained with the commonly found feldspars (sodic and potassic). The strongest response to stimulation is in regions of the infra-red (circa 800–1000 nm), with lower efficiency between the green and red wavelengths (circa 550–700 nm). The infra-red and green stimulation bands are believed to be associated with the excitation of the same type of crystal defect – this is important because the zeroing for alluvial sediments is likely to have been achieved by visible radiation. A full characterization of the response of potassium feldspar (Hütt *et al* 1988) yielded both a model for the process and the necessary detail to show that infra-red wavelengths could be used to stimulate blue/uv luminescence (referred to as infra-red stimulated luminescence, IRSL); similar processes occur in the sodic feldspars (Bailiff and Poolton 1991). The possibility of using infra-red stimulation provided yet another important advance in technique since it offered immediate experimental advantages in detecting weak luminescence. The efficiency of response of the photomultiplier tubes used to detect the luminescence emission varies with wavelength – being highest in the blue/uv region and decreasing rapidly beyond the green to negligible levels in the infra-red. Thus discrimination against the considerably brighter stimulating light while detecting as much of the weak luminescence as possible is easier to achieve (Figure 3.3) in the case of IRSL compared with green stimulated OSL because of the transmission characteristics of currently available optical filters. Also, suitable light sources are readily available. The use of infra-red emitting diode arrays (Poolton and Bailiff 1989) soon followed, providing a cheap and simple means of observing IRSL using conventional TL equipment, and their use is now widespread. A luminescence decay curve obtained using a diode array and detecting the emission in the blue/uv part of the spectrum is shown in Figure 3.4. In this example the signal intensity has decayed by over 90% after about 200 seconds of infra-red stimulation. Although this is 3–4 times slower than to obtain the same result using green stimulation with quartz, nearly complete depletion can be achieved within minutes of exposure to direct sunlight.

An example of the contrast in the bleaching behaviour of IRSL and TL signals under simulated solar irradiation for similar sediment samples is shown in Figure 3.5. Under the conditions of this particular test,

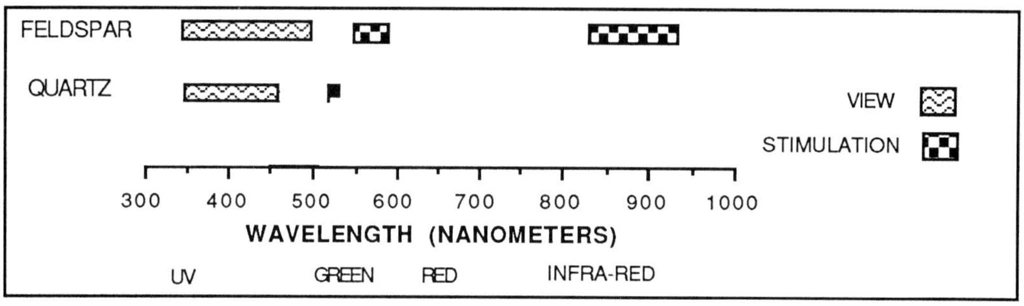

Fig. 3.3. Viewing and stimulation windows for optically stimulated luminescence. The limits of a viewing window are generally determined by the transmission characteristics of the optical filters used in the detection system. The stimulation window represents the limits of wavelengths selected for stimulation; in the case of quartz, the laser 'lines' are contained within a very narrow wavelength range.

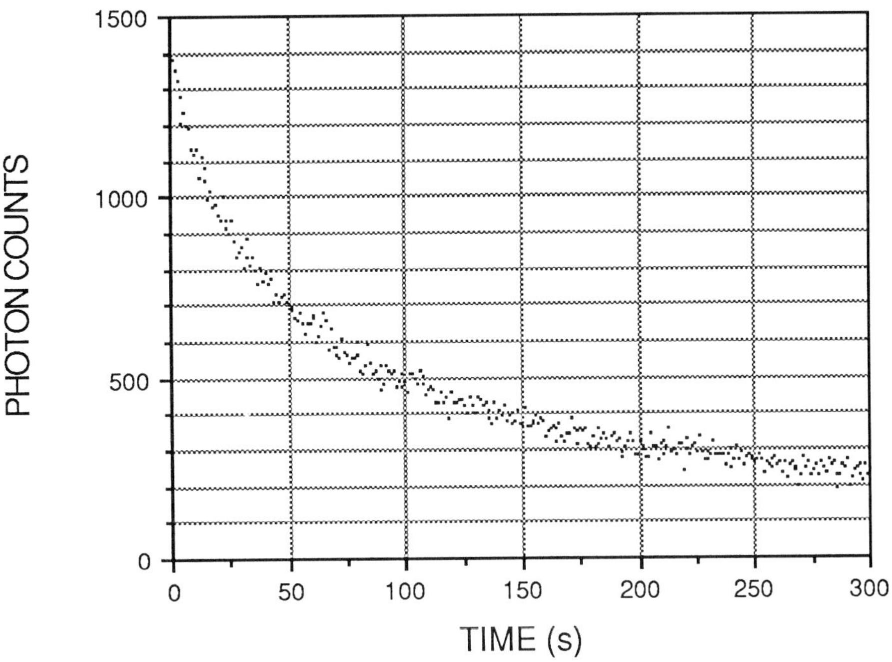

Fig. 3.4. Infra-red stimulated luminescence detected by a photon counting system and recorded as a function of time (seconds) for a sample of Holocene marine sediment using a diode array (860–900 nm). The decay curve reflects the depletion of stored energy (trapped charge) in the crystals; in the example shown, reduction to half the initial value is obtained in approx. 50 seconds (NB the decay is not a simple exponential). In principle the total quantity of light recorded during measurement is proportional to the energy stored in the crystals and the procedures used to evaluate the latter quantity are broadly similar to those developed for TL.

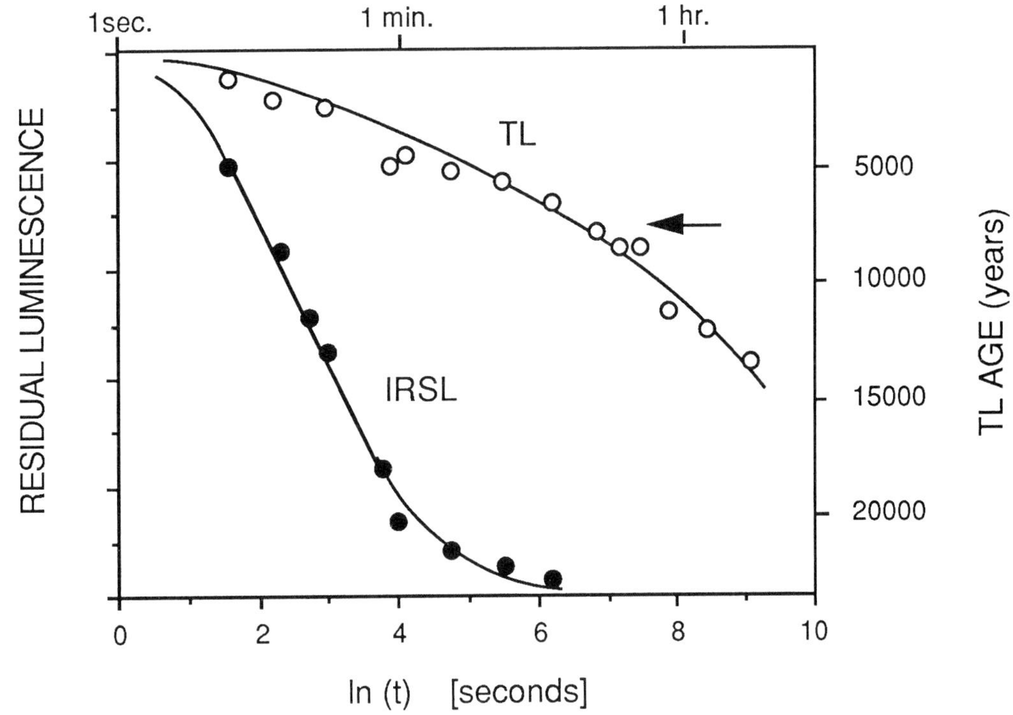

Fig. 3.5. Optical bleaching of a Holocene marine sediment; fine grain samples were placed approximately 0.5 meters from a 300 watt tungsten halogen filament lamp with quartz envelope for periods ranging from several seconds to several hours. The residual IRSL and TL signals were measured after the indicated periods in seconds (the values of time are plotted on a logarithmic scale, with equivalent values in solar units on the upper scale) of bleaching using a separate aliquot for each measurement. The ordinate scale on the right hand side represents the TL age (calculated using data gathered from the required dating measurements) for a given level of residual TL. Arrow refers to known age of sample (see text).

the IRSL was reduced to a negligible level after 20 mins of bleaching, whereas about 70% of the TL signal remained after the same period of exposure. Further reduction of the TL signal was obtained for longer bleaching periods; the data indicate that in excess of 1 day would be required to reduce the TL signal to a negligible level. A sample of this type illustrates the problem of incomplete zeroing discussed above. In Figure 3.5, the ordinate scale (RHS) gives the calculated age for a given intensity of residual TL signal (LHS) and the arrow indicates the age of the chronological control for this particular sample (a radiocarbon age of circa 6500–7000 cal BP for associated biogenic deposits). For the calculated TL age to agree with the radiocarbon age, 70% of the measured TL needs to be residual, ie the data indicate that the residual energy stored in the TL minerals immediately following burial accounts for 70% of the total measured in the laboratory. Under such circumstances the procedure for selection of the correct residual level without independent dating control would be arbitrary and thus inadequate.

There is, however, one aspect of the behaviour of OSL in quartz reported by Smith *et al* (1986) which could mar the dating of Holocene sediments from north-west European sites. Laboratory tests have indicated that following burial there may be a transfer of charge into trapping locations within the crystals (i.e. the energy storage system) which had been previously cleared of charge under sunlight exposure. This effect has been referred to as recuperation and can lead to an over estimation of the luminescence age. As Smith *et al* (1990) have commented in a subsequent paper, resolution of this problem requires the dating of known age deposits; such work is in progress (Stokes 1991). The magnitude of the effect depends on the conditions of bleaching (spectrum of incident light) before burial and is less severe for samples which have been thoroughly exposed to direct sunlight such as wind-blown deposits. Possible over-estimations of about 1000 years are suggested by Smith *et al* (*ibid*) for 'samples which did not receive extended bleaching when the OSL was set to zero at the event being dated' — evidently the problems related to the zeroing process encountered with TL have not yet been eliminated using OSL with quartz but nonetheless they have been substantially reduced. Limited IRSL tests with modern samples have been completed; Aitken and Xie (1991) report values of residual dose for flood silt and soil which correspond to ages of several hundred years.

Although part of the residual dose may be accounted for by a recuperative process, recuperation has yet to be clearly identified in feldspars (and has not been detected so far in this laboratory). If the absence of recuperation is substantiated, it represents a strong advantage of feldspars over quartz. Countering this advantage is the known susceptibility of feldspars to a phenomenon referred to as anomalous fading (Wintle 1977) which is a long-term

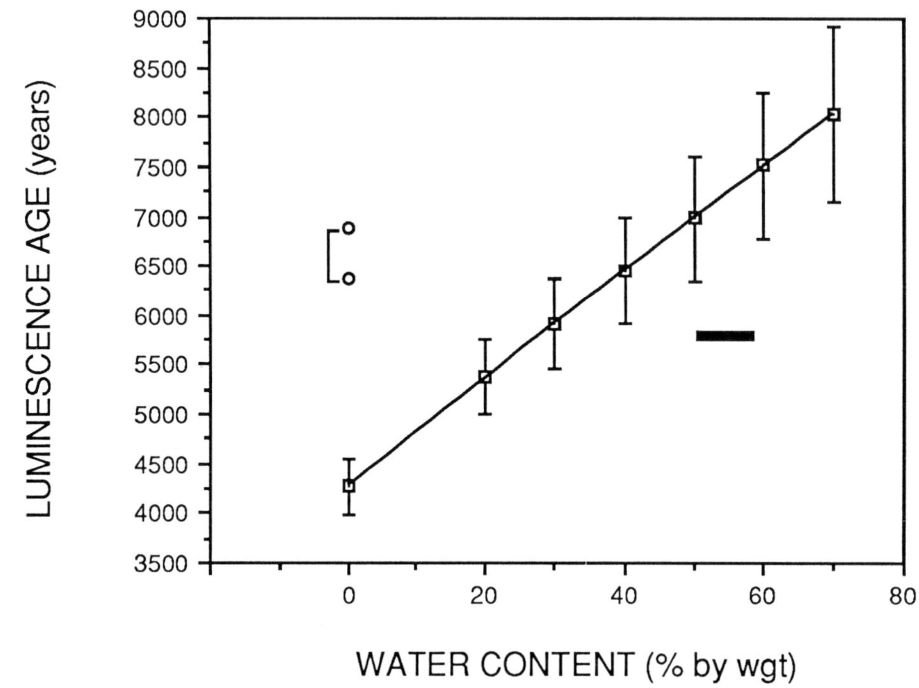

Fig. 3.6. The luminescence age calculated using various values for the average water content with overall error limits (for further discussion of the assessment of random and systematic errors, see Aitken 1985) incorporating an uncertainty of ±10% of the water content. The calibrated range of a carbon date for overlying organic materials is shown by linked circles and the measured in situ water content (50±5%) of the sediment is indicated by an horizontal bar. Errors limits represent the standard error on the mean given at the 68% level of confidence.

instability in the charge storage mechanism (resulting in an underestimate of the true age). However, laboratory procedures (Clark and Templer 1988) which were developed to accelerate the removal of unstable components in TL signals are also being used in IRSL/OSL procedures; so far they appear to offer a satisfactory treatment in certain cases, but considerably more detailed tests of known age *sequences* will be required for a full evaluation. The feldspar group of minerals are known to be highly complex and recent work by Spooner (1991), for example, further underlines the difficulties in establishing the degree of stability of the energy storage mechanism within samples, as probed by IRSL, which may contain a variety of different types of feldspar. Currently, the empirical approach is to exclude samples which exhibit persistent instability after performing long-term storage tests under various conditions, including 'aggressive' tests performed at elevated temperatures to remove unstable components of the luminescence signal. While this is an area of acknowledged complexity requiring more detailed study, so far results from 'known age' samples – with acceptable luminescent properties using current methods of assessment – have provided encouraging examples of agreement within the calibrated range of radiocarbon dating.

SAMPLING AND SITE RELATED FACTORS

Satisfactory samples for luminescence testing can be obtained without resorting to sorties on dark nights; monoliths or cores obtained with metal tins/tubes promptly double wrapped in black plastic after extraction can be used during daylight sampling. Where there is a risk of penetration of light (sand deposits), the sample should be large enough to ensure a 'dark' sample is available within the interior (suggested minimum volume 400 cm^3 where the cross-sectional area is greater than 20 cm^2). Where a face is being sampled, the sampling area should be cleaned up shortly before insertion of tin/tube and shaded from direct sunlight. It is essential that the field samples remain under dark conditions until they are excavated under subdued red light conditions in the laboratory. Retention of moisture is very important, particularly where the water uptake is high (see below).

The choice of technique for dating alluvial deposits may be dependent upon factors which are related to the dynamics of deposition and subsequent burial processes and careful examination of the morphology of the facies is strongly advised before considering dating tests. For example, where alluvial sands have been severely abraded the feldspar content is likely to be insufficient and this will confine tests to the use of quartz OSL (so far, requiring expensive laser systems).

The luminescence from feldspars is generally much stronger than from quartz under the same conditions and both luminesce under green stimulation. Consequently the observation of quartz OSL requires the use of mineral separation and this is most simply achieved using inclusions. Fine-grain samples comprise the range of minerals present within that grain size range and thus, under green stimulation, the OSL is expected to be dominated by feldspar emission. On the other hand currently available experimental data indicate that feldspars in the presence of quartz may be selectively stimulated using IR. Where the characteristics of response to stimulation wavelength for the two mineral groups are contrasting, quartz samples can be checked for the presence of feldspar by using IR stimulation – this has been used to check mineral separation procedures.

Since the behaviour of optically stimulated luminescence from quartz and feldspar differ, procedures are used to ensure that the emissions detected are dominantly from one mineral component. Feldspars are normally present in measurement aliquots as either fine grains (4–11 μm) or inclusions (90–150 μm) and use of these two grain size ranges allow optimization of the radiation dose-rate measurements which are performed to determine the annual dose-rate (as required to evaluate the Age equation – see Aitken 1985). Work with intermediate and larger grain sizes has also been reported, but the majority of dating work is currently conducted with fine grains and inclusions.

POST BURIAL PROCESSES

The other, and no less complex, aspect of age evaluation is the determination of the annual radiation dose which grains of dating mineral receive within their burial medium. The decay of naturally occurring radioactive isotopes of the parent elements uranium, thorium and potassium within the burial medium provides a weak flux of alpha, beta and gamma radiation which, as discussed above, is the driving mechanism of the 'luminescence chronometer'. This flux will be essentially constant where the decay chains (uranium and thorium) are in equilibrium and where there is no transport by chemical action of contributing members of the decay series. The presence of water within the voids of the burial medium acts as an attenuator and reduces the proportion of dose available to the dating minerals. For most waterlogged deposits, in particular unconsolidated alluvial sediments, the saturation water content is likely to be in excess of 100% by weight of dry sediment for a given volume.

The estimated uncertainty in the water content during the burial period is reflected in the error associated with the luminescence age and the magnitude of the effect depends on the mineral and grain size selected for measurement and the contributions of the different forms of radiation (α, β, γ, and cosmic) to the total dose-rate. Providing a steady state prevails (i.e. in semi-arid or permanently waterlogged conditions) or regular short-term change has occurred (e.g. seasonal) and the dose-response of the minerals used for dating is linear, an average dose-rate can be used in the calculation of the age.

The calculation of the luminescence age is slightly less sensitive to water content variation for techniques employing sand size (90–150 μm) compared with silt size (2–11 μm) grains and differences are frequently not significant. However, the change in calculated age with average water content needs careful examination, as shown in Figure 3.6 for a fine grain sample of inter-tidal sediment which has been dated using IRSL. The measured *in situ* water content (50 ± 5%) of the sediment is indicated by an horizontal bar, and the luminescence age calculated using this value (ca. 7000 ± 650 years) is consistent with a radiocarbon date for material from an overlying deposit with calibrated range as shown (linked circles). If the sediment had been dry throughout the burial period, the quantity of stored energy (accrued dose) would be some 30% more. In such cases the calculated error limits, without the uncertainties due to variations in water content, are impressive, reflecting the high precision which can be obtained with IRSL — an advantage which could be exploited with sediments from semi-arid regions.

Since alluvial deposits from north-west European sites are likely to have a porosity which: a) is currently high and b) may have undergone long-term changes (e.g. consolidation) which cannot be reconstructed, the effects of water content on the calculation of age and its associated error should be examined at an early stage of luminescence dating.

SEEKING CHRONOLOGICAL CONTROLS

At the current stage of development of the method the sampling of well characterized sequences which include material for which there are independent dating controls is essential. In the Durham laboratory we are currently studying the potential of the method for dating coastal archaeological sites. Well characterized sequences of coastal marine sediments of Holocene age (Tooley 1978) have been selected where a chronological framework is available from radiocarbon dates for intercalated peats. In addition to the aspects discussed above (Figs.3.3 and 3.4), which emerged in testing these sediments, dates were sought for samples at the interface of minerogenic/biogenic facies. An exploration of the variation in luminescence/unit weight of sediment revealed marked reductions in intensity for material underlying the peat interface. Part of this effect can be accounted for by contact with a much less radioactive medium and a significant increase in water content (and therefore reduction in dose) — providing good reasons to move further from the interface for dating evaluations. However, during further investigations with estuarine and coastal sediments from other sites, strong evidence is emerging to suggest that the luminescence from feldspar minerals can be affected by chemical weathering (R. Parish, poster to this conference). Fine grain samples are particularly susceptible to this problem and, in addition to basic mineralogical assessment of facies, detailed analysis using a scanning electron microscope is likely to be necessary to establish the extent of weathering.

Smith *et al* (1990) report several OSL (514 nm stimulation) dates for alluvium one of which is a date ($6.2\,^{+0.9}_{-1.2}$ ka, Ox735b1; the error term was expressed in this way by the laboratory because tests indicated that the increase in stored energy with time followed an exponential function, rather than the linear case illustrated in Figure 3.2) for sediment from LesTares, France. This agrees with a TL date for the same layer and is thought to be of Early Holocene age based on geomorphological evidence. Tests on Holocene alluvial deposits from the Tyne Valley (in collaboration with Dr M Macklin) in this laboratory using TL and OSL (514 nm) showed that the TL signal had suffered very little bleaching at deposition but a preliminary OSL age of 2.5 ± 0.5 ka is consistent with other dating evidence (Macklin, pers. comm.). Also, Aitken and Xie (1991) have obtained a series of IRSL ages for alluvial and colluvial deposits from British archaeological sites ranging from circa 2 to 5 ka. The reader is referred to forthcoming volumes of *Quaternary Science Reviews* and *Nuclear Tracks and Radiation Measurements* which will include a number of papers concerning the study of optically stimulated luminescence presented at the 1990 International TL and ESR Seminar, held in Clermont Ferrand; further papers on the use of TL for dating alluvial sediments from Australia can be found in proceedings edited by Gillespie (1990).

CONCLUSIONS

The luminescence dating of alluvial sediments from archaeological sites is emerging as a technique of great promise and value since it offers the significant advantage of dating directly the deposition of sediments.

The method is being developed in a number of laboratories along different lines of inquiry. Although optically stimulated luminescence is not a new phenomenon in physics, little is known of the specific physical mechanisms of the phenomenon in minerals and an understanding of such mechanisms will form an essential part of developing an accurate dating technique. Nonetheless, substantial progress has been made using more empirical approaches and luminescence ages obtained for known-age deposits have been extremely encouraging.

Thermoluminescence is by no means eclipsed by the arrival of optically stimulated luminescence and continues to play an important role both in luminescence studies and in the dating of sediments which have received a high level of insolation before burial (Roberts *et al* 1990). It is worth noting that the uncertainty in the degree of effectiveness of optical bleaching achieved before burial is significantly reduced using optically stimulated techniques but not eradicated, and it is likely that sediments of alluvial origin with an archaeological context will be found where zeroing was incomplete. One approach in such situations is to test whether consistent ages can be obtained with mineral components of sediments which have different bleaching characteristics (e.g. OSL vs TL).

An understanding of the process of deposition and incorporation of luminescent minerals in the medium under study, including any physical and chemical changes in the burial medium (e.g. water content, consolidation), is important because such factors may affect the accuracy of the luminescence dates produced. Constituents of many types of sedimentary deposit will have undergone and may continue to undergo processes of mixing, abrasion and exposure which need careful examination, particularly in the case of exposed coastal sites. While it appears that satisfactory results can be obtained with waterlogged sediments there are obvious attractions in studying sediments from semi-arid regions.

The use of optically stimulated luminescence techniques has induced a great invigoration in the field of luminescence dating; whereas there had previously been calls for the use of the term 'apparent' age there is now cautious optimism that this qualification will soon be dropped.

Acknowledgements

The author holds a Science and Engineering Research Council Advanced Fellowship and is pleased to acknowledge additional grant support from the Science-based Archaeology Committee of SERC and the University of Durham for the development of luminescence techniques for dating coastal archaeological sites. This work has been conducted in collaboration with Dr M.J. Tooley and Dr I. Shennan of the Dept. of Geography and with the valued assistance of Dr. N.R.J. Poolton. The author is grateful to the Research Laboratory for Archaeology, Oxford University, for providing access to their laser system for feasibility tests and to Mr S. Stokes for his help in completing them.

References

Aitken, M.J. 1985. *Thermoluminescence dating.* London: Academic Press.

Aitken, M.J. and Xie, J. 1991. Optical dating using infra-red diodes: young samples. *Nuclear Tracks and Radiation Measurements,* in press.

Bailiff, I.K. and Poolton, N.R.J. 1989. Charge transfer in quartz and feldspar; factors affecting the dating of young sediments. In *Synopses from a Workshop on Long and Short Range Limits in Luminescence dating, April 1989.* Oxford University: Research Laboratory for Archaeology and the History of Art, Occasional Publication No. 9.

Bailiff, I.K. and Poolton, N.R.J. 1991. Charge transfer studies in feldspars. *Nuclear Tracks and Radiation Measurements,* in press.

Berger, G.W. 1988. Dating Quaternary events by luminescence. In D.J.Easterbrook (ed.) *Dating Quaternary Sediments; Special Papers of the Geological Society of America,* 227, 13–50.

Berger, G.W. 1990. Effectiveness of natural zeroing of the thermoluminescence in sediments. *Journal of Geophysical Research,* 95(88), 12375–97.

Clark, P.A. and Templer, R.H. 1988. Thermoluminescence dating of materials which exhibit anomalous fading. *Archaeometry,* 30, 19–36.

Gillespie, R. (ed.) 1990. *Proceedings of a Quaternary Dating Workshop 1990.* Department of Biogeography and Geomorphology, Australian National University.

Godfrey-Smith, D.G., Huntley, D.J. and Chen, W-H. 1988. Optical studies of quartz and feldspar sediment extracts. *Quaternary Science Reviews,* 7, 373–80.

Huntley, D.J. 1984. On the zeroing of the thermoluminescence of sediments. *Physics and Chemistry of Minerals,* 12, 122–127.

Huntley, D.J., Godfrey-Smith, D.G. and Thewalt, M.L.W. 1985. Optical dating of sediments. *Nature,* 313, 105–107.

Hütt, G., Jaek, I. and Tchonka, J. 1988. K-feldspars optical response stimulation spectra. *Quaternary Science Reviews,* 7, 381–385.

Poolton, N.R.J. and Bailiff, I.K. 1989. The use of LEDs as an excitation source for photoluminescence dating of sediments. *Ancient TL,* 7, 18- 20.

Roberts, R.G., Jones, R. and Smith, R.G. 1990. Thermoluminescence dating of a 50,000 year old human occupation site in northern Australia. *Nature,* 345, 153–156.

Smith, B.W., Aitken, M.J., Rhodes, E.J., Robinson, P.D. and Geldard, D.M. 1986. Optical dating: methodological aspects. *Radiation Protection Dosimetry,* 17, 229–33.

Smith, B.W., Rhodes, E.J., Stokes, S., Spooner, N.A. and Aitken, M.J. 1990. Optical dating of sediments: initial quartz results from Oxford. *Archaeometry,* 32(1), 19–31.

Spooner, N.A. 1991. Optical dating: preliminary results on the anomalous fading of luminescence from feldspars. *Quaternary Science Reviews,* in press.

Stokes, S. 1991. Optical dating of young (modern) sediments using quartz: results from a selection of depositional environments. *Quaternary Science Reviews,* in press.

Tooley, M.J. 1978. Sea-level changes. *NW England during the Flandrian Stage.* Oxford.

Wintle, A.G. 1977. Detailed study of a thermoluminescent mineral exhibiting anomalous fading. *Journal of Luminescence,* 15, 385–393.

Wintle, A.G. and Huntley, D.J. 1980. Thermoluminescence dating of ocean sediments. *Canadian Journal of Earth Sciences,* 17, 348–360.

Wintle, A.G. and Huntley, D.J. 1982. Thermoluminescence dating of sediments. *Quaternary Science Reviews,* 1, 31–53.

4

Magnetic dating of alluvial deposits

A. J. Clark

Summary: *Detrital remanent magnetism, which retains a record of the geomagnetic field direction, can be used to date sediments including alluvium. A calibration for Britain, back to 10,000 cal BP, has been obtained from a combination of the archaeomagnetic record and cores taken from three lakes. Sediments can either be sampled as columns which can show the time span of deposition and variations in its rate, or alternatively by multiple sampling at one level to produce a single date. Dry-land samples are normally taken in a series of 5 cm deep plastic tubes, the columns being taken from sections. Some of the better data obtained are contributing improvements to the original calibration. Improvements will also come from the wider use of objective matching procedures. When this method of dating is combined with magnetic intensity and susceptibility measurements, there is potential for extracting fuller information about the history and nature of the sedimentation process.*

DETRITAL REMANENT MAGNETISM AND MAGNETOSTRATIGRAPHY

It is well known that the direction and intensity of the geomagnetic field are continuously changing with time. Practical magnetic dating depends upon mechanisms which are able to preserve a record of the direction at a particular time. The best-known example is the thermoremanence of fired clay, which makes possible the dating of structures such as kilns. Another mechanism is the detrital remanent magnetisation which is acquired by sedimentary deposits such as alluvium: magnetised particles, predominantly iron oxides, which these can be expected to contain, tend to align with the geomagnetic field during the process of settlement, so that the deposit retains a record of the changing field direction – the secular variation – as it builds up (Creer *et al* 1983).

The quality of the magnetic record retained by a sediment is adversely affected by turbulence and flow during deposition, post-depositional disturbances such as bioturbation, and by chemical changes which cause redistribution and alteration of the iron compounds. These, and slumping, can be especially troublesome if the environment of the deposit changes from its underwater equilibrium to dry land conditions with such exacerbating problems as oxidation and fluctuating water-tables. It is fortunate, however, that the most common chemical alteration, the development of rusty-looking lepidocrocite mottling, is not necessarily disastrous, because this compound is very weakly magnetic.

SEDIMENT MEASUREMENTS IN THE ARCHAEOLOGICAL CONTEXT

Sediment measurements in Britain are normally made as a vertical sequence, so that the pattern of changing declination and inclination values can be matched to the separate reference curves mentioned below. A widely used instrument is the Molspin fluxgate spinner magnetometer. This can be adapted to take long cores, but most land-based measurements are made on columns of individual samples taken from sections. These can be collected in specially made non-magnetic containers (Austin & Baldwin 1984), but in the method used by the writer and others they are economically encapsulated in approximately 5 cm diameter x 5 cm long tubes cut from uPVC drainpipe. If the deposit is soft enough, the tubes can be pushed horizontally into the section, or a 'shelf' can be prepared into which they are pushed vertically. If the tubes cannot be pushed in because the deposit, or part of it, is too hard or stony, samples can be taken from a shelf by cutting away material to leave a short 'pillar' or stump, which is fixed with plaster of Paris into a levelled tube dropped over it. The levelling gives a reference for the angle of dip or declination, while a north reference line is marked on the tube or plaster. This reference may be obtained simply with a magnetic compass or, if there is magnetic interference, from a timed sun observation or a gyro theodolite (Clark *et al* 1988). After removal, samples are normally completely encapsulated with plaster and stored in sealed polythene containers or cling-film wrapping to prevent drying out and shrinkage.

To improve the quality of any series, especially if the material sampled is of dubious stability, two or more samples can be taken at each level and their results averaged. An alternative column sampling approach has been designed by Austin and Baldwin (1984), consisting of a frame with a controlled sample container insertion mechanism which is clamped to a section face. A number of samples can be taken for each position of the frame, and the results show excellent consistency.

In the laboratory, the measurement process includes testing magnetic stability by stepwise partial demagnetisation in a progressively increased alternating field: any viscous magnetic components – acquired under the influence of the changing geomagnetic field after deposition – are removed by this means, or by prolonged storage in zero field.

These methods have been used successfully on estuarine clays at Westward Ho!, Devon (David & Clark 1987) and at the mouth of the river Crouch in Essex (Austin 1991), as well as dry-land alluvial deposits, examples of which are discussed in this paper. Some success has even been obtained with soils that have never been submerged (Clark *et al* 1988).

An alternative type of measurement, especially relevant to thin deposits that may represent a brief inundation, is multiple sampling at a single level. Measurements from such a group of samples are averaged to produce a single high quality reading, a strategy similar to that adopted for the archaeomagnetic dating of a feature by its thermoremanence. So far, some suitable sediments have been found in archaeological features such as ditches and water-holes, and the method has also proved effective, alongside the column method, for dating estuarine clays (David & Clark, 1987), and alluvial deposits resulting from brief deposition episodes such as a major river flood. For instance at Monmouth, a deposit of this kind found in excavations in Monnow Street by the Monmouth Archaeological Society gave a date of AD 1315–1345 at the 68% confidence level (AJC-67), which was in accord with the archaeological evidence. For periods earlier than the beginning of the archaeomagnetic curve, difficulties arise from uncertainties in the absolute values of declination and inclination in the master curves derived from the lake sediments (discussed below) and, furthermore, the occurrence of periods when changes are small or values are similar at different times. The choice between the alternatives arising from this latter problem, a major one in magnetic dating, can usually be resolved from auxiliary evidence in later times, but with increasing difficulty in more remote periods because of the paucity of such evidence. The problem is much less important with columns or cores, because the patterns of relative change can be matched with less dependence on absolute values. Nevertheless, the situation can only improve with the increasing accumulation of data.

CALIBRATION

A prerequisite for magnetic dating is a calibration of the variations in declination and inclination with time. These follow no calculable pattern and, prior to the beginning of direct recording in 1576, the calibration has had to be built up from measurements of two types: archaeomagnetic measurements mainly on burnt features, including a proportion of known date, and palaeomagnetic measurements on cores taken from lake-bottom sediments. The burnt features may be dated by historically documented events, dendrochronology or high quality radiocarbon determinations, while the lake cores depend mainly on radiocarbon measurements. A further problem is that the master curves obtained are only valid for an area of about 500 km radius, and a normalisation procedure must be applied to readings even within this area before they are truly comparable. Calibrations for Britain have been established from the present to 2000 cal BP, using the first method combined with directly recorded data from AD 1576 (Clark *et al* 1988), and back to about 10,000 cal BP (Fig. 4.1) using the second (Turner & Thompson 1982), with a useful overlap between the two. The archaeomagnetic calibration is normally presented as a combined plot of inclination against declination – a Bauer plot – which is most useful for spot readings from individual features, while separate curves of declination and inclination are easier to use for columns of sediment readings.

Archaeomagnetic data are on the whole the more satisfactory for the construction of master curves. They are affected by the magnetic distortions that occur in such generally strongly magnetised material, and by the unsystematic way in which the curve has to be built up as sites are discovered, as well as the inexact reference dating that many of them provide. But new points are constantly being added, and the curves can only improve. The quality of the lake record is variable. The fine and closely reproducible detail found in a lake such as Elk Lake, Minnesota (Sprowl & Banerjee 1985), with yearly varves against which the magnetic record can be calibrated, has not yet been equalled by any lake in Britain, where the best record so far obtained has been in Loch Lomond. In the almost complete absence of varves in Britain, calibration has had to depend upon radiocarbon measurements on the organic content of the sediment, which must be done judiciously to avoid the apparent age being increased by the inwash of 'old' carbon; and even when the age offset is thought to be minimal, there remains an inherent imprecision in the radiocarbon measurements (expressed by the error term) and possible further loss of precision in the calibration process into calendar date bands.

The archaeomagnetic master curve is constantly supplemented by new data from individual features, but the programmes of core sampling and measurement involved in obtaining reference data from lake beds are substantial exercises which cannot be repeated often. However, land-based sampling is contributing to our knowledge of the curve.

A difficulty with lake-bed cores is orientation, which cannot be directly established, and to a lesser extent levelling. Turner and Thompson could only orientate their lake cores by matching their declination measurements with a turning point at AD 1820 on the directly recorded part of the curve, at a level just below the surface of the sediment where their data were rather poor, probably because of disturbance during sampling. Subsequent increases in our knowledge of the archaeomagnetic curve have made an improved estimate possible. A provisional correction was based on a

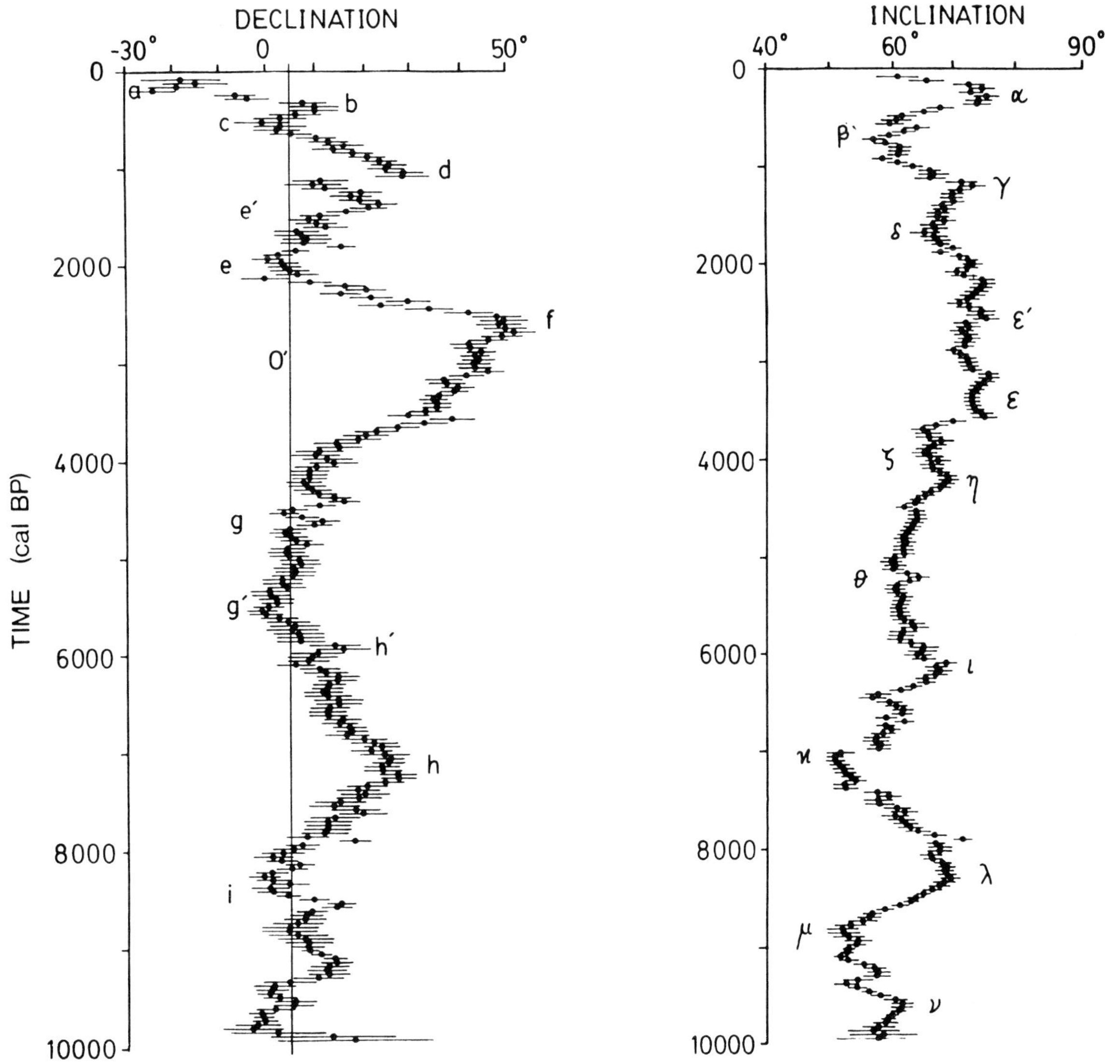

Fig. 4.1. The curves of Turner and Thompson, based on 10 cores from Loch Lomond, Lake Windermere and Llyn Geirionydd. The calibration is in calendar years. Mean magnetic directional values are plotted at 40–year intervals with 95% confidence limits for 80–year windows. Prior to 2000 BP, the 'preferred time' scale is based on radiocarbon determinations on the cores, and its error limits cannot be precisely quantified. The Roman lower case and Greek letters mark significant turning points useful for matching. The easterly excursion marked f, the largest in the 10,000-year span, is a particularly valuable benchmark in sediment column interpretation.

comparison of the mean direction of this with that of the lake data for the part of the Roman period up to cal AD 250, during which the average declination was close to due north, with little variation. This indicated that the lake curve needed to be moved 6.8° west (Clark *et al* 1988); but comparing the means of the lake and archaeomagnetic data over the longer period of 350–2000 cal BP gives a difference of only 4.4°. There is also a less reliable value of 3.3° from Brean Down (see below). Allowing appropriate weighting for these figures, a provisional declination zero correction of 5°E is proposed. There remains the proviso that the declination correction need not be constant along the length of the curve: this is because of the varying quality of the data and the possibility of some twisting during core extraction.

Inclination measurements are also likely to be erroneous because of deviations of cores from vertical. However, this is normally a small effect, which should have been largely removed by the use of the average of measurements from 10 cores. More fundamental is a general tendency for shallowing of inclination angles.

A series of dry land measurements obtained from buried sediments at Brean Down, Somerset (AJC-19), gave high quality results probably because they had remained waterlogged since deposition, covering the period 5450 to 3630 cal BP. Compared with the lake master curves, this indicates a smoother calibration, and a zero correction of only 3.3°, but inclinations 5.4° steeper on average (Clark 1990). Results from alluvium at Runnymede Bridge, on the River Thames, although not of such high quality, probably because of greater exposure to water flow and subsequent chemical changes, also show steeper inclinations as well as supporting the declination correction, as exemplified by the first of them to be published (Clark & David 1991). A rare example of this effect with accurate control data was provided by a column from the silt of the millpond at the Medieval village of Wharram Percy, South Yorkshire. The silt was deposited over the last 400 years, and reproduced the shape of the directly recorded curve for that period, though with inclination reduced by an average of about 6° when normalised (Tarling 1983: 154–5).

The explanation for shallowing could lie in the geometry of the magnetised particles in some sediments: if the particles are generally elongated rather than approximately spherical, their magnetisation is likely to align with the long axis. On settlement, such particles tend to lie flat, reducing the apparent inclination. This effect could have been stronger in the lake sediments used for the calibration than in the material used for the measurements discussed above. However, the comparison already mentioned between the lake and archaeomagnetic data from 350 to 2000 cal BP showed the mean inclination of the lake sediments to be only about 2° less than the archaeomagnetic mean. Thus the deviation may increase with age, which would imply a change in the nature of the sediments with time, perhaps from predominantly mineral grains to cultivated soils with more nearly spherical particles — although this is not in keeping with the Wharram Percy evidence. A much more satisfactory explanation is that the inclination shallowing is due simply to compaction, which would tend to increase with depth (Blow & Hamilton 1978; Tucker 1983).

Support for a shallowed inclination in the sediment reference data comes from two clay hearths at Runnymede, from a context with a core period dated by radiocarbon to about 3800–3500 cal BP. Their mean declination is in keeping with the reference curve with the modified zero proposed above; but the mean inclination, taking into account the error limits of the curve and of the hearth samples, and including an allowance for magnetic refraction in these, is steeper than the average of the calibration curve over this period by 10.9° ± 3.3° at the 95% confidence level. The Brean Down column showed a smaller difference, presumably because it was also affected by some shallowing.

The hearths were small and broken up, and required unusually careful sampling and analysis. The precious material has been retained for further study, which could determine the result more precisely.

There is a need for more evidence for the inclination deviation and its possible variation before the calibration master curves can be confidently modified. Even when this problem is satisfactorily resolved, it will still be uncertain to what extent inclination shallowing affects a column being measured — Wharram Percy, for instance, seems unusually steep for a young sediment. Thus, although the modified declination values can probably be used with reasonable confidence, the inclination matching of columns is likely to depend more on the general pattern of changes than on absolute values. Corrected reference curves will be most useful for dating fired single-date features such as hearths, for which inclination values can be more reliably estimated.

MATCHING COLUMN DATA TO THE REFERENCE CURVES

This is comparable with the matching procedure in dendrochronology, but complicated by additional variables: deposition rates are changeable, with no year markers, and deposition may be interrupted by periods of stasis or even erosion. So far matching has been largely confined to visual comparison, checked by correlation coefficient testing. Matching involves cognisance of absolute values, overall trends — especially in view of the uncertainty of the inclination reference data discussed above — and turning points in the data, and is much assisted by a computer program able to stretch and compress the data alongside the reference curves. Progress is being made with the introduction of objective statistical matching methods, as discussed by Clark (1983), and Thompson and Clark (1989).

INTERPRETATION OF A COLUMN

An example of the matching process is given in Fig. 4.2, which shows a column through 2.5 m of alluvium beside the Severn estuary, taken at Rumney Great Wharf on behalf of the Glamorgan-Gwent Archaeological Trust (AJC-56). The plots are of the raw data; this is 'noisy', but it was found that any smoothing tended to suppress turning points important to the matching process; nor could the results be improved by partial demagnetisation treatment. The sample column is shown schematically, with an indication of the materials encountered on the left and dates derived from the interpretation on the right. In the data plots, the first point to observe is the overall rise and fall of the inclination, a westerly trend in the declination with advancing time, and a series of easterly readings around the level of the thick peat band (with one relatively very low reading perhaps due to bioturbation in the peat), likely to represent the distinctive easterly excursion marked f on the reference curve (Fig. 4.1). The peat implies a period when practically no accretion took place, and, judging from the narrowing of the excursion, this could have lasted for as long as 1300 years. Further brief periods of stability are indicated by thin peat bands just before and during the Roman period. The lesser excursion at about 1200 cal BP (d) is recognisable but diminished, whereas the westerly extremum at 2000 cal BP is exaggerated.

The plot of magnetic intensity shows high readings coinciding with two periods of rapid accretion. The lower group seems to have passed its peak below the bottom of the column in the Neolithic; the upper is the material with which accretion recommenced in the Late Iron Age, overwhelming the peat. The magnetic enhancement of both could be due to run-off from upstream ground clearance involving fire — which increases magnetic susceptibility — or the erosion of highly magnetic soil, from the Cotswolds for instance.

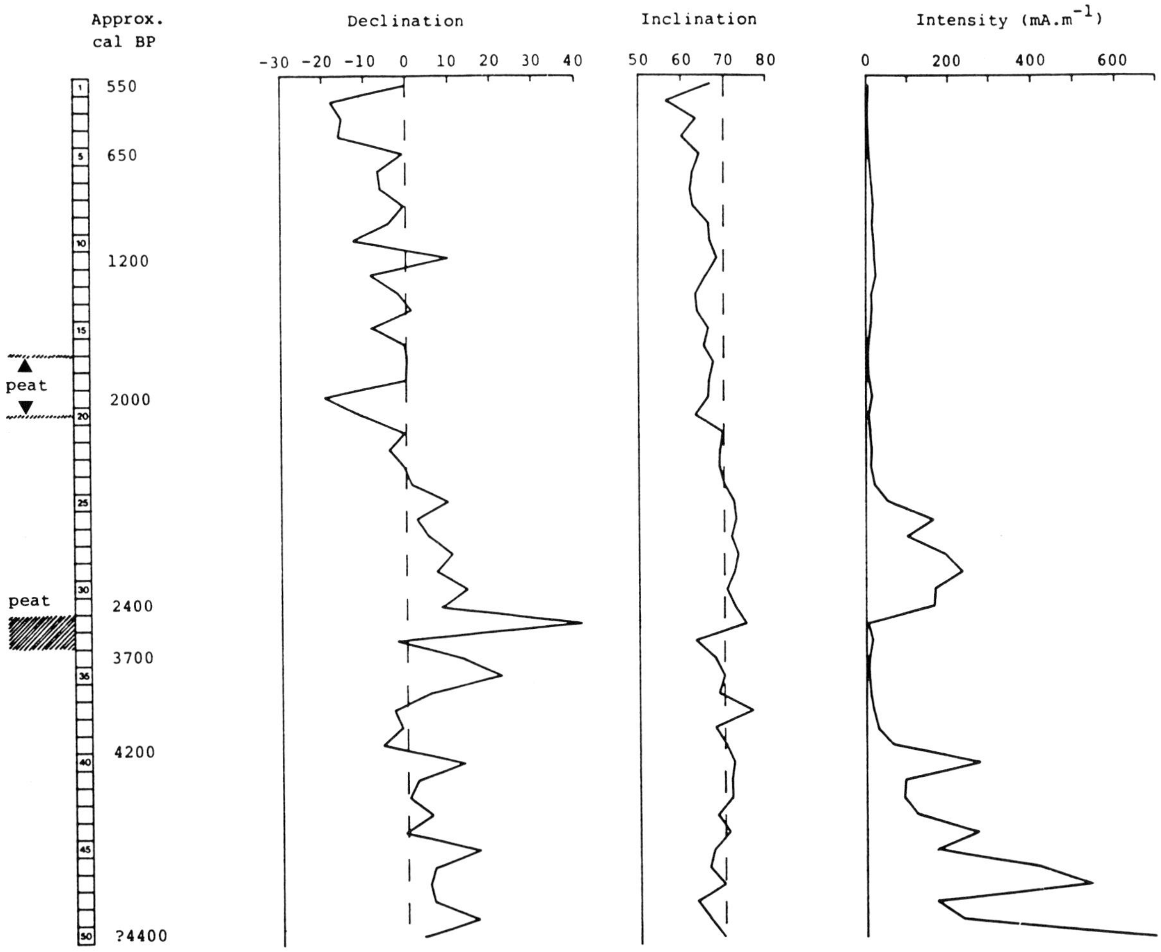

Fig. 4.2. Data from the 2.5 m column taken at Rumney Great Wharf, Gwent.

Dual frequency magnetic susceptibility measurements on the alluvium would probably help to resolve this.

Overall, the interpretation of the imperfect directional magnetic record of this column remains tentative, especially at its ends where the measurements are not supported by data on both sides as are those closer to the centre. The lower limit was particularly difficult to be certain about. Radiocarbon determinations for the peat are planned, and will provide a most valuable check — yet another way in which on-shore measurements can augment the original lake-derived calibration curves (Austin, 1991).

Acknowledgements

I am most grateful to Roy Thompson and Gillian Turner, and the Royal Astronomical Society, for permission to reproduce, not for the first time, their lake sediment magnetograms; and to the Glamorgan-Gwent Archaeological Trust to adapt the information and figure from the report on Rumney Great Wharf. Any errors in the interpretation of this are my responsibility. I am also grateful to my anonymous referees for helpful suggestions.

References

Austin, C.J.F. 1991. Preliminary magnetostratigraphic dating of inorganic estuarine clay sequences along the estuary of the River Crouch, Essex. *Quaternary Newsletter*, 64, 29–40.

Austin, C.J.F. and Baldwin, C.T. 1984. A fast and accurate palaeomagnetic sampler for field use. *Boreas*, 13, 273–277.

Blow, R.A. and Hamilton, N. 1978. Effect of compaction on the acquisition of a detrital remanent magnetization in fine-grained sediments. *Geophysical Journal of the Royal Astronomical Society*, 52, 13–23.

Clark, A.J. 1990. Archaeomagnetic measurements. In M. Bell, *Brean Down Excavations 1983–1987*, 113–116. London: English Heritage.

Clark, A. and David, A. 1991. Palaeomagnetic dating. In S.P. Needham, *Excavation and Salvage at Runnymede Bridge, 1978*. London: British Museum Press.

Clark, A.J., Tarling, D.H. and Noël, M. 1988. Developments in archaeomagnetic dating in Britain. *Journal of Archaeological Science*, 15, 645–667.

Clark, R.M. 1983. Statistical analysis of palaeomagnetic data. *In* K.M. Creer, P. Tucholka and C.E. Barton (eds.) 1983, 249–261.

Creer, K.M., Tucholka, P. and Barton, C.E. (eds.) 1983. *Geomagnetism of Baked Clays and Recent Sediments*. Amsterdam: Elsevier.

David, A.E.U. and Clark, A.J. 1987. Magnetic measurements. In N.D. Balaam, B. Levitan and V. Straker (eds.), *Studies in Palaeoeconomy and environment in South West England*, 249–253. Oxford: British Archaeological Reports, British series, 181.

Sprowl, D.R. and Banerjee, S.K. 1985. High-resolution palaeomagnetic record of geomagnetic field fluctuations from the varved sediments of Elk Lake, Minnesota. *Geology*, 5, 531–533.

Tarling, D.H. 1983. *Palaeomagnetism*. London: Chapman & Hall.

Thompson, R. and Clark, R.M. 1989. Sequence slotting for stratigraphic correlation between cores: theory and practice. *Journal of Paleolimnology*, 2, 173–184.

Tucker, P. 1983. Magnetization of unconsolidated sediments and theories of DRM. *In* K.M. Creer, P. Tucholka and C.E. Barton (eds.) 1983, 9–19

Turner, G.M. and Thompson, R. 1982. Detransformation of the British geomagnetic secular variation record for Holocene times. *Geophysical Journal of the Royal Astronomical Society*, 70, 789–792.

5

Archaeogeophysical prospecting on alluvium

A. J. Clark

Summary: *The three most useful ground-based prospecting methods for the rapid evaluation of the archaeological content of a landscape are magnetometer survey using the fluxgate gradiometer, magnetic susceptibility measurement and phosphate sampling, which all respond sensitively to anthropogenic effects. The first is especially effective for the detailed planning of individual burnt structures and cut features, while the others are generally more suitable for the broader definition of former occupation areas. Electrical resistivity, conductivity and ground penetrating radar have more limited application because they are less specific in response and generally slower. In searching for archaeology under alluvium, the limit of detection of minor features by gradiometer survey is about 1 m depth of deposit, and for stronger features such as kilns about 3 m. Total field instruments such as proton magnetometers, and Overhauser and optically pumped instruments, have rather better sensitivity, but generally slower ground coverage. Magnetic susceptibility may require boreholes for alluvium deeper than 0.5 m, depending on the type of instrument used. For small features, the limit of resistivity is generally about 1 m, but it and larger conductivity instruments can define substantial features at greater depth and, together with radar, are useful for planning subsurface topography, which can provide a guide to the most likely settlement areas.*

THE AVAILABLE TECHNIQUES

Magnetic methods

Modern exploitation of the landscape demands rapid and economical archaeological evaluation of large areas. Archaeogeophysical specialists have risen to the need (Clark 1990). Magnetic techniques are especially powerful because of their speed and response to burnt structures and anthropogenic enhancement of soil magnetic susceptibility. The mainstay is magnetometer survey with the fluxgate gradiometer, which can produce detailed, high-sensitivity surveys at a rate of about two hectares per day. The type of instrument assumed here has a 0.5 m detector separation and a resolution of about 0.1 nT (nanotesla, equivalent to the formerly used gamma), for instance that of Geoscan Research. For standard surveys, the instrument is carried at a steady but brisk walking pace along traverses a metre apart and generally 20–30 m long, with readings automatically logged at 2–4 per metre. Using markers at either end of the traverses and matching pacing to a timing device reduces marking out to a minimum. Results are monitored on a computer screen as the survey progresses, and later processed into sophisticated profile and half-tone plots.

Magnetometers respond to local modification of the geomagnetic field by magnetic iron oxides in archaeological features, either due to the thermoremanent effect in fired structures, or the silting of more magnetic soil into features cut into the natural. The magnetism of the soil is enhanced by activities associated with human occupation, especially burning. Magnetic susceptibility measurement, in contrast to magnetometers, responds to the general enhancement of soil magnetism by these activities. Used for detailed surveys, it can sometimes extract remarkable archaeological detail from the topsoil, but it is mostly used for the broad definition of areas of former occupation, either in its own right, or as a precursor to fluxgate coverage, or to complement field walking. Such a survey can be effective with readings on a grid as coarse as 15 m or even 30 m, enabling large areas to be assessed with great economy. A square kilometre can be surveyed in this way on a 15 m grid by a small team in about two weeks.

A number of ingredients contribute to the interpretation of a susceptibility survey. An essential preliminary is a knowledge of the underlying geology, variations of which will affect the absolute level of background readings, and also a geomorphological appreciation of relatively recent soil accretion or erosion likely to have affected the survival or accessibility of archaeological features. The basic survey can be carried out with field sensors, or samples can be taken for more accurate and sophisticated measurements in the laboratory. In particular, a dual frequency laboratory sensor will measure the frequency dependence of a susceptibility measurement. A relatively low reading at the higher frequency reveals the presence of a narrow range of fine particle sizes, the production of which is favoured by human activities ranging from burning to cultivation (Thompson & Oldfield 1986). This helps to distinguish anthropogenic susceptibility enhancement from changes of geological origin. Another useful test is to determine the potential susceptibility enhancement of a few background samples, which can be done by measuring the samples before and after a simple heat treatment. This needs to be repeated for any variation in underlying geology. If this test gives strong enhancement in an area where field measurements are all low, it can then be stated with some confidence that prolonged human settlement is unlikely to have existed in the area. If the test gives little increase, it means that former occupation areas will be difficult or impossible to detect by susceptibility. It will also suggest that any considerable increases that do occur in the area are probably of natural rather than human origin. An adequate method for this test is to place the soil sample in a lidded crucible and bring it to red heat with a gas flame: a crude version of an experiment developed by Tite (1972).

It seems possible to define some human artefacts even in such coarse surveys. In a recent survey on a 15 m grid, negative lynchets appeared to be represented by dilution of topsoil by less magnetic subsoil, and it was also possible to identify such features as the lines of old hedgerows, partly from the tendency of magnetic rubbish to collect along them, and partly because of a pedological enhancement effect that can occur under trees (Thompson & Oldfield 1986). An area of former woodland was tentatively identified on the same basis.

The discussion above assumes the use of equipment such as that of Bartington, which requires intimate contact between the sensor and the measured soil. Another class of magnetic susceptibility instrument makes use of a transmitter and receiver separated by a horizontal boom, typically 1–1.5 m in length, and carried above the ground. A sophisticated version, developed for archaeology, is the SH3 of Tabbagh (1986). This will be discussed below, but good results, at least with fairly shallow features, can also be obtained with, for instance, the Fisher M-Scope, designed as an inexpensive metal detector for finding pipes and cables. Although such instruments can have quite deep penetration, sensitivity falls off as the sixth power of the depth, more rapidly than a magnetometer. They can be carried continuously, with automatic recording, but accuracy and resolution depend critically upon height control. As they respond to the overall magnetic susceptibility of the soil, and increase rapidly in sensitivity near the surface, they do not clearly differentiate between overall magnetic enhancement of soil over a site, thin occupation layers and deeper cut features. However, this lack of discrimination is a positive advantage in the initial location of occupation sites.

Instruments of this type, when operated at a suitable frequency, will respond to soil conductivity rather than susceptibility, and are then equivalent to resistivity meters. The latest version of the long baselength, deeply penetrating Geonics EM31 is able to record simultaneously both susceptibility and conductivity measurements.

Phosphate

This type of geochemical survey is another technique well suited to large areas (Craddock *et al* 1985). Like magnetic susceptibility, it responds to the broader effects of human occupation, but indicates additions to the soil from the detritus of living rather than the alteration of soil properties detected by magnetic susceptibility sensors. Because of this, detailed studies applying both techniques can provide subtle complementary information by identifying different areas of a site: susceptibility will be biased towards living areas, but phosphate towards areas of rubbish disposal. In landscape assessments, they can be assumed to respond broadly to the same occupation areas, providing mutual confirmation. Similar grid intervals are appropriate; several hundred phosphate samples can be processed per day, so that the rate of working is also comparable. The addition of phosphate survey to susceptibility must increase the reliability of the archaeological deductions made, but there seems little doubt that its much rarer use is due to the inhibiting effect of the wet chemistry involved, including the frequent need to make up fresh reagents!

Resistivity

This venerable method of archaeological prospecting has been little used for evaluating large areas because of its relative slowness and lack of specificity of response to the anthropogenic alterations of soil properties to which the magnetic and phosphate methods are sensitive. It tends to be regarded as a secondary technique for obtaining additional information about a buried site. It is, however, more powerful as a method of detecting and defining building foundations and similarly solid structures, and long, widely spaced search traverses are sometimes used for the location of, for instance, Roman buildings and roads. There is evidence that the approximate location of a building can be ascertained by magnetic susceptibility, which is enhanced by the effects of its occupation – and sometimes a magnetometer can define the walls effectively as negative anomalies. However, electromagnetic conductivity instruments, as well as the tractor-drawn 'rateau' resistivity system of Hesse (Scollar *et al* 1990) on suitably accessible sites, have the scarcely realised potential to make this type of measurement competitive in speed.

Ground probing radar

The development of ground penetrating radar for archaeology is receiving much attention at the time of writing (Scollar *et al* 1990). It has the attractive ability, in favourable conditions, to give quite detailed images in profile to considerable depths. It is relatively slow to use, and for immediately subsurface sites with minor features it can be less satisfactory than the techniques described above. Although most effective in penetrating dry soils, it could be useful for seeing through alluvial deposits, as will be discussed below.

THE SEARCH FOR ARCHAEOLOGY UNDER ALLUVIUM

Magnetometry

The techniques described above are effective for locating the majority of archaeological sites superficial to the local lithology and not far from the surface. If archaeology lies on top of an alluvial deposit, all is well providing suitable susceptibility contrasts exist between the alluvium and the features or their fillings; but the masking effect of later alluvium (or colluvium) can be troublesome. Figure 5.1 demonstrates how the magnetic plan of a site can be distorted by alluvium deposits. Another example is provided by Vespasian's vexillation fort at Wimborne, Dorset (Clark 1990: 139–41). The defences of this cease to be directly traceable where they pass beneath alluvium deposited by the River Stour, although the situation is redeemed by a line of strong peaks probably caused by hearths, kilns and furnaces sited just behind the defences, away from the internal buildings. These peaks allow the limits of the fort to be continuously traced through the deepest part of the alluvium.

The magnetic signal of a feature falls off approximately as the inverse cube of the depth for discrete features such as pits and kilns, and as the inverse square for linear features such as ditches (Clark 1990). The sensitivity fall-off with a gradiometer is more rapid than with a total field instrument such as a proton magnetometer because of the subtractive effect of the upper sensor. A second sensor is also required for total field instruments, to subtract out diurnal geomagnetic variations, but this can be remotely placed so that it does not enter into the sensitivity equation. For technical reasons, this is not possible with the fluxgate detector. Let us consider an archaeological feature with its magnetic centre a typical 0.5 m below the surface of the natural, under 20 cm of soil, giving an effective separation of 1 m between the feature and a sensor carried 30 cm above the ground surface. For a discrete feature at this depth, a typical 5 nT anomaly obtained with a 0.5 m fluxgate gradiometer would be increased to 7.5 nT with a total field instrument, or 9.6 nT for a linear feature. Table 5.1 shows the reduction in this anomaly size with the addition of increasing depths of alluvium cover. The 5 nT initial gradiometer anomaly could be caused, for instance, by a pit of 1 m depth and diameter with a magnetic susceptibility contrast of 137×10^{-8} SI/kg between its filling and the surrounding ground. The lower part of the table shows the effect on a relatively strongly magnetic discrete feature such as a kiln or a bloomery hearth, giving a 100 nT gradiometer signal at normal depth.

These figures show that, for relatively minor features, the effective depth of alluvium with which the instruments can cope is not much more than 1 m. An exception is the response of the total field instrument to a linear feature, which is still appreciable under a 2 m (1.1 nT) or even 3 m (0.7 nT) deposit. However, it must be borne in mind that the spatial definition of anomalies also declines with depth.

The use of highly sensitive magnetometers of the optically pumped or Overhauser types in differential configuration might be beneficial. A common argument is that these expensive instruments would provide no advantage because of limitations to their effective sensitivity set by 'noise' generated by soil irregularities and magnetic contaminants such as igneous stones. These problems tend to be minimal in a homogeneous and relatively young deposit such as alluvium, at least in the absence of surface irregularities such as furrows. On the other hand, the response to archaeological features embedded in underlying gravel can be affected by natural magnetic inhomogeneity, channelling, and low magnetic contrast with the gravel itself. In lowland England, generally good results are obtained from the gravel terraces of rivers rising on or close to the Jurassic ridge, whose strongly magnetic soils contribute to an alluvium giving clear responses where it has silted into cut features, and tending to dominate any gravel problems. For instance, in a survey at Radley on the upper Thames, ring ditches gave gradiometer anomalies of about 15 nT at normal depth, which would have been a readily detectable 2.3 nT under a 1 m alluvium

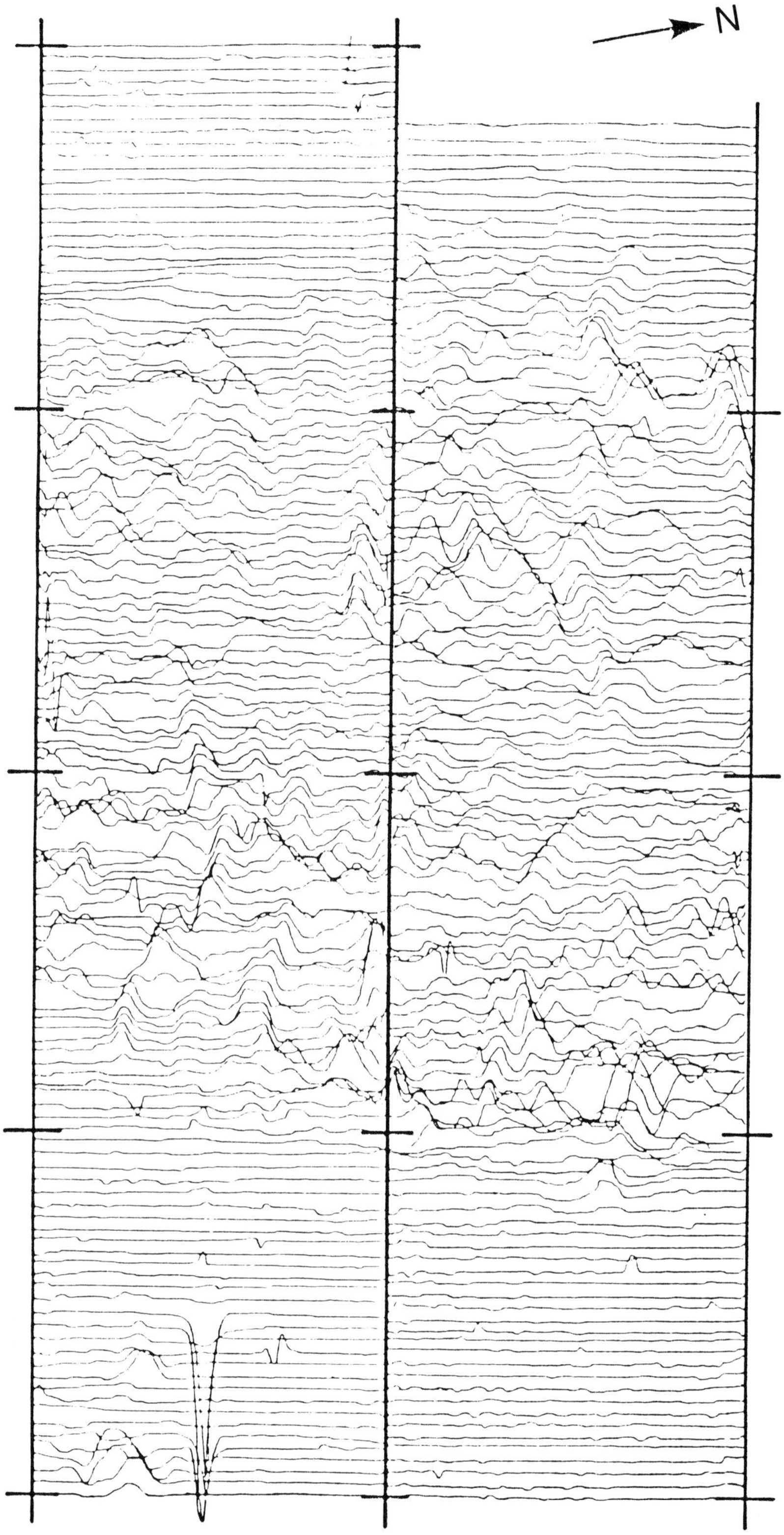

Fig. 5.1. Directly recorded fluxgate gradiometer survey at Stanwick, Northants: a complex of magnetic anomalies representing a Romano-British site. The ground rapidly shelves down beneath Medieval alluvium deposits on the west and east sides, masking the anomalies, whose rapid diminution may also be due to preceding erosion and also the possibility that occupation was more concentrated on the 'island'. On the east side, some anomalies due to strongly magnetised features such as kilns are apparent through the deposit (the strong negative features are probably superficial iron). The survey strips are 30 m wide, with traverses at 1 m intervals. Sensitivity is 7.5 nT/traverse interval. With modern digital recording and computer plotting, more information might have been obtainable from the alluviated areas.

deposit (Clark 1990: 135). Conditions are less favourable down river from the Goring gap, where presumably the alluvial soil is diluted by contributions from tributaries originating on less magnetic soils. In the Highland Zone, the presence in some areas of igneous boulders in gravel underlying alluvium can be even more disturbing.

Thus the use of total field instruments, including the high sensitivity types, is in principle advantageous for deeper deposits, although some ground conditions may make it difficult to realise their full potential. The relatively low sensitivity of the fluxgate gradiometer is compensated for by its rapid data-gathering power, which gives the opportunity for increasing the effective signal-to-noise ratio at the data processing stage. To gain full advantage from their sensitivity, the total field instruments need to be provided with comparable data recording systems.

There is also likely to be some advantage, where alluvial soils cause little surface 'noise', in lowering the detector of any magnetometer from 30 to 10 cm or less from the ground. For example, with a fluxgate gradiometer, the 5 nT discrete feature anomaly would be reduced to 1.8 rather than 1.1 nT with 0.5 m of alluvium; 0.6 rather than 0.4 nT with 1 m. Thus the signal can be increased to about half as much again if the instrument is carried at a height of 10 cm or less.

Magnetic susceptibility and phosphate

As discussed above, the standard equipment for this method requires direct contact with the material being measured. Thus, for deep alluvial deposits at least, it is necessary to use boreholes. At present measurements must be made on samples taken from the hole, but a borehole sonde is proposed by Bartington which will be suitable for measuring rapid profiles in 3 cm holes conveniently made with a coring auger. Boreholes do demand some effort, but an initial grid as coarse as 30 m is adequate, only narrowed to 15 m or less where

Depth of alluvium (metres)	Signal (nanotesla)			
	Discrete feature		Linear feature	
(5 nT feature)	Grad	Total	Grad	Total
0.0	5.0	7.5	5.0	9.6
0.5	1.1	2.2	1.7	4.3
1.0	0.4	0.9	0.8	2.4
2.0	0.1	0.3	0.3	1.1
(100 nT feature)				
0.0	100.0	154.0		
1.0	8.6	12.2		
2.0	2.0	5.9		
3.0	0.8	2.5		

Table 5.1. *Examples of signals detectable by a 0.5 m baselength fluxgate gradiometer and a total field instrument such as a proton gradiometer from discrete and linear features under different depths of alluvium. The assumption is made that the magnetic centre of the feature would be 1 m below the instrument in the absence of the alluvium layer. The figures are derived from the curves of Clark (1990: 79).*

indications of occupation are detected; they also give the bonus of directly examinable cores from which samples for dual frequency and phosphate measurement can be taken. Because of the spreading of activity around occupied sites, it is likely that susceptibility will be more sensitive than direct observation of cores around the periphery of a site.

The SH3 transmitter-receiver instrument discussed above has a penetration of 70 cm, so this could also be effective for alluvial depths up to about 0.5 m, without the need for boreholes and with the facility for rapid traversing. The EM38, with a claimed maximum penetration of 6 m, should also be very useful in its newly developed susceptibility mode, although it must be borne in mind that such an instrument looks at a very large volume of soil, so that a site represented by thin stratigraphy at considerable depth might be hardly detectable.

Either of these approaches to susceptibility measurement would be most usefully complementary to the 'island searching' mentioned in the radar and resistivity sections below, as a method of identifying any of these that were actually occupied. The borehole method has been used by Yates (1989) under peat.

A remarkable observation by Craddock *et al* (1985) was that occupation beneath 0.5 m of alluvium could be detected by phosphate measurements on surface soil, presumably due to root or worm action or both. Thus, to this depth at least, magnetic susceptibility may be similarly dispersed vertically, so that direct contact sensors could be effective at the surface, and the response of the transmitter-receiver type of instrument enhanced.

Ground probing radar

Although the uniformity of alluvium should make it a suitable subject for this method, attenuation of signals by waterlogging has been a problem, at least with older equipment. An experiment was carried out some years ago by the Ancient Monuments Laboratory (now part of Science and Conservation Services, English Heritage) over alluvium on gravel in the valley of the River Nene (cf. Fig. 5.1), to see whether radar could be used to locate once habitable gravel 'islands' amongst a complex of palaeochannels. The alluvium proved to be wet enough to absorb too much of the signal, so that the experiment was not successful.

Radar is best adapted to the production of individual profiles rather than extensive area coverage; it is uneconomical as a method for producing detailed plans. Its greatest value in the present context, apart from the broad topographical use, will probably be in confirming and adding detail to indications obtained by other methods. A proviso is that it is likely to be less sensitive than magnetic susceptibility or phosphate to subtle sites that survive as no more than thin occupation spreads on the original ground surface.

Resistivity

An alluvium deposit 0.5 to 1 m deep should not prevent the detection of walls by resistivity. A powerful feature of the technique is that its penetration can be increased almost indefinitely by expanding electrode spacing, although this is accompanied by a decrease in response to small features because of the wide spread of the current, so that depths of more than a metre can present problems. The use of multiprobe methods to construct resistivity pseudo-sections, and resistivity tomography which is at present being developed, share some of the advantage of radar in their ability to produce information in profile form.

Larger conductivity instruments would seem to have potential. The well-known Geonics EM31 can penetrate to a depth of 6 m. It is not suitable for resolving other than large archaeological features, but is a good instrument for mapping underground topography in the search for the gravel 'islands' discussed above.

A method has been developed by the OYO Corporation specifically for the detailed description of resistivity profiles in alluvium. It consists of a single probe, incorporating two ring electrodes separated by only about 2 cm, with two fixed probes at some distance to complete the circuit (Clark 1990). The probe is power driven into the ground, producing a high resolution profile which is automatically logged and printed out. A successful application observed by the writer in Japan was in tracing the extent of a dark layer associated with occupation, about 1 m below the surface and 25 cm thick. The purpose of the exercise was to extend the evidence of initial trenching which located the layer and provided a 'calibration' for the resistivity work. The layer was widespread and lacking in specific features; the search for the focus of the occupation would probably have benefited from the use of magnetic susceptibility in addition to the resistivity measurements.

CONCLUSION

The following is a summary of prospecting approaches that might be used for finding archaeology under alluvium.

1. *Borehole data.* Obtain borehole data for the depth of the alluvium cover over the area to be examined. If data are not already available (for instance from contractors' boreholes), a few holes will have to be made, preferably with a 3 cm corer which is economical to use and enables the soil profile to be examined and magnetic susceptibility samples to be taken.

2. *Magnetometry.* A prerequisite of any magnetometer work involving ditches and pits is to measure the susceptibility of a number of samples of the alluvium and of the underlying soil and natural, to ensure that adequate contrasts exist, especially if the alluvium has silted into open features. If the alluvium is strongly magnetic, it may be possible to show masonry surviving within it as negative magnetic anomalies. The thermoremanent magnetism of substantial fired clay structures such as kilns, and its contrast with surrounding material, can normally be relied upon without preliminary tests; thus, for example, a Roman or Iron Age industrial site should be detectable under 2–3 m of alluvium.

Although some information may be gained by manual scanning with the fluxgate gradiometer, modern recorded survey methods are so fast that search surveys are now normally made as detailed exploratory strips of traverses 1 m apart. Alternate strips of this type 20–30 m wide give a good sample of an area of ground quite economically.

3. *'Island' location.* Resistivity or conductivity measurements, and possibly ground penetrating radar, are suitable for locating 'islands' in original ground or underlying gravel which might have been particularly attractive for early occupation. Test traverses 10–20 m apart are appropriate, depending on the expected scale of such features, while the new EM31 with built-in logging seems suitable for recording area surveys. It also has the facility to simultaneously record susceptibility, although the archaeological response at depth could be marginal, and a follow-up borehole strategy may be preferable.

The following notes give an indication of the response of the different techniques with increasing depths of alluvial cover. They should be used alongside Table 5.1.

4. *0.5 m alluvium.* With a shallow deposit to this depth, the area may be treated as a normal site for magnetometer and resistivity prospecting. Carrying the magnetometer closer to the ground than normally can be especially helpful at this depth and at 1.0 m. If a general indication of areas of occupation is required, magnetic susceptibility or phosphate samples, or both, should be taken on an initial grid preferably not larger than 30 m. Surface soil may be effective, but samples from old soil beneath the alluvium will be more reliable. Dual frequency susceptibility measurements should be made if possible. This is a good depth range in which to use the smaller transmitter-receiver susceptibility instruments such as the SH3.

5. *1 m alluvium.* Ditches and pits tend to give very small magnetometer signals, many less than 1 nT. Although an originally 5 nT ditch will still give 2.4 nT signal with an absolute instrument, this is reduced to 0.8 nT with the gradiometer. Boreholes are essential for standard susceptibility measurements, but the larger transmitter-receiver instruments may be effective at the surface.

6. *2 m alluvium.* A marginal response obtained from both types of magnetometer to many ditches and pits, although subtle signal processing could produce a reasonable picture of some sites. Stronger discrete features such as kilns are reliably detectable. Transmitter-receiver susceptibility instruments may be effective, but borehole susceptibility may be essential.

7. *3 m alluvium.* Only the strongest features detectable by magnetometers. Resistivity, electromagnetic instruments and radar possibly able to map sub-alluvial topography, but boreholes necessary to obtain detailed information.

Acknowledgements

I am most grateful to the Archaeometry Branch, Science and Conservation Services Division, English Heritage, for supplying and allowing me to use the survey plot in Fig 5.1, and also to David Neal, excavator of the site, for permission to use it; also to Anthony Johnson of Oxford Archaeotechnics for allowing me to benefit from his experimental design and experience in magnetic susceptibility surveying.

References

Clark, A.J. 1990. *Seeing Beneath the Soil*. London: Batsford.

Craddock, P.T., Gurney, D., Pryor, F. and Hughes, M.J. 1985. The application of phosphate analysis to the location and interpretation of archaeological sites. *Archaeological Journal*, 142, 361–376.

Scollar, I., Tabbagh, A., Hesse, A. and Herzog, I. 1990. *Remote Sensing Techniques in Archaeology*. Cambridge: Cambridge University Press.

Tabbagh, A. 1986. Applications and advantages of the Slingram electromagnetic method for archaeological prospecting. *Geophysics*, 51, 576–584.

Thompson, R. and Oldfield, F. 1986. *Environmental Magnetism*. London: Allen & Unwin.

Tite, M. S. 1972. *Methods of Physical Examination in Archaeology*. London: Seminar Press.

Yates, G. 1989. Environmental magnetism applied to archaeology. Ph.D.thesis, University of Liverpool, unpublished.

SECTION 2

Characterisation of soils and biota in river depositional environments

6

Micromorphological studies of buried soils and alluvial deposits in a Wiltshire river valley

Susan Limbrey

Summary: *Micromorphological characteristics of soils and deposits in the valley floor of the Winterbourne/Upper Kennet in the Avebury and West Overton area of Wiltshire, England, are discussed. Evidence is given for cultivation of decalcified brown earths in the Neolithic period. The alluvial deposits which overlie these soils are described as 'calcite loam' and discussed in terms of environment of deposition and pedogenesis, and the contribution of biomineralization processes to their development is considered. After a period of stasis with the development of soils on the floodplains and their cultivation in the Medieval period, the resumption of alluvial deposition is associated with the construction and management of a water-meadow system. The deposition and characteristics of a distinctive non-calcareous clay under this regime are described.*

INTRODUCTION

Numerous studies in recent years have established that Holocene alluvial sequences in lowland Britain have in common a transition from earlier to later complexes of deposits of distinctive and contrasting kinds (Limbrey 1983; Limbrey and Robinson 1988). The earlier of them is characterised by materials laid down in active and abandoned channels of anastomosed systems, including sands and gravels, silts, tufas, marls and peats, and these channels interlaced and were bordered by ground showing soil development. The later complex consists of rather uniform bodies of fine silty and clayey alluvium deposited from a simplified channel system. They may or may not show evidence of periods of stasis marked by buried horizons of soil development, but they are uniformly topped by well developed humic soils. Such sequences present problems of chronology and of causation in terms of climatic change and human impact, which are addressed elsewhere in this volume (e.g. Robinson; Macklin, Passmore and Rumsby). From the point of view of human communities, it is the effects of the changes that are important, and some of the contributions which soil micromorphology can make in this field will be discussed here.

Soil micromorphology is a technique which, by displaying the effects of pedogenetic processes involving development and modification of soil structure, internal organization of soil constituents, and translocation of materials in solution and in suspension, can allow us to identify sequences of processes. Thin sections of resin-impregnated blocks of soil, taken so as to cover the whole depth of the profile to be studied, are examined under a polarizing microscope. Study of thin sections follows detailed field examination of exposed profiles, and is complemented by mineralogical and particle size analysis as appropriate.

Buried soils

In interpretation of features observed in soils buried beneath alluvial deposits, it is necessary to consider whether the phenomena observed belong to the period before the change in hydrological regime, or are the result of rise of groundwater, flooding and processes associated with deposition of alluvium and subsequent pedogenesis. In particular, it must be determined whether the presence of microlaminated and layered coatings in soil pores, in which clay, coloured yellow to brown by iron oxides and humus and mixed and interstratified with variable amounts of fine silt, should, in these circumstances, be interpreted only as evidence of cultivation. These 'dusty clay coatings' (Jongerius 1970; Bullock *et al* 1985) have been much studied by micromorphologists. Kwaad and Mücher (1979) review the evidence for their mode of formation, which is also outlined and illustrated in Courty *et al* (1989). Deposition of these coatings is explained in terms of disruption of soil peds exposed at the soil surface by rain impact, so that fine material, silt and clay, is washed into spaces in and below the cultivated horizon. The 'dustiness' of the clay is in part due to a conspicuous content of fine black particles, usually thought to be charcoal broken up in a cultivated soil after deposition on the land in manuring materials (Courty *et al* 1989), but I think it may also have an origin in soot, deposited from smoke as well as forming a substantial component of occupation deposits used as fertilizer.

The disruption of the exposed peds happens more readily if soil structure is weak or poorly developed. Some of the soils considered here were decalcified and show signs, in their clay mineralogy, of degradation of clay minerals. In such circumstances, they would be dispersed easily by rain. Other soils which show the same phenomenon, however, must have been base-rich, and probably calcareous, at the surface when the coatings formed, and these more strongly structured soils should have been much more resistant to disruption by rain impact.

Coatings of mixed silt and clay in gleyed alluvial deposits have been interpreted in terms of the slaking of the soil in the surface by floodwater, the fine material being deposited in soil pores as the flood subsides (Brammer 1971). In this case, where gleying renders the colloid susceptible to dispersion, the coatings are not found to be microlaminated, and the clay is poorly orientated. Coatings answering to this description have not been seen in the soils considered here, but the contribution of flooding to the process cannot be excluded. People might have continued cultivating soils as they became more susceptible to flooding. The processes need not be mutually exclusive, therefore, and flooding might be the additional factor promoting dispersion in calcareous soils which would otherwise resist it.

Fine-grained alluvial deposits

The study of the second of the riverine complexes outlined above, the fine-grained rather uniform alluvium, presents problems of interpretation which cannot be resolved by sedimentological studies alone, because at the macroscopic scale little sedimentary structure is detectable, and the pedological processes of mineral alteration (particularly in a situation where source materials are calcareous), translocation and faunal turbation have modified the particle size distribution of the material since it was laid down. At the same time, soil profile characterization for the purpose of soil classification and mapping can only suggest answers to historical questions concerning the later stages of a floodplain's development. Micromorphology can be used to search for remaining traces of sedimentary structure and to investigate the processes of translocation of silt and clay and of dissolution and redeposition of calcium carbonate which, while modifying the particle size distribution also illuminate the pedological processes which have affected the deposit during and after deposition. Evidence for vigorous earthworm activity in the development of soil structure accounts for obliteration of sedimentary structure as well as indicating the pedogenic environment affecting the material. The size and morphology of calcareous particles may be interpreted in terms of autochthonous mineralization rather than a sedimentary origin.

THE STUDY AREA

The examples to be discussed come from a collaborative study of the soils and alluvial sequences of the Winterbourne/Upper Kennet valley in the Avebury and West Overton areas, Wiltshire. Results of preliminary studies, giving an outline of the stratigraphy, molluscan and ostracod sequences and archaeological materials, are published in Evans *et al* 1988. Publication of the full results is in preparation.

The soil parent materials and sources of sediment in the locality are provided by the whole range of chalkland solid and drift geology: Upper, Middle and Lower Chalk, Clay-with-Flints, periglacial deposits forming the mixtures of cryoturbated and soliflucted chalk, flint and loess known as Coombe Rock, and gravels formed of flint and chalk. The Early Holocene deposits are tufaceous gravels and sands, calcareous silts and marls. With the exception of Clay-with-Flints, these materials are all calcareous, and the groundwater which issues from springs at the junction of the clay-rich Lower Chalk and the more permeable Middle Chalk is charged with calcium. Present soils of the area are: calcareous and pelo-calcareous alluvial gley soils of the floodplains; typical brown calcareous earths on the clay-rich Lower Chalk and on the gravels and Coombe Rock; rendzinas on Middle and Upper Chalk; and palaeoargillic brown earths on Clay-with-Flints (Findlay *et al* 1984).

The Winterbourne rises close to the top of the dip slope in the Lower Chalk about 10 km north of Avebury and flows mainly over Lower Chalk in a wide, shallow valley floored by periglacial gravels. It is augmented by springs at the foot of the subsidiary scarp formed by the Middle and Upper Chalk of the Marlborough Downs, and by a major tributary in the Lower Chalk from the north-west which passes south of Windmill Hill to a confluence on the west side of Avebury village. One of the localities studied here lies just south of this confluence, between Avebury and Avebury Trusloe. In summer the stream here forms only a grassed-over, cow-trampled depression, but its channel remains better defined where cattle are denied access.

The Swallowhead Springs lie almost 2 km south of Avebury. Here perennial flow is established from the base of the Middle Chalk, though borehole extraction now causes failure in dry summers, and the stream, now called the river Kennet, flows eastwards through Middle and Upper Chalk. 3.5 km downstream, the second locality studied lies at the foot of a steep bluff on the north side of the valley opposite the village of West Overton.

Avebury, Avebury Trusloe and West Overton are shrunken villages, with abundant evidence of Medieval house platforms between the surviving nuclei and the floodplain. At both localities, remains of channels in the floodplain attest the former existence of water-meadow systems. It is worth noting that in the eighteenth century the farm at West Kennet, immediately downstream of the springs, had an exceptionally high proportion of its area in water meadows (Wordie 1984).

STRATIGRAPHY

Avebury: 7 cuttings, up to 15 m long, were distributed on both sides of the stream across the 200 m of floodplain width and along 300 m of its length. Buried soils were exposed at depths of 0.6 to 0.8 m, on Coombe Rock, on gravel of flint and chalk, and on calcareous sands and marl which had infilled a hollow in the surface of the periglacial deposits. No certain evidence was identified for the existence of the channel of the Winterbourne in this early valley floor. The soils contained artifacts of Mesolithic and Neolithic types. The soils were buried by some 20–30 cm of highly calcareous silty clay, an alluvial deposit, upon which formed a calcareous alluvial gley soil, its humic horizon 10–15 cm, into the upper part of which was incorporated Medieval debris, augmenting its thickness to form up to 35 cm of a predominantly anthropic horizon. This soil was buried by 30–40 cm of black clay, upon which the present pelo-alluvial gley soil developed.

West Overton: 7 cuttings extended up to 50 m out from the north side of the floodplain and were distributed along 150 m of its length. Flint gravels were exposed at 1.8 to 2.5 m depth, lapping over Coombe Rock at the valley side. Channels in the gravels contained marls, tufacous gravels and calcareous silts, within which a humic horizon had developed, though apparently without a complete cessation of silt deposition. Further calcareous silt merged laterally with slope-derived flint and chalk gravel banked against the foot of the bluff. The buried soil, containing Mesolithic and Neolithic material, was exposed at depths of 1.0 to 1.5 m, and was formed on the valley-side gravels and on the silts and tufaceous deposits of the channel complex. The buried soil was overlain by about 1.0 m of calcareous silty clay, which merged and interleaved with gravels and colluvial soils derived from the valley-side bluff. The fine alluvium was interstratified with thin layers of flint and sarsen gravels which spread from a buried channel and petered out laterally. A coarse colluvium with flint and chalk and containing Romano-British and Anglo-Saxon material spread across the alluvium, extending further from the valley side than earlier colluvial deposits, and merged laterally and vertically, at about 0.5 m depth, with a black clay deposit on which the present pelo-alluvial gley soil had developed.

The common features of the two localities are the two components of the upper alluvial complex: the calcareous silty clays and the black clays. The contrasting features are the shorter period of development of Mesolithic/Neolithic buried soils at West Overton, where they formed on the Early Holocene channel and valley-side deposits, whereas, except in the case of the soil on the marl-filled hollow, those at Avebury formed on the periglacial substrate, and the Medieval soil at Avebury, whose features demonstrate a period of stasis in alluviation, whereas at West Overton such a period is marked by an extension of colluvial material across the floodplain without clear evidence for substantial *in situ* soil development.

THE BURIED SOILS

At Avebury, buried soils of three kinds were seen:

1. An argillic brown earth, formed on Coombe Rock. The soil, which was exposed and excavated over more than 100 square metres, showed considerable complexity of its A horizon. In its simplest development it had a brown A horizon of up to 20 cm relatively stone-free non- calcareous clay over a horizon of very stony clay, and a stony, light brown calcareous Bt horizon, 10 to 30 cm, tonguing down into the C horizon. A and B horizons both show clay coatings. In some places it was truncated to the stony layer. Elsewhere, the stone-free and stony horizons had been mixed together, and in other places a double stone- free horizon occurred, with a stone line in between. The appearance was that of a cultivated soil, cultivation depth being variable and soil being redistributed laterally as well as mixed vertically.

2. A black clay soil, formed on the marl-filled hollow. The profile comprised an A horizon of black clay, 10–15 cm, with an abrupt boundary to a brown clay B horizon, 8 cm but tonguing down a further 20 cm, both horizons non-calcareous and exhibiting ferruginous and manganiferous mottling, and an abrupt boundary to white calcareous marl. This profile merged rapidly laterally into a truncated form of the brown earth at the margin of the hollow, and a spread of flints from that profile extended across it and was overlain by a further spread of less stony brown clay.

3. A soil formed on chalk gravel. This soil had characteristics of a weakly developed rendzina, with a greyish brown A horizon, 20 to 40 cm, highly

calcareous silty clay with a darker, stony band within it in its deeper expression, and an abrupt boundary to small chalk gravel with flint concentrated at its surface. Archaeological material occurred in both the stony level within the A horizon and at its base. The duplication of stone-free and stony horizons here reflects that seen in soil no. 1, and may again be the produce of a sequence of worm sorting, disturbance and localised truncation.

At West Overton, buried soils within the valley floor channel deposits were variably developed, with evidence for transient soil development on tufaceous gravels in the form of slightly humus-stained horizons and iron staining and manganese mottling developing in downward swathes from particular stratigraphic horizons, suggesting erosion of a temporary surface subsequent to its functioning as a focus of soil development in a marshy situation. Only in the case of a deep calcareous silt deposit in a channel did the humus staining develop markedly, and its protected position within a silting up channel allowed it to accumulate over a depth of 25–30 cm before more rapid fluvial deposition covered it with clean silt.

The main buried soil complex, equivalent in stratigraphic position and archaeological content to that at Avebury, was developed on the valley side deposits and the silts and tufaceous gravels, extending about 15 to 20 m across the valley floor before suffering truncation, being represented further out by only a gravel spread associated with iron and manganese mottling. As in the case of Avebury soils 1 and 2, this soil was decalcified and could be classified as a brown earth with some macroscopically visible clay coatings. Its depth was less here, however, and the B horizon poorly developed. Depth of the A horizon, a brown clay, was ca. 20 cm, with an abrupt boundary to chalk gravel, or a variable boundary to strongly mottled tufaceous gravel or silt. The profile was more complex where deepest, and stone lines occurred above and within it. these stone lines merged along a line parallel to the valley side, so as to cut out an upper part of the A horizon and form a step down to a thinner soil. The step was in places marked by larger stones. The soil in this region contained abundant Neolithic archaeological material, and the field interpretation was of a soil forming part of a field which at some time had been cultivated, the step marking a boundary of cultivation. As at Avebury, a sequence of periods of worm sorting and disturbance is suggested.

Micromorphology

In the case of Avebury soil no.1 (Fig. 6.1), field evidence left little doubt that the soil had been

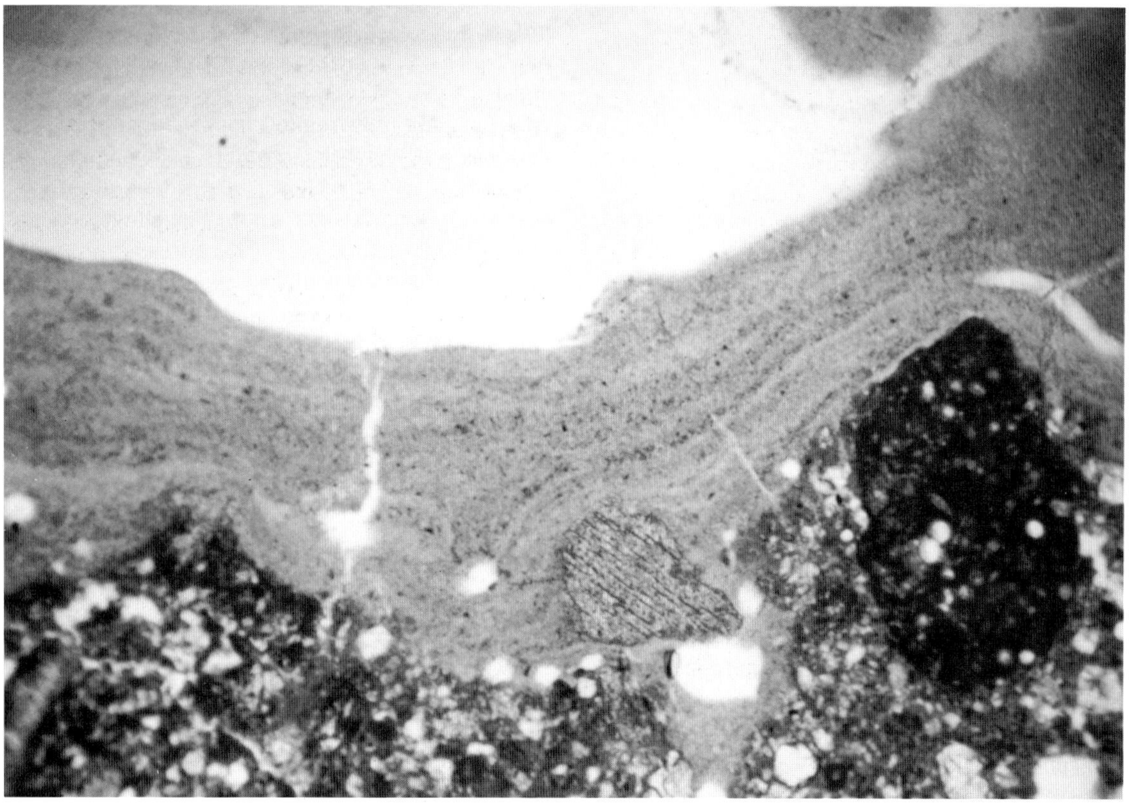

Fig. 6.1. *Buried soil on chalk gravel at Avebury, A horizon. Microlaminated dusty clay coating in which some of the coarser laminae are rich in calcite crystals. Plane polarized light; frame length 3 mm.*

cultivated. The contribution of soil micromorphology, therefore, was to illustrate the surface conditions brought about by cultivation and to determine the sequence of cultivation and the gleying affecting the black clay soil. A vertical sequence of thin sections was taken which covered the black clay profile and the redistribution brown soil which overlay it. The soil fabric of the brown soil displayed abundant evidence of faunal activity, being dominated by the crescentic and mammilated form of earthworm faecal material. Rounded peds of contrasting patterns of birefingence were incorporated into the soil matrix, and voids showed both coatings and hypocoatings. Dusty clay coatings were layered, with abundant fine black particles in the coarser layers. Some of the coatings were strongly iron stained, others less so. A hierachy of features confirmed the sequence of development of faunal structure, disturbance and mixing of soil material, inwash of dusty clay from an exposed surface, and gleying.

In contrast to the decalcified soils, soil no.3 was calcareous throughout as observed macroscropically, but in thin section it was found that there was abundant infilling and lining of voids with non-calcareous or only partly calcareous dusty clay. The soil fabric was dominated by faunal excrement structure showing vigorous earthworm activity prior to development of the coatings.

The clay in these coatings was distinctive in being very pale yellow under plain polarized light, and having low birefringence, but showing continuous orientation, under crossed polarizers. Some of the coatings were entirely non-calcareous; others had calcite crystals of fine silt grade scattered through the clay laminae or forming part of the silt-dominated laminae. Evidence of gleying following emplacement of the coatings was present as iron depletion features and staining affecting the surfaces of the coatings in those pores which were not completely filled. The coatings were present throughout the buried A horizon and were particularly abundant in the upper part of the C horizon.

A problem with this profile is the origin of the non-calcareous clay. While decalcification *in situ* after emplacement of the coatings is a possibility, the calcareous components of the coatings do not show signs of dissolution. Chalk fragments in the stone line, and in the C horizon, however, do show thinning and development of porosity suggesting that the profile as a whole had at some time been subjected to decalcification, and it is hypothesised that a decalcified A horizon had developed but has been lost by erosion, perhaps by the floods which initiated the alluvial deposition which buried the profile. The distinctive appearance of the clay suggests neoformation of kaolinitic clay in the development of a soil on the gravel. That the coatings were not formed entirely by the flooding itself is suggested by their layered and laminated form, with variable 'dustiness' showing repeated small episodes of locally variable activity, but in any case for flooding to result in inwash of dispersed soil, a disrupted surface such as that produced by cultivation is implied. There is no evidence in the distribution of iron and manganese oxides that the coatings were formed after gleying had occurred. The presence of calcite in the coatings, increasingly so with depth, suggests a more complex history of soil disturbance. There was no evidence in this profile of tree clearance pits, but in the lower part of the soil micromorphology shows areas of disrupted pedality and a sort of slurry of the dusty clay and fragments of soil peds and free sand grains filling larger pores (Figure 6.2), which is consistent with deep disturbance of some kind. Features interpreted as tree clearance pits were identified at this stratigraphic level elsewhere on the floodplain.

The West Overton buried soil was studied by micromorphology where its substrate was the valley-side gravels. Although decalcified, some recalcification had occurred, there being patchy impregnation of the fabric with micritic calcite and some local micritic coating to pores. In the field, it was observed that clusters of small stones including chalk occurred, the remains of aestivation chambers of earthworms coming down through the overburden. The recalcification was clearly later than the formation of structure in the soil fabric by faunal activity and the development of dusty clay coatings. It must therefore relate to the period after burial by colluvium containing chalk.

Here there was not such clear macroscopic evidence of cultivation as at Avebury. Micromorphology again showed abundant coatings of dusty clay, and though there was some development of the 'limpid' coatings which develop where clay translocation is not a product of rain impact on an exposed soil, subsequent to the dusty coatings, this was slight, and faunal activity visible in the thin section as infillings of burrows was prior to the development of the dusty coatings. There is, however, evidence in the incorporation of fragments of earlier dusty clay coatings into the matrix that there had been a sequence of cultivation and reworking of the soil by worms, which supports the stratigraphic evidence for a number of episodes.

CALCITE LOAM

Radiocarbon dating shows that the soils of the valley floor began to be overwhelmed by alluvial deposits in the earlier part of the second millennium BC (uncalibrated).

The alluvium which covered the buried soils in the Winterbourne/Upper Kennet valley has an extremely high calcium carbonate content, 45–70%. particle size analysis after decalcification gives silt and clay contents each of 40–50%, but the almost entirely silt grade of the

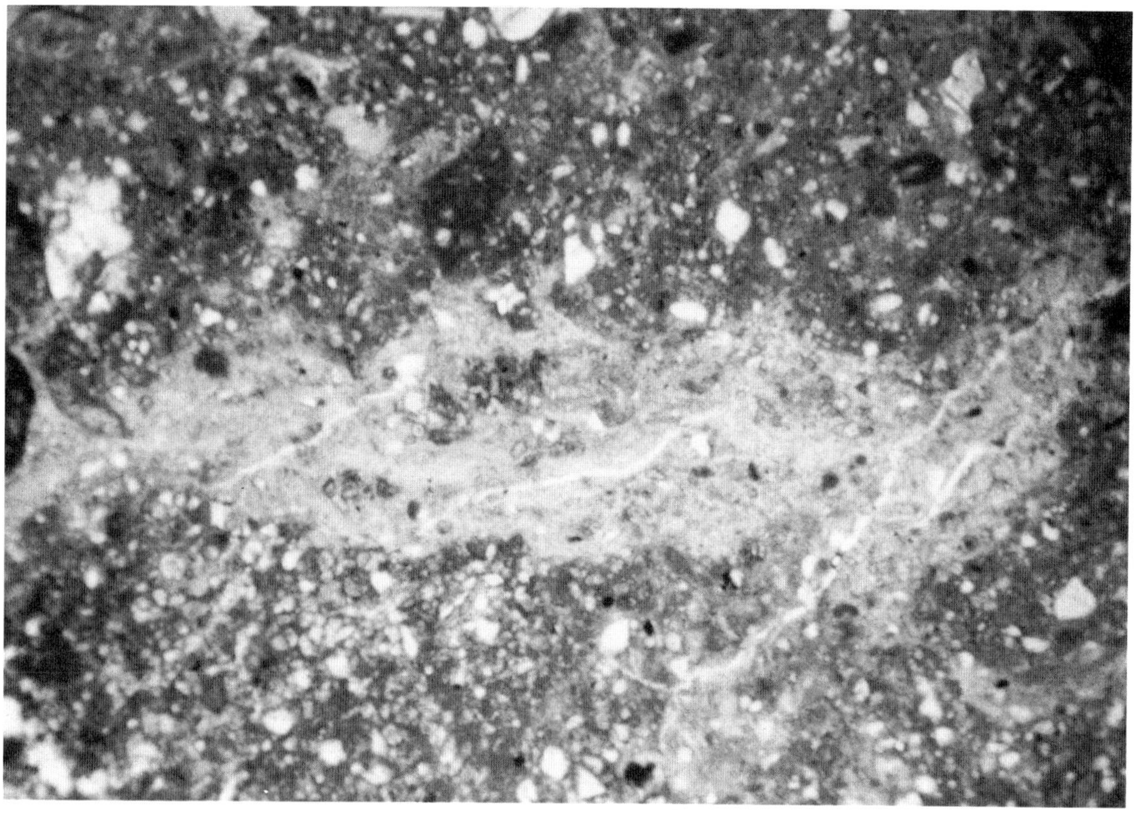

Fig. 6.2. 10 cm deeper in the same soil as Figure 6.1, below base of A horizon. Infill of pore with a mass of dusty clay containing disrupted soil and disaggregated material. Plane polarized light; frame length 3 mm.

calcite makes the texture a loam rather than a clay or clay loam. Clay percentage is of the order of 15% in undecalcified samples, but without decalcification it is probable that some of the clay remains in silt-sized aggregates. Since the clay is unlikely to have been in dispersed form when deposited, the silt-dominated texture has greater sedimentological reality than the actual clay content would suggest, even without consideration of the source of the calcite silt. The term *'calcite loam'* has been coined.

Rapidity of accumulation of the calcite loam is suggested by the failure of significant humus accumulation to occur, and by the lack of evidence of decalcification in the bulk of the deposit. The gravel layers which occur at intervals, fanning out from a buried channel running parallel to the valley side and petering out into barely detectable scatters of small stones and eventually traces of sand, define lobes of sediment, building up and extending out across the floodplains over a period of two thousand five hundred years. This sequence suggests alternation of gentle deposition and more violent, erosive, flooding. The gravels, then, are lag deposits, distributed when stony colluvial soils encroaching upon the floodplain from the valley side were eroded, and left behind as finer material was winnowed away.

Micromorphology

Most of the calcite is in the form of crystals which fall in the coarse and medium silt grades (sparite) and finer (micrite, less than 4 microns). At Avebury there is a higher proportion of sparite to micrite than at West Overton, and though the particle size distribution of the non-calcareous fraction is similar at the two sites, at the former, the high sparite content gives a coarser texture as seen in thin section.

There are abundant snail shells and shell fragments and frequent composite granules of sparitic calcite which fall in the sand grade, and indeed sometimes exceed 2 mm diameter. Where there is a substantial sand fraction, shell and granules are particularly abundant. Elsewhere, shell fragments and granules constitute the only particles of their size grade, with a large size gap between them and the silt-dominated sediment, so they are clearly formed *in situ*.

The vigour of faunal activity in the deposit is shown by its microstructural characteristics, with channel fills of crescentic form (Figure 6.3), intercalations of micritic calcite stained with humus (Figure 6.4), and, a reverse effect, intercalations of sparite and the non-calcite silt component in a more micrite-rich texture. Some of these structures are still clearly the product of faunal

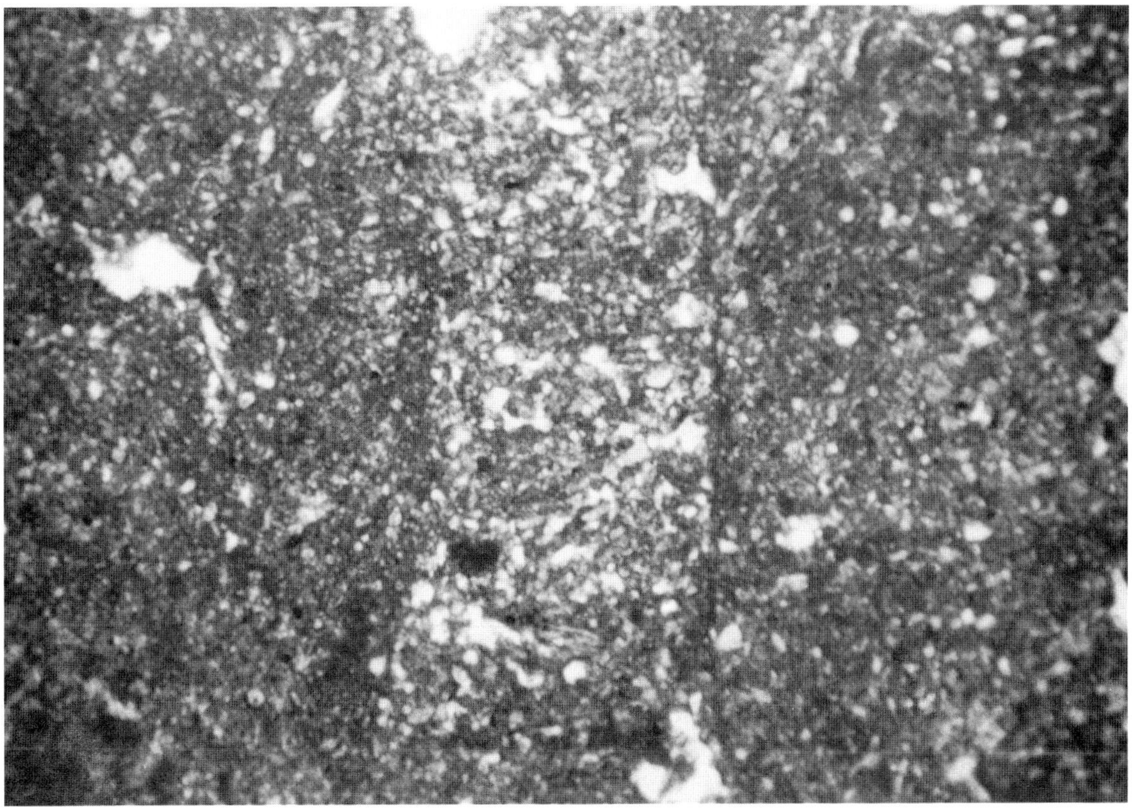

Fig. 6.3. *Calcite loam. Vertical section through a worm channel with fill of faecal deposits in bow-like form and having high porosity. The edge of the channel is darkened by humus. Plane polarized light; frame length 3 mm.*

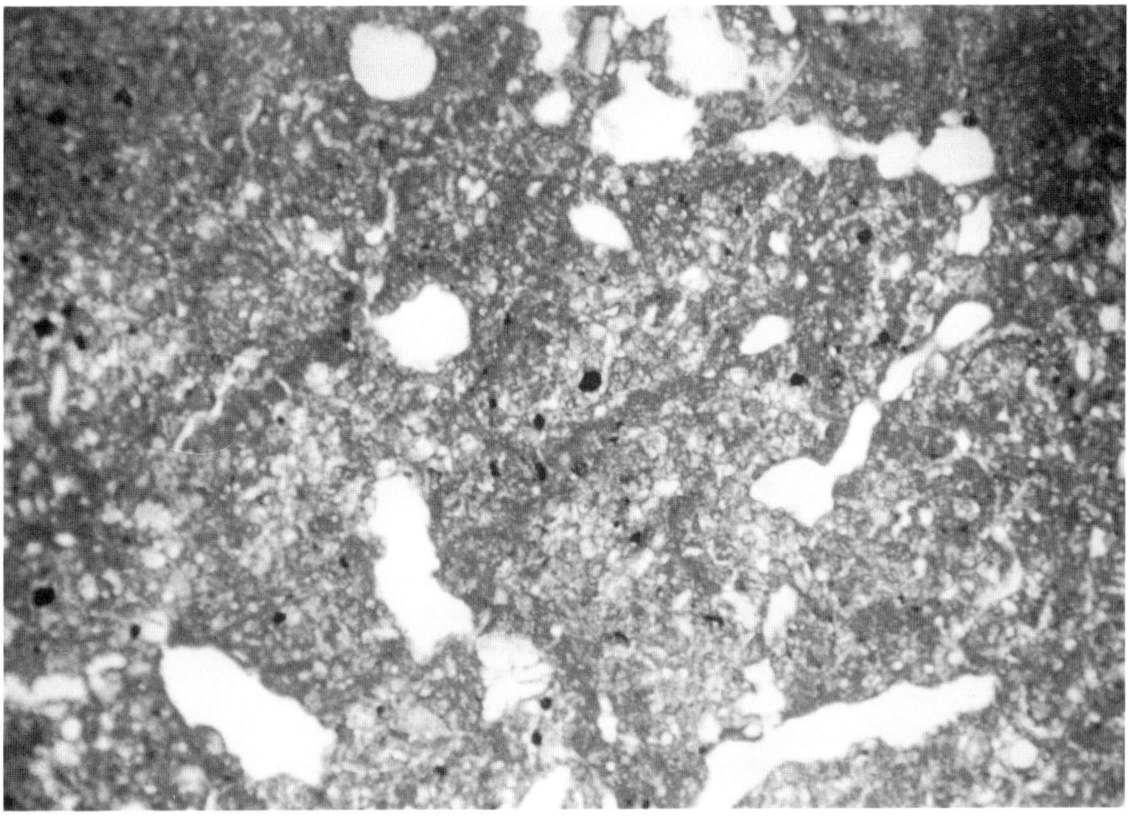

Fig. 6.4. *Calcite loam. Interlaced intercalations of humus-stained micritic calcite delineating coalesced earthworm faecal deposits. Subsequent porosity has thin dusty clay coatings. Plane polarized light; frame length 3 mm.*

excretion; others appear to be the result of compaction and coalescing of a structure of burrows and excretion. The organic intercalations discussed by Thompson *et al* (1990) and shown to be the humus-rich outer zone of earthworm deposits, has an exact parallel here in the humus stained micritic calcite intercalations which are a conspicuous element of the structure of this deposit (Figure 6.4).

While there are calcareous void coatings in the calcite loam, there is no evidence of decalcification in the bulk of the deposit. If the redistribution of calcium carbonate represented by these coatings derives from decalcification phases during accumulation rather than at its culmination, these incipient soils were destroyed in the erosive events. Non-calcareous dusty clay coatings occur throughout, in variable quantity, being particularly associated with the sandy horizons of the lag deposits. If the lag deposits do indeed represent the remains of colluvial soils, then the coatings could derive from these. Micromorphological studies of the colluvium at the valley side show similar coatings within the deposit, consistent with its deposition and exploitation under arable use, and their occurrence in very similar form on the floodplain itself also suggests that cultivation extended from the slope to the valley floor during periods when risk of flooding was reduced.

The problem in understanding the origin and nature of this deposit is that it is not seen forming today, and its characteristics have been modified by soil process during and after its deposition. There are three possibilities for the origin of the calcite: derivation directly by erosion of chalk or combe rock, precipitation around springs or within stream channels, and precipitation on the floodplain itself. The sparsity in the sediment of chalk fossils of appropriate size grade (and they are abundant in the chalk itself) seems to rule out the first. The greater proportion of sparite in the calcite loam at Avebury, however, could be accounted for by the presence in the chalk gravels at that location of chalk fragments having a distinctive large crystal size and readily disaggregating; this form of chalk was not seen at West Overton. Precipitation within the stream channels certainly accounts for some of the calcite in the lower part of the calcite loam at West Overton, since tufa fragments derived from the underlying deposits occur in the sand grade, but they die out upwards. Precipitation around the springs, producing loose crystals rather than tufaceous encrustations, is a possibility, but the Avebury location is further from the springs feeding the Winterbourne, than West Overton is from the Swallowhead springs, and the Winterbourne must always have had a lower flow than the Kennet. It is surprising, therefore, to find a coarser element in the deposits of the former.

Biogenic sources of calcite

An hypothesis presented here is that a large part of the calcium carbonate component of the deposit crystallised *in situ* on the floodplain, through biomineralization processes. In the Early Holocene, biologically mediated precipitation formed tufa within the aquatic system (cyanobacteria, *Rivularia* sp., have been identified by B. Whitton, of Durham University, on the internal structure of nodules). Once the stream channels became inhospitable to the cyanophyts, in terms of seasonal extremes of flow and sediment load, and the waters which were formerly retained within the fluvial system were spread out into the sub-aerial zone, the calcium charge of the waters was, in part, handed over to the terrestrial regime for precipitation.

Biomineralization activity involving calcite is known in a very large number of terrestrial organisms, from higher plants and molluscs to bacteria (Watabe and Wilbur 1976; Leadbeater and Riding 1986). A common feature of drying-out flood deposits is the growth of an algal mat, and there are some micromorphological indications that calcite precipitation is associated with algal mats in these deposits, as well as with the tufa deposits of the earlier regime. Again, species of *Rivularia* can colonise soils in this way. Such mats are ephemeral, and since the deposits are thoroughly reworked by soil fauna there are only traces of mat-like intercalations of micritic calcite, darkened by organic matter, in the thin sections.

Many invertebrates produce calcite granules of many sizes in their soft tissues (Watabe *et al* 1976), and though most of these are probably dissolved during decay of the body or digestion in a predator's gut, some are probably released into the soil. Lumbricid earthworms produce calcite spheruliths in their calciferous glands, and these may aggregate in the sinus of the gland to form composite granules (Bevelander & Nakahara 1959; Nakahara & Bevelander 1969). Such spheruliths have been observed under SEM in earthworm aestivation chambers (Kretzschmar 1987), but not identified in the general soil matrix. Since their size is of the order of a few microns, they would form a component of the micritic calcite of soils, but under light microscopy the details of their morphology would not be apparent. Robertson (1936) described the aggregate granules forming in the cavity of the calciferous glands and stated that they were voided at intervals. The later writers cited above concern themselves with the processes of excretion in the glands and stop short of describing the final form and the process of voiding of the aggregate granules, but the stages observed by Bevelander and Nakahara (1959) do appear to be leading towards Robertson's granules. These granules have not, however, been clearly identified in soil micromorphology and distinguished from those produced by Arionid slugs. Examined macroscopically as extracted from slugs, it is clear that the latter can be larger and more irregular in shape than the former, but the granules which occur abundantly in calcareous soils and sediments have a range of size, shape and internal structure which probably includes both origins; those that are regular ovoids of 0.3 to 0.4 mm across could well be earthworm granules. A scatter of spherical bodies with walls formed of micritic calcite which are observed in these deposits are probably eggs of snails which failed to develop; the calcite is used to form the shell of the embryo and so does not persist in this form if development is completed (Tompa 1975).

Calcification of decaying plant roots is a well established phenomenon (Thompson *et al* 1990; Jaillard *et al* 1991). Although there are traces of calcite structures in these deposits which could derive from the subsequent disruption of calcified roots, faunal reworking has been too thorough to allow certain identification. Whereas some of the large granules of circular or ovoid section and regular surface display internal structure reminiscent of the cell structure of fine roots seen in cross-section, they are never seen as the longitudinal sections which might be expected to predominate in soil sections orientated vertically. Moreover, there is no evidence of the zone of calcium carbonate depletion around the structures which has been observed in the case of calcified roots (Jaillard & Callot 1987), and no cortical zone to the root is seen.

Mites, some of which concentrate calcium in their tissues, contribute to the destruction of roots and litter, and their faecal pellets have a distinctive size and form. In a study of richly organic soils, formed where archaeological deposits have been incorporated into soils on a very similar calcite loam at Runnymede on the Lower Thames floodplain (Needham & Sorensen 1988), ovoid bodies of appropriate size composed of a mass of mictritic calcite crystals slightly darkened by humus are very abundant and can be seen as loose fillings to voids as well as a component of the soil matrix, and an origin as the excreta of mites feeding on calcium rich-debris is entirely plausible. In the calcite loam of the Avebury/West Overton sequence, these bodies are less distinct but do appear to form a component of the soil matrix.

The implication of a biogenic origin for a substantial part of the calcite component of the loam is that the erosion of potential source soils is not so drastic as the volume of the sediment would imply, and that the particle size distribution of the sediment is no guide to the fluvial conditions of its deposition. A further implication is that the initial sediment load of any flood event was unlikely to be so great as to spoil the sward and delay exploitation of the fodder or grazing resource of the floodplains. The characteristics of each year's deposition would develop over the season as the plants grew and died and the grazing animals and soil fauna incorporated the residues, so that newly formed calcite in excreta and tissue remains was mixed with sediment to form the porous, well structured, but not richly organic, soil.

Stasis

Deposition of the calcite loam was terminated with the development of a brown earth which, at Avebury, contains both Romano-British and Anglo-Saxon material and is associated with features of Medieval date. Accumulation of occupation debris of the Medieval period and its incorporation into a cultivated soil as it built up, partly as a result of the debris itself and partly by colluviation from the valley side, is demonstrated stratigraphically and by both macroscopic and micromorphological characteristics. At West Overton, the soil on the calcite loam merges laterally with colluvium from the valley side which incorporates material from Romano-British to Medieval age.

The soil which developed on the calcite loam at Avebury is non-calcareous in its matrix but contains abundant calcareous particles, including snail shells and granules. The transition between the calcite loam and the soil shows two process: decalcification, with the development of infillings and coatings to pores in the upper part of the calcite loam, and the diminution of input of calcite, the finer component being lost before the coarse silt sized crystals, but eventually these go too. The implication is that cessation of flooding interrupted the input for many of the biomineralization processes, while there was still abundant calcium in the soil/plant systems to support slugs and snails and maintain pH high enough to prevent dissolution of their shells and granules.

Micromorphological evidence was hardly needed to demonstrate cultivation and incorporation of domestic refuse in these soils, given their depth, abundance of pottery, bone, chalk and stones, and their high organic content. The structure seen in thin section amply demonstrates vigour of faunal activity, and there is an abundance of dusty clay coatings both disrupted by cultivation subsequent to their formation and *in situ* in the later pore system. Floodplains were again an arable resource during the period from Roman to Medieval times. At Avebury, there is archaeological evidence for occupation of the floodplain itself during the Medieval period, and at both localities the present villages are the remnant of much larger Medieval villages.

WATER-MEADOW CLAY

Resumption of alluvial deposition in Late- or Post-Medieval times saw a very different regime to that responsible for the calcite loam. A richly humic clay was deposited, and its culmination in the present pelo-alluvial gley soils is associated with the surface topography of abandoned channels of a water-meadow system.

Clay content after decalcification is 60–70% at Avebury and 55–65% at West Overton, with sand content a few percent only. The matrix is non-calcareous but there are abundant shell fragments and particularly abundant calcite granules, which are thus by far the commonest mineral particles of sand grade. It is rich in charred particles and humified remains (Figure 6.5) and has a large number of thick-walled spherical organic bodies which appear to be fungal chlamydospores. Phytoliths also occur. Figure 6.5 shows the abundance of organic and charred material.

Observation of the water meadows south of Salisbury, the remaining part of a very extensive system which is still being managed in the traditional way, provides an analogy, albeit on a larger scale and fed in part by effluent from a modern urban area. Nevertheless, the enthusiastic welcome for 'nice and dirty' water by the farmers today emphasises the fertilizing benefits, in addition to those of the soil warming which produces early grazing in spring, and irrigation in summer. Sedimentation in the present-day water meadows occurs by the flocculation of black, organic-rich clay within the feeder channels during the 'drowning' phases, the water which spills down the slopes being clear. This material is cleaned out each year during the winter and spread down the slope between feeder and drain, where it is incorporated by faunal activity during the following year. The clay texture therefore does not necessitate a standard sedimentological interpretation of deposition directly from quiet floodwater spread over the floodplain. The clay deposit observed in the Winterbourne/Upper Kennet floodplain is consistent with the water-meadow regime. The visible remains of the channels represent only the last stage of the system, which may have held sway for at least 150 years.

Unlike the soils of the stasis period, it is not possible now to invoke lack of fluvial input for absence of all fine calcareous material, whether detrital or biogenic, since regular drowning by stream water is implied, and quartz-dominated silt is present. A factor which should be considered is the source of the water in the streams, and in particular in relative contribution of surface run-off and groundwater from the chalk. Water meadows are a component of an integrated system of land use (Kerridge 1967; Bowie 1987). Sheep grazed the meadows, following the sequence of drowning and draining, from early spring until the grass on the hills had grown enough to provide grazing. The meadows were then used to grow hay. In the autumn the sheep were brought down

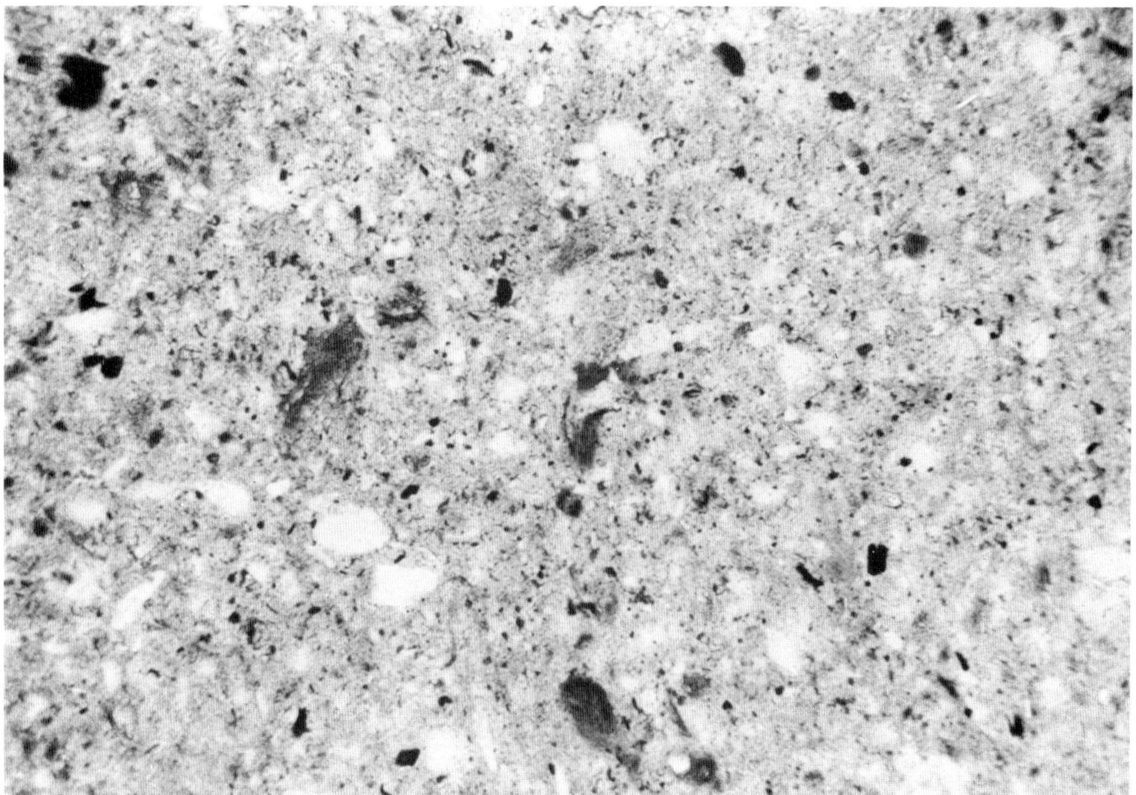

Fig. 6.5. Water-meadow clay, showing abundance of fine charcoal and humified particles. Plane polarized light; frame length 0.75 mm.

and folded on the arable land through the winter. Both heavily grazed hills and trampled arable would suffer impedence to infiltration and so encourage run-off. The snag with this hypothesis, however, is that it is the warmth of groundwater which stimulates the early growth on the water meadows. If run-off predominated, this benefit would have been reduced.

CONCLUSIONS

Though micromorphology has illuminated a number of aspects of the soils and deposits of the valley floor, it raises questions concerning others. In the localities referred to here, we can confirm that arable agriculture was practised on valley floors during the Neolithic period and again between Romano-British and Medieval times. That the alluvial deposits which accumulated between these two periods provided a rich substrate for biological activity even though they did not support the development of humic topsoils, can perhaps be interpreted as showing that plant growth continued and provided a grazing resource. There is also some indication that these floodplains were cultivated at intervals during the accumulation of the calcite loam. The nature of the water-meadow clay is illuminated by micromorphology, and the contrast between this and the earlier phase of active deposition of material on the floodplain is brought out particularly clearly. Further consideration is needed of the contrasting conditions of arable land in the catchment and of the regime of alluvation which produced such contrasting deposits during the two periods when deposition on the floodplains was substantial.

Acknowledgements

This work was funded by SERC Research Grant No. GR/D 35407. Supported by this grant, Ian Mâté carried out much of the field description and sampling, performed the chemical and physical analyses, and prepared most of the thin sections. The other component of the collaborative project, under John Evans at University of Wales, College of Cardiff, assisted by Rosina Mount, as well as dealing with the biological aspects of the research, provided, with the labour and skills of students, all the excavation and the archaeological analysis. To all these people I am most grateful.

References

Bevelander, G. and Nakahara, H. 1959. A histological and cytological study of the calciferous glands of *Lumbricus* terrestris. *Physiological Zoology*, 32, 40–46.

Bowie, G.G.S. 1987. Watermeadows in Essex — a re-evaluation for the period 1640–1850. *Agricultural History Review*, 35, 151–158.

Brammer, H. 1971. Coatings in seasonally flooded soils. *Geoderma*, 6, 5–16.

Bullock, P., Federoff, N., Jongerius, A., Stoops, G. and Tursina, T. 1985. *Handbook for Soil Thin Section Description.* Albrighton: Waine Research Publications.

Courty, M.A., Goldberg, P. and Macphail, R., 1989. *Soils and Micromorphology in Archaeology.* Cambridge: University Press.

Evans, J.G., Limbrey, S., Mâté, I. and Mount, R.J. 1988. Environmental change and land-use history in a Wiltshire river valley in the last 14000 years. In J.C. Barrett and I.A. Kinnes (eds.), *The Archaeology of Context in the Neolithic and Bronze Age: Recent Trends*, 97–103. Sheffield: Department of Archaeology and Prehistory, University of Sheffield.

Findlay, D.C., Colborne G.J.N., Cope, D.W., Harrod, T.R., Hogan, D.V. and Staines, S.J. 1984. *Soils and their Use in South West England.* Harpenden: Soil Survey of England and Wales, Bulletin No.14.

Jaillard, B. and Callot, G. 1987. Action des racines sur la segregation mineralogique des constituents du sol. In N. Federoff, L.M. Bresson, and M.A. Courty (eds.), *Micromorphologie des sols*, 371–375. Paris: Association Française pour l'étude du Sol.

Jaillard, B., Guyon, A. and Maurin, A.F. 1991. Structure and composition of calcified roots and their identification in calcareous soils. *Geoderma*, 50, 197–210.

Jongerius, A. 1970. Some morphological aspects of regrouping phenomena in Dutch soils. *Geoderma*, 4, 311–331.

Kerridge, E. 1967. *The Agricultural Revolution.* London: Allen and Unwin.

Kretzschmar, A., 1987. Characterisation microscopique de l'activité des lombriciens endoges. In N. Fedoroff, L.M. Bresson and M.-A. Courty, (eds.), *Micromorphologie des sols*, 325–330. Paris: Association Française pour l'étude du Sol.

Kwadd, F.J.P.M. and Mücher, H.J. 1979. The formation and evolution of colluvium on arable land in northern Luxembourg. *Geoderma*, 22, 173–192.

Leadbeater, B.S.C. and Riding, R. 1986. *Biomineralization in Lower Plants and Animals.* Oxford: Systematics Association/Clarendon Press.

Limbrey, S. 1983. Archaeology and Palaeohydrology. In K.J. Gregory (ed.), *Background to Palaeohydrology*, 189–212. Chichester: Wiley.

Limbrey, S. and Robinson, S. 1988. Dry land to wet land: soil resources in the Upper Thames valley. In P. Murphy and C. French (eds.), *The Exploitation of Wetlands*, 129–141. Oxford: British Archaeological Reports, International Series, 186.

Nakaharsa, H. and Bevelander, G. 1969. An electron microscope and autoradiographic study of the calciferous glands of the earthworm, *Lumbricus terrestris. Calcified Tissue Research*, 4, 193–201.

Needham S. and Sorensen, M.-L. 1988. Runnymede refuse tip: a consideration of midden deposits and their formation. In J.C. Barrett and I.A. Kinnes, (eds.), *The Archaeology of Context in the Neolithic and Bronze Age: Recent Trends*, 113–26. Sheffield: Department Archaeology and Prehistory, University of Sheffield.

Robertson, J. 1936. The function of the calciferous glands of earthworms. *Journal of Experimental Biology*, 13, 279–297.

Thompson, M.L., Federoff, N. and Fournier, B. 1990. Morphological features related to agriculture and faunal activity in three loess-derived soils in France. *Geoderma*, 46, 329–349.

Watabe, N., Meenakshi, V.R., Kurtz, E.M. and Dunkelberger, D.G. 1976. Calcareous spherules in the gastropod *Pomacea paludosa*. In N. Watabe and K.M. Wilbur (eds.), 283–308.

Watabe, N. and Wilbur, K.M. (eds.) 1976. *The Mechanisms of Mineralization in the Invertebrates and Plants*. Colombia, S. Carolina: University of S. Carolina Press.

Wordie, J.R. 1984. The South: Oxfordshire, Buckinghamshire, Berkshire, Wiltshire and Hampshire. In J. Thirsk (ed.), *The Agrarian History of England and Wales, Vol.V*, 1640–1750. I Regional Farming Systems, chapter 10. Cambridge: University Press.

7

Molluscan taxocenes from Holocene overbank alluvium in central southern England

J. G. Evans, P. Davies, R. Mount and D. Williams

Summary: *Although Mollusca are abundant in fine-grained, minerogenic alluvium of Holocene age, there have been few investigations, especially of overbank flood deposits. This is regrettable because of the importance of the floodplain environment as a human resource. This paper is concerned with problems of interpretation. These are due to the partial inapplicability of modern analogue data and to features of floodplains and their faunas which cause mixing, such as fluvial processes (taphonomy) and molluscan behaviour (different faunas living in the same place at different times of the year). The importance of using autochthonous assemblages is stressed. A key unit of analysis is the taxocene, a recurring collection of species in particular proportions which has environmental implications even where modern parallels are unknown. Ways of characterising and environmentally assessing taxocenes are itemised, and the distinction is made between properties that do not relate to named species (e.g. diversity) and those that do (e.g. relative abundance). The approach is thus a partial analogue one. Examples of floodplain taxocenes from central southern England are listed and their environmental significance discussed.*

INTRODUCTION

In the British Isles, fine-grained, overbank alluvium of Holocene age is widespread in river valley bottoms. It is often rich in the shells of land and freshwater molluscs, but there have been few studies, in spite of early recognition (Kennard and Woodward 1903; 1906). Most work has been done in association with archaeological sites (Evans 1986; French 1983; Horsman and Davison 1989; Jackson and Ambrose 1976; Lambrick 1985; Lambrick and Robinson 1979; Thomas *et al* 1986; Needham 1991) or with problems of human land use in mind (Robinson 1988; Shotton 1978); other investigations include those of Sparks and Lambert (1961) and Preece and Robinson (1982).

In this paper we discuss problems of studying the Mollusca from deposits which have formed largely by overbank flooding. We detail (tentatively) a number of recurrent, distinctive, assemblages (taxocenes) which we consider to have relevance in environmental reconstruction. The paper takes further the initial, more general, statement by Evans (1991).

At the outset, the distinction between "assemblage" and "taxocene" should be clarified. "Assemblage" refers to a collection of shells without implications as to its environmental significance or to the relationships (living or dead) between its individuals – e.g. the raw data of all the shells from a soil sample. "Taxocene" refers to particular proportions of species of a particular taxonomic group, either as a part of a larger living community or, if subfossil, as reflecting a former life situation. A taxocene is more of an interpretative statement, for example referring to a group of stratigraphically contiguous assemblages which as a whole have particular characteristics *but which individually need not do so.* "Taxocene" is also better than "community" (Pielou 1975: 270) because it stresses the artificial, taxonomic, bias of studying just one group of organisms, and it does not imply any form of intra- or inter-specific interaction which "community" often does.

PROBLEMS WITH MODERN ANALOGUE DATA

In using molluscs as indicators of past environments, we are hampered not only by a paucity of modern comparative data but also by their partial inapplicability. Matching of subfossil and modern data, in terms of species-abundance compositions is often impossible (Cameron 1978; Thomas 1985). This is a temporal reflection of the fact that spatially most species are eurytopic, occurring in a wide variety of habitats when their entire geographical range is considered, but on a local scale — e.g. in individual lakes or rivers — being often more narrow. Because of this, and because mollusc species do not occur in interactive communities, there is not precise covariation of different species on a regional or countrywide scale, or, by the same token, through time.

Differences between taxocenes relate to changes in large-scale, all-pervasive factors like climate, the cumulative impact of man and biogeography, to factors of a more localised and immediate nature like vegetation, hydrology, water chemistry and human management, and, on an even smaller scale, to lack of adjustment through the youthfulness of environments. It is a question of disentangling the different scales of factor and of establishing which differences are due to local factors and which to regional ones. A start is to build up the data in small areas and short time blocks and then make comparisons at larger scales.

The applicability of modern data is also limited because of non-matching of modern and past environments. Modern floodplain environments in the British Isles are artificial and generally managed, so the use of terms like "fen", "carr", "water-meadow" or "reed swamp" are to be discouraged unless used very specifically and with good cause.

FLOODPLAINS AND THEIR FAUNAS

Floodplains often have more extreme annual variations in properties such as flooding/drying, temperature, parching and vegetation structure than places which are not flooded. In consequence, the molluscan faunas tend to be specialised ones, often consisting of species which can invade and breed rapidly or which can survive drought or flooding (e.g. Williams and Hynes 1977). *Lymnaea truncatula*, for example, one of the commonest floodplain gastropods, has enormous reproductive potential; under laboratory conditions in one observation it reached maturity in 21 days, and a single snail gave rise to 25,000 individuals in 12 weeks (Kendall 1953), although in nature an 8- to 10-fold seasonal increase is more realistic (Morphy *et al* 1969). Some snails and bivalves can survive in damp mud (Kenk 1949; Bevercombe *et al* 1973).

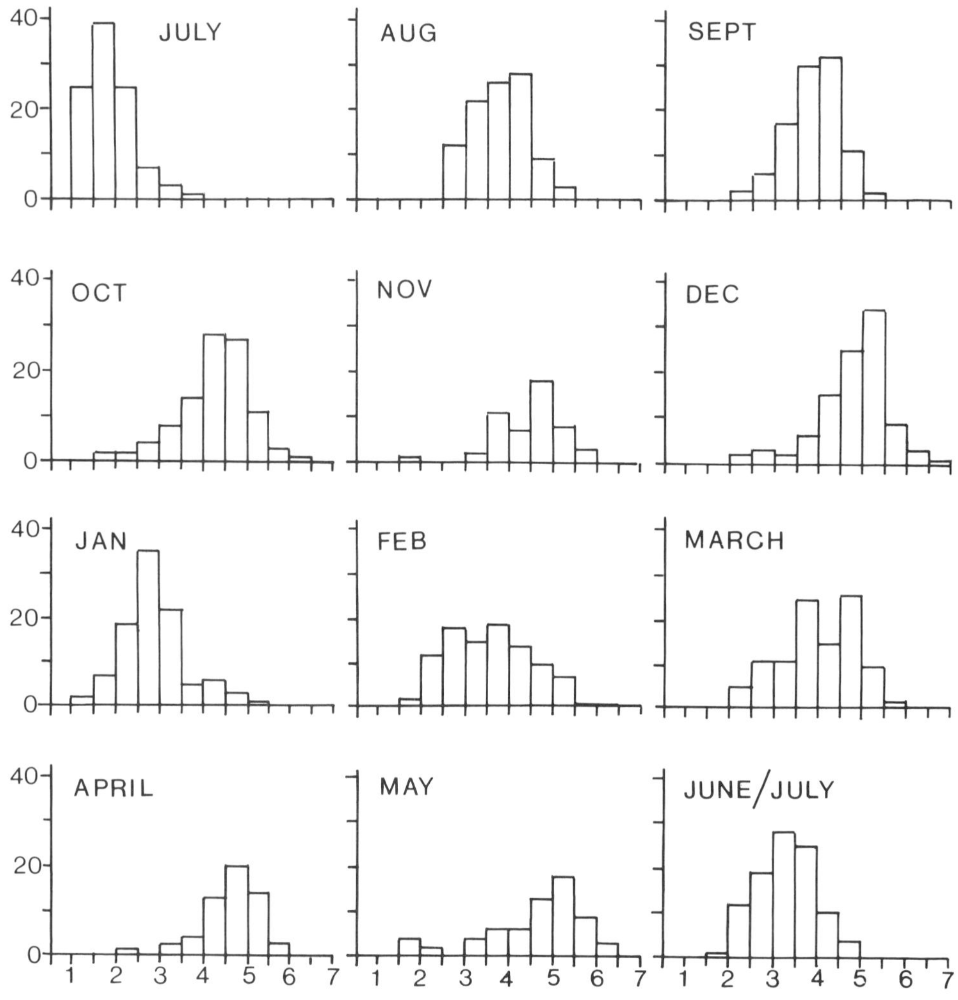

Fig. 7.1. *River Kennet, West Overton, Wiltshire. Measurements of shell diameter of modern Anisus leucostoma, 1987–88. Vertical axis = numbers of shells; horizontal axis = shell diameter in mm. (Measurements, R. Mount.)*

An important characteristic is that there are sometimes different life assemblages in the same place at different times of the year. This has been demonstrated by Robinson (1988) for hay meadows where there is a winter breeding population of amphibious species and a summer breeding one of land species. In the fossil record, such separation is unlikely to be preserved so that one ends up with a mixed assemblage. This is a general feature of molluscan palaeoecology in that, however good the preservation, much small-scale temporo-spatial behaviour is lost, as for example, seasonal or short-term annual variations in abundance, month to month variations in size distributions, shown here for Anisus leucostoma (Fig. 7.1), or local seasonal migrations.

THE IMPORTANCE OF AUTOCHTHONY

Ideally, wholly autochthonous assemblages should be used. Autochthony is judged from the topographical context, lithopedology, and the molluscan assemblages themselves. In fine-grained alluvium, the proportion of land to freshwater species is a useful guide. The only possible contexts in overbank alluvium for fully autochthonous assemblages are the tops of palaeosols and fine-grained overbank deposits which lack true freshwater species.

In dealing with taphonomically (as opposed to behaviourally) mixed assemblages, the land and freshwater components can be considered separately, as done by Kerney (1971) for the interglacial fauna at Swanscombe. This approach works well for channel deposits where the distinctions between the two components are clear. However, with overbank deposits, it is not so simply applied because it is not always clear which species were living on the floodplain and which are derived from the river. This especially applies to amphibious species such as *Anisus leucostoma* and *Pisidium personatum* and to certain freshwater species which might breed on well-watered floodplains and in adjacent drainage ditches, such as *Valvata cristata* and *Lymnaea peregra.*

There are degrees of autochthony. For example, in mixed assemblages of land and freshwater shells, some degree of integrity may yet be preserved and detected by examining intra-specific variation, as in the case of the two assemblages of *Carychium minimum* from Neolithic and Bronze Age deposits at Runnymede (Needham 1991) shown here (Fig. 7.2). In this diagram, each point represents the measurements of an adult shell. There is approximate correlation of two different

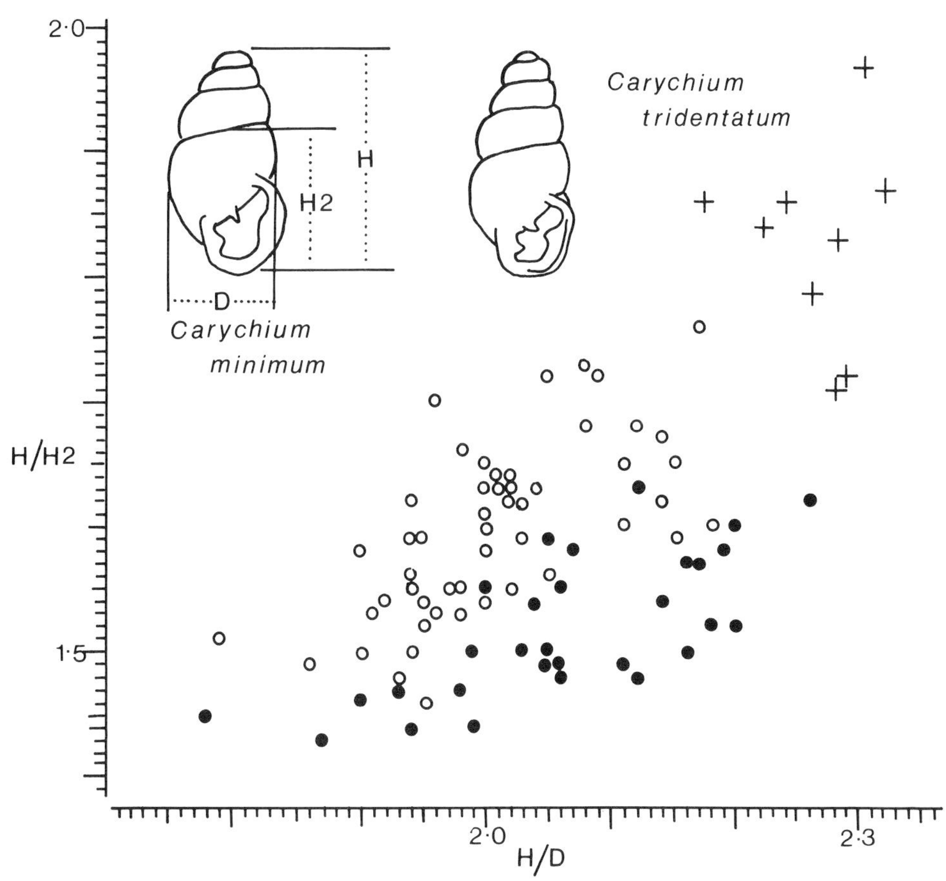

Fig. 7.2. Runnymede Bridge, Berkshire. Measurements of three populations of Carychium from Neolithic and Bronze Age deposits. Open circles = C. minimum (series SS1); closed circles = C. minimum (series WF1b); crosses = C. tridentatum. (Analysis, J.G. Evans.)

forms of *C. minimum* with two separate sample series respectively even though the assemblages are partly allochthonous. This demonstrates, too, the point made by Thomas (1978) that it is important to examine intra-specific variation.

CHARACTERISATION AND ASSESSMENT OF MOLLUSCAN DATA

There are several ways in which molluscan data can be characterised and assessed. The first five (below) do not use frequencies and relative abundances of *named* species. The species concept is accepted, but there is no one-to-one matching between named species and environment. You can say something about the fauna, and therefore the environment, on the basis of relations of relevance between fauna and environment that apply — and have always applied — throughout ecology.

(1) *Total numbers of individuals.* This gives information about the suitability of the environment for molluscan life, e.g. its calcium carbonate content, but will also be related to sedimentation rate.

(2) *Total numbers of species.* This is more useful because it gives information on the number of microhabitats and niches. The number of species may also be a function of the size of the habitat (e.g. Boycott 1936: 126; Baur 1987).

(3) *Distribution of numbers of individuals among species* — diversity or evenness (Magurran 1988) — as used by Pryor and French (1985) on freshwater slum assemblages from Iron Age and Roman ditches. Diversity gives an indication of the structure of the environment and the extremeness of conditions. It is best applied when there are high or low values or when there are marked trends in sequences — either temporal or spatial.

(4) *Biomass.* Because there are considerable size differences between species, biomass estimates might be more reliable than numbers of individuals in giving information on aspects such as ecosystem productivity (e.g. Mason 1970; Cameron 1982; Holopainen and Jónasson 1989; Russell-Hunter 1978).

(5) *Age structure.* Although, as indicated above, age composition is unlikely to be detectable in the subfossil record, it is worth bearing in mind as a possible environmental indicator. Types of reproductive strategies could reinforce the information from diversity indices, suggesting the severity or favorability of an environment for Mollusca.

The next five aspects use named species, and recurrence of characteristics is important in establishing the significance of the assemblages. Modern and subfossil assemblages are used alike. "Today" is only one point in a continuum with the past.

(6) *Intra-specific variation.* The use of different forms of a species which are related to environment helps to refine interpretation, as shown by Coope *et al* (1961) on *Gyraulus laevis*, from floodplain pools. Work has been done on modern populations of certain species. In one study on *Lymnaea peregra*, differences of size, growth pattern and reproductive pattern were related to environment, with snails from exposed places being smaller, initiating reproduction earlier and putting more effort into it than individuals from sheltered habitats (Calow 1981). Similar differences might be expected between adjacent populations of species common to both rivers and their floodplains (e.g. *Lymnaea peregra* in taxocene 7 — see below). There has been little work on this aspect of subfossil assemblages, however, in spite of pleas by Thomas (1978; 1985).

(7) *Stenotopy.* Stenotopy is a function of the context and the species present. Most narrowly, stenotopic species of known ecology — in the subfossil context, "indicator species" — are those which have a restricted habitat over their entire range. Unfortunately they are few and mostly rare species such as *Vertigo moulinsiana* which usually lives on the stems of *Carex* spp. or *Glyceria maxima* in fen environments (Butot and Neuteboom 1958; Bishop 1974) and *Valvata macrostoma* which lives in ditches (Boycott 1936). By their very rareness, however, these are of limited value. Perhaps the commonest stenotope is *Vallonia pulchella*, on floodplains probably confined to grassland rather than coarse sedge or reed vegetation.

A few species, although having a wide range of habitats, behave more like stenotopes than others in individual instances, in showing peaks of abundance and strong antipathetic relationships with other species. *Valvata cristata*, for example, is typical of ditches (Boycott 1936; Sparks 1961). It also occurs in the vegetated littoral of lakes around 2 m, rarely below 5 m (e.g. Berg 1938), sometimes showing a completely antipathetic spatial distribution with the deeper-water species, *Valvata piscinalis* (e.g. Odhner 1929), and likewise through time (e.g. Alexandrowicz *et al* 1984). Occasionally, however, it is common in the profundal zone, e.g. between 14 and 22 m (Ekman 1915). In our work, we see it as the most tolerant to drying out on floodplains of the true freshwater species (e.g. Fig. 7.3b).

In the genus *Pisidium* from the upper reaches of chalk streams, Ham and Bass (1982) have shown strong associations between intermittent flow and *P. personatum* and between permanent flow and *P. milium.*

More broadly, *P. casertanum* and *P. personatum* were recorded most frequently from intermittent headwaters, while *Sphaerium lacustre*, *S. corneum*, *P. henslowanum* and *P. hibernicum* showed a preference for channel widths exceeding 25 m.

(8) *The behaviour of pairs of species.* This is illustrated by assemblages from floodplain deposits from Ergolding in Bavaria (Evans, unpublished), in which individual species are arranged in rank order and values for another species from the corresponding assemblage plotted against them. Interestingly, there are two types of behaviour, mutually exclusive, as with *Zonitoides nitidus* and *Vallonia pulchella*, and negatively correlated, as with *Vallonia costata* and *Succinea oblonga* (Fig. 7.3a).

(9) *Ecological groups.* These are groups of species of similar ecology but that are not necessarily associated in life. They are useful in certain circumstances. For example, the ratio of land to freshwater shells can give an indication of the taphonomy of an assemblage. Sparks (1961) used four freshwater groups in the interpretation of Pleistocene assemblages, and in one instance for the Holocene (Sparks and Lambert 1961). Robinson (1988) has used proportions of land, amphibious and freshwater species in floodplain alluvial open-country assemblages as an indicator of the status of grassland, whether meadow or pasture, although since these are from largely autochthonous contexts they perhaps belong in the next section.

(10) *Associations of distinctive composition.* Unlike the ecological groups, these, at least in theory, are life associations and close, if not equivalent, to taxocenes. They can be identified subjectively, as we have done on the whole, or by multivariate analyses such as Principal Components Analysis (e.g. Bishop 1981), Detrended Correspondence Analysis (e.g. Bush 1988), TWINSPAN (Hill 1979) or CLUSTAN (e.g. Young and Evans, 1991). The assemblages from Ergolding, referred to above, were grouped by Ward's method (Ward 1963; Wishart 1978) (Fig. 7.3b), and although these groupings may not reflect former life assemblages, they do provide a basis for further consideration. The problem with rigid numerical methods of analysis is that they do not take the stratigraphical context of an individual assemblage into account. In our view it is perfectly legitimate to include an individual assemblage in a taxocene by the nature of its context rather than its faunal characteristics alone.

Equally, the separation of taxocenes must take trends in time and horizontal space into account, that is the time-space context of particular species and associations. Small sequences are useful, and again, recurrence is an important criterion of validity. External contexts such as geomorphology, lithopedology, archaeology and biology are also used, although prone to problems of analogy just as are molluscs. With all these aspects, it is a question of building up groups of similarities and differences.

SOME EXAMPLES

We are concerned with wet ground taxocenes from fine-grained alluvium of Holocene age in central southern England. Other contexts with similar environmental parameters and molluscan assemblages might be considered, such as dune slacks (e.g. Bevercombe *et al* 1973), tufa (e.g. Preece 1980), various slum habitats (e.g. Pryor and French 1985; O'Connor 1988) and wetlands generally (e.g. Bishop 1981; Bush 1988), but in the context of these proceedings and our research we have restricted this paper to floodplains. Seven wetland taxocenes and one freshwater taxocene are proposed, and of all the features used in establishing their validity the most important is recurrence. Although we have not applied multivariate analysis (except for TWINSPAN, which being based on presence-absence does not appear to be suitable), they appear to be distinct and not, for example, to seriate or occur at random. None, except assemblage 8, has true aquatic shells in more than 10% abundance. The proposal of discrete taxocenes does not mean that these are the only possibilities but is a pointer to those that are best defined. Itemisation and characterisation are the key aims at this stage in the work. Environmental interpretation has been kept to a minimum because it requires detailed knowledge of contexts; where it has been attempted it is at the level of vegetation structure and hydrology rather than at the broader scale of biogeography and climate.

Taxocene 1

Low diversity, with open-country land species, mainly *Vallonia pulchella* agg., *Trichia hispida*, *Cochlicopa* and *Limacidae*; amphibious and marsh species in very low abundance or absent. Structurally low diversity, open and relatively dry environment – meadow or pasture, with occasional winter flooding. The taxocene often occurs in the uppermost parts of sequences, e.g. in Medieval levels in the Upper Kennet (Evans *et al* 1988) and in the Anton at Fullerton (Fig. 7.4), although the latter is approaching taxocene 2 in having a marsh component. At the simplest end of the range of structural diversity (e.g. short grassland), it is likely that *Vallonia pulchella* agg. and Limacidae would be the sole components.

Taxocene 2

Taxocene 1 + marsh/swamp species, *Lymnaea truncatula*, *Carychium minimum* and Succineidae. Open environment, wetter and structurally more diverse than

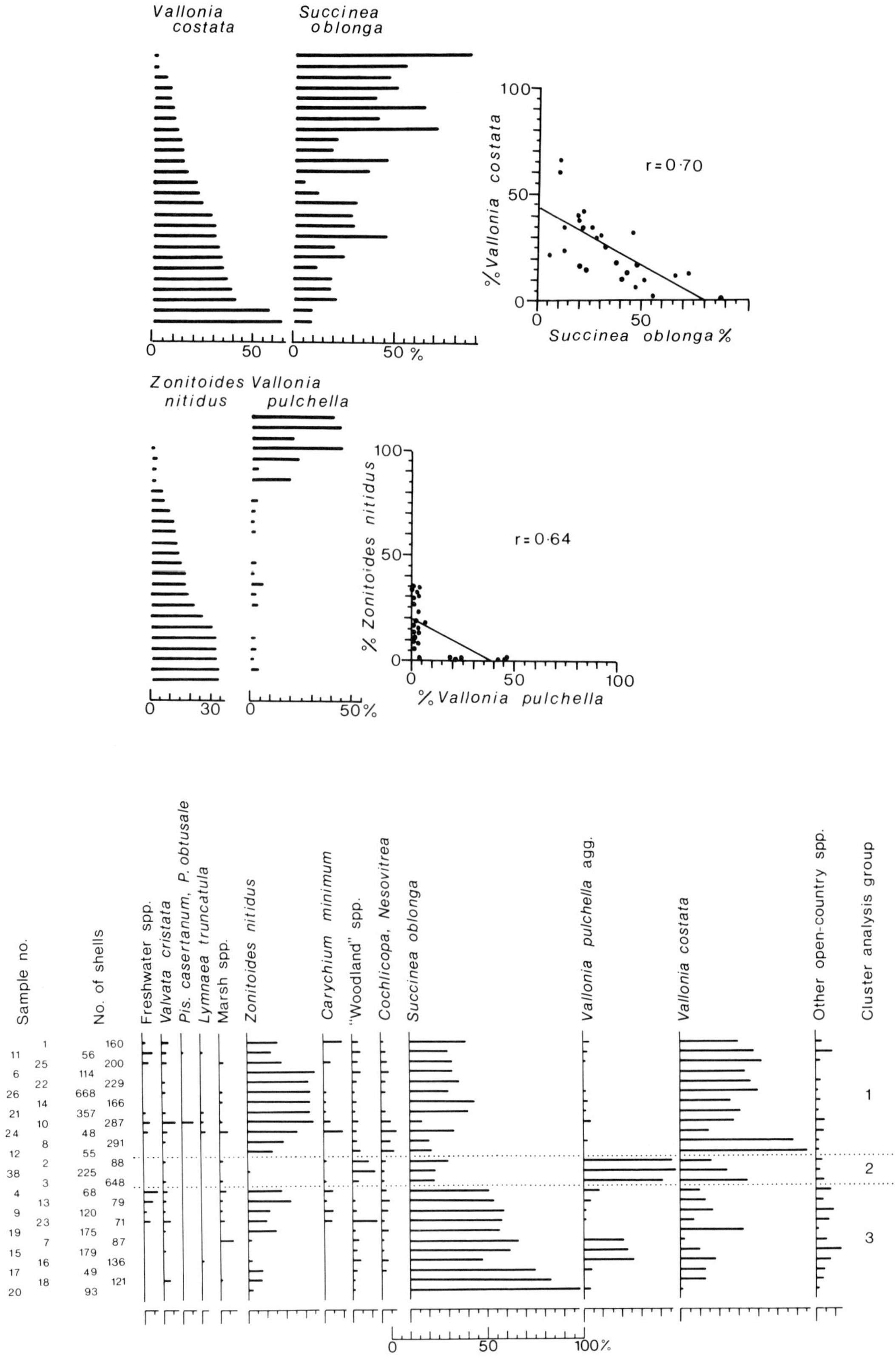

Fig. 7.3. *Ergolding, Landshut, Bavaria.* ***(a)*** *(top) Rank order curves and correlation coefficients for two pairs of Mollusca from Neolithic to Carolingian (Early Medieval) deposits.* ***(b)*** *(bottom) Assemblages seriated according to the order established by cluster analysis, using Ward (1963) and Wishart (1978). (Note that the order of each sample, not just the groups, is based on the cluster analysis and is significant in terms of seriation, although we have not considered it to be so ecologically.) (Analysis, J.G. Evans.)*

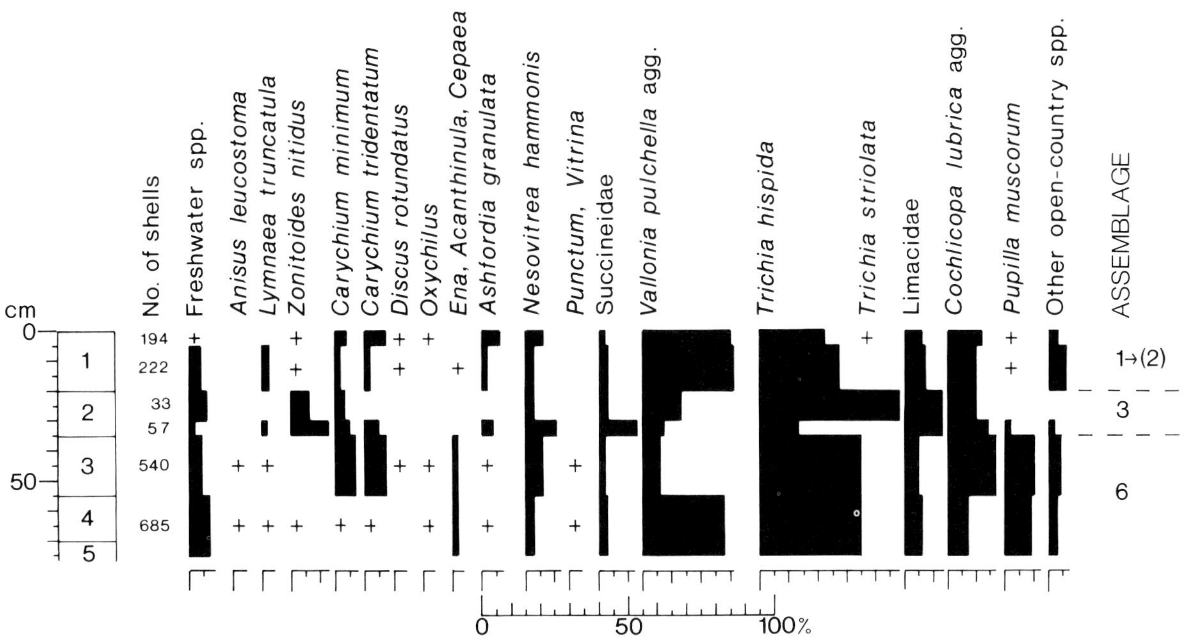

Fig. 7.4. River Anton, Fullerton, Hampshire. Molluscan diagram (FAH 9) from fine alluvium. 0 cm = modern surface; 1, topsoil; 2 and 3, brown silty clay loam; 4, greyish brown clay; 5, silty clay loam overlying peat. (Analysis, D. Williams.)

taxocene 1. Meadow or lightly grazed pasture, winter-flooded. This is the situation in the upper part of the sequence at Kingsmead Bridge in the Wylye (Fig. 7.5).

Taxocene 3

Marsh and sometimes woodland species, and sometimes of high diversity; marsh species, *Zonitoides nitidus* and *Carychium minimum*, distinctive, while *Vallonia pulchella* and Succineidae are poorly represented. There is an antipathetic relationship between *Z. nitidus* and *V. pulchella* (e.g. Fig. 7.3) which probably relates to the structural diversity of the habitat, although it should be noted that *Z. nitidus* is a carnivore, being used to control *Lymnaea truncatula* (Rondelaud 1982). Environment of high structural diversity such as woodland or tussocks of sedges; winter-flooded.

Taxocene 4

Lymnaea truncatula predominant, plus Succineidae sometimes. Low diversity environment, lower than that of taxocene 1. One possibility is mudflat. At Anslows Cottages in the Lower Kennet (Evans, unpublished) it occurs in two separate alluvial sequences just below a change to peat in the sequence "Taxocene 2" to "Taxocene 4" to "Peat".

Taxocene 5

Amphibious species — *Anisus leucostoma*, *Lymnaea truncatula* and *Pisidium personatum* — distinctive; land species, woodland or open-country, abundant. Environment flooded for longer than in those of taxocenes 2 and 3, and/or with permanent pools. This is the situation in the lower part of the Kingsmead Bridge sequence (Fig. 7.5), although tending towards taxocene 8 between 60 and 65 cm. Where the land component is open-country, the taxocene is similar to the meadow assemblages of Robinson (1988).

Taxocene 6

Taxocene 1 + *Pupilla muscorum*. Grassland, perhaps long and rank with a thick thatch (i.e. of loose, dead stolons and leaves), as in the early stages of abandoned pasture; but it could also be where alluviation is starting, with drying out presenting a baked mud surface of high insolation which the species favours. The problem is that *P. muscorum*, while typically a xerophile, also occurs in wetlands (Ant 1963). This is a case where modern analogues from floodplains might be useful. Also, the wetland form is likely to be morphologically different from the dry-ground form, so investigations along these lines — e.g. size and shape measurements or denticle counts — might help. The taxocene is typical of the *early* part of alluvial sequences in the Upper Kennet (Evans *et al* 1988) and occurs at Fullerton (Fig. 7.4) and Kingsmead Bridge (Fig. 7.5).

Taxocene 7

Lymnaea peregra distinctive, together with other amphibious species; land component open country or woodland. This taxocene has only been found in the Upper Kennet (Evans *et al* 1988), where it occurs especially in deposits of silty loam. It is not known whether the shells are from animals living on the floodplain or are derived from overbank flooding, and the situation is especially tricky because the Kennet is a winterbourne in this locality so that the river and floodplain faunas are likely to have been similar. One approach would be to see if two different forms were present or if they represented a uniform population, because it is possible that river and floodplain forms would differ (cf. Calow 1981; Lam and Calow 1988). Wright *et al* (1984) showed an interesting two-locational distribution of *L. peregra* in a winterbourne, with a high frequency in the intermittent part, then a gap of low frequency/absence in the upper reaches of the permanent part, and then another high frequency in the lower reaches of the permanent part. It would be useful to know if the two high frequency faunas were of different forms.

Taxocene 8

Freshwater species of Sparks' (1961) slum group, especially *Valvata cristata* and *Pisidium casertanum.* The taxocene reflects the transition from aquatic to wetland habitats. It is indicated within taxocene 5 at Kingsmead Bridge between 60 and 65 cm (Fig. 7.5). *Valvata cristata* is often the only freshwater species consistently present in terrestrial-amphibious assemblages (e.g. Fig. 7.3). At Apethorpe, Northamptonshire (Sparks and Lambert 1961: Table 4), it shows an antipathetic distribution with *V. piscinalis*, and at Farmoor in the Upper Thames (Lambrick and Robinson 1979: Table 18), it occurs in the upper levels of alluvium with only amphibious and marsh/swamp species. The environment indicated is permanently wet, but subject to stagnation, as in ditches.

RECOMMENDATIONS FOR FUTURE WORK

(1) More data are needed, both modern and fossil, the latter from autochthonous contexts, so that recurrent associations can be built up. Associations need to be both "molluscan – molluscan" and "molluscan – other" (e.g. lithopedostratigraphical, archaeological). Small recurrent sequences or trends in time and space should also be sought.

(2) More data are needed on the relationship between molluscan assemblages and environmental structure and diversity at different scales: species (inventory), structural or architectural, small-spatial and landscape.

(3) An implication of the changing environmental significance of a species or group through time is that the traditional histogram method of presenting results needs re-assessing.

(4) Studies of intra-specific variation are needed to refine environmental interpretation.

(5) More understanding is needed of wetland environments on river floodplains, and specifically the possibilities in the past which may be rare or non-existent today.

Acknowledgements

We are grateful to the Science and Engineering Research Council for supporting some of this work. For the use of data prior to full publication we thank Barbara Ottaway (Ergolding).

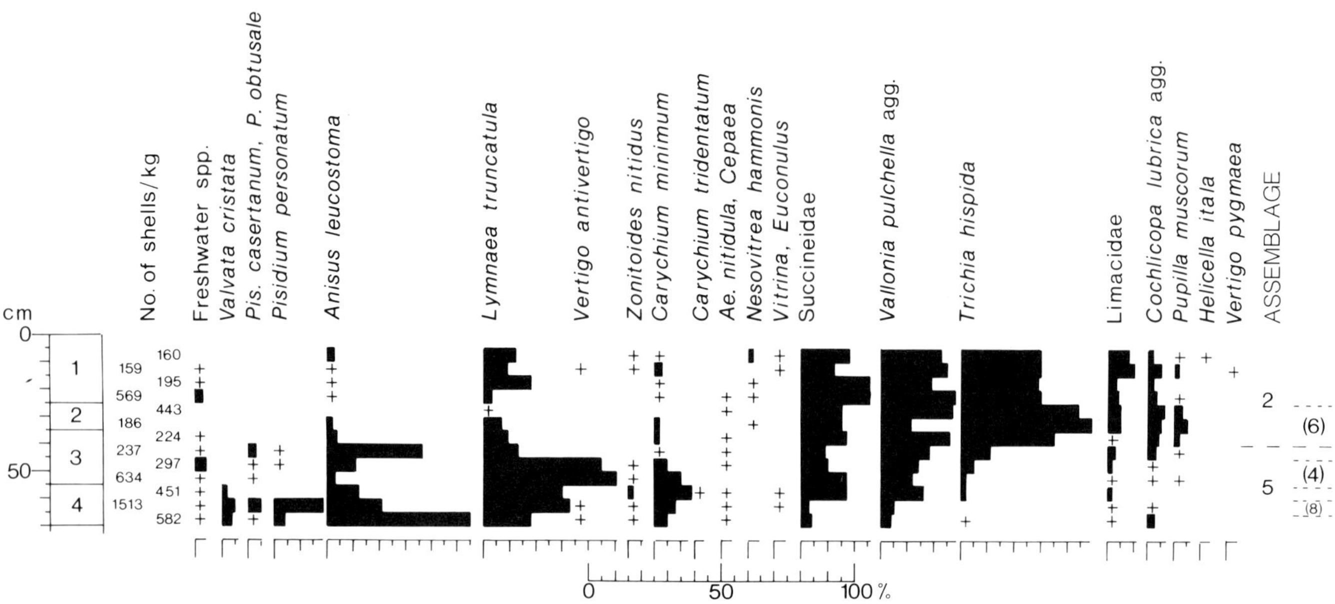

Fig. 7.5. River Wylye, Kingsmead Bridge, Wiltshire. Molluscan diagram (KB, F) from fine alluvium. 0 cm = modern surface. 1, topsoil; 2, inorganic clay; 3, orange-mottled, inorganic clay; 4, organic clay, overlying peat.

References

Alexandrowicz, S.W., Snieszko, Z. and Zajaczkowska, E. 1984. Stratigraphy and malacofauna of Holocene deposits in the Sancygniówka valley near Działoszyce. *Quaternary Studies in Poland*, 5, 5–28.

Ant, H. 1963. Faunistische, ökologische und tiergeographische Untersuchungen zur Verbreitung der Landschnecken in Nordwestdeutschland. *Abhandlungen aus dem Landesmuseum für Naturkunde zu Münster in Westfalen*, 1–125.

Baur, B. 1987. Richness of land snail species under isolated stones in a karst area on Öland, Sweden. *Basteria*, 51, 129–133.

Berg, K. 1938. Studies on the bottom animals of Esrom Lake. *Det. Kgl. Danske Videnskabernes Selskabs Skrifter, Naturvidensk og Matematisk Afd.*, 9, Raekke 8, 1–255.

Bevercombe, A.M., Cox, N., Thomas, M.P. and Young, J.O. 1973. Studies of the invertebrate fauna of a wet slack in a sand dune system. *Archiv für Hydrobiologie*, 71, 487–516.

Bishop, M.J. 1974. The Mollusca of Wicken Fen with some additional records. *Cambridgeshire and Isle of Ely Naturalists Trust Annual Report 1974*, 1–4.

Bishop, M.J. 1981. Quantitative studies on some living British wetland mollusc faunas. *Biological Journal of the Linnaean Society*, 15, 299–326.

Boycott, A.E. 1936. The habitats of freshwater Mollusca in Britain. *The Journal of Animal Ecology*, 5, 116–186.

Bush, M.B. 1988. The use of multivariate analysis and modern analogue sites as an aid to the interpretation of data from fossil mollusc assemblages. *Journal of Biogeography*, 15, 849–861.

Butot, L.J.M. and Neuteboom, W.H. 1958. Over Vertigo moulinsiana (Dupuy) en haar voorkomen in Nederland. *Basteria*, 22, 52–63.

Calow, P. 1981. Adaptational aspects of growth and reproduction in *Lymnaea peregra* (Gastropoda: Pulmonata) from exposed and sheltered habitats. *Malacologia*, 21, 5–13.

Cameron, R.A.D. 1978. Interpreting buried land-snail assemblages from archaeological sites – problems and progress. In D.R. Brothwell, K.D. Thomas and J. Clutton-Brock (eds.), *Research Problems in Zooarchaeology*, 19–23. London: Institute of Archaeology.

Cameron, R.A.D. 1982. Life histories, density and biomass in a woodland snail community. *Journal of Molluscan Studies*, 48, 159–166.

Coope, G.R., Shotton, F.W. and Strachan, I. 1961. A late Pleistocene fauna and flora from Upton Warren, Worcestershire. *Philosophical Transactions of the Royal Society* B, 244, 379–421.

Ekman, S. 1915. Die Bodenfauna des Vättern, qualitativ und quantitativ untersucht. *Internationale Revue der gesamten Hydrobiologie*, 7, 146–204.

Evans, J.G. 1986. Mollusca from the Iron Age sites at Meare, Somerset. *Somerset Levels Papers*, no. 12, 97–102.

Evans, J.G. 1991. An approach to the interpretation of dry-ground and wet- ground molluscan taxocenes from central-southern England. In D.R. Harris and K.D. Thomas (eds.), *Modelling Ecological Change*, 75–89. London: Institute of Archaeology.

Evans, J.G., Limbrey, S., Máté, I. and Mount, R. 1988. Environmental change and land-use history in a Wiltshire river valley in the last 14,000 years. In J.C. Barrett and I.A. Kinnes (eds.), *The Archaeology of Context in the Neolithic and Bronze Age: Recent Trends*, 97–103. Sheffield: Department of Archaeology and Prehistory, University of Sheffield.

French, C.A.I. 1983. An Environmental Study of the Soils, Sediments and Molluscan Evidence Associated with Prehistoric Monuments on River Terrace Gravels in North-west Cambridgeshire. University of London, unpublished PhD thesis.

Ham, S.F. and Bass, J.A.B. 1982. The distribution of Sphaeriidae in rivers of central southern England. *Journal of Conchology*, 31, 45–55.

Hill, M.O. 1979. *TWINSPAN – A FORTRAN Program for Arranging Multivariate Data in an Ordered Two-way Table by Classification of the Individuals and Attributes*. Ithaca, New York: Cornell University.

Holopainen, I.J. and Jónasson, P.M. 1989. Bathymetric distribution and abundance of *Pisidium* (Bivalvia: Sphaeriidae) in Lake Esrom, Denmark, from 1954 to 1988. *Oikos*, 55, 324–34.

Horsman, V. and Davison, B. 1989. The New Palace Yard and its foundations: excavations in the Palace of Westminster 1972–4. *Antiquaries Journal*, 69, 279–97.

Jackson, D.A. and Ambrose, T.M. 1976. A Roman timber bridge at Aldwincle, Northamptonshire. *Britannia*, 7, 39–72.

Kendall, S.B. 1953. The life history of *Lymnaea truncatula* under laboratory conditions. *Journal of Helminthology*, 27, 17–28.

Kenk, R. 1949. The animal life of temporary and permanent ponds in southern Michigan. *Miscellaneous Publications of the Museum of Zoology*, University of Michigan, 71, 1–66.

Kennard, A.S. and Woodward, B.B. 1903. The non-marine Mollusca of the River Lea alluvium at Walthamstow, Essex. *Essex Naturalist*, 13, 13–21.

Kennard, A.S. and Woodward, B.B. 1906. On sections in the Holocene alluvium of the Thames at Staines and Wargrave. *Proceedings of the Geologists' Association*, 19, 252–258.

Kerney, M.P. 1971. Interglacial deposits in Barnfield Pit, Swanscombe, and their molluscan fauna. *Journal of the Geological Society of London*, 127, 69–93.

Lam, P.K.S. and Calow, P. 1988. Differences in the shell shape of *Lymnaea peregra* (Müller) (Gastropoda: Pulmonata) from lotic and lentic habitats; environmental or genetic variance? *Journal of Molluscan Studies*, 54, 197–207.

Lambrick, G.H. 1985. Further excavations on the second site of the Dominican Priory, Oxford. *Oxoniensia*, 50, 131–208 and Microfiche 1.

Lambrick, G.H. and Robinson, M. 1979. *Iron Age and Roman Riverside Settlements at Farmoor, Oxfordshire*. London: Council for British Archaeology.

Magurran, A.E. 1988. *Ecological Diversity and its Measurement*. London: Croom Helm.

Mason, C.F. 1970. Snail populations, beech litter production, and the role of snails in litter decomposition. *Oecologia* (Berlin), 5, 215–239.

Morphy, M.J., Ross, J.G. and Taylor, S.M. 1969. Problems of *Lymnaea truncatula* ecology in investigations of fascioliasis. *Malacologia*, 9, 127–134.

Needham, S.P. 1991. *Excavation and salvage at Runnymede Bridge 1978: the Late Bronze Age Waterfront site*. London: British Museum Press.

O'Connor, T. 1988. Slums, puddles and ditches: are molluscs useful indicators? In P. Murphy and C. French (eds.), *The Exploitation of Wetlands*. Oxford: British Archaeological Reports, British Series, 186.

Odhner, N.H. 1929. Die Molluskenfauna des Tåkern. *Sjön Tåkerns Fauna och Flora*, 8, 2–129.

Pielou, E.C. 1975. *Mathematical Ecology*. New York: Wiley Interscience Publication.

Preece, R.C. 1980. The biostratigraphy and dating of the tufa deposit at the Mesolithic site at Blashenwell, Dorset, England. *Journal of Archaeological Science*, 7, 345–362.

Preece, R.C. and Robinson, J.E. 1982. Mollusc, ostracod and plant remains from early Postglacial deposits near Staines. *The London Naturalist*, no. 61, 6–15.

Pryor, F.M.M. and French, C.A.I. 1985. *Archaeology and Environment in the Lower Welland Valley*. Cambridge: Cambridge Archaeological Committee.

Robinson, M. 1988. Molluscan evidence for pasture and meadowland on the floodplain of the upper Thames basin. In P. Murphy and C. French (eds.), *The Exploitation of Wetlands*, 101–112. Oxford: British Archaeological Reports, British Series, 186.

Rondelaud, D. 1982. Le contrôle biologique par prédation de *Lymnaea truncatula* (Müller). Etude expérimentale de la dynamique de cinq espéces de mollusques après arrêt du traitment. *Malacologia*, 22, 697–700.

Russell-Hunter, W.D. 1978. Ecology of freshwater pulmonates. In V. Fretter and J. Peake (eds.), *Pulmonates. Vol.2A. Systematics, Evolution and Ecology*, 335–83. London: Academic Press.

Shotton, F.W. 1978. Archaeological inferences from the study of alluvium in the lower Severn-Avon valleys. In S. Limbrey and J.G. Evans (eds.), *The Effects of Man on the Landscape: The Lowland Zone*, 27–32. London: Council for British Archaeology.

Sparks, B.W. 1961. The ecological interpretation of Quaternary non-marine Mollusca. *Proceedings of the Linnean Society of London*, 172, 71–80.

Sparks, B.W. and Lambert, C.A. 1961. The Postglacial deposits at Apethorpe, Northamptonshire. *Proceedings of the Malacological Society*, 34, 302–315.

Thomas, K.D. 1978. Population studies on molluscs in relation to environmental archaeology. In D.R. Brothwell, K.D. Thomas and J. Clutton-Brock (eds.), *Research Problems in Zooarchaeology*, 9–18. London: Institute of Archaeology.

Thomas, K.D. 1985. Land snail analysis in archaeology: theory and practice. In N.R.J. Fieller, D.D. Gilbertson and N.G.A. Ralph (eds.), *Palaeobiological Investigations: Research Design, Methods and Data Analysis*, 131–156. Oxford: British Archaeological Reports, International Series, 266.

Thomas, R., Robinson, M., Barrett, J. and Wilson, B. 1986. A Late Bronze Age riverside settlement at Wallingford, Oxfordshire. *Antiquaries Journal*, 143, 174–200.

Ward, J.H. 1963. Hierarchical grouping to optimize on objective function. *Journal of the American Statistical Association*, 58, 236–244.

Williams, D.D. and Hynes, H.B.N. 1977. The ecology of temporary streams. II. General remarks on temporary streams. *Internationale Revue der gesamten Hydrobiologie*, 62, 53–61.

Wishart, D. 1978. *Clustan Users Manual*. (3rd edn.) Edinburgh: Programme Library Unit, Edinburgh University.

Wright, J.F., Hiley, P.D., Cooling, D.A., Cameron, A.C., Wigham, M.E. and Berrie, A.D. 1984. The invertebrate fauna of a small stream in Berkshire, England, and the effect of intermittent flow. *Archiv für Hydrobiologie*, 99, 176–199.

Young, M.S. and Evans, J.G. 1991. Modern land mollusc communities from Flat Holm, South Glamorgan. *Journal of Conchology*, 34, 63–70.

8

Archaeological inferences from alluvial sediments: some findings from southern England

Robert G. Scaife and Paul J. Burrin

Summary: *The reliability of pollen analyses from inorganic alluvial sediments in reconstructing paleoenvironments has long been questioned. This paper examines the results of research undertaken within four valleys in southern England which demonstrates that such concerns appear unwarranted in this region. The mirroring of pollen profiles between peats and inorganic fills is demonstrated and aspects of the complex pollen taphonomy are discussed. Some archaeological inferences are also presented from consideration of the nature and spatial patterns of the valley fills.*

INTRODUCTION

Fluvial responses to environmental stimuli are recognised as being spatially and temporally variable (e.g. Burrin and Scaife 1984). Early models (Davis 1899) argued that cycles of erosion and base-level changes were the underlying causes. Other ideas followed which focussed on glacio-eustasy controlling the long profiles of river systems (Baulig 1935) and hence phases of incision or aggradation. Subsequently, climatic change (Zeuner 1959; Dury 1964) was advocated as the principal mechanism: warmer, wetter periods caused erosion; colder or dryer periods aggradation. More recent advances in geomorphology have demonstrated however that fluvial systems often respond to environmental changes in a complex, episodic manner involving thresholds (Schumm 1979). Furthermore, it is only relatively recently that the impact of prehistoric cultures on valley fill sedimentation has been fully appreciated.

There is now widespread and increasing evidence of the significant role played by prehistoric cultures in causing or exacerbating inorganic inwash from valley sides to valley floors. From such studies, it is possible to date the alluvium by identifying the age of the buried site(s) and associated artifacts; it is also possible to infer estimated rates of subsequent alluviation, although such attempts are fraught with potential errors. Other approaches have also demonstrated a less apparent and more complex interaction between man and his environment. Such evidence comes from research on chalkland dry valley colluvial fills in southern England (Kerney *et al* 1964; Bell 1981, 1982, 1983; Allen 1986) and in floodplain studies at point locations (e.g. Shotton 1978; Limbrey 1978; Hazelden and Jarvis 1979; Brown 1982). In these circumstances, anthropogenic activity is often recorded as episodes of accelerated sedimentation within bog or floodplain lithostratigraphy (Scaife 1980, 1988).

Initial studies of such alluvial architectures led to the suggestion (Limbrey 1978, 1983; Bell 1982) that many British lowland river valleys exhibit a basically comparable sequence of Holocene sediments. However, this now appears to be a premature conclusion, for our knowledge of many British inland floodplain fills is limited and in many instances, only part of the alluvium has been studied (often that exposed only in river banks by fluvial incision). Furthermore, relatively little is known about the deeper and presumably older Holocene deposits found beneath contemporary inland floodplains and there have been few detailed systematic studies into these sediments.

Consideration of these and other factors led to a model (Burrin & Scaife 1984) focussing some of the difficulties of correlating alluvial fill sequences with exogenic or endogenic factors. Likewise, the application of theoretical models such as catastrophe theory to the problem of valley alluviation (Burrin and Scaife 1988) has also exemplified some of the ways in which fluvial systems can respond to environmental changes, including the impact of man. It was also advocated that in order to understand how fluvial systems have responded to such stimuli, studies have to be instigated throughout the catchment (along the lines suggested by the IGCP Correlation Programme 158A), for single site studies are not necessarily indicative of wider responses within the valley. This has subsequently been demonstrated further by Starkel *et al* (1991).

In spite of these significant advances in the understanding of valley fill development, there are still a number of uncertainties which require further examination. For instance, models dealing with pollen transport and sediment flux are often inappropriate in considering the origins of pollen found in floodplain sediments and its palaoenvironmental significance. There is also some concern in the literature regarding the reliability of pollen found within inorganic, fluvially derived sediments when used in palaeo-environmental reconstruction. Perhaps somewhat surprisingly, the interpretation of peat accumulation within valley fills with respect to environmental conditions (stability or instability?) still remains unclear. There is also, disappointingly, a paucity of detailed independent archaeological evidence which might be used to support or question some of the interpretations offered in palynological studies of alluvium in which prehistoric man appears to be an instigator of significant environmental change.

In an attempt to address some of these problems, research has been undertaken since the mid seventies in a number of river valleys in southern England. The aim has been to record systematically throughout the valleys the stratigraphical character of their alluvial fills; to date these deposits; and to identify the causal mechanisms involved in producing these accumulations. This paper will synthesise some of the results of these endeavors (for more details see Burrin 1988, Burrin and Jones 1991, Scaife 1987, 1988, and Scaife and Burrin 1983, 1985, 1987) and discuss a number of the environmental archaeological issues which have still not been adequately considered or explained.

THE RESEARCH AREA

Research has focused on the Medina valley on the Isle of Wight and the Ouse, Cuckmere and Rother drainage basins of East Sussex. The Medina is the principal river on the island, rising in the Lower Cretaceous Wealden Series, before flowing northward through the Lower Greensand ridge at Gatcombe. It breaches the chalk downlands south-east of Newport, at which point the river becomes tidal until entering the Solent at Cowes.

The Sussex basins are three relatively small, southward oriented catchments of size 450, 150 and 700 km^2 respectively. The headwaters rise in steep-sided narrow valleys (ghylls) dissecting the sandy/clayey southern flanks of the Ashdown Forest in the Central or High Weald. After leaving this region the rivers meander southwards across the low-lying Vale of Sussex towards the South Downs chalklands. The Ouse and Cuckmere breach the chalk scarp by means of spectacular gaps, with local relief rising to over 200 metres through which the rivers pass before entering the English Channel. In the case of the Rother, the valley fill deposits extend out from the Low Weald into the marshes of Romney and Walland (Eddison and Green 1988), with the river draining into the sea south of Rye.

Extensive subsurface investigations have been undertaken by auger surveys throughout all four river valleys. Over 450 boreholes sunk through the valley fill deposits have enabled the construction of numerous litho-stratigraphical cross and long profile sections. This field work has been coupled with detailed sedimentological, mineralogical and biostratigraphic analyses. As a result, a detailed understanding of the character, age and probable causes of alluviation has been produced. For the first time, a regional picture is now beginning to emerge of the Holocene development of some of the river valleys and their associated fills in this part of Britain. A summary of some of the salient research results will be presented here for each valley before focussing on a range of issues which still require discussion.

THE MEDINA VALLEY

An almost continuous floodplain extends the length of the Medina valley from its source at Chale Green northwards to the tidal reaches between Newport and Cowes. Augering has been undertaken along the length of the valley, which has revealed that the inland alluvium comprises up to 3 metres of peats with thinner bands of interbedded inorganic sediments. The principal site detailed here is the Withy Bed at Gatcombe (SZ 502858), a topogenous, soligenous mire typical of valley fen carr and sedge fen communities existing in a topographical location dominated by high groundwater tables. This Withy Bed is one of a number of typical valley peat sequences which have been investigated in the Isle of Wight (Scaife 1980, 1982, 1987, 1988) and which are now becoming rapidly depleted through post-war changes in land management.

The cross valley stratigraphy and relevant pollen data are summarised in Figure 8.1. Full descriptions of the pollen assemblages and stratigraphy are given (Scaife *op cit*) and the site is used here to illustrate the principal causes of peat development and/or sedimentation typical of that occurring in other southern English valley peat mires.

A pollen and stratigraphical sequence dating from the Late Devensian through the full Holocene has been constructed and this provides an insight into the principal periods and causes of valley alluviation. Complete Holocene valley fill sequences are rare in southern England and consequently, this is an important site. From the study of these sequences, it has become clear that the valley fill began accumulating at different times as a result of both human and natural causal

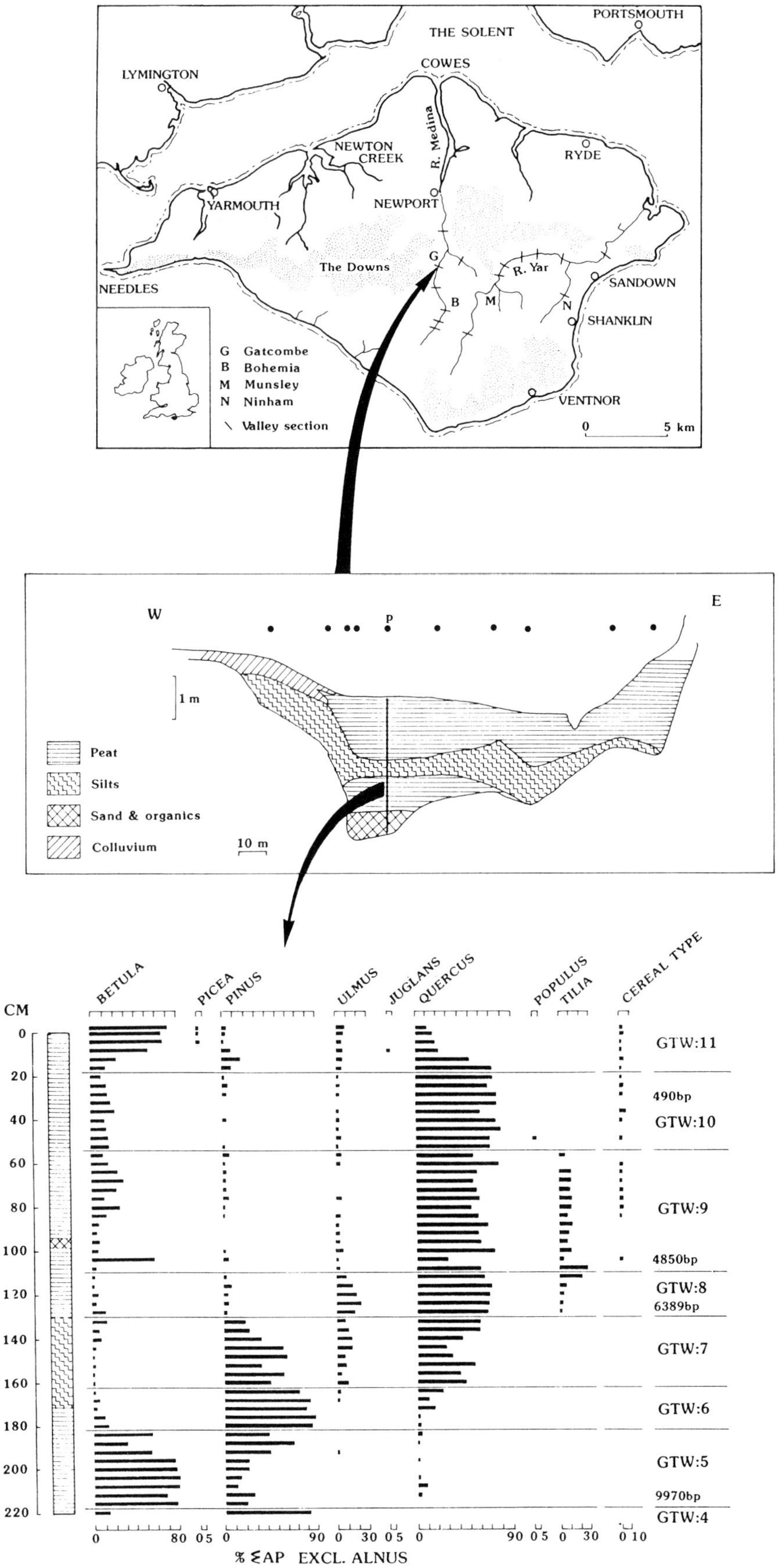

Fig. 8.1 The river Medina, Isle of Wight: summarised stratigraphy and selected pollen taxa.

factors. Three major phases of peat initiation have been recognised:

Early Holocene: The transition from stadial to interglacial conditions at 10,000 BP (Late Devensian to Holocene) was marked by a transition from minerogenic sediments typical of Late Devensian regimes to a period of peat accumulation in valley bottoms indicating waterlogged conditions. This was the response to climatic change, increased humidity and melting permafrost. Peat accumulation in topographical basins such as described at Gatcombe Withy Bed also occurred at other sites in the Isle of Wight. This increase in water availability was also coupled with the stabilisation of interfluves by colonising vegetation. This was initially dominated by juniper and rapidly followed by birch, pine and hazel (Scaife 1982) which had a stabilising effect on valley side soils.

Atlantic: Changing climate at 7000 BP is widely documented and marks the onset of the often described Mid Holocene climatic optimum (Flandrian chronozone II), being represented by evidence of increased oceanicity (the Atlantic; Godwin's pollen zone VIIa). This is manifested in a marked increase in the pollen record of thermophilous vegetation (e.g. *Tilia* and *Ilex*) and by substantial increases in the values of *Alnus* pollen. The latter has been seen as an indication of much greater water availability (Godwin 1975a, 1975b). The consequences of this climatic change were a further period of increased water-tables in valley bottoms and subsequent peat growth. Although the rise in importance of *Alnus* is diachronous, this phenomenon has been recognised at a number of sites in southern England (Brown 1988). It is as yet unclear whether rapidly rising sea levels during the preceding Boreal and attainment of present levels shortly after 7000 BP was partly responsible for this climatic amelioration. Similarly, rising sea levels may also have had a ponding back effect on river systems with consequent effects on peat initiation. This cause may not have been wholly responsible, however, because peat accumulation has been shown to occur in basins outside the reach of such eustatic effects at this time. Furthermore, this phenomenon will also depend on valley gradient and, in cases where this is pronounced, the effects may only be seen in the coastal reaches.

Post-Atlantic: Pollen analyses of soligenous valley peats, as at Gatcombe, strongly indicate that in many cases post-Atlantic peat formation was initiated directly as a result of woodland clearance on interfluves by man. This is likely to have resulted in a reduction of evapotranspiration rates and increased surface run-off which in turn gave rise to higher ground water tables and waterlogging of valley bottoms (Moore and Willmot 1976). This possible anthropogenic causation is evidenced by a sharp increase in herb pollen characteristic of anthropogenic activity. Basal levels may contain sediments and tree pollen representing woodland characteristics prior to deforestation.

Episodes of inorganic sediment accumulation have also been recognised within the valley fill at Gatcombe. These are as follows:

Late Devensian: A phase of sedimentation under periglacial conditions where the environment was shaped by processes including freeze-thaw, solifluction and seasonal meltwater runoff. The potential for higher sediment transport occurred at this time because of restricted vegetation cover.

Boreal: Flandrian Ib, Ic. The identification of a period of Boreal sedimentation within the valley peats is of significance for it indicates that destabilisation of valley side soils and vegetation occurred at this time. Mechanisms for this are, however, enigmatic. The possible effects of Mesolithic communities on the environment have been the focus of much debate (e.g. Dimbleby in Rankine *et al* 1960, and in Keef *et al* 1965; Smith 1970, 1981) and it has been suggested that anthropogenic factors (fire) may have been the causal factor at this location. Of further interest is that the sedimentology of this inwashed material is similar to that of the reworked loess identified in the Weald (Burrin 1981, 1983). These aeolian deposits which were originally deposited during the Devensian cold stage were thus reworked at a later date consequent upon destabilisation of the interfluves perhaps by prehistoric activities.

Sediment inwash resulting from later prehistoric woodland clearances: Localised sediment inwash into peat basins has been discussed in a number of studies of upland peat basins resulting from the ephemeral clearances of Mesolithic communities. It is with the arrival of Neolithic agricultural communities after ca. 5000 BP that more sustained soil erosion, sediment transfer and deposition occurred. Early Neolithic activity in southern England of typical landnam character was responsible for relatively minor quantities of inorganic sediments found in otherwise organic/peat profiles (Scaife 1988). Local changes in pH status of the mire at Gatcombe Withy Bed and consequent changes in vegetation status attest to such activity and also a local origin of the sediments. Subsequent intensification of woodland clearance and more efficient and extensive agricultural practices from the Early Bronze Age produced increases in the inorganic component identified within the peat (Scaife *op cit*). This is regarded as typical of lowland valley peat sequences dating to the later prehistoric period.

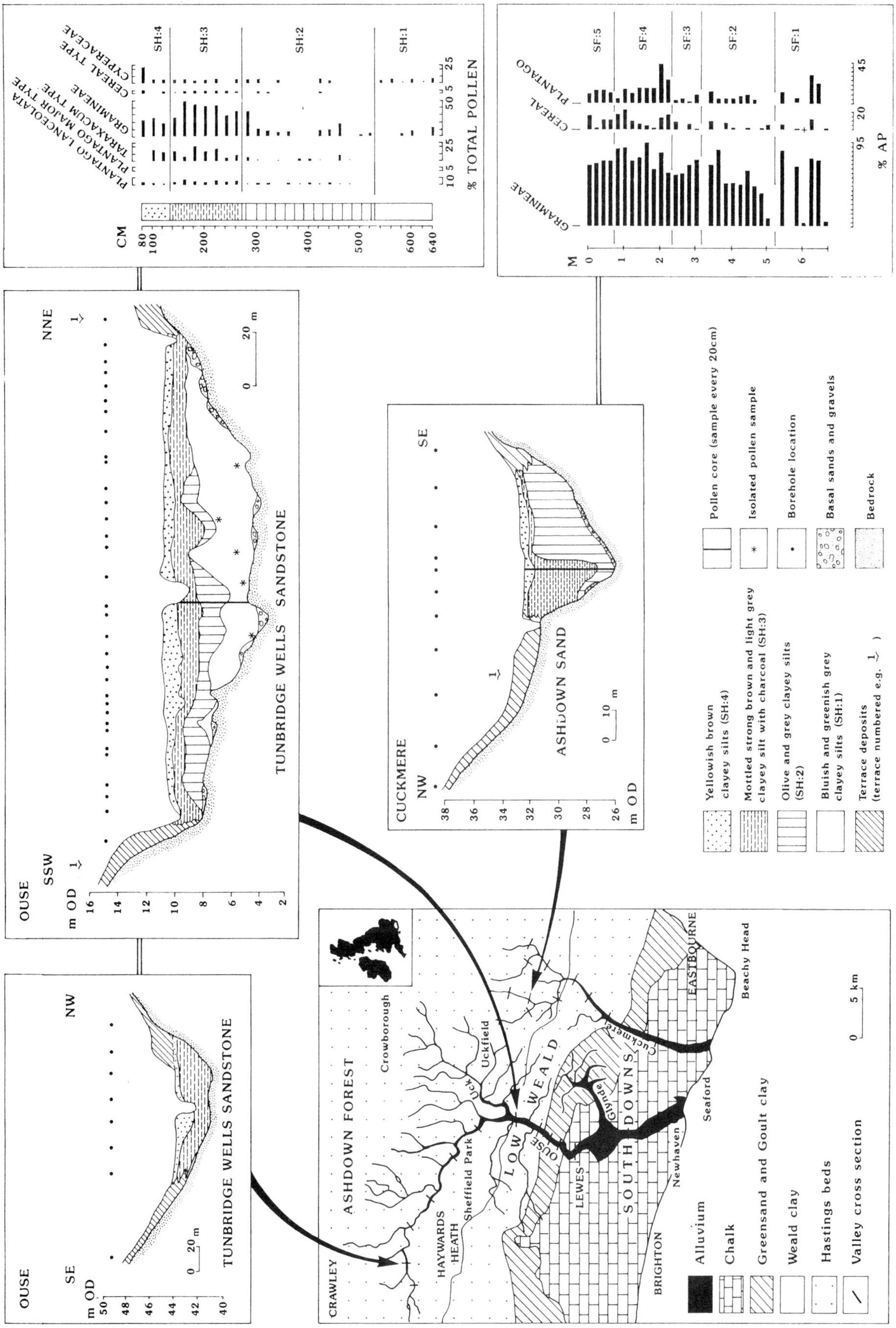

***Fig. 8.2.** The rivers Ouse and Cuckmere, Sussex: summarised stratigraphy and selected pollen taxa.*

THE RIVER OUSE VALLEY

Investigation of the alluvial fill deposits of the Ouse has been undertaken at a number of locations throughout the valley (Burrin & Jones 1991). Attention is focused here on the results of subsurface investigations at Sharpsbridge (TQ 444208) located on the southern margins of the High Weald, and at Wellingham in a Low Weald tributary valley (Brooks 1983).

Auger survey at Sharpsbridge revealed a complex polycyclic, suballuvial surface cut into the Hastings Beds. The alluvial fill (Figure 8.2) has been described in detail in Burrin 1983, 1985; Burrin and Scaife 1984. Four lithostratigraphical units have been recognised. The rockhead is partly mantled by a thin, residual clayey sand and gravel deposit which is buried by up to 7 metres of mainly inorganic silts, clayey silts and silty clays. Despite the occurrence of occasional thin interbedded peaty laminations, the predominantly inorganic nature of the alluvium precluded the possibility of obtaining an accurate radiocarbon chronology. Consequently, a palynological investigation was carried out in these inorganic sediments in the hope that some initial idea of the associated palaeo-environments and probable age of these sediments could be gained.

Alluvium has long been regarded as of little significance in palaeobotanical reconstruction because of the problems in interpreting the more complex taphonomy of the sub-fossil pollen and spores. However, this study enabled the approximate dating of periods of sedimentation and erosion at Sharpsbridge as well as providing useful information regarding the environmental history of the Ouse catchment. Full pollen data have been presented in more detail elsewhere (Scaife and Burrin 1983; Burrin and Scaife 1984) and only selected taxa are presented in Figure 8.2 which illustrate the depths of alluvium which have largely accumulated since arable practices were introduced in this locality (i.e. Neolithic onwards).

A good correlation appears to exist between the lithostratigraphy and the pollen stratigraphy. The lowest pollen zone SH:1 is characterised by dominance of *Pinus* and *Corylus* type. The dating of this zone is dependent on the suggested time of entry of *Pinus* into southern England in the Late Devensian and Early Flandrian. The domination of arboreal pollen by *Pinus* and the relatively low numbers of herb pollen taxa, especially those often indicative of Late Devensian conditions (Scaife 1980, 1982), suggest that an Early Flandrian age is most likely for the earliest sediments of the Ouse valley.

A sharply defined bio-stratigraphical discontinuity occurs between pollen assemblage zone SH:1 and SH:2. *Quercus*, *Tilia*, *Ulmus* and *Fraxinus* were dominant during sedimentation. Of significance is a major hiatus in the sediment continuity and there occurs a range of herb taxa indicative of anthropogenic activity and environments (some are represented in Figure 8.2 and include ruderal and segetal taxa). It has been suggested (Burrin and Scaife 1984) that sedimentation in zone SH:2 was initiated during the Neolithic period after the 'primary *Ulmus* decline' and it is this which makes this site especially interesting, since there are some 6 metres of sediment overlying the beginning of this pollen zone. In the upper part of pollen zones SH:2 and in zones SH:3 and SH:4 a wide range of herb taxa (including cereals) occur which are typical of the prehistoric period. Also noted is the '*Tilia* decline', a phenomenon widely seen in pollen data from southern England and attributable to anthropogenic activity (Turner 1962).

Further pollen samples were collected from the valley fill near Cuckfield in the upper Ouse catchment. Augering revealed (Figure 8.2) an almost symmetrical trough-like suballuvial surface cut into Lower Cretaceous Tunbridge Wells Sands with a minimum elevation of 40.3 metres O.D. and a maximum proven fill thickness of 3.2 metres. Two lithostratigraphic units have been identified. A deeper fill consists of mottled strong brown and light grey clayey silt with charcoal inclusions and an overlying yellowish-brown clayey-silt. These are similar to those described in the upper part of the valley fill further downstream at Sharpsbridge. The pollen from the basal sediments also correlate closely with zone SH:3 at Sharpsbridge, with *Quercus*, *Alnus* and *Corylus* type dominating. There are also notable quantities of *Fagus* and herb pollen types. These results suggest a Late Bronze Age/Early Iron Age for the initiation of alluvium within the upper part of the valley. It is noteworthy that no evidence of Early Flandrian or Neolithic sedimentation has been identified in the most inland valley tracts.

Additional palynological data have also been produced from Wellingham bog (Brooks 1983), located in one of the tributary valleys of the Ouse. These complement that produced from the Sharpsbridge site. Analysis of a 6 metre core through peats dated to ca. 7000 BP revealed high woodland pollen, with *Quercus*, *Alnus*, *Fraxinus*, *Corylus*-type and *Salix* all present. The *Ulmus* and *Tilia* declines are both recorded, with *Fraxinus* increasing as *Ulmus* decreases. Periods of decline, accompanied by increases in Gramineae, are followed by periods of recovery. This can be interpreted as phases of anthropogenic disturbance and clearance alternating with periods when the woodland vegetation was able to re-establish itself. A significant expansion of Gramineae, *Plantago lanceolata*, *Rumex* and *Artemisia* following the *Tilia* decline attests to subsequent cultivation. There is also evidence of *Pinus* afforestation in the top 20 cm of the fill. Hence, a similar response to environmental change is recorded in very different types of alluvial fill deposits within the valley, with peat profiles mirroring those identified within inorganic sequences.

THE CUCKMERE VALLEY

Similar geomorphological investigations have been undertaken along the length of the Cuckmere valley. The floodplain site chosen for detailed palynological study is at Stream Farm (TQ 557157) in the headwaters of the upper Cuckmere where the system drains part of the Dudwell anticline. Subsurface investigations revealed the presence of a buried, narrow and relatively flat rockhead bench cut into the Lower Cretaceous Ashdown Sand. A deep polycyclic gorge or channel with a minimum elevation of ca. 26 metres O.D. has been cut into this bench. This rockhead or suballuvial surface is partly veneered by a thin accumulation of clayey sands and gravels, which is buried in turn by up to 6.6 metres of relatively homogeneous fine sandy and clayey silts much of which is reworked loess (Burrin 1981, 1983). This fill is a significant accumulation in this small catchment given the inland location of the site, being only a few kilometers from the watershed.

Three lithostratigraphic units have been recognised (Figure 8.2) which are of similar texture and character to those recorded in the valley of the Ouse at Sharpsbridge. Full pollen analysis was carried out on these sediments the results of which are presented in Scaife and Burrin (1985). Here, only Gramineae, cereal type and *Plantago lanceolata* are reproduced to illustrate the pertinent fact that anthropogenic factors are evident at the base of this remarkably deep sediment sequence.

The basal pollen zone SF:1 is characterised by high percentages of deciduous tree taxa (*Quercus*, *Tilia*, *Alnus*, *Betula* and *Corylus*) reflecting substantial woodland cover. In addition, however, is a diverse assemblage of herbs dominated by Gramineae, cereal type, *Plantago lanceolata* and other herbs indicative of arable and pastoral land use. This zone has a somewhat disjunct/mingled appearance which possibly indicates destabilisation of the valley side soils and rapid inwash. This may be attributable to the woodland clearance and agriculture during the Neolithic-Early Bronze Age.

Subsequent pollen zones SF:2 to SF:5 show the increasing effects of anthropogenic activity in the form of woodland clearance and agriculture. This appears to have resulted in a more continuous rate of sedimentation but with some periods of more rapid inwash characterised by fewer and more degraded pollens, plus a taxonomically more diverse range of herbs. Of note also are phases of standing water characterised by aquatic weeds (e.g. *Myriophyllum*, *Sparganium* and *Typha*). This is possibly attributable to higher groundwaters similarly caused by continued anthropogenic forest clearance resulting in reduced evapotranspiration and surface run-off from the interfluves causing locally higher water-tables and waterlogging in valley bottoms (Moore and Willmott 1979; Scaife 1980). This cannot be proven as yet since other factors, such as natural influences (e.g. beaver damming), human management of stream channels and anthropogenic interference downstream causing ponding back, or climatic change and local base level fluctuations, must also be considered.

Undoubtedly, the most significant point resulting from this analysis is the presence of agricultural plant taxa at the base of this extremely thick sediment sequence within the headwater region of this small catchment. The inference is that some 6.6 metres of sediment has accumulated since the Neolithic period, that is, since the introduction of arable agricuture. Some reworking of the original loess mantle, local soils and bedrock has occurred. Consequently, the alluvium also contains some pollen from earlier periods which are predominantly palynomorphs derived from Lower Cretaceous strata. Earlier Holocene pollen is largely absent which may be attributed to what must have been the high biological activity on the earlier well drained nutrient-rich soils of the interfluves.

THE RIVER ROTHER

Other investigations have also been undertaken in the eastern Rother valley (Figure 8.3). As with the Ouse and Cuckmere, subsurface investigations were carried out throughout the valley (Burrin 1988). Sites at Mayfield (TQ 604250) and Robertsbridge (TQ 743240) were selected for more detailed investigation including sedimentological and palynological analyses (Scaife and Burrin 1987; Burrin 1988).

The section to the south-east of the village of Mayfield (Figure 8.3) revealed a maximum fill depth of 5.87 metres overlying a polycyclic sub-alluvial surface cut into Lower Cretaceous Wadhurst Clay with a minimum elevation of 39.6 metres O.D. Overlying and partly mantling the rockhead is a silty sand and gravel deposit which is irregular in thickness and extent. This is buried by fine-grained alluvium with a cross sectional area of 422 square metres which has a significant loessal component (Burrin 1988). Three lithostratigraphic units have been identified. The deepest unit extends from 39.6 metres O.D. to 44 metres O.D. and contained burnt wood remains.

The valley fill at Robertsbridge overlies an undulating polycyclic suballuvial surface in the Lower Cretaceous Ashdown Beds, with a minimum elevation of 1.7 metres O.D. Here, the rockhead is also partly covered by a thin (0.25 metres) silty, sandy gravel (Figure 8.3). This is usually buried by a bluish and greenish-grey clayey silt. Peat with wood charcoal and other inclusions was found between 7.6 metres and 5.8

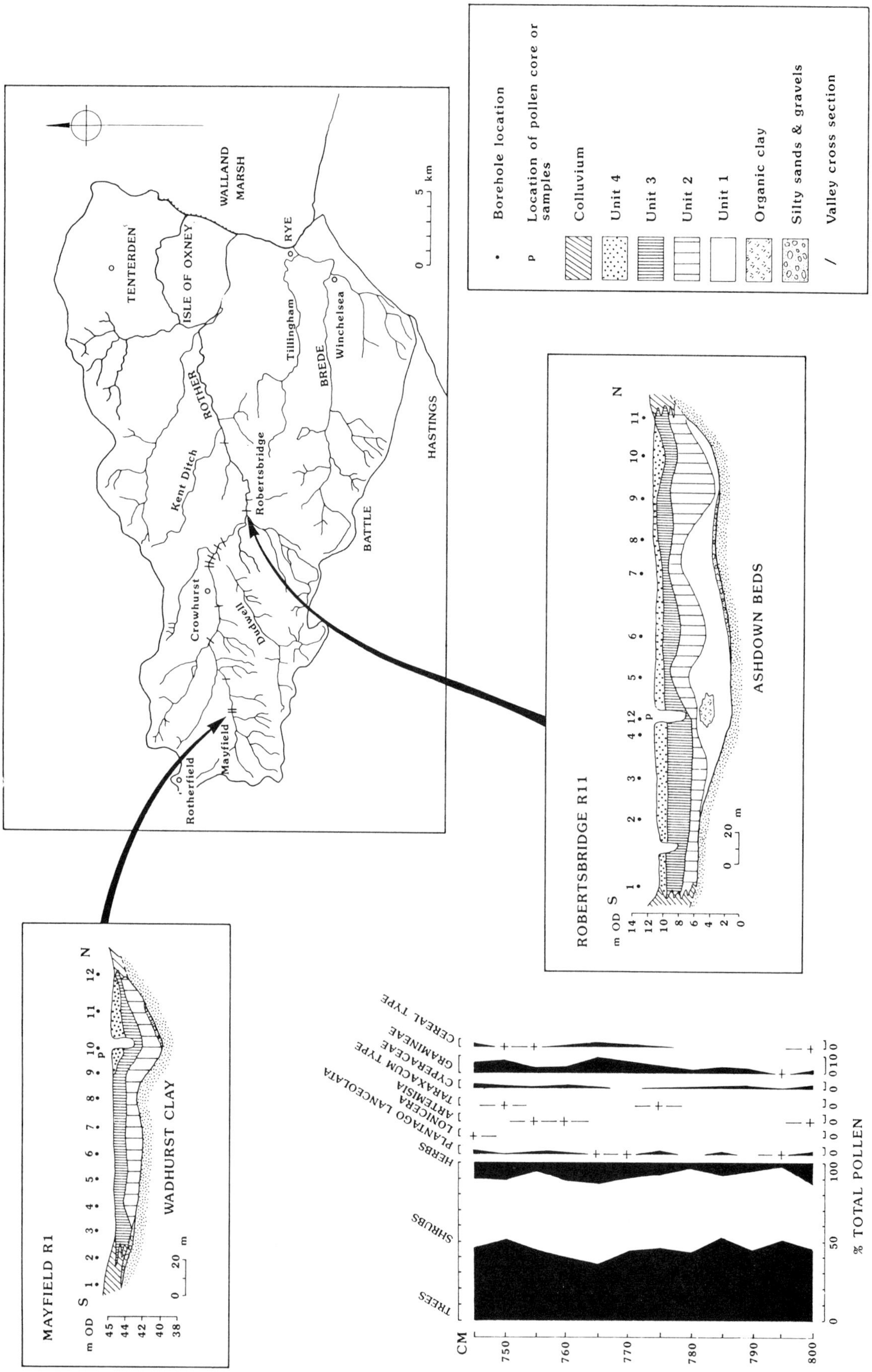

Fig. 8.3. The river Rother, East Sussex: summarised stratigraphy and selected pollen taxa.

metres O.D. Above these basal fills are light grey and strong brown mixed and mottled silt with occasional charcoal and manganese inclusions from 6.1 metres to 11.4 metres O.D. The uppermost fill deposit recognised is a pure brown silt which ranges in elevation from 9.9 metres O.D. to 12.4 metres O.D.

Pollen analyses were carried out on a core from Mayfield and on the organic peaty clays interbedded with the lowest unit at Robertsbridge. The pollen spectra are dominated throughout by arboreal (*Tilia*, *Quercus*, *Alnus* and some *Ulmus*, *Betula* and *Fraxinus*) and shrub pollen (*Corylus*, *Salix*, *Viburnum*) in approximately equal proportions (45% total pollen). Pollen of herbs is present but in much lower proportions ranging from 2% to 14% of total pollen. The latter are, however, relatively diverse with most taxa only occurring sporadically. Again these include cereal pollen and weeds of cultivation such as *Plantago lanceolata* from which it may be suggested that some anthropogenic activity was taking place in the catchment at the time of accumulation of these basal sediments. However, these data indicate the relative absence of non-wooded areas at least in proximity to the site. From the relatively high values of *Tilia* pollen found, it is shown that this taxon was probably dominant on the interfluves of this catchment.

There is now much evidence for the importance of *Tilia* in the Atlantic Flandrian chronozone II/post-Atlantic period/Flandrian chronozone III (e.g. Scaife 1980; Thorley 1981; Greig 1982a; Moore 1977; Waller 1987). Clusters of pollen of *Corylus* and *Alnus* were noted in the pollen preparations and it is likely, therefore, that these taxa formed the major part of the 'on-site' vegetation. These fill sequences have yet to be radiocarbon dated but there are now sufficient pollen data available to suggest that the sequence described from Robertsbridge is of a Mid Holocene (Flandrian II) or later (probably Neolithic) date. This probability is based on the prevalence of mixed deciduous woodland and/or dominance of *Tilia*. The peaty clay at Robertsbridge must, therefore, pre-date the decline of *Tilia*, which has been dated to ca.3700 B.P. in this part of the Weald (Waller *et al* 1988). Furthermore, the presence of cereal pollen from a depth of 8 metres and *Plantago lanceolata* throughout, allows a Neolithic to Bronze Age date to be postulated.

This interpretation is also corroborated by the relatively low values of *Ulmus*. Pollen percentages of *Ulmus* prior to its decline (at 5040 ± 80 BP in the Brede tributary; Waller *et al* 1988) usually attain 10–15% throughout southern England during the Atlantic (Flandrian chronozone II; Scaife 1988). Here *Ulmus* values are significantly lower throughout, suggesting a post-*Ulmus* decline date for the basal pollen spectra. This is also commensurate with the range of anthropochorous pollen taxa present in the spectra which are normally associated with Neolithic and later activities.

As with the sites analysed in the Ouse and Cuckmere valleys, the pertinent point here is that the lowest levels contain such evidence of Neolithic (or later) activity. The subsequent burial of this organic matter by 7.45 metres of inorganic alluvium, is almost certainly the result of widespread deforestation and movement of surface soils off the interfluves into the valley systems. As such, this represents a very substantial impact of prehistoric man on the hydrology of this part of the Weald.

TAPHONOMY OF POLLEN IN ALLUVIAL SEDIMENTS

Pollen analyses have rarely been carried out on alluvial sediments although a number of researchers have noted the potential value of these deposits where lacustrine or peat deposits are not available (Grichuk 1967; Hall 1981, Burrin and Scaife 1984; Hunt 1987). This is not surprising in view of the interpretative problems and dangers inherent in analysing minerogenic sediments of allochthonous character (Faegri and Iversen 3rd edit. 1974; Cundill and Whittington 1987; Hunt 1987). Difficulties are expected to occur for many reasons. These include, for example, the possibility of pollen being eroded from older sediments and re-incorporated into younger materials, oxidation caused by fluctuating moisture regimes, biological reworking, surface weathering where burial is not rapid, and the pH (especially alkalinity) of the water entering fluvial the catchment. Preservation of pollen in mineral sediments is frequently inferior to that of acid peats. Interpretation of pollen data is thus fraught with problems. Nevertheless, it has been demonstrated that the application of pollen analytical methods to the detailed analysis of inorganic alluvial deposits can produce valuable palaeoenvironmental data.

Many models of pollen production, transfer and deposition are typically confined to a consideration of woodland environments. This is perhaps because most palynological research has concentrated on the history and development of such communities. Typical are the often quoted transfer models from continental Europe including Tauber (1965, 1967) and the relative pollen production information of Andersen (1970, 1973). These studies seek to relate absolute pollen frequencies to the differing modes of pollen production and dispersion. Notable are the more recent studies by Bradshaw (1981a, 1981b), who has researched the question of pollen representation in England, and Edwards (1982), who has considered in particular the conceptual aspects of pollen representation in relation to human activity and the woodland edge. Moore *et al* (1991) give a valuable summary of the mechanisms of pollen transport.

It is now increasingly clear that models which discuss only woodland systems are of only partial value when considering the Holocene (Flandrian III) anthropogenic effects on the environment. Prehistoric forest clearance and agriculture have produced a diverse range of vegetation communities and ecotonal environments too numerous to itemise. In considering the nature and character of pollen transfer in alluvial sediments, the origin of the pollen in them has been found to be intricately linked with the activities of prehistoric cultures on the adjacent interfluves. Typical sources of pollen in such rural anthropogenically affected catchments may be briefly summarised as follows (Figure 8.4):

i) That which arrives via airborne means such as originally discussed by Tauber (1967) and subsequently by Jacobson and Bradshaw (1981). This may relate to woodland components described by Tauber, including: (a) rain component with long distance transport (Cr), (b) canopy (Cc) and (c) trunk space components (Ct).

ii) Pollen derived from fen carr woodland growing on, or adjacent to river floodplains, typically *Alnus* and *Salix* but other taxa may be significant (e.g. *Corylus*, *Myrica*, *Cyperaceae*) depending on local hydrological conditions.

iii) The local component (Cl) that is, pollen derived directly from vegetation growing actually on the site of interest. This is usually considered as being aquatic and mire taxa (iii.a.) since these are the traditional sites sampled for pollen analysis. Jacobson and Bradshaw (1981) have described this as the gravity component. However, in floodplain situations a range of additional taxa (iii.b.) may be present (for example wet meadow and tall herb communities). In the case of buried soils (e.g. archaeologically buried under mounds and banks) from terrestrial vegetation growing on dry soils.

iv) Pollen regarded as contemporary but transported from upstream environments of the fluvial catchment(Cw). Peck (1973) has clearly illustrated the importance of this component.

The above are factors usually taken into account in the interpretation of 'normal' pollen data. However, a number of other pollen sources must also be considered in the majority of cases where anthropogenic environments and/or sediment systems are being considered. Such sources are:

v) Pollen produced from pastoral (v.a) and arable (v.b.) agricultural environments which is dispersed aerially and can be deposited on the valley floor.

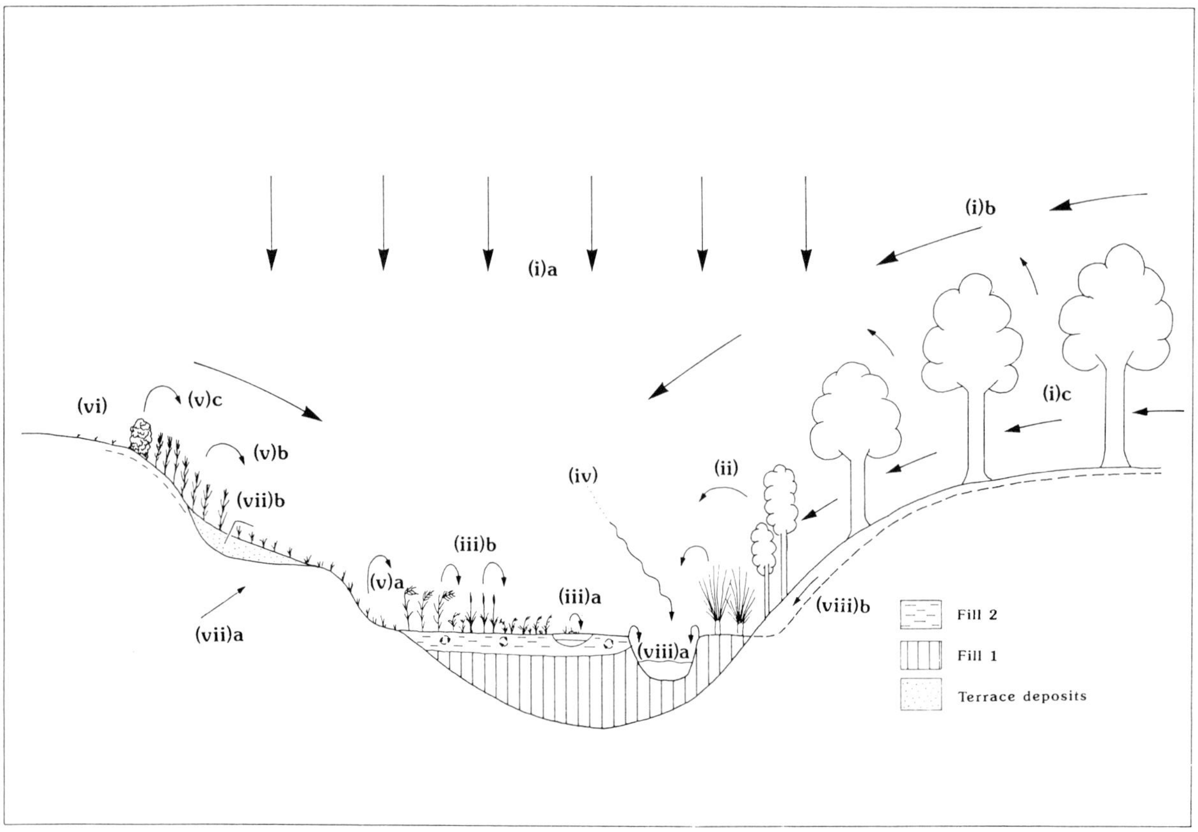

Fig. 8.4. Sources of pollen in alluvial sediments.

Field boundary environments consisting of hedgerows (Groenmann-Van Waateringe 1978) and weed communities should also be considered.

vi) Pollen produced from secondary anthropogenic sources. There are numerous ways in which sometimes significant numbers and types of pollen can become incorporated into sediments. The extreme cases are indicated in the urban pollen models of Greig (1982b) and Krywinski (1983), and the liberation and subsequent dispersal of pollen trapped in the husks of cereal inflorescences (Robinson and Hubbard 1977) by cereal processing (e.g. threshing and winnowing). Other crop processing activities such as flax retting should also be considered.

vii) Palaeopollen; a term used to denote pollen (palynomorphs) which is undoubtedly eroded from geological sediments of substantially earlier age. This includes:

a) palynomorphs from geological bedrock formations which are frequently encountered in large numbers within alluvial deposits (e.g. Robinson and Lambrick 1979; Scaife 1987 and in Burrin 1988; Hunt 1987).
b) those of more recent Pleistocene superficial deposits such as from adjacent river terraces.

viii) Pollen derived from the erosion of older alluvial deposits (viii.a) and soils of the interfluves (viii.b).

Some studies have sought to address the problems of the reincorporation of older pollen in sediments through erosion of soils on interfluves. Peck (1973) studied the frequencies of pollen transported in a small fluvial system and depositionary environment in Yorkshire and concluded that considerably higher frequencies of pollen and spores entered the system at times of higher discharge and were incorporated into the lake basin sediments. This situation is, however, somewhat different to that encountered in river floodplain alluvium which may be regarded as a more open system and consequently, a wider range of pollen sources needs to be considered. Sources may include all of those noted in those categories described above including contemporaneous pollen originating from plant taxa growing on the floodplain and pollen derived from secondary sources. Whilst there is an undoubted primary component, it is the secondary, derived component which may be a major source of concern when interpreting pollen spectra in alluvial sediments.

This secondary transport of pollen raises a number of taphonomic problems. It is accepted that pollen in lacustrine sediments may be derived from the drainage basin (Peck 1973) and can undergo degradational changes (Birks 1970; Lowe and Walker 1984). Pollen is potentially more vulnerable to degradation in colluvial environments. Consequently, much pollen subsequently incorporated within alluvial sediments will inevitably be substantially degraded. This raises the question of the differential preservation and/or destruction of taxa with differing exine robustness. This situation may be further enhanced by the fact that pollen in the alluvium may also be derived from the eroding soil of the interfluves which may have a very substantial pollen component. This is especially so in acid soils such as podzols. The taphonomy of pollen in soils is complex and in most soil palynological analyses, the effects of differential preservation are fundamental to the interpretation of such pollen spectra. Reworking of the upper soil horizons where absolute pollen frequencies are highest (Dimbleby 1961,1985) could impart substantial quantities of pollen into an alluvial system (Peck 1973). Pollen derived from soils and especially the lower soil horizons may also contain substantial numbers of (geological) palynomorphs and earlier Quaternary grains (e.g. from river terraces).

There is also a possibility of time-lag between pollen production and its incorporation into the sediments analysed. These time-lags involve pollen derived from erosion of soils on the interfluves which may be of greater age, impeded downslope movement of soil by field boundaries and other obstructions, pollen eroded from earlier fluvial sediments and the time taken in transport. It might be argued that this negates the value of the stratified pollen and its interpretation. Such a time-lag is, however, implicit in any study of mineral sediments of allochthonous origin such as lacustrine and marine sediments. These factors have not prevented palynological work on alluvial sediments; on the contrary they have been presented as the most ideal of pollen preserving situations in spite of the problems of transport, and erosive phases causing the mixing of otherwise stratified sediment. It is only in the case of pure and usually ombrotrophic peats that pollen is shed directly onto the site being studied. Here, the attendant problems of the autochthonous component become paramount. Once pollen is incorporated into alluvial sediments, phases of river incision may subsequently erode the sediments and transport them and their contained pollen further down the river valley where they may be re-deposited thus producing a second time-lag effect. Studies of river alluvium in the United States of America (Hall 1981) and in Southern Russia (Grichuk 1967) have indicated that there is a strong possibility that pollen from different ecotypes may be transported substantial distances and may give problems in interpreting the environment local to where the pollen is recovered. Whilst this may be the case in large catchments, this is not felt to be the case in the small valley systems studied here where the pollen contained in these sediments is almost certainly largely derived from the fluvial catchment although as noted above, an air-borne, extra-regional element may be present. The bulk of the pollen recovered from alluvial sediments is derived from a possibly diverse range of local vegetation communities and from pollen falling on to the soils of the catchment.

ALLUVIAL RESPONSES AND ARCHAEOLOGICAL INFERENCES

It has been demonstrated that phases of woodland clearance by prehistoric communities can cause a significant movement of sediments from the interfluves into the river valleys. This appears to have been a relatively continuous process with phases of enhanced/pulsed sediment flux. In this way the transfer of pollen directly from the interfluves to the valley floors was facilitated with perhaps some intermediate time-lag from soils/sediments held back by field banks and/or hedges. Processes of colluviation (*sensu stricta*) grading downslope into alluvial processes, would certainly have been an effective means of transporting overland, the pollen falling on to the soils of the interfluves.

Once the sediment/pollen flux reached the valley floor, sedimentation appears to have been relatively continuous through time, although there is some indication of periods of incision within the lithostratigraphy. Alluviation may have occurred by a range of processes including overbank deposition although other processes must also have occurred, such as by anastomosing channels, for it is considered unlikely that overbank deposition would have been a significant process in the more inland valley tracts. Anthropogenic changes to river channels may also have important consequences on the hydrology such as for example, the transition from anastomosing systems to single channel as a consequence of alluvial sedimentation (Limbrey and Robinson 1988).

The stratified nature of the pollen in these thick sediment accumulations suggests that alluviation appears to have taken place through time with little significant pollen reworking. Pollen profiles from these inorganic deposits appear as 'normal' pollen sequences such as might be obtained from peat accumulations; that is, with progressive changes in different plant taxa through time in the pollen profiles. This gives, therefore, an apparent mirroring of the pollen profiles of peats and inorganic accumulations. This finding is both surprising and unexpected given the different types of deposits and the nature of the depositionary environments. However, it is pertinent to note that periods of dominant woodland with soil stability (for example the Atlantic, Godwin's pollen zone VIIa/Flandrian Chronozone II), are not be represented in alluvial sediment sequences at the sites noted above. Thus, in extreme cases there is an inverse relationship between these lithostratigraphical units of the floodplain alluvium and the biostratigraphy of peat accumulations which are present. As with peat accumulation, discontinuities and phases of more rapid deposition may also be evidenced. Such interactions result from natural physical processes such as climatic changes and from human activities affecting vegetation, soils, erosion rates and hydrology.

Within-valley responses can also be as variable as those between valleys. For example, the inland trunk of the Ouse valley consists predominantly of inorganic material (Burrin & Jones 1991), yet significant accumulations of peat have been identified in the short left-bank Wellingham tributary (Jones, pers. comm.; Brooks *op cit*). Why peat was accumulating in some locations but not others is still not clear. Furthermore, the environmental interpretation of peat development is also problematic.

The environmental interpretation of peats

It is a commonly held belief that peats and lacustrine sediments are the most satisfactory media for the study of preserved subfossil pollen and spores when attempting to reconstruct past vegetation and environments. Peat accumulation has long been interpreted as being indicative of periods of relative environmental stability (characterised by waterlogging of valleys and stability of soils on surrounding interfluves). This provides a contrast with inorganic alluvial sediments which are more likely to be indicative of periods of environmental instability. Interestingly, such sediment regimes may provide a more reliable means of assessing the anthropogenic impact on an environment than organic accumulations.

This interpretation of peat indicating environmental stability has, however, been challenged by Moore and Wilmott (1979) who have suggested that forest clearance by prehistoric communities may have resulted in the reduction of evapotranspiration rates and increased surface run-off resulting in higher groundwater tables and waterlogging of valley bottoms. Consequently, peat accumulations need not be indicative of stable environmental conditions, particularly when there are anthropogenic indicators (e.g. pollen from weeds of cultivation) contained within the peats. Hence, the role of prehistoric man in determining the type of sedimentation in valley bottoms remains unclear, for peats as well as minerogenic accumulations can develop in response to the impact of man within catchments.

Inorganic accumulations

The accumulation of relatively thick inorganic fills in the Sussex valleys contrasts with the thin inorganic bands identified in the Gatcombe Withy Bed. Why there should be such variation in the thicknesses of these inorganic accumulations considering the proximity of their geographical locations is also an interesting point for debate.

The presence of significant quantities of loessal derived sediment in the southern Weald may have been a major contributary factor in providing such copious volumes of inorganic material to the valley bottoms. Consequently, these Sussex valleys may be atypical of Holocene fluvial responses resulting from the impact of prehistoric societies in that the soils therein were very different from those of today, being highly susceptible to erosion. It should be noted however, that sediments of a similar character have been identified within the inorganic component of the Gatcombe deposits. Hence, there may have been some (thin?) loess deposited on the Isle of Wight during the Late Devensian or very early Holocene.

A further factor to be considered in attempting to explain these large inorganic accumulations is the environmental sensitivity and associated thresholds within the valleys which might account for such complex and variable responses. The thin bands of sediment input into the Gatcombe Withy Bed peats resulting from Neolithic and later anthropogenic activity could be viewed as a proportional response to possibly lower population densities within this catchment. However, to view increased anthropogenic activity resulting in proportionally greater sediment movement is of course unrealistic. Sites discussed above in the Sussex river valleys illustrate thicknesses of sediments of up to 10 metres of largely inorganic silts of secondary loessic origin which have moved off the interfluves within the last 5000 years. This fact suggests that the character of valley side soils has undergone very substantial changes since the introduction of agriculture. It is unlikely that such great thicknesses of sediment are a proportional response to anthropogenic activity (i.e. the thicker the fill the greater the former population numbers or densities). Consideration of environmental sensitivity to man's activities is of fundamental significance in this context, a theme discussed in considerably more detail in Burrin and Scaife (1988). Whatever the merits or otherwise of these considerations, one fact remains paramount, *viz.* that there is a need for archaeologists to provide independent evidence concerning the size and significance of former societies within these catchments.

Impact of soil erosion on prehistoric communities

Of further note are the wider types of environmental change that must have ensued at this time. For example, the relative relief in the valleys studied has clearly reduced through time as they silted up and the landscape became more subdued. Not only has the vegetation cover been repeatedly cleared and modified, but the soils have also undergone marked changes as indeed have the floodplains. If the sediment flux from the valley sides to the valley floor was contemporaneous with deposition therein, the rate of inwash must have exceeded the capacity of the system to remove it. The clearance of woodland for agriculture must have caused periods of significant soil erosion, with rill and gully erosion the principal mechanisms. These processes are evident in the Weald today. It is, however, unclear how frequently woodland and scrub was cleared from the catchment during the prehistoric period. It would seem that the ground cover certainly regenerated at intervals, indicating short periods of soil stability. Arable cultivation contrasting with pastoral activities and the differing technologies and socio-economic character of prehistoric and later cultures undoubtedly had an impact on the extent and intensity of soil erosion and sediment transport from the interfluves. The impact of soil instability on prehistoric communities is an interesting point for debate. It might well have been of such significance as to cause disruption of communities in the valleys at times when soil erosion was a potentially serious environmental problem.

Long profiles

Long profiles of the valley fills reveal a spatial and temporal ordering (Figure 8.5). It appears that the fills become thinner and younger in an upstream direction. In the case of the Ouse valley, the deepest unit identified (Burrin & Scaife 1984) in the Sharpsbridge fill (SH:1) was not identified in the upper catchment within the High Weald. Similar results have been obtained in the fills from the Cuckmere (Burrin 1983) and the Rother (Burrin 1988).

The spatial distribution of the various fills throughout the valleys suggests sedimentation may have taken place in much of the catchments once valley side vegetation clearance(s) had been initiated. It is inferred from the fills that soil erosion was not confined to small local sites, but was occurring throughout large areas of the catchment, except perhaps for the innermost ghylls in the High Weald. Hence, alluviation was not seemingly confined to single sites or within local valley reaches, but was probably occurring throughout a large area within the catchments. It is also pertinent to note that no evidence has yet been identified in the fills for significant secondary reworking of the sediments within the valley bottoms. Any secondary time-lag effect in the deposition of the sediment fills and their contained pollen is therefore thought to be minimal.

Whether this conclusion has an archaeological significance remains unclear. However, if the rivers were unable to transport material and rework the fills

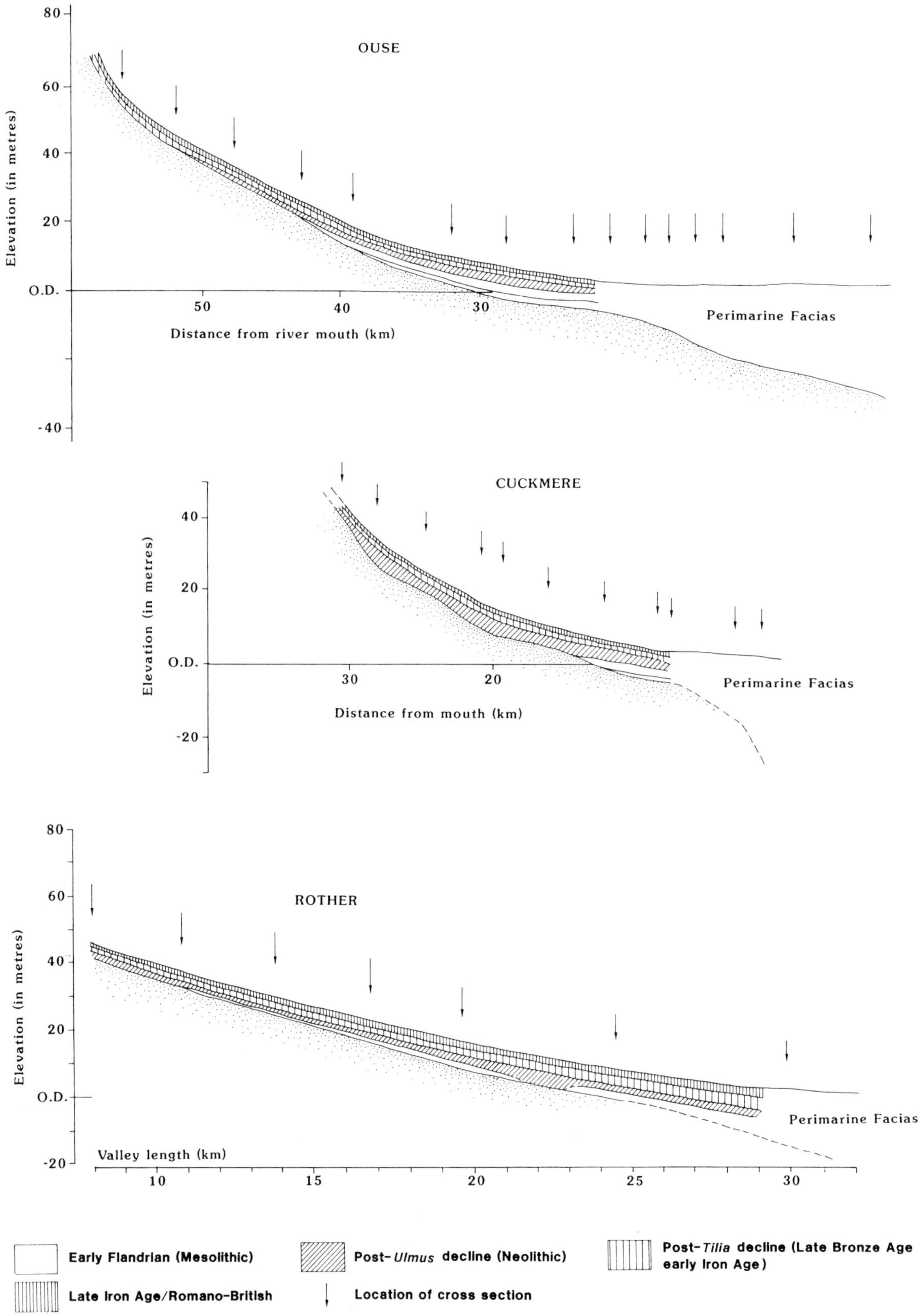

Fig. 8.5. The long profiles of some Sussex river valleys.

significantly, then the fact that the fills become both thinner and younger inland might indicate a slow but gradual penetration of Neolithic and later cultures into the High Weald. Traditionally, the Weald during the Holocene has been regarded as a densely forested region, which would have been inaccessible and inhospitable in comparison to the better drained terrain offered by the chalklands. While this rather simplistic view has now been challenged (Scaife and Burrin 1985, 1987; see also Gardiner 1990), it is plausible that parts of the central Weald (particularly the deeply incised narrow valleys or ghylls that characterise this part of south-east England) may have remained relatively untouched until the Late Holocene.

The extent to which the valley fills discussed above can yield further archaeological inferences remains to be seen. To date, no artifacts have been discovered within the alluvial fills although the augering techniques used are considered to be inappropriate for identifying such relics should they remain buried within the alluvium. Only in the lower Rother valley is there any documented evidence of an archaeological site now buried by alluvium (Lemmon and Hill 1966). This is the Romano-British horizon identified in the vicinity of Bodiam Castle which lies beneath a blanket of alluvium almost 2 metres thick. Similarly the burial of Roman roads by ca. 15–30 cm of alluvium in parts of the Ouse valley and elsewhere (Margary 1965) attests to later periods of sedimentation although the impact of the wealden iron industry on valley alluviation now appears to have been less than previously thought (Burrin 1985). There is a clear need for further investigations in these valleys in an attempt to find other independent archaeological evidence which may help to corroborate, or otherwise, some of these inferences. This is further emphasised by the substantial archaeological evidence recovered from other alluvial sequences in southern England. An example of particular note in this respect is the Thames site of Runnymede Bridge (Needham 1991).

CONCLUSIONS

This paper has focussed on the interpretation of pollen within alluvial deposits and its value in assisting in the reconstruction of past fluvial and archaeological environments. The pollen taphonomy of fluvial domains has been discussed and a range of sources for pollen identified. The problem of the palaeo-environmental significance of peats within Late Holocene fills has also been considered. Views that these are indicative of environmental stability are suspect, particularly where the peats contain pollen which is attributable to anthropogenic activity. A clear need has been identified for independent archaeological research to substantiate some of the research findings presented for areas of southern England. In spite of some obvious difficulties, the challenge nevertheless remains.

Acknowledgements

The authors would like to express sincere thanks to Ann Burrin for her assistance in the preparation of various drafts of this paper and to Linda Coleman for production of the illustrations.

References

Allen, M.J. 1986. Archaeological and environmental aspects of colluviation in South-East England. In M. Robinson and W. Groenman-Van Waateringe (eds.), *Man Made Soils*, 69–94. Oxford: British Archaeological Reports, International Series, 410.

Andersen, S. Th. 1970. The relative pollen productivity and pollen representation of North European trees, and correction factors for tree pollen spectra. *Danmarks Geologiske Undersøgelse*, Ser. II, 96.

Andersen, S. Th. 1973. The differential pollen productivity of trees and its significance for interpretation of a pollen diagram from a forested region. In H.J.B. Birks and R.G. West (eds.) *Quaternary plant ecology*, 109–15. Oxford: Blackwell.

Baulig, H. 1935. *The changing sea-level.* Institute of British Geographers Publication No. 3.

Bell, M.G. 1981. Valley sediments and environmental change. In Jones, M. and Dimbleby, G.W. (eds.) *Climatic change in later European prehistory*, 75–91. Oxford: British Archaeological Reports, British Series, 87.

Bell, M.G. 1982. The effects of land-uses and climate on valley sedimentation. In A.F. Harding (ed.) *Climatic change in later Prehistory*, 127–47. Edinburgh: University Press.

Bell, M.G. 1983. Valley sediments as evidence of prehistoric land-use on the South Downs. *Proceedings of the Prehistoric Society*, 49, 119–150.

Birks, H.J.B. 1970. Inwashed pollen spectra at Loch Fada, Isle of Skye. *New Phytologist*, 69, 807–820.

Bradshaw, R.H.W. 1981a. Modern pollen-representation factors for woods in south-east England. *Journal of Ecology*, 69, 45–70.

Bradshaw, R.H.W. 1981b. Quantatative reconstruction of local woodland vegetation using pollen analysis from a small basin in Norfolk, England. *Journal of Ecology*, 69, 941–955.

Brooks, A. 1983. Untitled contribution. In *Sussex: environment, landscape and society*. University of Sussex, 118.

Brown, A.G. 1982. Human impact on the former floodplain woodlands of the Severn. In M. Bell and S. Limbrey (eds.) *Archaeological aspects of woodland ecology*, 93–104. Oxford: British Archaeological Reports, International Series, 146.

Brown, A.G. 1985. The potential use of pollen in the identification of suspended sediment sources. *Earth Surface Processes and Landforms*, 10, 27–32.

Brown, A.G. 1988. The palaeoecology of *Alnus* (alder) and the Postglacial history of floodplain vegetation. Pollen percentage and influx data from the West Midlands, United Kingdom. *New Phytologist*, 110, 425–436.

Burrin, P.J. 1981. Loess in the Weald. *Proceedings of the Geologists Association*, 92, 87–92.

Burrin, P.J. 1983. The nature of Wealden floodplain alluvium and some possible implications. *Geologisches Jahrbuch*, Series A, 71, 237–263.

Burrin, P.J. 1985. Holocene alluviation in south-east England and some implications for palaeohydrological studies. *Earth Surface Processes and Landforms*, 10, 257–271.

Burrin, P.J. 1988. The Holocene floodplain and alluvial fill deposits of the Rother valley and their bearing on the evolution of Romney Marsh. In J. Eddison and C. Green (eds.) *The development of Romney Marsh*, 30–52. Oxford: University of Oxford Archaeological Committee.

Burrin, P.J. and Jones, D.K.C. 1991. Environmental processes and fluvial responses in a small temperate zone catchment: a case study of the Sussex Ouse valley, south-east England. In L.Starkel, K.J. Gregory and J.B. Thornes (eds.), *Temperate Palaeohydrology*, 217–51. Chichester: John Wiley and Sons Ltd.

Burrin, P.J. and Scaife, R.G. 1984. Aspects of Holocene valley sedimentation and floodplain development in southern England. *Proceedings of the Geologists Association*, 95, 81–96.

Burrin, P.J. and Scaife, R.G. 1988. Environmental thresholds, catastrophe theory and landscape sensitivity: their relevance to the impact of man on valley alluviation. In J. Bintliff (ed.) *Conceptual Issues in Environmental Archaeology*, 211–32. Edinburgh: University Press.

Cundill, P. and Whittington, G. 1987. Flood-plain sedimentation in the Upper Axe valley, Mendip, England: a comment. *Transactions of the Institute of British Geographers*, 12(3), 360–362.

Davis, W.M. 1899. The geographical cycle. *Geographical Journal*, 14, 481–504.

Dimbleby, G.W. 1961. Soil pollen analysis. *Journal of Soil Science*, 12, 111.

Dimbleby, G.W. 1985. *The palynology of archaeological sites*. London: Academic Press.

Dury, G.H. 1964. *Subsurface exploration and chronology of underfit streams*. United States Geological Survey Professional Paper 452–B.

Eddison, J. and Green, C. 1988. *Romney Marsh – Evolution, Occupation, Reclamation*. Oxford: Oxford University Committee for Archaeology Monograph No. 24.

Edwards, K.J. 1982. Man, space and the woodland edge: speculation on the detection and interpretation of human impact in pollen profiles. In M. Bell and S. Limbrey (eds.) *Archaeological aspects of woodland ecology*, 5–22. Oxford: British Archaeological Reports, International Series, 146.

Faegri, K. and Iversen, J. 1975. *Textbook of pollen analysis*. Oxford: Blackwell (3rd. edition).

Gardiner, M. 1990. The archaeology of the Weald – a survey and a review. *Sussex Archaeological Collections*, 128, 33–53.

Godwin, H. 1975a. *The History of the British Flora*. Cambridge: University Press (2nd. edition).

Godwin, H. 1975b. History of the natural forests of Britain: establishment, dominance and destruction. *Philosophical Transactions of the Royal Society*, London, B271, 47–67.

Greig, J.R.A. 1982a. Past and present lime woods of Europe. In M. Bell and S. Limbrey (eds.), *Archaeological Aspects of Woodland Ecology*, 23–55. Oxford: British Archaeological Reports, International Series, 146.

Greig, J.R.A. 1982b. The interpretation of pollen spectra from urban archaeological deposits. In A. Hall and H. Kenward (eds.), *Environmental Archaeology in the Urban Context*, 47–65. London: Council for British Archaeology Research Report 43.

Grichuk, M.P. 1967. The study of pollen spectra from recent and ancient alluvium. *Review of Palaeobotany and Palynology*, 4, 107–112.

Groenman-Van Waateringe, W. 1978. The impact of Neolithic man on the landscape in the Netherlands. In S. Limbrey and J.G. Evans (eds.), *The effect of man on the landscape: the lowland zone*, 135–46. London: Council for British Archaeology Research Report 21.

Hazleden, J. and Jarvis, M.G. 1979. Age and significance of alluvium in the Windrush valley, Oxfordshire. *Nature*, 282, 291–292.

Hall, S.A. 1981. Deteriorated pollen grains and the interpretation of Quaternary pollen diagrams. *Review of Palaeobotany and Palynology*, 32, 193–206.

Hunt, C.O. 1987. Comment: the palynology of fluvial sediments: with special reference to alluvium of historic age from the Upper Axe valley, Mendip Hills, Somerset. *Transactions of the Institute of British Geographers*, 12(3), 364–367.

Jacobson, G.L. and Bradshaw, R.H.W. 1981 The selection of sites for palaeovegetational studies. *Quaternary Research*, 16, 80–96.

Keef, P.A.M., Wymer, J.J. and Dimbleby, G.W. 1965. A mesolithic site on Iping Common, Sussex, England. *Proceedings of the Prehistoric Society*, 31, 85–92.

Kerney, M.P., Brown, E.H. and Chandler, T.J. 1964. The lateglacial and post-glacial history of the chalk escarpment near Brook, Kent. *Philosophical Transactions of the Royal Society*, London, B248, 135–204.

Krwynski, K., Fjelldal, S. and Solvedt, E.C. 1983. Recent palaeoethnobotanical work at the Medieval excavations at Bryggen, Norway. In B. Proudfoot (ed.), *Site, Environment and Economy*, 145–69. Symposia of the Association for Environmental Archaeology No 3. Oxford: British Archaeological Reports, International Series, 173.

Lemmon, C.H. and Hill, J.D. 1966. The Romano-British site at Bodiam. *Sussex Archaeological Collections*, 104, 86–102.

Limbrey, S. 1978. Changes in quality and distribution of the soils in lowland Britain. In S. Limbrey and J.G. Evans (eds.), *The effect of man on the landscape: the lowland zone*, 21–6. London: Council For British Archaeology Research Report 21.

Limbrey, S. 1983. Archaeology and palaeohydrology. In K.J. Gregory (ed.), *Background to palaeohydrology*, 189–212. Chichester: Wiley.

Limbrey, S. and Robinson, S. 1988. Dry land to wet land: soil resources in the upper Thames valley. In P. Murphy and C.A.I. French (eds.), *The exploitation of wetlands*, 129–44. Oxford: British Archaeological Reports, British Series, 186.

Lowe, J.J. and Walker, M.J.C. 1984. *Reconstructing Quaternary Environments*. London: Longman.

Margary, I.D. 1965. *Roman ways in the Weald*. London: Phoenix House.

Moore, P.D. 1977. Ancient distribution of lime trees in Britain. *Nature*, 268, 13–14.

Moore, P.D. and Willmott, A. 1976. Prehistoric forest clearance and the development of peatlands in the uplands and lowlands of Britain. *VIth International Peat Congress*, 1–15. Poznan, Poland.

Moore, P.D., Webb, J.A. and Collinson, M.E. 1991. *Pollen analysis*. Oxford: Blackwell Scientific Publications (2nd edition).

Needham, S. 1991. *Excavation and salvage at Runnymede Bridge, 1978: the Late Bronze Age waterfront site*. London: British Museum Press.

Peck, R.M. 1973. Pollen budget studies in a small Yorkshire catchment. In H.J.B. Birks and R.G. West (eds.), *Quaternary Plant Ecology*, 43–60. Oxford: Blackwell.

Rankine, W.F., Rankine, W.M. and Dimbleby, G.W. 1960. Further excavations of a Mesolithic site at Oakhanger, Selborne, Hants. *Proceedings of the Prehistoric Society*, 26, 246–262.

Robinson, M. and Hubbard, R.N.L.B. 1977. The transport of pollen in bracts of hulled cereals. *Journal of Archaeological Science*, 4, 197–199.

Robinson, M.A. and Lambrick, G.H. 1984. Holocene alluviation and hydrology in the upper Thames Basin. *Nature*, 308, 809–14.

Scaife, R.G. 1980. Late-Devensian and Flandrian Palaeoecological Studies in the Isle of Wight. PhD thesis. King's College, University of London.

Scaife, R.G. 1982. Late-Devensian and early Flandrian vegetation changes in southern England. In M. Bell and S. Limbrey (eds.), *Archaeological Aspects of Woodland Ecology*, 57–74. Symposia of the Association for Environmental Archaeology No.2. Oxford: British Archaeological Reports, International Series, 146.

Scaife, R.G. 1987. Late-Devensian and Flandrian vegetation of the Isle of Wight. In K.E. Barber (ed.), *Wessex and the Isle of Wight*, 156–80. Cambridge: Quaternary Research Association Field Guide.

Scaife, R.G. 1988. The Elm Decline in the pollen record of South East England and its relationship to early agriculture. In M. Jones (ed.), *Archaeology and the flora of the British Isles*, 21–33. Oxford: Oxford University Committee for Archaeology Monograph No.14.

Scaife, R.G. and Burrin, P.J. 1983. Floodplain development in, and the vegetational history of the Sussex High Weald and some archaeological implications. *Sussex Archaeological Collections*, 121, 1–10.

Scaife, R.G. and Burrin, P.J. 1985. The environmental impact of prehistoric man as recorded in the Upper Cuckmere valley at Stream Farm, Chiddingly. *Sussex Archaeological Collections*, 123, 27–34.

Scaife, R.G. and Burrin, P.J. 1987. Further evidence for the environmental impact of prehistoric cultures in Sussex from alluvial fill deposits in the eastern Rother valley. *Sussex Archaeological Collections*, 125, 1–9.

Schumm, S. A. 1979. Geomorphic thresholds: the concept and its applications. *Transactions of the Institute of British Geographers*, New Series, 4, 485–515.

Shotton, F.W. 1978. Archaeological inferences from the study of alluvium in the lower Severn-Avon Valleys. In S. Limbrey and J.G. Evans (eds.), *The effect of man on the landscape: the lowland zone*, 27–33. London: Council for British Archaeology Report No. 21.

Smith, A.G. 1970. The influence of Mesolithic and Neolithic man on British vegetation: a discussion. In D. Walker and R.G. West (eds.), *Studies in the Vegetational History of the British Isles*, 81–96. Cambridge University Press.

Smith, A.G. 1981. The Neolithic. In I.G. Simmons and M.J. Tooley (eds.), *The Environment in British Prehistory*, 125–209. London: Duckworth.

Starkel, L., Gregory, K.J and Thornes, J.B. 1991. *Temperate Palaeohydrology*. Chichester: John Wiley and Sons.

Tauber, H. 1965. Differential pollen dispersion and the interpretation of pollen diagrams. *Danmarks Geologiske Undersøgelse*, Ser.II, 89, 1–69.

Tauber, H. 1967. Investigation of the mode of pollen transfer in forested areas. *Review of Palaeobotany and Palynology*, 3, 277–287.

Thorley, A. 1981. Pollen analytical evidence relating to the vegetation history of the chalk. *Journal of Biogeography*, 8, 93–106.

Turner, J. 1962, The *Tilia* decline: an anthropogenic interpretation. *New Phytologist*, 61, 328–341.

Waller, M. 1987. The Flandrian vegetational history and environmental development of the Brede and Pannel valleys. PhD thesis, Polytechnic of North London.

Waller, M., Burrin, P.J. and Marlow, A. 1988. Flandrian sedimentation and palaeoenvironments in Pett Level, the Brede and Lower Rother Valleys and Walland Marsh. In J. Eddison and C. Green (eds.), *Romney Marsh: evolution, occupation, reclamation*, 3–29. Oxford: University of Oxford Archaeological Committee Monograph no.24.

Zeuner, F.E. 1959. *The Pleistocene Period*. London: Hutchinson.

9

Further thoughts on the impact of beaver on temperate landscapes

Bryony Coles

Summary: *Beaver have great potential as changers of the landscape, and it is argued that more attention should be paid to their possible effect on the Postglacial environment of Britain, especially in and adjacent to running water. Their relevant activities are briefly described, and the archaeological evidence for similar activity in southern Britain reviewed, with a consideration of the implications for prehistoric landscapes and the people who inhabited them.*

The occasion of papers focussed on the relationships between people and rivers provides almost ideal conditions to draw attention to an important third force: beaver. Anyone studying landscape use and environmental change in a context where beaver were present should be aware that this relatively small animal can effect significant alterations to the environment, and it is the aim of this paper to demonstrate their possible impact on past conditions, especially in southern Britain.

My interest in beaver dates back to a childhood Christmas present from grandparents, Grey Owl's "Sajo and her Beaver People" (Grey Owl 1951), but it was not until the 1970s that some inkling of their archaeological significance emerged. Excavation of a Neolithic site in the Somerset Levels exposed apparently worked wood (the Baker Platform – Coles, Fleming and Orme 1980) which in due course was identified as beaver-felled. There followed some intensive research by J.M. Coles on beaver in present-day Canada, and a search of the archaeological literature for references to beaver, which resulted in two papers 'Beaver in the Somerset Levels' (Coles and Orme 1982) and '*Homo sapiens or Castor fiber?*' (Coles and Orme 1983). Some of the content of these papers will be repeated in what follows, as it is my contention that in Britain, at least, there is still insufficient attention paid to beaver and the havoc they may cause in landscapes with running water.

This lack of interest may perhaps be due to our general unfamiliarity with beaver, which became extinct in Britain in the course of the Middle Ages. There is a late reference to their presence in the 12th century *Itinerary through Wales* of Giraldus Cambrensis:

> The Teivi has another singular particularity, being the only river in Wales, or even in England, which has beavers; in Scotland they are said to be found in one river, but are very scarce. (Llewelyn Williams 1930)

The inference is that beaver were already a rarity in the early 2nd millennium AD, and by the time antiquaries began to speculate on prehistory, beaver were a forgotten force in the land.

Today, general histories of environmental change in England or in Britain pay scant attention to the animal. *England before Domesday*, for example, mentions bear and wolf and even bison several times, but not beaver (Jones 1986: 143). Archaeological surveys can be equally devoid of reference to the animal (e.g. Darvill 1987), although John Evans and John Wymer both commented that beaver may have caused a lake at Thatcham, where early Mesolithic peoples camped (Evans 1975: 88; Wymer, pers.comm.). Generally, however, there is little consideration of the environmental impact of the animal, although a range of excavation reports notes the presence of beaver bone, and a trickle of beaver-cut wood has recently started to flow (see below). As a matter of interest, geologists have also noted the presence of beaver and mentioned its possible influence on the landscape, although I do not know of any detailed investigation of the evidence (e.g. Thomas 1961: 425; Worsley 1978).

On the continent, beaver still exist in the wild in various regions, particularly but not only in northern and eastern Europe. In some cases they have been reintroduced, in others it is possible that their occupation has been continuous since the retreat of the last ice sheets. Perhaps because they are a familiar animal to workers in those regions, they are more often acknowledged as agents of change than is the case in Britain. The possibilities may be summed up by one recent example, the landscape reconstruction for the Skateholm Project, an investigation of Mesolithic settlements, cemeteries and their surroundings on the south Swedish coast (Larsson 1988). Jonsson, reporting on the vertebrate remains, includes a beaver dam and meadow in one of his reconstruction diagrams (Jonsson 1988: fig. 12) and writes as follows:

> The beaver has played an important role in the creation of biotopes for many other animals. The tree-felling activity of the beavers opened the close and shady forest to the advantage of grasses and herbs. Here elk and aurochs were able to find temporary habitats. These 'beaver meadows' also attracted rodents and birds of prey, as well as wild boar and smaller carnivores. Old beaver dams, once deserted and drained were transformed into vast open grass lands before the forest was once again re-established.

In order to make the case for beaver as agents of change in the British landscape, I will first summarise the salient points from earlier papers as to their behaviour and environmental impact. Then, the likelihood of the prehistoric beaver in Britain following similar behaviour patterns needs to be examined, which means turning to the archaeological evidence for their presence and activities. When these have been established, the repercussions will be briefly noted, whether in terms of environmental change or in terms of the benefits and disadvantages for the humans who co-habited with the beaver.

Fieldwork in Ontario, mainly carried out by J.M. Coles, along with some observations in other parts of Canada and on this side of the Atlantic in Sweden and recently in Poland, has been supplemented by various written accounts of beaver behaviour (e.g. Anon. 1978; Novak 1976). Adult beaver are less than 1 m long and weigh 15–30 kg. Trees are felled for food. Using its teeth, a beaver can fell a sapling in seconds and a 12 cm diameter willow in three minutes, and it will tackle trees up to about 1 m in diameter although normally it concentrates on smaller trees. Preferred food species are poplar, willow and maple, with quite a wide range of other species eaten if necesssary, along with aquatic vegetation.

The beaver live in small colonies consisting of a pair and two successive litters; in temperate and northern forested lands they dam running water to create a pond where they build a lodge with an underwater entrance. Both dam and lodge are built from the branches of beaver-felled trees, along with other fallen timber as long as it is not rotten.

Depending on local topography, the dams may be 2–3 m high and a few metres long or up to several hundred metres long in broad shallow valleys if a colony is long resident. If the beaver are undisturbed, dams are constantly repaired and frequently extended, and they may be stabilized by growing vegetation. This comes about because quantities of mud and stone, as well as wood, are used in dam building, and the mud contains seeds which germinate and bits of plant which root, and some of the wood species such as willow also root very readily.

Dams are built across the slow-flowing watercourses of lowland marshes, across the sluggish side branches of major rivers, across streams and across tiny rivulets in hill country: if there is a flow of water and trees to block it with, the beaver will build to create their pond. Water floods the land to either side of a blocked watercourse, and any vegetation is killed. Trees destabilised by the rising water keel over in winds; those left standing soon die. As the years pass, the tree-free expanse of water spreads where previously was forested dry land, and these open areas can be quite large. One beaver-created lake in the Algonquin in Canada covers about 20 hectares, and it is retained by a dam that is no more than 50 m long.

If flooding is the main mechanism causing deforestation, tree-felling for food and for building materials nevertheless contributes to the process. Trees are normally felled within 50 m of the pond, for ease of transport of the wood, and so that the beaver can quickly escape to water if predators appear. In the autumn, large surpluses of wood are accumulated for winter food, stored at the bottom of the pond, and in this season beaver may rapidly create clearings (and chaos) as they fell tree after tree and carry off some of the branches. At times, they dig a canal from their pond towards the food source, which helps with transport as it easier for them to tow floating logs by water than to drag the logs over rough dry land, and so in this way some clearings are made which are not immediately adjacent to a pond.

One side-effect of the beavers' attack on trees is natural coppicing, in that the 20–50 cm high stump which they leave behind may sprout and grow a crop of long straight shoots, just as happens with deliberate coppicing by humans. If the beaver are active when there is snow on the ground, the trees are felled at 20–50 cm above the surface of the snow thereby leaving, after the thaw, a stump approaching a pollard in height.

The further effects of beaver activity are those described by Jonsson and quoted above, namely the development of a habitat which favours a variety of wild life. The pond is colonised by aquatic plants and animals, and other animals such as elk, bear and birds of prey come to exploit them. Cleared dryland provides browse from the natural coppice, and grasses and other plants flourish and yield fruit, all of which attracts herbivores and rodents, which in turn attract predators. Abdandoned beaver ponds silt up and develop into wet meadows, providing a similar attraction for herbivores though perhaps at a different season of the year. Humans are amongst the predators who may benefit from these after-effects of the beavers' presence, and they may also appreciate the abundant supplies of firewood which the animals litter around the ground, sometimes ready-cut into burning lengths. One of our Scandinavian volunteers in the Somerset Levels reported that his grandparents had regularly collected beaver-felled wood for fuel, thereby saving themselves a certain amount of hard work.

A visit to the Narev river valley near the village of Drozdowo outside Łomsa in eastern Poland in July 1991 provided the opportunity to observe the effects of beaver in a rural European landscape. Along the broad and strongly-flowing Narev, the beaver live in burrows in the river bank and do not build dams. In 1979 one family had moved into a small stream flowing across the floodplain and into the main river. Their many descendants co-exist with the farmers in an area of hay meadows and small patches of woodland, and unlike the neighbouring river beaver they build dams and dig canals with great vigour. Along one field boundary ditch dams were no more than 10 m apart, and one had been rearranged by the farmer to make a crossing for his tractor. Apparently the farmers found some dams useful in controlling the spring floodwaters of the Narev, and these they left. Others caused water to stand where it was not wanted, and these they broke. Dams in the woods were generally left intact, resulting in the water table in one wood being about 1 m above the adjacent dry hay meadows during our visit.

The high water table ensured that the beavers' canals stayed flooded. Some of these were open, some were virtually tunnels, and others were interrupted by short stretches of well-worn path where an obstacle had prevented the beaver from digging. The wood was criss-crossed with a dense network of these canals, used by the beaver for speedy and safe travel around their territory, and for carrying food to their stores.

Because of the beaver, the wooded land is now permanently waterlogged, whereas the adjacent hay meadows dry out seasonally. The wood itself is full of small bushy growth which has sprouted from felled stumps, and elk are attracted to browse (they like the floodplain copses in general). So far, although the beaver have altered the character of the woodland, they have not caused any noticeable clearings within it. This may be due to the very flat character of the floodplain, which means that the relatively short dams do not create huge ponds. What is most interesting about the Narev beaver is the way they have turned to dam-building and canal-digging as soon as they left the main river. Along the river, they could not be said to be significant agents of landscape change, but along its tributary they appear to have that potential.

What is the evidence that the beaver of Britain behaved in a manner to affect the landscape, that is, that they felled trees and built dams? This needs to be established, because the mere presence of beaver does not guarantee such activity, if, like those in the Narev main river, they live in burrows in the river bank instead of creating a pond and building a lodge. Evidence for the beavers' use of wood, whether for food or for building, consists of logs and branches of appropriate species bearing the distinctive facets left by beaver teeth. Fig. 9.1 illustrates a modern beaver-cut log and, beside it, a prehistoric example from the Somerset Levels. Wood with beaver facets has now been identified from several prehistoric sites in England and Wales, dating from the Early Mesolithic to the later Bronze Age at least. Amongst the relevant sites are Star Carr, Yorkshire (Early Mesolithic), the Baker Platform, Somerset Levels (Neolithic; Coles, Fleming and Orme 1980), Runnymede on the Thames (Beaker and Middle Bronze Age, information from Stuart Needham), Caldicot, South Wales (later Bronze Age, information from Caroline Earwood). In these cases, apart from Runnymede, the wood appears to have been collected and used by people, for example for building the Star Carr and Baker platforms, but it was clearly cut by beaver so we can be fairly sure that they were felling trees whether or not humans were also doing so.

The second question, did the beaver build dams?, is less easy to answer because, on the whole, archaeologists do not get funded to excavate and record non-human features. It is only when such are associated with human activity, or investigated up to the point where their non-human origin becomes clear, that they are likely to enter the archaeological literature. Yet various possible beaver constructions have now come to light, and I am grateful to colleagues, John Wymer in particular, for drawing my attention to relevant accounts.

Wymer's own excavations at Thatcham revealed what was possibly a beaver canal leading inland from the lake shore (Wymer 1962, 336; see also Coles and Orme 1983). Thatcham was situated in the Kennet valley and several earlier discussions of the Kennet peats and of archaeological finds from the peats refer directly or indirectly to beaver activity. Harold Peake (1937) drew together much of the evidence in a paper to the Newbury District Field Club. Beaver were often reported amongst other bones found in the peat near Newbury, and in 1825 Buckland noted abundant branches and other remains of trees, and bones from beaver and other animals. In the 1860s, both Palmer and Adams suggested that beaver dams could have obstructed drainage in the Kennet valley, leading to peat formation (Peake 1937: 121–122). Peake was of the

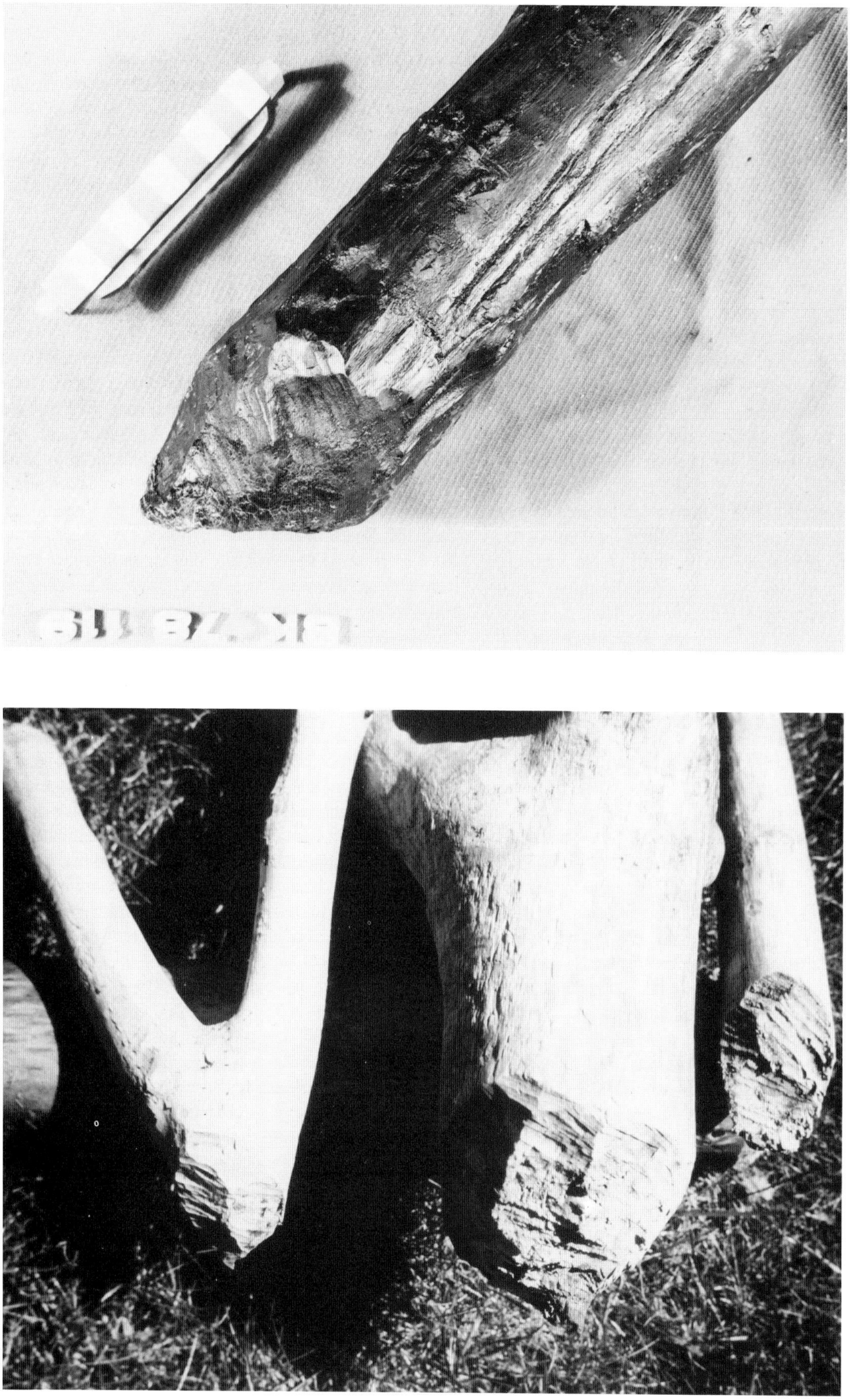

Fig. 9.1. *Ancient and modern beaver-cut wood. Left, from the Neolithic Baker platform, Somerset Levels, ca. 3000 cal BC. Right, from Ontario, Canada, ca. 1985 AD. Photos J.M. Coles.*

same opinion, and he drew attention to several long banks or mounds of "shell-malm" running across the valley, which he argued had accumulated against beaver dams and one of these, in Newbury itself, he suggested could have been used during the Iron Age as a causeway to cross the valley (1937: 125–126). But he also noted the need for excavation to find beaver-gnawed wood to confirm that the linear banks indicated the position of former beaver dams. As far as I know, no such excavations took place, and drainage has by now probably destroyed whatever evidence there may have been, whether for beaver-caused mounds or for palaeochannels as is now sometimes suggested.

It is worth turning back to an earlier account of the Kennet peats, quoted by Palmer and reprinted following his paper (Palmer 1875: 135–138). The account was written in 1756 by John Collet, who observed amongst other things that:

> Great numbers of trees are plainly visible in the true peats, lying irregularly one upon another; and sometimes even cart loads of them have been taken out, and dried for firing ... These trees are generally oaks, alders, willows and firs, besides some others not easily to be known. The small roots are generally perished; but yet have sufficient signs to show that the trees were torn up by the roots, and were not cut down, there being no sign of the ax or saw, which, had they been felled, would have been plainly visible ...

Collet continued with some very sound observations on stratigraphy, and the problems created by old peat-cuts which have subsequently filled-up. But what matters in the present context is that his description of uprooted trees fits well with the situation described above, when rising water levels follow the construction or extension of a beaver dam, and living trees are destabilised to such an extent that they fall over in quite moderate winds, creating the effect of the aftermath of a hurricane. Note that in this instance, as with Peake's long banks, there is no direct evidence that beaver were building dams, but there are good descriptions of phenomena which could be side-effects of dam building.

Downstream from Newbury and Thatcham, after the Kennet has joined the Thames, lies Runnymede, and from here Needham has recently reported "bundles of rods and branches, some with beaver-gnawed ends and all stripped of bark, from layer of organic mud in long stratified sequence with no associated cultural debris" (*Radiocarbon* vol. 31, pts 1–2, p 17, 1989). The wood gave radiocarbon dates of 3770 ± 60 BP (BM-2435) and 3740 ± 80 BP (BM-2436). Further wood, from a "dense mat of twiggy material, including many pieces with signs of beaver chewing" gave a date of 2900 ± 50 BP (BM-2661). Runnymede may therefore have seen beaver active in side channels or backwaters of the Thames and the accumulations of wood suggest lodge, winter food store or dam and although the latter seems the least likely of the possibilities there may have been a dam just downstream to pond up the water and provide conditions both for the beaver to live in safety and for their wood to remain preserved in waterlogged conditions for several millennia (see also Needham, this volume).

Further on downstream near the Thames estuary, Wilkinson's excavations across Mar Dyke revealed, amongst other things a layer of:

> Dark reddish brown peat containing abundant mainly horizontal wood fragments ranging from twigs to trunks ...layer present throughout length of section 575 and also formed a 1.8 m thick deposit in section 273. Deposition took place between c 4950 and 3550 bp (3000–1600 bc) ... (Wilkinson 1988: 106).

From Section 575 to Section 273 is a distance of nearly 60 m, covering the middle portion of a valley 100–200 m wide at the time of wood deposition. Wilkinson's field notes include reference to "one stem c 5 cm diameter which had the characteristic teeth marks of beavers", and other less clear pieces and several wood chips, and he considered the following possible interpretations for the layer: "a) valley floor woodland, b) drifted wood, c) beaver dam." The thickness of the wood layer reported from Section 273, the possibility that the wood extended for at least 60 m across the valley, and the positive identification of beaver tooth marks, make this one of the best candidates for a dam, although it remains unverified.

Recent discoveries in Shropshire and in Wales may in due course provide relevant evidence. The Shropshire discovery comes from peat from a small bowl-shaped hollow immediately adjacent to the course of a river on the Welsh border, and it consists of wood with definite beaver tooth marks (M.D. Watson, Shropshire Senior Archaeologist, pers.comm.). The Welsh find was made by Mark Redknap of the National Museum of Wales, who unearthed beaver-felled wood in the course of an excavation to check on the findspot of a stray artefact. The evidence in the light of the local topography suggests a line of wood crossing fairly narrow waters between islands of dry land, perhaps a beaver dam. But as yet neither of these two discoveries has been fully investigated, and it is premature to speculate further.

Thus we have sound evidence for the presence of beaver in Postglacial England and Wales, we can be sure that they felled trees, and it seems probable that they were building dams. Knowing some of the present-day effects of these activities, the implications for prehistoric landscapes and the people who inhabited them can now be considered briefly. Some of these points have been further elaborated in Coles and Orme 1982 and 1983.

It is uncertain whether or not pollen analysis will identify beaver clearings in a forested landscape, and

10

Alluvial sedimentation style and archaeological sites: the lower Vyrnwy, Wales

John Lewin

Summary: *Alluvial sedimentation in upland Britain occurs in a variety of environments which give vertical or lateral ageing sequences in river deposits. These are associated with contrasting river patterns and incision/aggradation sequences which are themselves conditioned in part by Quaternary climate changes and by human activity. Major geomorphological themes explored to date in upland Britain include contemporary channel changes, progressive Holocene incision, colluvial and fluvial adjustment to glacially modified relief, and disturbance phases effected by deforestation, mining and extreme flood events. Themes which have received less attention so far are extensive documentation of Holocene palaeochannels (cf Poland) and, with a few exceptions, an appreciation of the complexity of sedimentary environments on a local scale.*

The lowest 20 km of the Vyrnwy provide some useful additional evidence. The area contains four major morphological elements: (a) A complex of glacial outwash and fan gravels, with closed depressions and disordered drainage patterns. (b) A series of Holocene incision terraces with an impressive set of palaeochannels of braided and meandering patterns. The terraces are associated with an early Roman military site, though incision has occurred subsequently. (c) A currently actively migrating gravel river showing historical movement over some 200 years, including a major cutoff in 1947. (d) A stable reach of imperceptible gradient, with overbank vertical sedimentation dominant.

The actual and potential archaeological significance of these sites differs in response to their contrasting sedimentation styles, a phenomenon which is to be expected more widely in upland Britain.

INTRODUCTION

Fluvial deposition results from the major processes of *sedimentation* of fine material which has been suspended in flowing or ponded water, and from *accretion* where material travelling along a river bed, which is usually coarser, comes to rest. The latter commonly occurs in bedforms or bars, as in point bars on the inside of meander bends. Lag deposits consisting of larger rock particles which the river is unable to transport are also found. It is usual to associate suspension deposits (sometimes known as 'vertical-accretion' deposits) with overbank flows on floodplains and accretionary ones with within-channel flow, but it is also possible for flood gravel accretion to take place overbank while fine sedimentation can occur in slacker water quite commonly within the channel zone. Particular channel patterns (meandering, braiding, anastomosing) are also sometimes believed to have characteristic sets of sedimentary units associated with them, but here again much variety is possible and research in recent years has shown that mistaken interpretations based on simple premises are quite easily made.

Of particular interest archaeologically is the association of repeated sedimentation, *sensu stricto*, with a vertical ageing of deposits. With stable channels, this may persist on floodplains for a very long time burying successive landsurfaces with an increasing thickness of alluvium. By contrast, preserved accretionary deposits formed in-channel may be associated with lateral channel shifting, thus giving a lateral ageing pattern. Here deposits at any one level may be diachronous, with a spatial age pattern of some intricacy depending on the history and pattern of channel shifts across their floodplains.

Extended aggradation may lead to vertical stacking of sediments in age sequences; as well as simply involving units of suspended sediment, this may incorporate sets of coarse and fine materials or other components of both suspension and accretion deposits. Overall river incision, on the other hand, may leave no direct and local sedimentary record, excepting in the important context of terrace formation and the preservation of former alluvial units on valley floor margins, possibly in a staircase with both lateral and vertical age sequencing on each of the 'treads'.

A dominant factor affecting these processes, though one which is often neglected, is the calibre and volume of sediment actually being supplied to rivers from hillslopes or upstream channel erosion. Depending on patterns of climate-influenced weathering and vegetation cover, the soil and bedrock present, and the degree of human activity, there may be very considerable differences in both space and time between supply-conditioned alluvial sedimentation styles. Hydrological conditions can also fluctuate, affecting the ability of streams to transport sediments. Competence-limitation is especially important in the case of coarser bed sediments, whilst supply- limitation is generally the more important with finer materials.

THE ALLUVIAL CONTEXT IN UPLAND BRITAIN

For upland Britain, the pattern apparent in preserved alluvial deposits has been strongly conditioned by the fluctuation in Late-Quaternary climates and the sequences in landforming processes that resulted. In particular the alternation of cold and warm phases has produced recognisable sequences, with earlier events conditioning later processes, and hence deposits. Cold climate phases were associated with glacial deposition, in the form of tills or outwash deposit assemblages and enhanced slope erosion under periglacial conditions. Eustatic and glaciotectonic processes also altered sea levels and river gradients. Warmer interglacial stages, by contrast, have been broadly associated with greater slope stability, river incision under conditions of lower sediment supply and possibly discharge, and the reworking of earlier alluvial and other valley deposits. Soil development, and then the surface erosion of these finer materials, has given greater relative importance in the present Holocene to the finer components of total river sediment loads and deposits, though at least on steeper valley gradients coarser materials have continued to move and laterally to accrete even under interglacial conditions. There is therefore an important contrast between environments where the basal coarser units underlying finer alluvial ones are largely relict from cold phases, as is common in lowland Britain, or ones where both parts of a superficially similar couplet have continued to be actively deposited, as at many upland valley sites.

Sedimentation in the Holocene has, of course, also been rendered more complex by human activity. Through deforestation, agriculture or mining activity this may have released sediments, or, more recently through reservoir construction and river bank protection, may have retarded them. The precise causal mechanisms which link human actions and sedimentation are more complex than may be generally appreciated. Deforestation affects the water balance, with decreased evapotranspiration, more runoff, and a more rapid and considerable response to the rainfall of individual storms. This may, according to site, increase hillside erosion, channel erosion or alluvial sedimentation. This hydrological effect may be regarded as distinct from that of sediment release because of enhanced soil erodibility from the bare ground of individual plots cleared for cultivation. A third and again different effect is that of removal of riparian trees. This affects bank stability and eventually channel dimensions, and may be associated with the increased mobilization of river bank and bed sediments. Clearly these several mechanisms require careful evaluation. It is not easy to distinguish climatic and anthropogenic effects in the sedimentary record, especially when they are interacting.

A final important factor lies in the physical controls on river channels and sedimentation. River discharges, valley slopes and sediment size may vary in a systematic manner and in a down-valley direction to produce distinctive alluvial environments. These range from upland boulder-bed streams through braided and actively meandering patterns, to inactive sinuous rivers. Location within river systems may affect sedimentation styles considerably.

The combination of sedimentary processes and environments, contrasting Quaternary history, human activity, and indeed unique location within particular drainage basin sedimentary systems, means that particular sites have also an unique assemblage of alluvial sediments — variations on a few themes, perhaps, but likely to be different in important detail from sites elsewhere. Previous detailed study in upland Britain and its margins has explored four phenomena :

1. Progressive Holocene incision into Late-Quaternary glacigenic valley fills, with the partial preservation of sets of Holocene incision terraces (e.g. Macklin and Lewin 1986).
2. Fluvial adjustment, by localized incision or infilling, of glacially created relief involving also lacustrine and backswamp (organic) deposits (Brown 1987).
3. Phases of disturbance effected by both anthropogenic activity and extreme flood events (e.g. Harvey 1986; Harvey *et al* 1981; Lewin *et al* 1983).
4. The quite considerable mobility of rivers at some sites, involving active floodplain reworking over recent centuries (e.g. Lewin 1987; Macklin and Lewin 1989).

Despite the relatively good understanding of river channel processes in Britain (see Ferguson 1981), there also remain some significant omissions from published academic analyses of the U.K. upland sedimentary record. Complex palaeochannel systems have been described in outwash deposits in Scotland (Maizels 1983), but not generally on the spatial scale and covering an extended timescale on through the Holocene of the now classic Polish work (see, for example, Starkel 1982). Do such channels, and the temporal variety in environmental controls that their variable patterns and sizes suggest, not so generally exist in the U.K. uplands? A second problem lies with the often unperceived complexity of alluvial sediments, frequently within a small area, and this could well form a barrier to fuller exploration and understanding of alluvial sites from an archaeological point of view. Simplistic views on the origin and dating of coarse and fine alluvial units may be particularly unhelpful, for example.

In the light of both these points, a discussion of a small area on the Welsh borderland between the confluences of the Vyrnwy and Tanat, and Vyrnwy and Severn, may be useful. Within a direct distance of 10 kms there is here a large range of alluvial environments, channel patterns, palaeochannels and terraces. These are associated with archaeological sites, so that the potential value to archaeologists of an appropriate appreciation of the nature of alluvial sedimentation and some of its controls may be evaluated. Recent alluvial history in the area has been briefly discussed in Lewin (1987), prehistory by Limbrey (1987), and the Roman archaeology by Jones (1990).

THE LOWER VYRNWY

The Afon Vyrnwy has a catchment area of 878 km^2, its major tributary being the Tanat (229 km^2). Figure 10.1 shows river location and long profile, together with the present study area. River discharges have been partially regulated since the construction of the Vyrnwy reservoir, but gauging records at Llanymynech show a mean flow of 20 m^3 s^{-1} rising to a 10–year flood of 276 m^2 s^{-1} . Gradients within the study reach have declined considerably from those in the headwaters (Figure 10.1), but the study reach is situated where the lateral confinement of a relatively narrow glaciated valley is replaced by the broad Shropshire-Cheshire plain.

In the study area discussed here, which is shown in Figure 10.2, alluvial deposits comprise four major elements:

1. A complex of outwash and fan gravels in the Four Crosses and Carreghofa areas, along the valley and river margins and between the Vyrnwy and Severn. These gravels may extend up to 10 m above present river level, and north of the Vyrnwy their surface slopes relatively steeply eastwards (> 0.005 m m^{-1}). The plain itself is thickly covered with 'blue clay', according to bore hole logs up to 56 m in thickness at Trepannal Farm on the River Morda (SJ 287220). The gravels overlie blue clay just east of Llanymynech (SJ 277209), with clay and gravels inter-digitating in some boreholes (e.g. at Wern SJ 257205).

   The surface of these gravels has a mottled appearance on air photography (see Figure 10.3), with closed depressions (as at Carreghofa Farm) and deranged channel forms of small dimension. They may be interpreted as outwash and ice marginal deposits of Late Devensian age, comparable with deposits of similar origin which have been studied further north in the Wrexham area (Thomas 1985).

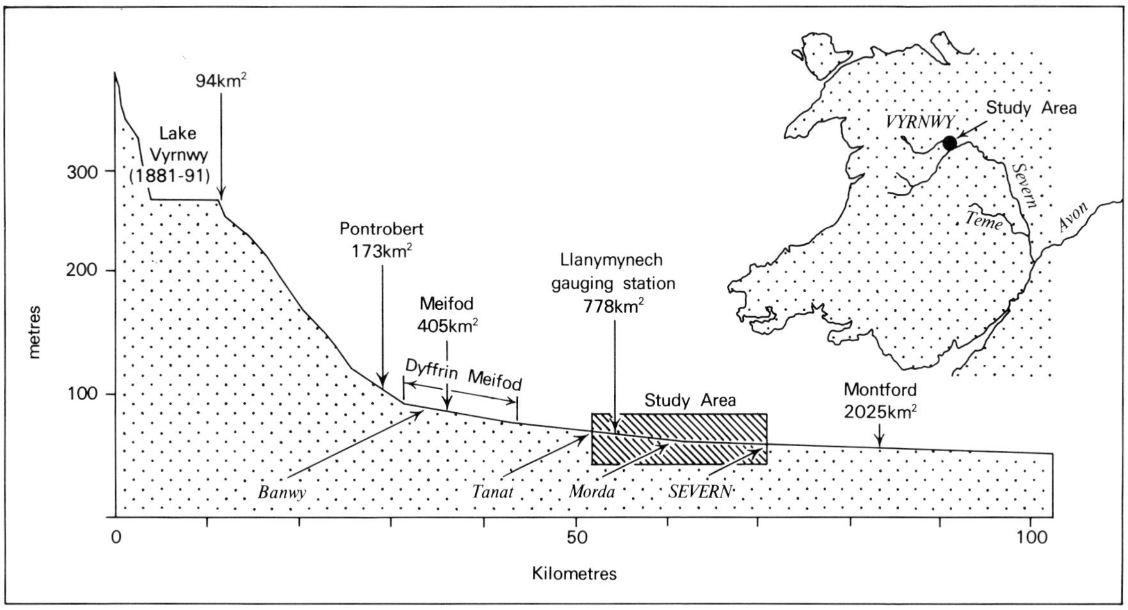

Fig. 10.1. The study area and long-profile of the Afon Vyrnwy.

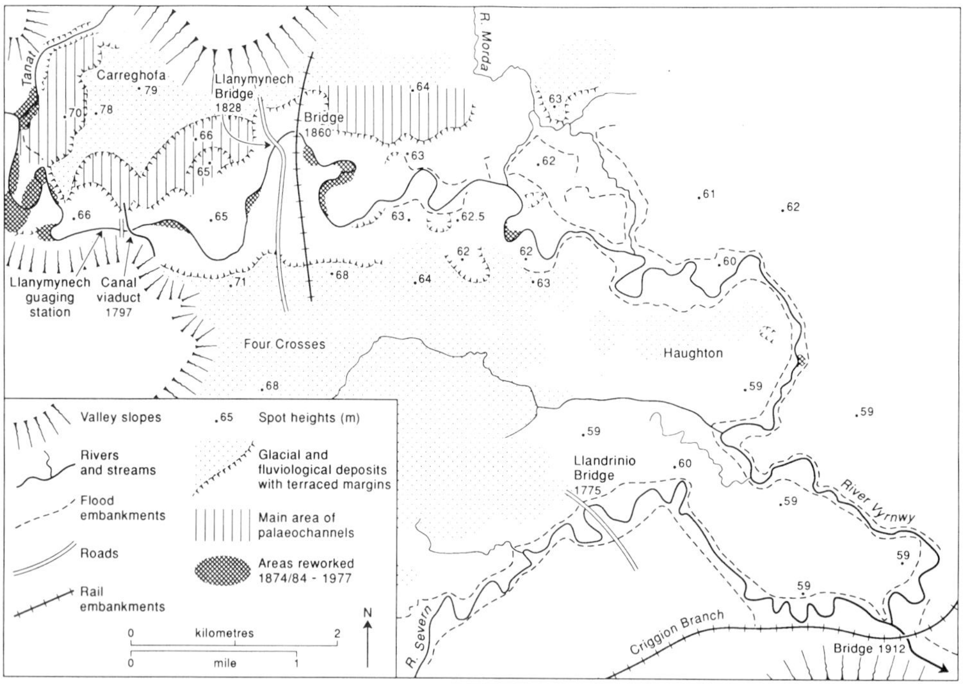

Fig. 10.2. The Lower Vyrnwy, showing alluvial sedimentation elements, elevations, and construction features.

Fig. 10.3. The Tanat palaeochannels and Carreghofa outwash deposits. The line of trees from middle left (North) to bottom right (South) follows a 6 m bluff dividing the Holocene palaeo-channels (element 2, bottom) from the outwash sediments (element 1, top). Photo by Chris Musson, Clwyd-Powys Archaeological Trust.

2. A series of incision terraces is cut into these Devensian deposits, and at the lowest levels within the Tanat, into bedrock. There is a veneer of fluvial gravels on the terraces together with a well developed set of palaeochannels many of which are only revealed as crop marks on suitable air photography (Figure 10.3). The palaeochannels are mostly cutoff meander arcs, but they can also indicate accretionary bar complexes in places. These terraces are currently being mapped and studied in detail.

   The lowest of these terrace levels is occupied by an early Roman military site, which is under investigation by Professor G.D.B. Jones following identification of a titulum on air photography. Part of the Roman site contains ditches infilled, apparently naturally, with fine fluvial gravel showing that the Roman encampment was still on the active river floodplain of the Tanat at the time. Subsequently (after the 1830s according to map evidence) the Tanat was artificially straightened and it is now incised some 2 m below its Roman level. Some of the incision was probably pre-nineteenth century in date.

   The excellent and unusual preservation of the palaeochannels calls for some explanation, particularly in view of the paucity of such Holocene features being recorded in Britain. This probably arises for a number of reasons: the rapid incision and lateral migration of the Tanat and Vyrnwy at this site, and their possible bedrock 'fixing' once the gravels and (?) clay had been cut through, and the relative thinness of a suspension-sediment blanket masking their existence, are two reasons. However, it is also quite possible, and ongoing research elsewhere seems to indicate this, that palaeochannel systems of Holocene age in upland Britain are in fact quite widespread, but that they remain to be recorded and fully studied.
3. The present Vyrnwy river is actively migrating, particularly above its junction with the Morda. Comparisons of historical maps and air photographs show the extent and rate of river migration and active lateral accretion. Active meandering is associated with accretionary channel bars, cutoffs and cutoff infilling as is indicated by photography of a then very recent cutoff on the Vyrnwy on 16 April 1947 (probably formed during the exceptional 1947 spring floods, whose discharge has not been locally reached since that date); the cutoff channel has been infilled subsequently, with overbank sedimentation gradually masking the relief of the accretionary channel bars in time (see Figure 10.4).

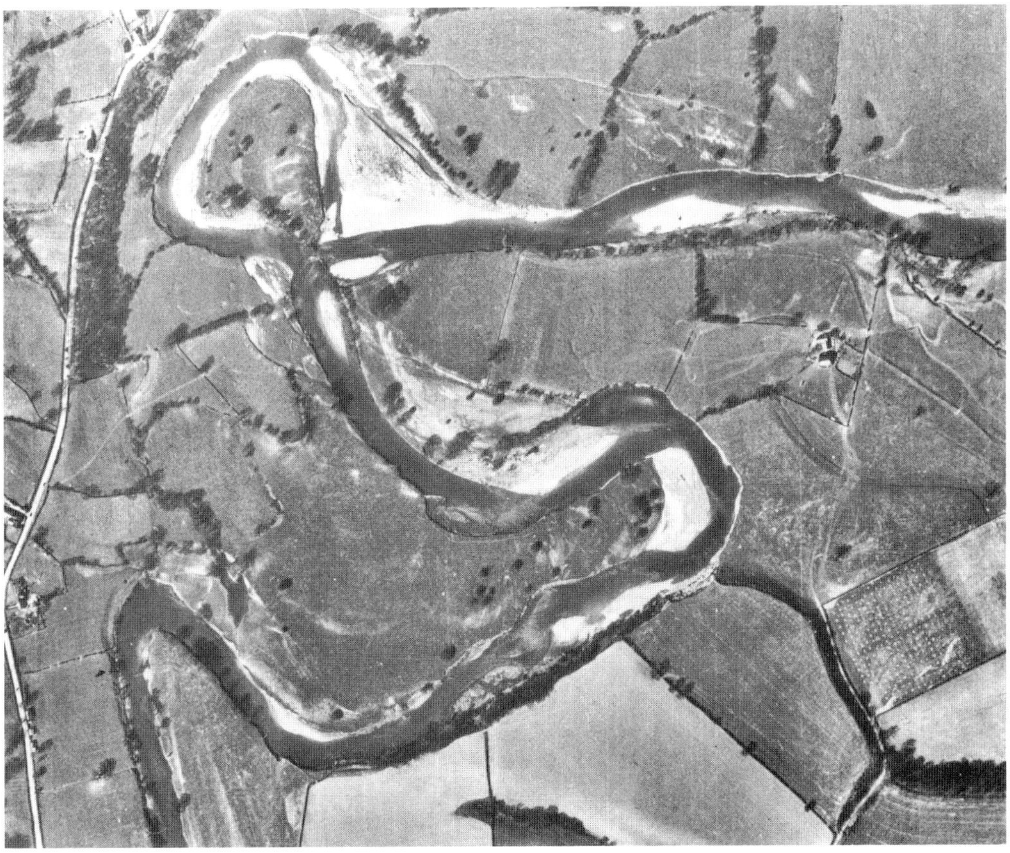

Fig. 10.4. *An alluvial cutoff on the Vyrnwy, 16 April 1947. This reach of the river is accretion-active with gravel deposition under the present river regime (SJ 2419). British Crown Copyright, reproduced with the permission of the Controller, Her Majesty's Stationery Office, London.*

4. The lowest reach of the Vyrnwy (below Houghton and the Morda junction) has an imperceptible gradient over the thick glacigenic clays. Very little lateral activity is apparent: the sixteenth century Melverley church is immediately adjacent to the river which is also closely approached by pre-enclosure ridge and furrow systems (Figure 10.5). Analysis of surveyed valley cross sections show between 0.23 and 0.84 m mean elevation differences along profiles between the (higher) parts of the profiles alongside the river and between the ca. 1800 enclosure flood banks, and the areas beyond. These data suggest a continuing build-up of sedimentation deposits, and the area is known for being exceptionally flood-prone. Burial of archaeological materials in this environment has been reported previously by Thompson (1982), and further literally in-depth exploration could be useful in this environment.

ALLUVIAL SEDIMENTATION AND ARCHAEOLOGY

Each of the four broad sedimentary environments so far described on the Lower Vyrnwy, all within about 10 kms of one another, provides a different context for the preservation of archaeological and environmental evidence. The outwash material preserves a landsurface of greatest antiquity, with its surface depressions potentially accumulating an extended environmental

Fig. 10.5. The Vyrnwy at the Severn junction (bottom). Note the flood embankments and pre-enclosure ridge and furrow, in places almost to the river's edge (SJ 3316). British Crown Copyright, reproduced with the permission of the Controller, Her Majesty's Stationery Office, London.

record in ponding sites. But the relatively old landsurface does form something of a palimpsest of human occupancy, with deposition a prelude rather than an accompaniment to human activity. The terrace sequences give a lateral and vertical ageing sequence of coarse/fine couplets deposited by braided and meandering streams. River depositional activity has here operated in association with human activity, and the Roman occupation involved valley floor conditions rather different from those now present. The third zone – that of actively migrating rivers today – appears to be destructive of archaeological remains older than those of the last few centuries, whilst the final zone, dominated by overbank sedimentation and a laterally inactive channel, is likely to preserve materials well below the present, young landsurface.

All this contrasting river erosional and depositional activity in this small area should lead to some caution in the interpretation both of sites and 'find' distributions. Burial by alluviation is important in upland valleys, whilst sites of a particular elevation above rivers today may have been either more or less so when occupied. The uplands can show quite considerable variety in sedimentation style, both in sequence downvalley and over short distances. The essential point to appreciate is that the dynamics of alluvial sedimentation must be considered alongside site archaeology so that misinterpretations are kept to a minimum, and useful opportunities for discovery and sequencing are fully realized.

References

Brown, A.G. 1987. Long-term sediment storage in the Severn and Wye catchments. In K.J. Gregory, J. Lewin and J.B. Thornes (eds.), *Palaeohydrology in Practice*, 307–332. Chichester: Wiley.

Ferguson, R.I. 1981. Channel form and channel changes. In J. Lewin (ed.), *British Rivers*, 90–125. London: Allen & Unwin.

Harvey, A.M. 1986. Geomorphic effects of a 100–year storm in the Howgill Fells, Northwest England. *Zeitschrift fur Geomorphologie*, 30, 71–91.

Harvey, A.M., Oldfield, F., Baron, A.F., and Pearson, G.W. 1981. Dating of post-glacial landforms on the central Howgills. *Earth Surface Processes and Landforms*, 6, 401–12.

Jones, G.D.B. 1990. Searching for Caradog. *Trivium*, 25, 57–64.

Lewin, J. 1987. Historical river channel changes. In K.J. Gregory, J. Lewin and J.B. Thornes (eds.), *Palaeohydrology in Practice*, 161–175. Chichester: Wiley.

Lewin, J., Bradley, S.B. and Macklin, M.G. 1983. Historical valley alluviation in mid-Wales. *Geological Journal*, 18, 331–50.

Limbrey, S. 1987. Farmers and farmland: aspects of prehistoric land use in the Severn Basin. In K.J. Gregory, J. Lewin and J.B. Thornes (eds.), *Palaeohydrology in Practice*, 251–267. Chichester: Wiley.

Macklin, M.G. and Lewin, J. 1986. Terraced fills of Pleistocene and Holocene age in the Rheidol Valley, Wales. *Journal of Quaternary Science*, 1, 21–34.

Macklin, M.G. and Lewin, J. 1989. Sediment transfer and transformation of an alluvial valley floor: the River South Tyne, Northumbria, U.K. *Earth Surface Processes and Landforms*, 14, 233–246.

Maizels, J.K. 1983. Proglacial channel systems: change and thresholds for change over long, intermediate and short time-scales. In J.D. Collinson and J. Lewin (eds.), *Modern and Ancient Fluvial Systems*, 251–266. Oxford: Blackwell.

Starkel, L. (ed.) 1982. *Evolution of the Vistula River Valley during the last 15000 years*. Geographical Studies, Polish Academy of Sciences Institute of Geography and Spatial Organization, Special Issue no. 1.

Thomas, G.S.P. 1985. The Late Devensian along the border of northeast Wales. *Geological Journal*, 20, 319–340.

Thompson, T.R.E. 1982. Soils – Powys 11 Sheet SJ 21 (Ardleen). *Soil Survey record* 75.

11

The determination of cause in the generation of major prehistoric valley fills in the Cheviot Hills, Anglo-Scottish border

Richard Tipping

Summary: *Radiocarbon-dating evidence from several fluvial aggradations in the upland Cheviot Hills is discussed, and two major prehistoric phases of landscape instability are identified. Attention is paid to the degree to which proxy palaeoenvironmental records (pollen diagrams) are of an appropriate scale to relate to the geomorphic activity, before making correlations which strongly imply an anthropogenic origin, through woodland clearance, for these aggradations.*

INTRODUCTION

The deposition of fluvial sediments as terrace fills (aggradation) is but one response by river systems to changes in catchment hydrology, but is that most likely to be preserved, and is the basis of most palaeohydrological research. Aggradation in itself, however, provides few clues to its cause. The explanation of causal events needs to be undertaken through correlation with proxy palaeoenvironmental records pre-dating aggradation. One of the most commonly utilized of these is pollen stratigraphy. This paper will examine some of the requirements for such correlations in relation to major prehistoric fluvial aggradations in valleys draining the Cheviot Hills (Fig. 11.1).

The present work developed from palaeoenvironmental reconstructions within the north-draining upper Bowmont Water (Fig. 11.1), involving analysis of several pollen diagrams (below) and geomorphological mapping, in conjunction with a detailed archaeological survey. Attempts to reconstruct the fluvial history of the valley were severely hampered by the lack of dateable sections. The best record of accelerated fluvial activity derives from a lacustrine sequence at Yetholm Loch (Fig. 11.1; below), where a very coarse inwash band was dated to around 2000 cal BP. The scale of the event at Yetholm Loch indicated that any fluvial aggradation would be of considerable topographic significance. To investigate whether this event could be recognized and dated within fluvial aggradations, other valleys were visited. Clapperton (1967) mapped coarse gravel aggradations in two east-draining valleys, the River Breamish and the Harthope Burn/Wooler Water systems (Fig. 11.1), and considered them to be of Devensian Lateglacial age. This interpretation was undermined in 1971 by his publication (Clapperton, Durno & Squires 1971) of pollen analyses from a peat beneath the gravel terrace fill within the Wooler Water catchment; these indicated a Mid to Late Flandrian age for the aggradation. No closer dating was possible from the pollen analyses, and this and the Breamish valley deposits were re-examined in order to obtain radiocarbon dating evidence.

This paper presents, in the first part, details of the radiocarbon ages of the coarse gravel aggradations in these valleys and one other, not mapped by Clapperton, the Halter Burn catchment (Fig. 11.1), in addition to the evidence at Yetholm Loch. As such, no attempt to reconstruct the complete fluvial history of any of these catchments has been made. In each catchment, the aggradations described are the largest and coarsest Flandrian valley fills. They are, therefore, in themselves exceptional, and clearly not typical of later sedimentary processes. In their considerable scale, however (below), they clearly represent major periods of instability in the upland Cheviot landscape. Their importance lies in their scale, which must imply a causal mechanism that should be readily discernible from pollen records, and the second part of the paper considers the possible causes, and discusses appropriate scales for correlating such proxy records with fluvial events. As such, the paper concentrates on evidence from one study area alone, since it is thought important to work from the particular to the general at this stage, and no attempt is made to

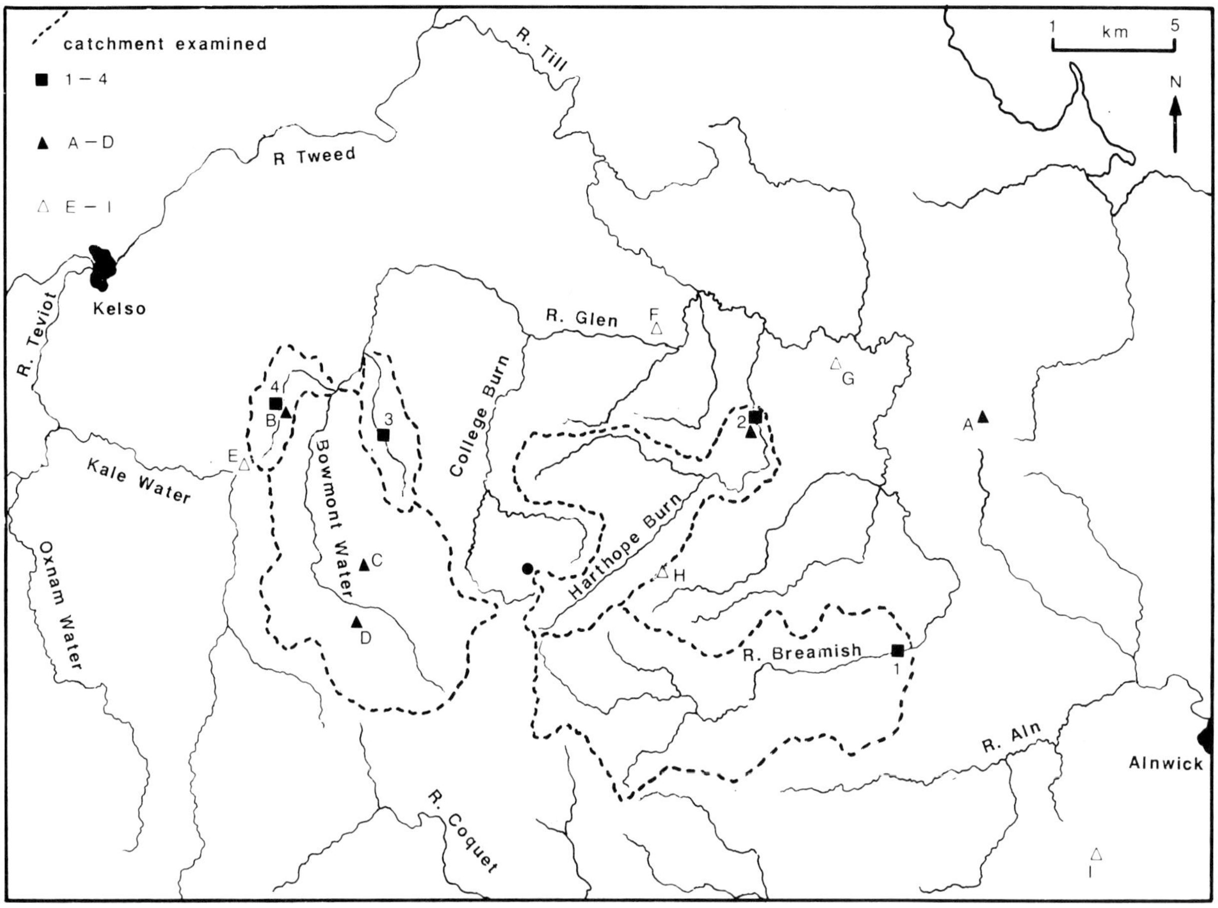

Fig. 11.1. Principal drainage of the Cheviot Hills (black dot = The Cheviot; 815 m OD), and the catchments and sites examined in the present study. Sites 1–4: fluvial sites described; 1 — River Breamish at Powburn; 2 — Harthope Burn/Wooler Water at Wooler; 3 — Halter Burn; 4 — Yetholm Loch. Sites A-D: radiocarbon dated pollen diagrams discussed; A — Camp Hill Moss (Davies & Turner 1979); B — Yetholm Loch ; C — Sourhope; D — Swindon Hill (all present study and Tipping forthcoming). Sites E-I: undated pollen diagrams; E — Linton Loch (Mannion 1978); F — Akeld Basin (Borek 1976); G — Trickley Wood (Turner 1968); H — Broad Moss (Davies & Turner 1979); I — Black Lough, Edlingham (Moyle 1980).

synthesize the relatively limited amount of work on Holocene upland fluvial activity (e.g. Brazier, Whittington & Ballantyne 1988; Brazier & Ballantyne 1989; Harvey *et al* 1981; Harvey & Renwick 1987; Macklin & Lewin 1986; Macklin *et al* 1991; Richards *et al* 1987; Robertson-Rintoul 1986).

THE STUDY SITES

The Breamish valley at Powburn

The River Breamish drains the east slopes of The Cheviot, the catchment rising to ca. 550 m OD. There is one prominent river terrace above the present-day floodplain between Ingram (NU 010163), ca. 16 km from the headwaters, and Powburn Bridge (NU 058170), a further 5 km downstream, up to 1.5 km wide. Above Ingram the deeply incised River Breamish leaves little room for terrace development. Below Powburn Bridge this terrace surface is eroded by a complex sequence of younger terraces, as yet undated.

A working gravel quarry at NU 047171, around 20 km from the headwaters, at ca. 90 m OD, exposed sections in the highest and oldest terrace fill, the surface lying ca. 4.3 m above the present river. The fill comprises in excess of 6 m (proved in boreholes) of coarse, probably braided-river, gravels, probably principally reworked from fluvioglacial deposits immediately upstream (Clapperton 1967).

The quarry has been excavated to the contact of the gravel with the underlying till. Fig. 11.2 shows schematically a plan of the locations, and the generalized stratigraphy beneath and within the gravel fill, of four radiocarbon-dated organic deposits (see also Table 11.1: radiocarbon determinations are expressed as radiocarbon years BP in Table 11.1, but as calibrated years BP in the text). Date SRR-3660 (sample 2: a — 2743 cal BP; b — 2356 cal BP) is from a peat-bed several metres in spatial extent, 3.40 m below the terrace surface, at or very close to the base of the gravel fill. The laterally extensive peat suggests the valley floor to have been marshy; Fig. 11.3 is a composite pollen diagram constructed from analyses of the organic sediments, and shows the valley floor here to have carried a damp woodland of alder (*Alnus*), birch (*Betula*) and possibly bog myrtle (Coryloid). The *in situ*

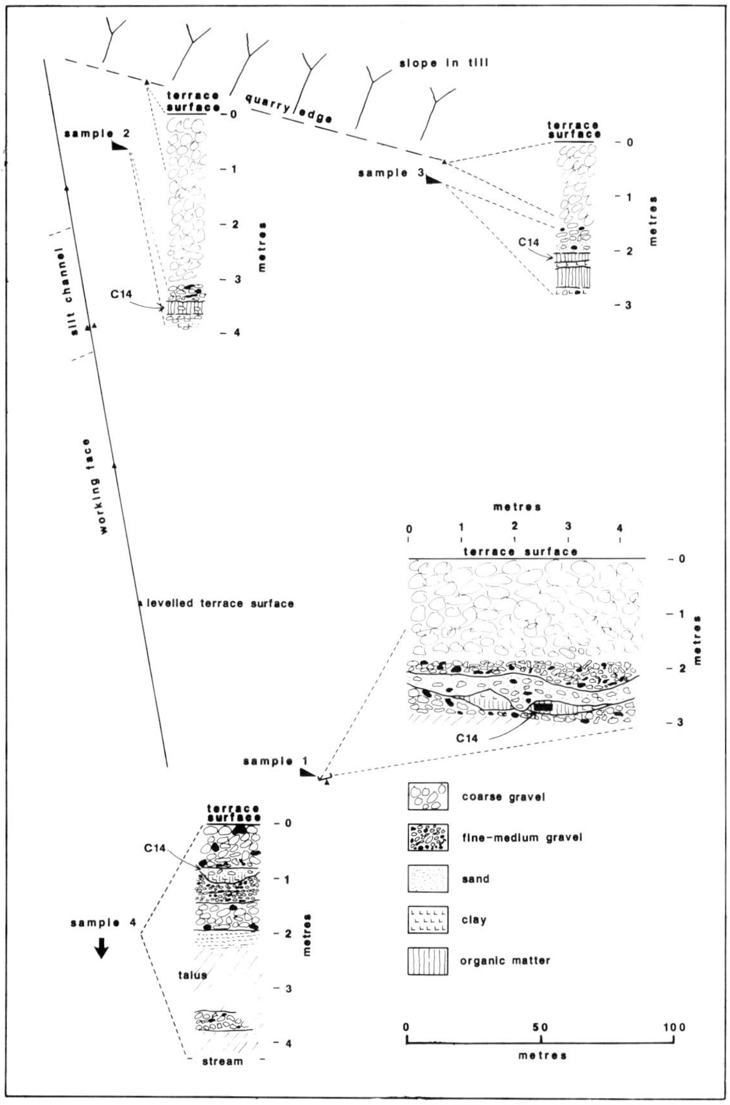

Fig. 11.2. Schematic plan of the western end of Powburn Quarry as at November 1989, with locations of samples 1–3 (Dates 1–3; Table 11.1); sample 4 is ca. 200 m south of sample 1. The sediment logs at each sampling point (with depths below terrace surface) are also depicted: dashed lines and detailed stratigraphies depict sediments preserved during quarrying; all sites are overlain by coarse gravels.

Table 11.1. *Radiocarbon dates relating to fluvial aggradations in the catchments examined.*

Site	Lab. no.	Fraction dated	Radiocarbon age (plus 1.0 σ)	δ13C content
Powburn	SRR-3660 (a)	'fine' organics	2565 ± 45	-29.6%
	SRR-3660 (b)	'coarse' organics	2390 ± 40	-30.1%
	SRR-3661 (a)	'coarse' rootlets	1800 ± 50	-27.0%
	SRR-3661 (b)	'humic 'carbon	2275 ± 40	-29.3%
	SRR-3661 (c)	'humin' carbon	1995 ± 50	-29.3%
	SRR-3659 (a)	'humic' carbon	1890 ± 45	-29.5%
	SRR-3659 (b)	'humin' carbon	1850 ± 45	-29.2%
	SRR-3662	'combined' sample	10025 ± 55	-30.4%
Wooler	SRR-3658	'coarse' rootlets	3695 ± 45	-28.7%
Halter Burn	SRR-3664	'coarse' roots	3940 ± 45	-30.5%

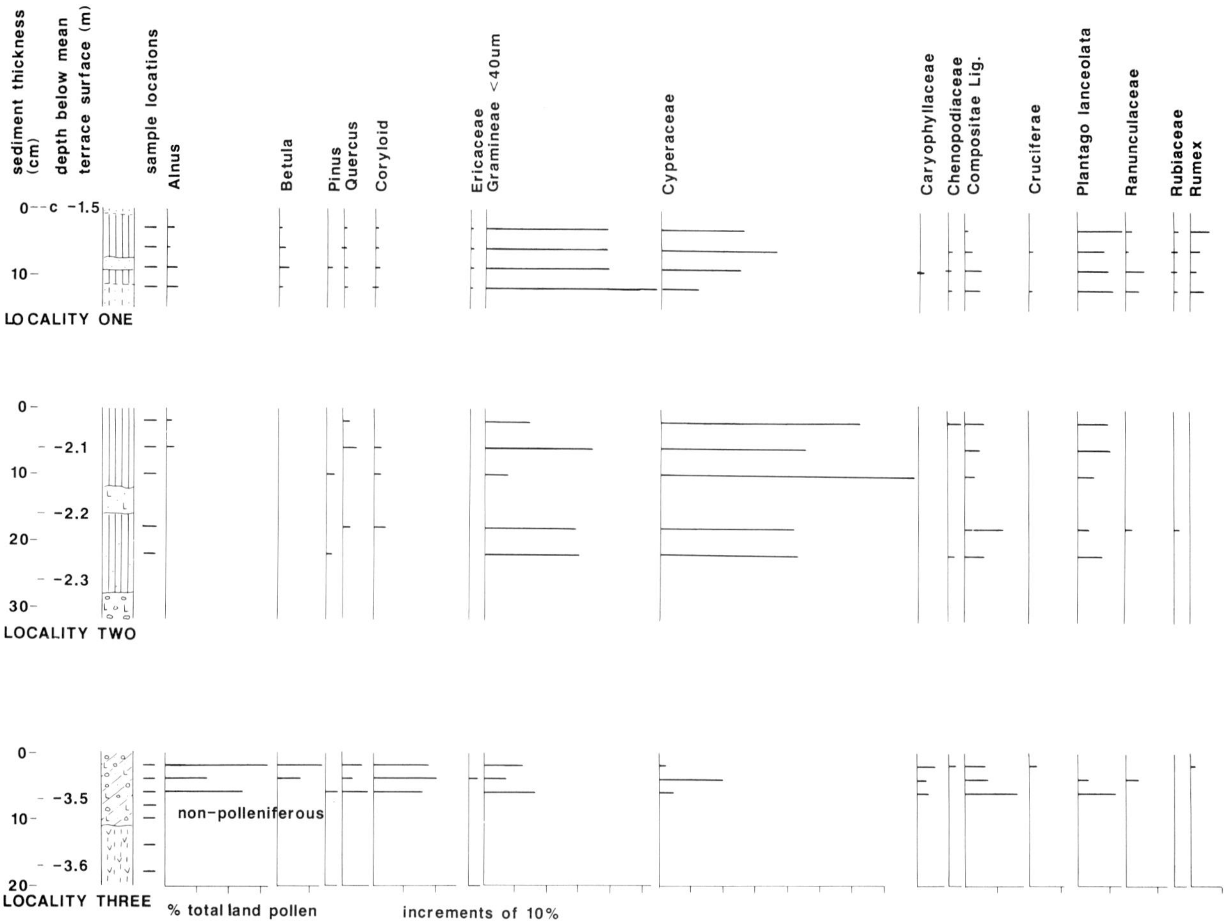

Fig. 11.3. Selected land pollen taxa from polleniferous spectra associated with the organic deposits of samples 1–3, Powburn Quarry, in stratigraphic order according to depth below the terrace surface.

peat is thin, only 24 cm, and the whole thickness was sampled for assay.

Date SRR-3661 (sample 3: a – 1723 cal BP; b – 2334 cal BP; c – 1946 cal BP) also lies close to the base of the gravel fill, against the till-covered hillslope at the quarry edge, 2.04 m below the terrace surface. A peat-bed of at least 3.5 m lateral extent is interbedded with thin fine sand stringers and a thicker (9–10 cm) silty clay; the overlying coarse gravel appears to lie conformably on the peat. The peat represents the *in situ* accumulation of organic deposits near the valley floor, but affected by occasional mineral washes, and very close in time to the initiation of the gravel fill.

Date SRR-3659 (sample 1: a – 1837 cal BP; b – 1816 cal BP) is a thin peat developed in still-water conditions at the base of an abandoned channel, 2.86 m below the terrace surface and within the sequence of aggrading coarse clast-supported gravels. Date SRR-3662 (sample 4: Table 11.1) is from a similar slack-water deposit within an abandoned channel, sandwiched between coarse to very coarse clast-supported gravels and cobbles, 0.80–1.10 m below the terrace surface. The deposit is only poorly organic.

SRR-3662 is clearly in error, and the poorly organic silts and clays may have contained non-contemporaneous or infinite-aged carbon. The good agreement between different fractions in the remaining dates indicates that they are securely dated and interpretable. SRR-3660 is stratigraphically the oldest, and differences between size fractions are insignificant at 2 σ. Of the three fractions dated for SRR-3661, there is no significant difference between the 'coarse' rootlets and 'humin' fractions, although the 'humic' fraction is significantly older. The rootlets are considered to provide the best estimate of age, since they are the least mobile fraction and, being capped by ca. 2 m of gravels, cannot be intrusive. Fluvial gravels covered this deposit at some time after ca. 1700 cal BP, but undoubtedly very soon after this date, since both fractions of SRR-3659 closely agree in providing an age for organic deposits within the aggradation of around 1800 cal BP.

The Wooler Water at Wooler

The headwaters of the Harthope Burn lie close under The Cheviot at ca. 800 m OD, and the river runs in a fault-controlled valley before debouching onto the lowland plain. The reach examined lies immediately south of Wooler, at ca. 70 m OD, some 15 km from the highest point of the catchment. The reach is mapped as Fig. 1 of Clapperton *et al* (1971), and is not reproduced here. Their 'gravel plain' is the highest preserved terrace surface of the Wooler Water, ca. 8 km in length, and a maximal 1.5 km in width. Clapperton (1967) and Clapperton *et al* (1971) described a thick peat-bed beneath up to 3.50 m of very coarse matrix-supported and imbricated gravels and cobbles, interpreted as an infilled kettle-hole overlain by fluvial deposits. The gravels again have their origin in reworked fluvioglacial deposits immediately upstream of the reach. Pollen analyses showed the top of the peat to be post-elm decline in age (post-ca. 5000 BP; Clapperton *et al* 1971).

The locality worked on by Clapperton was re-visited, but landscaping has removed much of the gravel fill. The peat was located at NT 99602730, an *in situ* woody and coarsely fibrous bed at least 45 cm thick, overlain by ca. 1.25 m of well-sorted structureless and poorly developed plane-bedded medium sand. The topmost ca. 10 cm of peat was excavated, sieved to concentrate the coarser, woody fragments, and sent for assay. Date SRR-3658 (Table 11.1; 4081 to 3995 cal BP (three intercepts)) indicates the age of the uppermost peat to be around 4000 cal BP. Such an age does not disagree with the pollen 'dating' of Clapperton *et al* (1971). There is no evidence at this locality, or in the interpretations of these earlier workers, for an erosional contact between the peat and overlying fluvial deposits.

The Halter Burn

This is a small stream flowing into the Bowmont valley north of Kirk Yetholm (Fig. 11.1) from a highest point at ca. 490 m OD. Three terrace surfaces have been mapped, of which the sampled left-bank section, at NT 84202685 (ca. 170 m OD, only 3 km from the headwaters), is in the highest and most extensive terrace fill, traceable along the lowermost 2–3 km of the valley, but less than 50 m wide. This section showed a complex stratigraphy of channel gravels and finer abandoned channel and overbank deposits overlying a well-developed and strongly gleyed soil profile derived from the underlying till. Within this soil were a large number of thin fibrous root remains; the similarity of the roots, their limited extent and positioning relative to one another strongly suggest the roots belonged to a single tree or shrub (species unidentified). The roots were truncated, probably by the overlying fluvial gravel, and preserved through rapid burial. They are thus *in situ* and closely related in age to the onset of deposition of the very coarse gravel and cobbles of the basal fluvial unit. Date SRR-3664 (Table 11.1; 4415 cal BP) gives the age of this gravel aggradation as close to ca. 4500 cal BP.

Yetholm Loch

This sequence is different from the other localities, in that the evidence derives not from mappable terrace fills but from within a continuously accumulating 3.10 m deep lacustrine sediment sequence at the southern, infill, end of a small loch within a very small catchment (Fig. 11.1), lying at about 100 m OD, and drained by a single small unnamed stream. A thin spread of alluvium, ca. 25 m wide, covers the valley floor, but is probably quite recent.

Fig. 11.4 depicts the sediment stratigraphy and loss-on-ignition values of contiguous 1.0 cm slices of the core, together with major changes in terrestrial pollen frequencies over the period from ca. 6000 BP to ca. 1000 BP (the pollen diagram is supported by eight radiocarbon dates); the elm decline at this site, at 165 cm (170.0–161.0 cm), has a radiocarbon age of 4690 ± 110 BP (GU-2513: 5450 to 5331 cal BP (three intercepts)), and a 6.0 cm slice centred on 122.0 cm has a radiocarbon age of 1750 ± 60 BP (GU-2514: 1697 to 1635 cal BP (three intercepts)). The palynological changes are discussed below, but attention is drawn to the sediment stratigraphy and loss-on-ignition values, which together indicate a phase, between ca. 138.0 and 130.0 cm, of coarse sediment inwashing. The sediments become too inorganic to date, and coarsen to predominantly sandy silt, and eventually to medium-coarse sand. This massive inwash band is not of lacustrine origin, but represents very substantial erosion within the catchment, its products probably carried to the lake by a stream choked with sediment. The time of the onset and cessation of coarse sediment inwashing is difficult to define, through the inability of radiocarbon to date critical horizons, the resolution of the available dates and the problems of interpolating between dates GU-2513 and -2514, where sedimentation rate changes probably make assumptions of linearity in interpolation unreliable. Date GU-2514, on sediment above the inwash band can only be a *terminus ante quem*, given the sediment thickness sampled, but an age of ca. 1700 cal BP can reliably be assumed for cessation of inwashing from linear interpolation of radiocarbon dates above and through GU-2514. The commencement of inwashing is more problematical, but similar interpolation results in a date of ca. 2200 cal BP.

Summary

Fig. 11.5 summarizes the radiocarbon dating evidence for the sites described above. Care has been taken to describe the context of the dates, since security of context needs to be demonstrated before correlations with other proxy records can be made.

It is unknown whether the data represent a complete record of prehistoric fluvial events, and at each locality there are limits to the dating evidence. Partly this

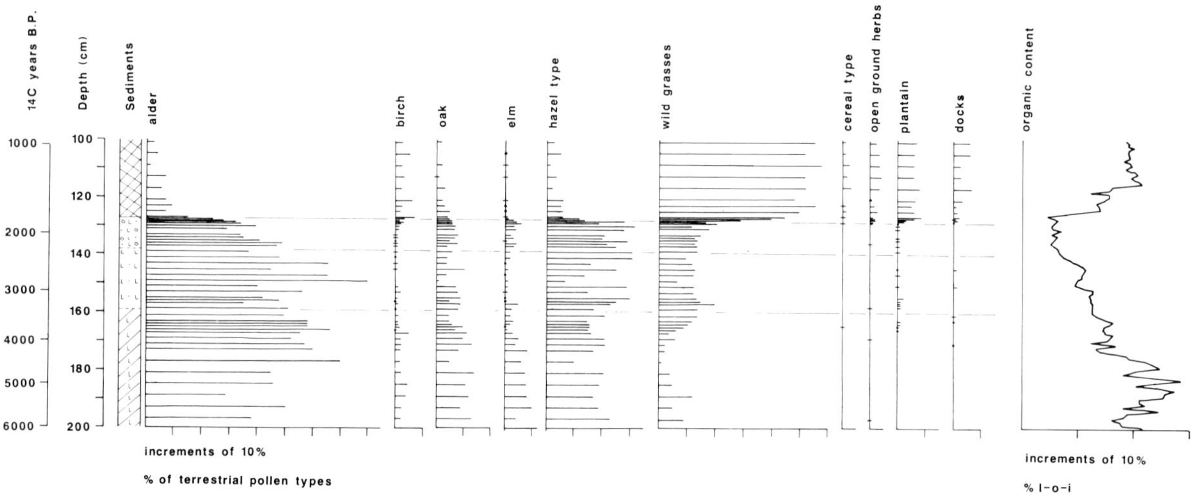

Fig. 11.4. Radiocarbon-derived chronology, sediment stratigraphy, organic content and selected land pollen taxa from the lacustrine sediments of Yetholm Loch, showing changes over the period ca. 6000–ca. 1000 BP.

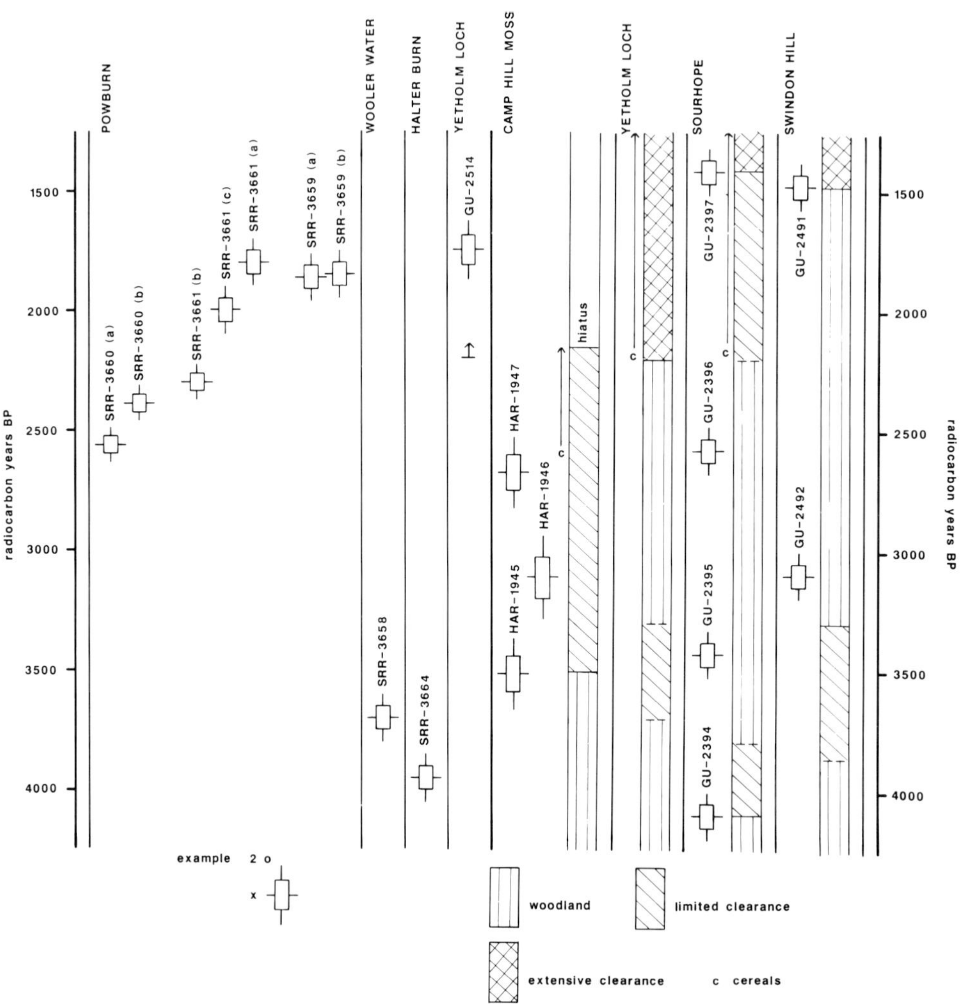

Fig. 11.5. Radiocarbon dates from the fluvial records described, and radiocarbon dates and major vegetation changes between ca. 4000 and 1500 BP in the dated pollen diagrams.

derives from the research design, in examining only particular reaches, and partly from the inability in such poorly exposed and dominantly inorganic sediments to date events; these are major difficulties in working on high-energy upland catchments. At Powburn the basal organic deposit, though on or close to till, lies away from the thalweg, and a further 2 m of gravel, not exposed, is known to exist below the thickest extent of exposed gravel, at sample 4. This basal gravel need not represent the same stratigraphic unit dated to the late prehistoric. Even were this the case, pre-existing terrace fills are perhaps unlikely to have survived the fluvial activity that accompanied the deposition of the dated fill, and the record is almost certainly incomplete.

This is less of a problem at the Halter Burn, where the date clearly pertains to the initial Flandrian aggradation, but here the clear evidence of later stages in valley-floor evolution cannot be dated, and at the Wooler Water, where the gravel, sedimentologically one unit, rests upon a peat-bed representing continuous accumulation since Devensian Lateglacial times, the ages of terrace fills later than ca. 4000 cal BP are unknown. There is now a need to investigate the ages of later valley fills in the Wooler Water and Till valleys. Only in the lacustrine sedimentary record at Yetholm Loch can a complete record be found.

Notwithstanding this, the dated fills are probably the most significant prehistoric fluvial stages, and this is demonstrable at Yetholm Loch. Fig. 11.5 suggests that there are two major phases of prehistoric fluvial aggradation affecting the catchments studied. At Halterburn the initiation of Flandrian fluvial activity occurred at ca. 4400 cal BP. In the Wooler Water catchment the earliest event is dated to ca. 4000 cal BP. The dates are sufficiently comparable to assume the event to be synchronous within these two catchments, but neither date provides a *terminus ante quem* for the event, and this remains unknown. At Yetholm Loch there is little indication in the sedimentological record for increased stream flow at this time; although organic content declines from the elm decline onwards (below), there is no single phase of inwashing comparable to that recorded later in the stratigraphy.

At Yetholm Loch, this later phase ends at ca. 1850 cal BP, and probably commenced at around 2350 cal BP. No dateable evidence was found in the Halter Burn, nor in the Wooler Water, although this may not be significant (above). In the Breamish valley, several dates concur in indicating the major gravel fill there to post-date ca. 2600 cal BP and to have been forming at ca. 1850 cal BP. Again, there is good reason to assume synchroneity in these two catchments.

THE IDENTIFICATION OF CAUSE

The only available alternative palaeoenvironmental proxy records in the region to correlate with the fluvial stratigraphy are pollen analyses. Fig. 11.1 locates the considerable number of pollen records covering the last ca. 5000 years that exist in and around the Cheviot Hills, within ca. 20 km of The Cheviot. However, the majority of sites (Sites E-I) have no radiocarbon dates, and cannot be considered further. This is unfortunate, since Site H (Broad Moss; Davies & Turner 1979) lies on the watershed between the River Breamish and Harthope Burn/Wooler Water catchments. Davies & Turner (1979) regarded the clearance phases at this site as synchronous with their dated sites at Camp Hill Moss (Site A) and as too far south to relate directly to the Cheviot Hills, Fellend and Steng Mosses. This correlation is considered below, but until the sequence at Broad Moss is dated little can be said with certainty.

Prior to the recent Bowmont valley survey only the diagram from Camp Hill Moss could be referred to, but this site lies on the Carboniferous Fell Sandstone, at ca. 205 m OD, and changes need not have been comparable to those on the granitic and andesitic Cheviot. This site can now be placed in context by comparison with three radiocarbon-dated but as yet unpublished diagrams from the central and northern Cheviot; Yetholm Loch (Site B; ca. 100 m OD), Sourhope (Site C; ca. 270 m OD) and Swindon Hill (Site D; 365 m OD)(Tipping forthcoming). In addition, the now radiocarbon-dated Wooler Water peat (cf. Clapperton *et al* 1971), together with the limited analyses from the Powburn section (above), can be considered. The chronology at Camp Hill Moss has been substantially revised, and is explained in the Appendix.

There is now a heightened awareness of the spatial resolution of different types of pollen site (cf. Jacobson & Bradshaw 1981). Sites represent vegetation changes at different spatial scales, such that in correlating geomorphic events with other records, there has in the past perhaps been too little concern with evidence specific to that locality, and in the case of fluvial changes, the catchment. In addition, the larger the catchment, and the greater the topographic diversity, the more need there is to fully summarize the probable diversity of past vegetation and land use.

The pollen sites considered in Fig. 11.5 are, with the exception of Yetholm Loch, relatively small soligenous mires, with basin diameters of ca. 100 to a few hundred metres. As a consequence, the area of the landscape which is represented in pollen spectra (the pollen recruitment area) is probably restricted, between perhaps a few hundred metres to one or two kilometres around each site. This is emphasized in the pollen records for Sourhope and Swindon Hill (below); the sites are 2 km apart, yet the higher-altitude site appears not to reflect in its entirety the prehistoric land-use record found at Sourhope, notably the period of 'mixed' farming at ca. 2200 cal BP. A valid criticism is that only one site,

Yetholm Loch, lies within the catchment from which fluvial records have been described, although that from the Wooler Water is, in its uppermost spectra (Clapperton *et al* 1971) relevant to the emplacement of the overlying gravel. The pollen spectra from beneath and within the gravel fill at Powburn are clearly relevant, but are limited in their interpretative value by their disjointed nature. Nevertheless, the remaining diagrams all lie within 20 km of all the fluvial sites, and with the possible exception of Camp Hill Moss, reflect vegetation changes relevant to the headwaters of all the catchments. In addition, given the relatively intensive cover of pollen sites, a great deal can be made of synchronous palynological events in correlations with the fluvial record.

The interpretations of the palynological changes can be summarized in diagrammatic form as in Fig. 11.5; interpretations of periods of predominantly undisturbed vegetation and anthropogenic disturbance are presented, with the intensity of disturbance (limited or large-scale) indicated. The subtler interpretations of land use cannot be considered here.

Only the diagram of selected taxa at Yetholm Loch (Fig. 11.4) is considered in detail, because it not only provides an excellent record of both changes in vegetation and erosion, but also because this combination of features necessitates a complex interpretation. Between 170.0 (the elm decline) and 145.0 cm the organic content (loss-on-ignition) declines in a step-wise pattern as river-borne silty clays become more prominent, a change associated with limited woodland clearance and the introduction of agriculture, albeit on an apparently modest scale, with resultant soil instability in the catchment. Between the elm decline and the onset of highly minerogenic sedimentation at ca. 140.0 cm the evidence for clearance is not conclusive, possibly due to the limited scale of the activity, but tentatively recognized is a fluctuation in dryland tree percentages, although with little response in the herbaceous element, between ca. 151.0 and 141.0 cm. There are difficulties in interpolating ages between the radiocarbon dates GU-2513 and -2514 (above), but approximate radiocarbon ages for the beginning and end of this clearance phase are ca. 4000 cal BP and ca. 3500 cal BP respectively.

Within the phase of coarse sediment inwashing (138.0–130.0 cm), the pollen diagram would appear to indicate that this accelerated erosion/sedimentation was unrelated to woodland clearance, since dramatic evidence for such anthropogenic activity is not seen until ca. 130.0 cm, when alder, elm and hazel/bog myrtle percentages are sharply reduced, and oak values decline less steeply, wild grass, plantain, dock, cereal and arable weed percentages all increase abruptly as the catchment was for the first time intensively farmed in an essentially treeless landscape. However, the interpretation of the diagram is not straightforward. Patterns of pollen deterioration in the major taxa, particularly by corrosion, a deterioration most closely associated with decay in aerobic environments such as soils (Havinga 1984), indicate that a large proportion of the pollen in the inwash band is derived from catchment soils, and is very likely to have been reworked, rather than being contemporaneous. The effects of reworked organic matter on radiocarbon dating in the later Flandrian have for long been recognized, and similar effects can in rare circumstances be seen in the pollen record; Hirons (1988) considered that up to 50% of grains of certain pollen types were non-contemporaneous at times of increased stream flow into one lake in Northern Ireland. The 'complacent' pollen record at Yetholm Loch, at a time of clear geomorphic disturbance, is similarly interpreted; prior to this period, the dominant vegetation types in the catchment were of wet and dry woodland, and the clearance phase is not recognized until around 130.0 cm because much of the pollen introduced with the coarse sediment inwashing is recycled, predominantly arboreal, pollen. The rapid palynological changes at 130.0 cm, where the devastating effect of this clearance activity is made clear, are regarded as a response to increased soil stability, waning coarse sediment supply and reduced reworking of pollen, which allow the true character of the vegetation to be seen.

The pollen diagram from Camp Hill Moss commences after, but probably only shortly after, the elm decline, and until 3510 ± 70 BP (HAR-1945: 3829 to 3778 cal BP (three intercepts)), a woodland cover existed around the site. Above ca. 3800 BP, limited clearance took place, initially for pasture alone, but later than 2670 ± 70 BP (HAR-1947: 2772 cal BP) cereals are present. Davies & Turner (1979) considered these to have been discrete episodes of activity, between ca. 3800 and 3350 cal BP, and later than ca. 2770 cal BP.

The Sourhope pollen diagram is currently only provisionally interpreted, but contains evidence of Early Neolithic pastoral activity, and after a break, intensified clearance for pasture between 4100 ± 50 BP (GU-2394: 4804 to 4568 cal BP (five intercepts)) and ca. 4200 cal BP. Sporadic cereal grains are recorded from around this time, but occur consistently only above ca. 2250 cal BP, with pastoral indicator herbs again well represented. This clearance phase was sustained until 1450 ± 70 BP (GU-2398: 1341 cal BP), when farming appears to have become significantly more intensive/extensive. There is some, though limited, evidence at Swindon Hill for woodland disturbance prior to 4500 cal BP, close to ca. 5350 cal BP, with cereal cultivation, but the first substantial clearance occurred at ca. 4300 cal BP, ending slightly before a horizon dated by assay GU-2492 to 3100 ± 50 BP (3355 cal BP). This appears to have been for pasture only; given the slow decline of dry-land tree pollen and the limited representation of microscopic charcoal, this may have been due to grazing pressure. In contrast to other sites in the region, there are very few indications of human presence on this upland plateau between ca. 3500 cal BP and ca. 1350 cal BP.

To these records can be added those at Wooler (Squires in Clapperton *et al* 1971; their fig. 3) and at Powburn (Fig. 11.3). Both these, though more limited in their applicability, show indications of woodland clearance immediately prior to the onset of gravel deposition. The Powburn sequence is particularly distinctive in this, showing above the basal sample (sample 2; above) the almost complete absence of wetland tree pollen (alder, birch possibly bog myrtle; above), and reductions in dry-land tree pollen (*Quercus* (oak)), associated with a rise in wild grasses (Gramineae <40µm) and some clearance indicators (particularly *Rumex* (dock family)). Given the depositional environment of these localities, many of these changes may be very local, for example, the removal of alder-birch woodland on the valley floor by gravel aggradation, so that the vegetation changes may be the result, and not the cause, of the geomorphic event. In addition, percentage distortions due to locally produced Cyperaceae pollen complicates interpretation. Nevertheless, there is considerable evidence to suggest that woodland clearance was contemporary with gravel aggradation.

DISCUSSION

To draw these threads together, there is thought to be good evidence for regarding the first significant human impact on the Cheviots to have been broadly synchronous, at around 4650–4200 cal BP. This phase is recognized at all the radiocarbon dated pollen sites (Fig. 11.5b) and at the Wooler Water. It is considered that this phase occurred throughout the study area. Indeed, it is very possible that this phase was synchronous across a much larger region, since the first clearance phases recognized at Fellend and Steng Mosses (Davies & Turner 1979) are dated to ca. 4300 cal BP and ca. 4400 cal BP.

Space is too limited to discuss other lines of evidence in detail, but the most recent synthesis of archaeological data pertaining directly to the Cheviots (Burgess 1984) indicates that this short time-period is of critical importance in human settlement patterns also. Prior to ca. 4500 cal BP settlement was in the lowlands. Between ca. 4500–3900 cal BP the population appears to have shifted to the higher ground of the Cheviots, and a period of "intensive and upland settlement", with areas "cleared and tilled up to 400 m" OD (Burgess 1984: 143) persisted until perhaps 3200 cal BP. The woodland clearance in the pollen diagrams is best seen as a reflection of this activity and suggests widespread human activity. Contemporaneous with this phase were the fluvial aggradations at the Wooler Water and Halter Burn (Fig. 11.5a). Contemporaneity need not be causally significant, but that is the assumption here.

The direct association of woodland clearance with accelerated fluvial activity is seen at Yetholm Loch in the deposition of the coarse sediment inwash band, between ca. 2250 and ca. 1850 cal BP. This clearance activity is near-total in the pollen recruitment area of this small loch, an area probably little larger than its catchment. In the uplands of the Bowmont valley this phase is clearly less pronounced, being recognized at Sourhope as one of limited intensity, and not having been discerned at all at Swindon Hill; but Turner (1979) has shown that the first intensive clearances in northern England were of dates comparable to that at Yetholm Loch. Following this clearance, agricultural activities were clearly at a higher level of intensity. Contemporaneous with this clearance activity is the massive coarse gravel aggradation in the middle reaches of the River Breamish (Fig. 11.5a) and, as at Yetholm Loch, clearance is detected in pollen counts from that catchment also (Fig. 11.3). Again, a causal relation is assumed.

It is argued that the two phases of prehistoric geomorphic activity recognized in the present study are causally related to anthropogenic activity, particularly woodland clearance for farming. Climatic controls have not been considered causal because there is very little evidence for climatic perturbations coinciding with the periods of geomorphic activity identified, either at the global scale (Lamb 1977; Harding 1982) or at the local level. The Sourhope pollen diagram provides evidence for increased precipitation dated to 2570 ± 50 BP (GU-2396: 2744 cal BP), but there is little evidence in the vegetation record for Bronze-Iron Age climatic deterioration ca. 3200–3000 cal BP (Burgess 1985, 1990; Higham 1986), probably the most significant climatic downturn in the period. Although terrace fills of this date and cause may exist, they are clearly not of geomorphic significance for the fills dated here, and at Yetholm Loch there is nothing to indicate changing catchment conditions at this time.

If we can develop the arguments beyond the basic climate-anthropogenic change models, there is the potential to explore these major landscape-forming events in more detail. The recognition that pollen sites with restricted pollen recruitment areas within one relatively small valley differ in the clarity with which clearance episodes are depicted, allows the possibility of reconstructing the loci of greatest anthropogenic activity. It is also possible to approach more subtle questions relating to landscape sensitivity and the mechanisms between clearance and accelerated geomorphic activity, such as the intensity of clearance needed to trigger major events (only at Yetholm Loch can this successfully be gauged, where the Late Iron Age clearance was due to near-total clearance of the woodland), the spatial extent of the clearance, the location of clearance within a catchment, the patterns of events prior to landscape instability, and the processes involved. These questions have not yet effectively been tackled and, apart from establishing their antiquity and suggesting by correlation their probable cause, we know little about the scales and the dynamics of these major changes in the landscape.

Acknowledgements

Grateful thanks go to the following for assistance with fieldwork during this and the Bowmont valley study; Geraint Coles, Nyree Finlay, Bill Finlayson, Peter McKeague, Carl Phillips, Jane Webster and undergraduates from the Department of Archaeology, Edinburgh University; to Gordon Cook (Scottish Universities Research and Reactor Centre) and Douglas Harkness (NERC Reactor Centre) and their support staff for providing radiocarbon dating facilities; to Chalmers Clapperton and Judith Turner for helpful correspondence; to Roger Mercer for his invitation to work in the Cheviot Hills, and to Noel Fojut and Patrick Ashmore of Historic Buildings and Monuments, Scotland for funding of the project. Grateful thanks go to Northern Aggregates Ltd (RMC Group) for access to Powburn Quarry.

References

Anderson, M.L. 1967. *A History of Scottish Forestry*. Edinburgh: Thomas Nelson.

Borek, M.J.E. 1976. Pollen Analysis and Vegetational History of the Akeld Basin. Durham: Unpublished M.Sc. Thesis.

Brazier, V. and Ballantyne, C.K. 1989. Late Holocene cone evolution in Glen Feshie, Western Cairngorm Mountains, Scotland. *Transactions of the Royal Society of Edinburgh, Earth Sciences*, 80, 17–24.

Brazier, V., Whittington, G. and Ballantyne, C.K. 1988. Holocene debris cone evolution in Glen Etive, Western Grampian Highlands, Scotland. *Earth Surface Processes and Landforms*, 13, 525–531.

Burgess, C. 1984. The Prehistoric Settlement of Northumberland: A Speculative Survey. In R. Miket and C. Burgess (eds.), *Between and Beyond The Walls: Essays on the Prehistory and History of North Britain in Honour of George Jobey*, 126–175. Edinburgh: John Donald.

Burgess, C. 1985. Population, Climate and Upland Settlement. In D. Spratt and C. Burgess (eds.), *Upland Settlement in Britain. The 2nd Millennium B.C. and after*, 195–219. Oxford: British Archaeological Reports, British Series, 143.

Clapperton, C.M. 1967. The Deglaciation of the East Cheviot area, Northumberland. Edinburgh: Unpublished Ph.D Thesis.

Clapperton, C.M., Durno, S.E. and Squires, R.H. 1971. Evidence for the Flandrian history of the Wooler Water, Northumberland, provided by pollen analysis. *Scottish Geographical Magazine*, 87, 14–20.

Davies, G. and Turner, J. 1979. Pollen Diagrams from Northumberland. *New Phytologist*, 82, 783–804.

Harding, A.F. (ed.) 1982. *Climatic Change in Later Prehistory*. Edinburgh: University Press.

Harvey, A.M., Oldfield, F., Baron, A.F. and Pearson, G.W. 1981. Dating of Post-glacial landforms in the Central Howgills. *Earth Surface Processes and Landforms*, 6, 401–412.

Harvey, A.M. and Renwick, W.H. 1987. Holocene alluvial fan and terrace formation in the Bowland Fells, Northwest England. *Earth Surface Processes and Landforms*, 12, 249–257.

Havinga, A.J. 1984. A 20–year experimental investigation into the differential corrosion susceptibility of pollen and spores in various soil types. *Pollen et Spores*, 26, 541–558.

Higham, N. 1986. The Northern Counties to AD 1000. London: Longman.

Hirons, K.R. 1988. Recruitment of cpr2 pollen to lake sediments: an example from Co. Tyrone, Northern Ireland. *Review of Palaeobotany and Palynology*, 54, 43–54.

Jacobson, G.L. Jr. and Bradshaw, R.H.W. 1981. The Selection of Sites for Palaeovegetational Studies. *Quaternary Research*, 16, 80–96.

Lamb, H.H. 1977. *Climate: Present, Past and Future. 2. Climatic History and the Future*. London: Methuen.

Macklin, M.G. and Lewin, J. 1986. Terraced fills of Pleistocene and Holocene age in the Rheidol Valley, Wales. *Journal of Quaternary Science*, 1, 21–34.

Macklin, M.G., Passmore, D.G., Stevenson, A.C., Cowley, D.C., Edwards, D.N. and O' Brien, C.F. 1991. Holocene alluviation and land-use change on Callaly Moor, Northumberland, England. *Journal of Quaternary Science*, 6, 225–232.

Moyle, D.W. 1980. Pollen Analysis of Peat Deposits near Edlingham, Northumberland. Durham: Unpublished M.Sc. Thesis.

Richards, K.S., Peters, N.R., Robertson-Rintoul, M.S.E. and Switsur, V.R. 1987. Recent valley sediments in the North York Moors: evidence and interpretation. In V. Gardiner (ed.), *International Geomorphology 1986*, Part 1, 969–883. London: Wiley.

Robertson-Rintoul, M.S.E. 1986. A quantitative soil-stratigraphic approach to the correlation and dating of post-glacial river terraces in Glen Feshie, western Cairngorms. *Earth Surface Processes and Landforms*, 11, 605–617.

Turner, C.H. 1968. Studies on a Small Post-glacial Peat Deposit in Northumberland. Durham: Unpublished M.Sc. Thesis.

Turner, J. 1979. The environment of north-east England during Roman times as shown by pollen analysis. *Journal of Archaeological Science*, 6, 285–290.

APPENDIX

Revision of the chronological framework at Camp Hill Moss (Davies and Turner 1979)

The pollen diagram is supported by five radiocarbon dates. Davies & Turner considered that errors had affected some of these dates, but this is disputed here. At ca. 30 cm, assay HAR-1949 gave a 'modern' age (0 ± 70 BP), which is thought unrealistic by the authors. However, at this depth there are increases in several taxa which are related here to post-Agricultural Improvement tree-planting, dated from documentary evidence (Anderson 1967) to 1775 + 25 AD (ca. 175 BP). In this interpretation, the peat at depth 30 cm is assigned an age of 175 + 25 BP. Davies and Turner cast doubts upon the radiocarbon age of the sample at ca. 53 cm, of 640 + 80 BP (HAR-1948), on the grounds that the peat accumulation rate inferred between that sample and that at ca. 80 cm (2670 + 70 BP; HAR-1947) was unacceptably slow, and that a date of around 1310 AD was unreasonable for the first major clearance. These arguments are accepted, but in the interpretation presented here, assay HAR-1948 is accepted as a valid age, separated from the lower samples by an hiatus immediately below the assayed peat at 54 cm, possibly caused by peat cutting in the 'high medieval' period. There is no apparent peat-stratigraphic change, and for many of the pollen taxa sharp changes in representation can be interpreted in terms of an intensifying period of anthropogenic activity, as Davies and Turner did. However, among the taxa local to the bog surface, those in their fig.14(c) plus *Potentilla*, the abrupt changes are perhaps less easy to explain in this way.

12

Climatic and cultural signals in Holocene alluvial sequences: the Tyne basin, northern England

Mark G. Macklin, David G. Passmore and Barbara T. Rumsby

Summary: *Investigations of British Holocene alluvial sequences have traditionally focused on assessing anthropogenic as opposed to climatic controls of river basin run-off and sediment yield. This problem has been addressed in the Tyne catchment, northern England, where Postglacial river sediments typical of those found in upland, piedmont and lowland rivers elsewhere in Britain can be studied in a single drainage basin. Until ca. 1000 cal BC valley floors in most of the Tyne were stable or slowly accreting. This was followed by major incision in the Middle and Lower Tyne between ca. 1350 cal BC and 550 cal AD, synchronous with climatic deterioration. Valley floor refilling after ca. 790 cal BC continued until ca. 680 cal AD and was partly coeval with forest clearance and expansion of agriculture, particularly in the Late Iron Age. Extensive valley floor erosion in the Tyne uplands did not begin until 550–680 cal AD. Erosion coincided with a shift to wetter conditions, although it may have been triggered by vegetation disturbance. Renewed valley floor alluviation ca. 1450 cal AD, and widespread channel degradation in the late eighteenth century, were associated with unusually high run-off and particularly severe floods. It is concluded that anthropogenic activity and climatic change need not be considered as competing hypotheses for explaining the timing and pattern of Holocene alluviation and river erosion. River response to environmental change should instead be viewed in terms of a continuum between these two factors, which, in particular circumstances and at certain times, can each control water and sediment yields and the alluvial record of a drainage basin.*

INTRODUCTION

Interpretation of Holocene river sequences by archaeologists and earth scientists in Britain, as elsewhere in Europe, has traditionally focused on evaluating anthropogenic as opposed to climatic factors that control catchment water and sediment yields, channel and floodplain morphology, and in turn the alluvial record of a drainage basin (Bell 1982; Burrin and Scaife 1988; Robinson and Lambrick 1984; Macklin and Lewin, in press). This important question needs to be addressed if we are to move from reconstruction to interpretation of Holocene river environments in general and, in particular, to model the response of human communities to changes in river regimes, as well as their impact (deliberate and inadvertent) on the fluvial system.

Climatic change has moved rapidly up the environmental research agenda, prompted primarily by growing concern over the effects of global warming (Houghton, Jenkins and Ephraums 1990). Alluvial valley floors are one of the most sensitive and susceptible environments in the British landscape to subtle shifts in hydroclimate and this has probably been the case throughout the Holocene. Current hydrological scenarios anticipate profound and widespread changes in patterns and rates of river sediment transport and floodplain deposition throughout much of upland and lowland Britain, together with significant disruptions of riparian ecosystems (Newson and Lewin 1991). However, to disentangle the respective roles and relative importance of human activity vis-à-vis climatic change it is necessary to assess and link a number of geomorphological, environmental and cultural elements which together provide historical insight to the development of a river basin. First, episodes of valley floor sedimentation and erosion need to be mapped and dated to determine their temporal and spatial pattern. Recent studies, both in Britain (Brown 1990) and in the U.S.A. (Patton and Boison 1986), have demonstrated that this is unlikely to emerge from investigations of single, discrete river reaches but requires evaluation at the scale of the river basin where the full range of Holocene depositional environments and fluvial stratigraphies are represented. Second, fluvial sediments should be provenanced, particularly when alluviation is believed to relate to catchment land-use change. Recognition of post-settlement alluvia associated with agriculturally-accelerated soil erosion (Brown and Barber 1985; Macklin and Lewin 1986; Shotton 1978) or mining activities (Macklin 1985) are especially important in this context. Third, independent proxy records of regional climatological/hydrological change with appropriate temporal resolution should be available. Finally, it is essential that prehistoric and historic catchment land-use histories are documented at least in general terms.

All four of these components have been studied, to a greater or lesser extent, in the Tyne basin, northern England, where since 1983 Holocene fluvial processes and stratigraphies have been investigated. The Tyne catchment is well suited to evaluate longer-term environmental controls of a river system, primarily because river dynamics and Postglacial alluvial sequences typical of those found in upland, piedmont and lowland rivers elsewhere in Britain can be studied in a single drainage basin.

GEOLOGY, PHYSIOGRAPHY AND ARCHAEOLOGY OF THE TYNE BASIN

The Tyne catchment (drainage area 2927 km^2, basin relief 893 m) is developed predominantly on Carboniferous sandstone, limestone and shales, with smaller outcrops of andesite in the headwaters of the North Tyne and dolerite along the lower part of the South

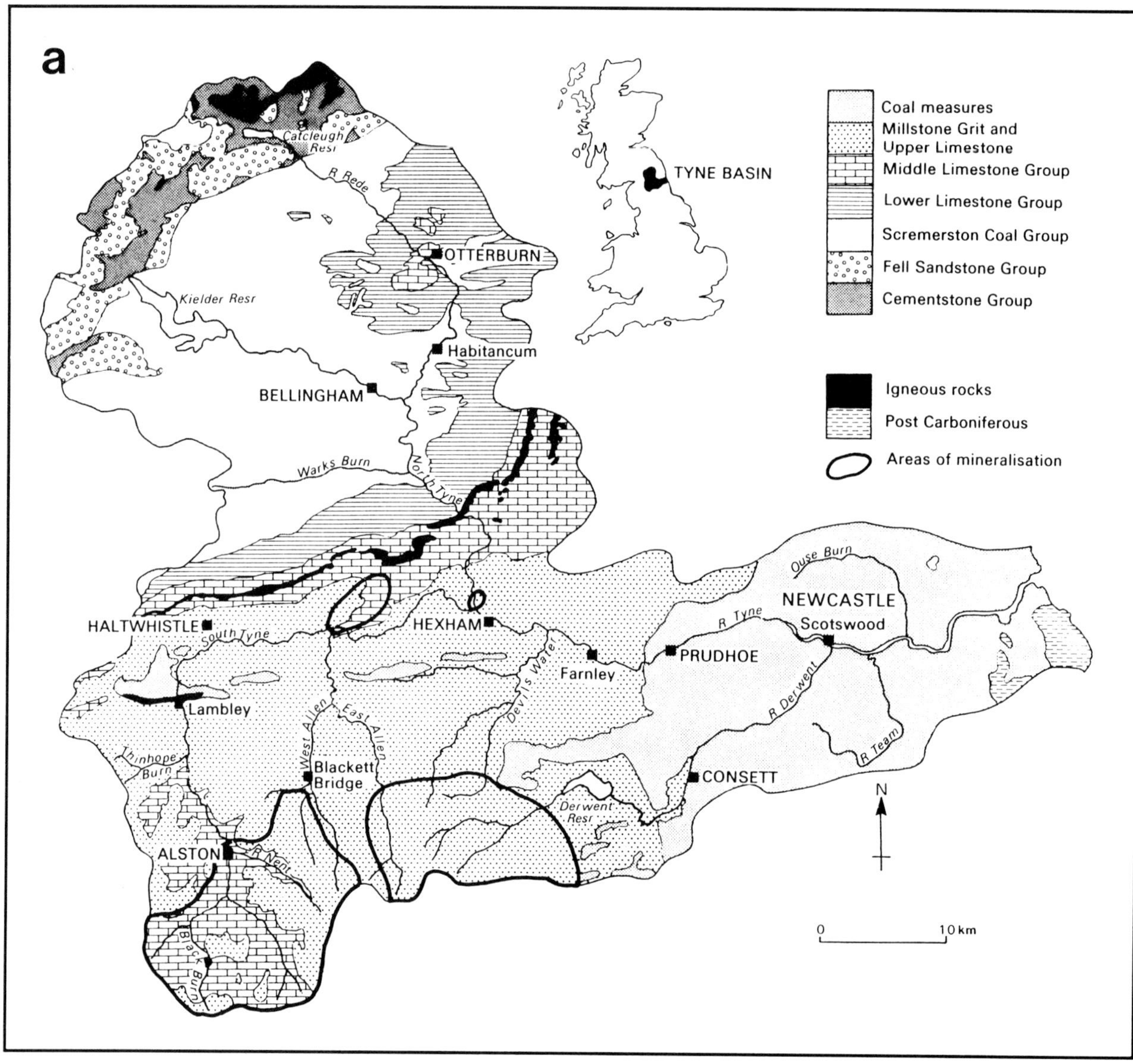

Fig. 12.1.(a) Geology of the Tyne basin.

Tyne valley (Fig. 12.1a). The geologies of the North and South Tyne catchments (the principal tributaries of the Tyne), however, differ in detail; the South Tyne and its upland tributaries drain substantial areas of limestone and the Northern Pennine Orefield, that was once the most productive lead and zinc mining area in Britain (Dunham 1990). Fine-grained sediment from these basins have distinctive chemical signatures that can be recognised in Holocene alluvium downstream in the Middle and Lower Tyne valley. Investigations of the dispersal of nineteenth century and early twentieth century mining waste in the region's rivers have been especially valuable in this context, enabling longer-term (Macklin, Rumsby and Newson 1992) and large-scale (Lewin and Macklin 1987; Macklin and Dowsett 1989) transport and storage of fine- grained alluvium to be examined. This has also been useful for interpreting the fluvial redistribution of eroded soil associated with prehistoric and later catchment disturbance.

The Tyne basin lies within the ice limits of the Late Devensian Dimlington Stadial, and deposits from this and earlier Pleistocene glaciations (extensively modified by solifluction processes) cover valley slopes and infill glacially-overdeepened valley floors. Present river channels are inset within Pleistocene glacial and glaciofluvial deposits, Holocene alluvium or bedrock. Holocene river sediments in the Tyne range from coarse gravels in the Northern Pennine uplands, deposited by laterally and vertically active wandering gravel-bed rivers (Macklin, in press), to fine-grained vertically accreted alluvium in the lower (Macklin, Rumsby and Newson 1992) and perimarine (Passmore

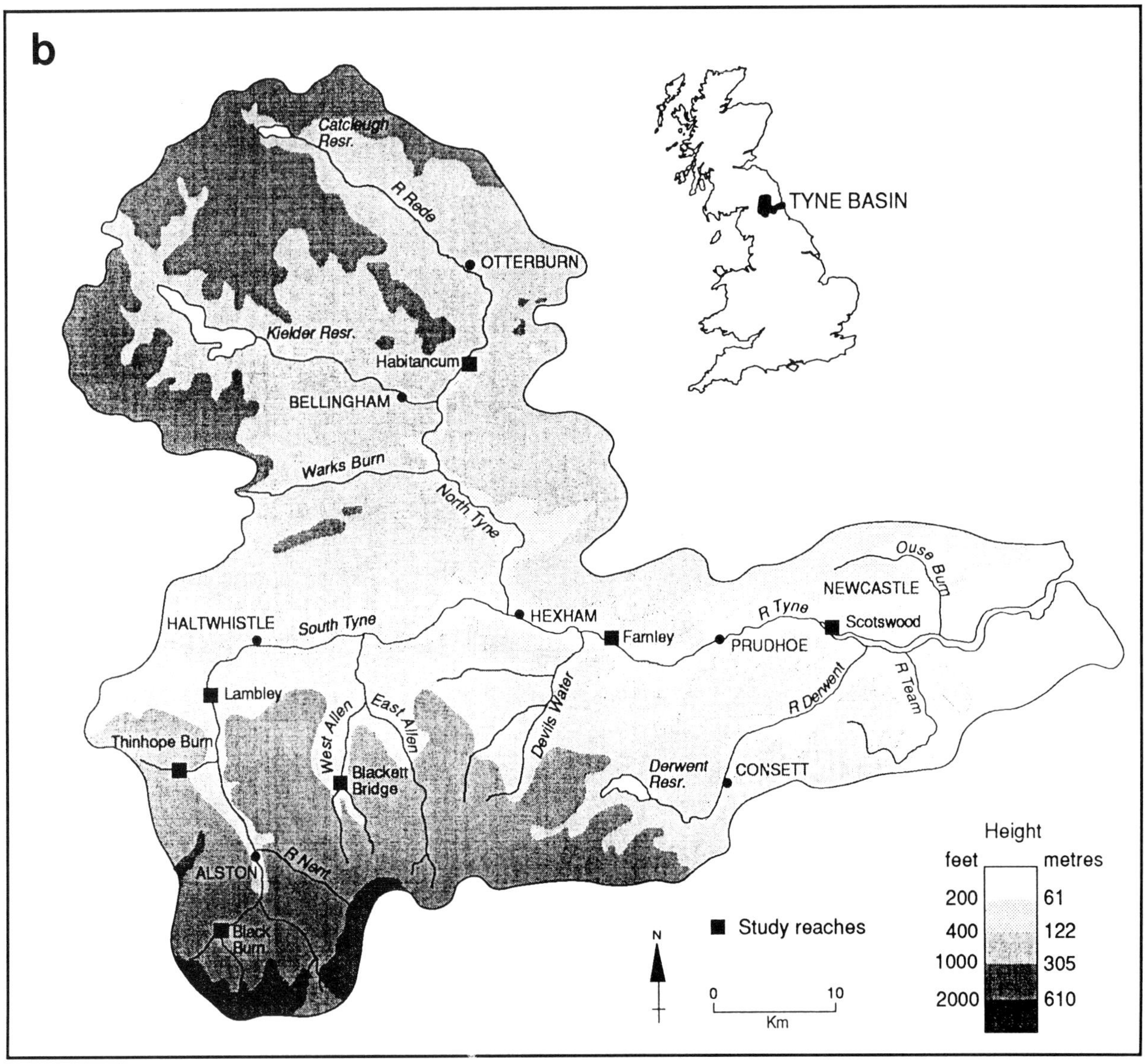

Fig. 12.1 (b) Drainage network and relief of the Tyne basin. Location of study reaches are indicated.

et al 1992) reaches of the basin, characterised by relatively low rates of Holocene channel migration.

Although the Tyne basin has an archaeological record extending from the later Mesolithic onwards (Weyman 1984), human activities only appear to have had a significant impact on the regional landscape in later prehistoric times. Episodes of small-scale and temporary forest clearance during the Bronze Age (associated with pasture and some arable cultivation) are evident in pollen diagrams from upland sites in the region (Davies and Turner 1979; Turner 1983; Rowell and Turner 1985), and can also be inferred from recovery of cereal grain from excavation of contemporary settlement (Gates 1983). However, the most dramatic phase of land-use change appears to have been extensive forest clearance during the later Iron Age and Romano-British periods (Davies and Turner 1979; Higham 1986). Subsistence strategies included both pastoral and arable components (e.g. Turner 1979; Gates 1981; Topping 1989) suggesting significant disturbance of catchment soils and primary forest cover. It is interesting to note that the vast majority of known Iron Age and Romano-British settlements in the Tyne basin upstream of Hexham (Fig. 12.1b) are located in the North Tyne, Rede and lower parts of the South Tyne valleys (Jobey 1977; 1978). By contrast, sites of this age are comparatively rare in the Northern Pennines, although this may in part reflect a greater intensity of archaeological survey in the central and northern areas of the catchment (Higham 1986).

Following a period of settlement contraction and forest regeneration in the latter part of the first millennium AD (Davies and Turner 1979; Higham 1986), this spatial imbalance had been partially redressed by the late thirteenth century with numerous villages and hamlets established throughout the valleys of the basin (Harbottle and Newman 1973; Jobey 1961; Northumberland Sites and Monuments Record). During the fourteenth century another phase of depopulation and settlement contraction, at least in North Tynedale and Redesdale, is evidenced by abandonment of farmsteads (Jobey 1977) and upland forest regeneration (Davies and Turner 1979).

By the mid sixteenth century, however, population had recovered in North Tynedale (Harbottle and Newman 1973; 1977) coinciding with renewed (and permanent) forest clearance with some arable production in the lower South Tyne valley (Davies and Turner 1979). Agricultural settlement reached its maximum extent in North Tynedale during the late eighteenth century as permanent farmsteads and year-round pastoral agriculture succeeded seasonal grazing patterns on the upper valley slopes (Harbottle and Newman 1973). Land was also brought into cultivation in the

Site	Lab.No. (Beta)	^{14}C Age BP	Calibrated age range (intercept 2 sigma)*
Scotswood	30348	6420 +/- 170	5640-4950 BC (1)
Scotswood	30347	6250 +/- 100	5450-4910 BC (1)
Scotswood	30346	4800 +/- 90	3780-3370 BC (1)
Clara Vale	43799	2590 +/- 90	970-410 BC (2)
Low Prudhoe	37797	400 +/- 50	1420-1640 AD (2)
Farnley Haughs	37060	5900 +/- 70	4940-4600 BC (1)
West Woodburn	37810	1300 +/- 70	600-890 AD (2)
Lambley	44527	310 +/- 50	1450-1660 AD (2)
Thinhope	37353	7910 +/- 80	7060-6500 BC (1)
Thinhope	37352	1670 +/- 50	240-530 AD (2)
Thinhope	33435	1230 +/- 60	660-980 AD (2)
Blackett Bridge	40171	1670 +/- 60	230-540 AD (2)

* Age ranges calibrated after (1) Pearson *et al* (1986) and (2) Stuiver and Becker (1986).

Table 12.1. *Uncalibrated ^{14}C dates and calendrically corrected ^{14}C ranges from the Tyne basin.*

South Tyne valley at this time to support the metal mining workforce (Hunt 1984).

STUDY REACHES AND DATING CONTROL

Three reaches (Fig. 12.1b) considered to be representative of recent and Holocene river development in the Tyne basin were selected for detailed investigations. These were:

1. Thinhope Burn, a small (drainage area 12 km^2) upland stream in the Northern Pennines 8 km north-west of Alston, (180–595 m OD).
2. A piedmont reach of the river Tyne at Farnley Haughs (drainage area 2000 km^2, 50 m OD), 10 km downstream of the confluence of the North and South Tyne rivers.
3. The Lower Tyne valley at Scotswood (drainage area 2337 km^2, 3–9 m OD), immediately west of Newcastle upon Tyne, located within the present tidal range.

Alluvial sediments at these sites together span most of the Postglacial period, ranging in age from Early Holocene to the present century. Alluvial units deposited between ca. 7060 cal BC and 1660 cal AD have been dated principally using ^{14}C dating of associated wood and peat, supplemented by optically-stimulated luminescence (OSL), palaeomagnetic techniques and archaeological evidence. All ^{14}C dates were calibrated to calendrical years according to calibration curves after Pearson *et al* (1986) and Stuiver and Becker (1986) (Table 12.1). In addition, recent vertically-accreted fine- grained alluvium in the Tyne valley and coarse flood sediments in Thinhope Burn, in both cases deposited over the last two hundred and fifty years, have been dated using trace metal chemistry and lichenometry respectively.

TYNE BASIN HOLOCENE ALLUVIAL SEQUENCES

Lower Tyne valley, Scotswood

Construction of the Newcastle Western By-Pass between 1987 and 1991 afforded an excellent opportunity to study Holocene fluvial sediments in the Lower Tyne valley 18 km upstream of the mouth of the river. Investigations focused on a 1700 m transect across the Tyne valley floor at Blaydon Haughs (NZ 195638), defined by the route of the by-pass and a new bridge across the Tyne at Scotswood. Geotechnical site evaluations prior to road building revealed up to 10 m of Holocene alluvium overlying up to 40 m of Pleistocene glacial and glaciofluvial deposits. Further site, sedimentological and stratigraphic details can be found in Macklin and Passmore (1988) and Passmore *et al* (1992).

Four major periods of Holocene sedimentation are evident (Fig. 12.2):

1. Between 5640–4950 and 3780–3370 cal BC nine metres of peaty and silty clays were deposited at

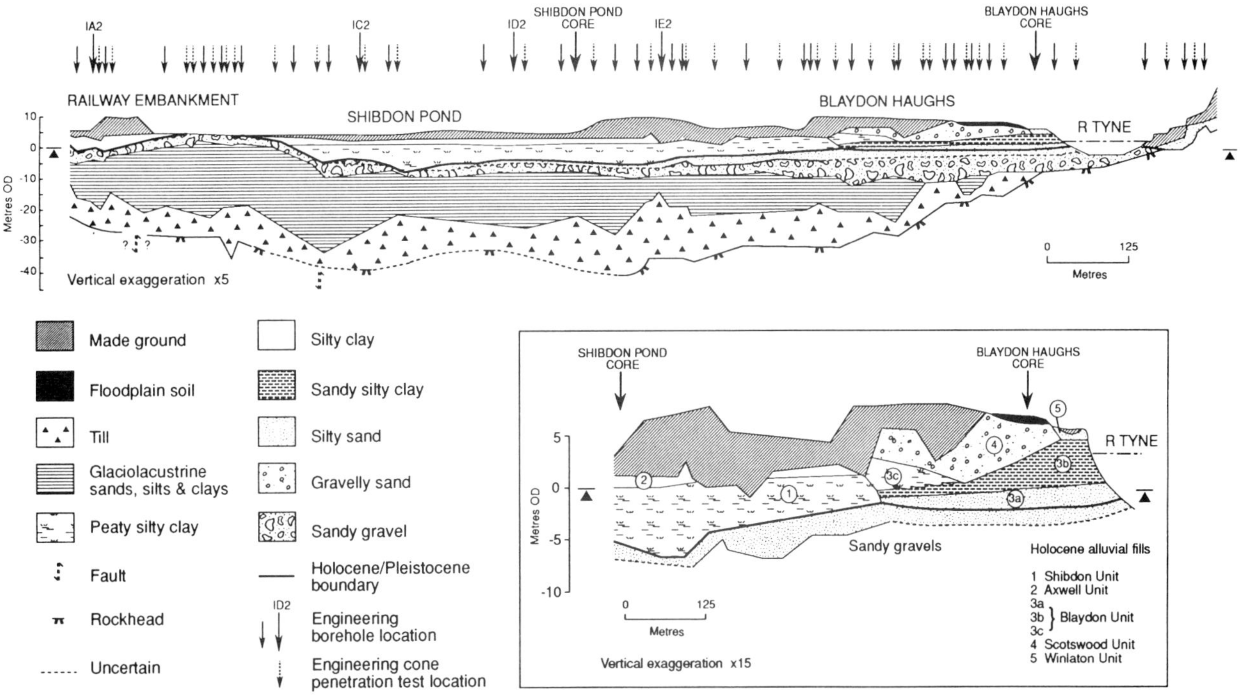

Fig. 12.2. Valley floor subsurface cross-section in the Lower Tyne at Scotswood showing Pleistocene and Holocene sedimentary sequences.

Scotswood primarily in response to increasing base levels associated with rapid sea-level rise. Pollen and sedimentological analyses show these sediments were deposited in a low-energy, freshwater floodbasin environment supporting alder carr (Passmore *et al* 1992). Early in this period (between 5640–4950 and 5450–4910 cal BC) there is evidence of floodplain alder forest disturbance which may reflect either increased flooding or possibly the activities of Mesolithic human communities during this period. This alluvial unit infills a broad trench eroded into Pleistocene glaciofluvial deposits, located on the south side of the valley, that was cut at a time of low sea level probably during the Lateglacial/Early Holocene transition.

2. Major avulsion occurred sometime after 3780–3370 cal BC but before 970–410 cal BC and resulted in the active channel zone of the Tyne switching from the south to its present location on the north side of the valley. This was followed by deposition of up to 4.6 m of laminated sands, silts and clays at Blaydon Haughs (adjacent to the present Tyne channel) and 1 m of silty clays across the southern part of the valley floor at Shibdon Pond (Fig. 12.2). Pollen analyses of these sediments shows both the local floodplain and wider catchment to be substantially deforested with extensive agricultural and pastoral grasslands. This suggests that deposition coincided with major anthropogenic disturbance of forest cover in the Tyne basin during the Late Iron Age and Romano-British periods. *Terminus post quems* for refilling of the entrenched valley floor comes from a large oak trunk radiocarbon-dated to 970–410 cal BC and a Late Iron Age/Romano-British wooden wheel (Piggott 1949) both discovered at Clara Vale, 6 km upstream of Scotswood, and buried beneath 5 m of sands and silts considered on lithological grounds to be analogous to those at Blaydon Haughs.
3. Fine-grained alluvium continued to be deposited at Scotswood probably until Medieval times. In the Post-medieval period, however, shortly after 1420–1640 cal AD there was significant channel aggradation of coarse sediment in the Lower Tyne valley (Rumsby 1991) with deposition of up to 6 m of gravelly sand at Blaydon Haughs (Fig. 12.2). Geochemical analysis of this latter unit indicates limited metal contamination, consistent with smaller-scale mining activities in the Tyne catchment before the early seventeenth century (Raistrick and Jennings 1965).
4. Major channel entrenchment in the eighteenth century was followed by deposition of up to 1.5 m of silty sands (Fig. 12.2). These sediments have a limited lateral extent alongside the present river and are inset within Post-medieval gravelly sands described above. Straightening and dredging of the Tyne channel in the mid-nineteenth century resulted in a further 2–3 m of river bed incision (Johnson 1895).

Middle Tyne valley: Farnley Haughs

Holocene alluvium is particularly well exposed in a sand and gravel pit at Farnley Haughs (NZ 064633) located at the eastern end of a large alluvial basin centred on Hexham (Fig. 12.1). Excavations have exposed a continuous 300 m section across the entire Holocene valley floor from its junction with Late Pleistocene coarse-grained alluvium on the south side of the valley (Fig. 12.3), northwards to the present river channel. Four Holocene alluvial units have been identified with younger fluvial deposits inset below and truncating older fills. This alluvial sequence is very similar to "row terraces" described by Schirmer (1983) in German valleys, produced by lateral channel shift and sedimentation separated by shorter periods of river cut and fill. Each Holocene fluvial unit at Farnley Haughs comprises two lithofacies; massive, imbricated sandy gravels overlain by sands and silts, that are either flat-bedded or display inclined heterolithic stratification (cf. Thomas *et al* 1987). Similar lithofacies associations are well-documented in meandering river systems (Jackson 1978; Miall 1985) with the coarse lower member representing river bed and bar material, and the finer upper member representing floodplain and channel fill sediments resulting from overbank and slackwater deposition.

The oldest Holocene alluvial fill at Farnley Haughs, termed the Willy Wood unit, has been dated to between 4940–4600 cal BC (radiocarbon-dated tree trunk recovered from within the lower gravel member) and 1350–550 BC (palaeomagnetic date on upper sands and silts). Sometime in the period 1350–550 BC major valley floor incision occurred that resulted in river bed level being lowered by more than 2 m (Fig. 12.3). An OSL date of 510 (± 375 years) BC from sandy silts at the base of a major channel fill within the Farnley unit brackets this phase of trenching to around the Sub-Boreal/Sub-Atlantic boundary. It also suggests incision and channel refilling was relatively rapid, and possibly was completed in less than 150–200 years. At present the top of the Farnley unit is undated, however, a ^{14}C date of 600–890 cal AD from wood within a palaeochannel on the surface of a stratigraphically equivalent unit in the Rede valley (Passmore *et al* 1991) indicates alluviation in other parts of the Tyne catchment, and probably at Farnley Haughs, had ceased by this time.

The Styford unit, separated from the Farnley unit by a phase of relatively minor downcutting, has been dated using trace metal chemical analysis. Heavy metal

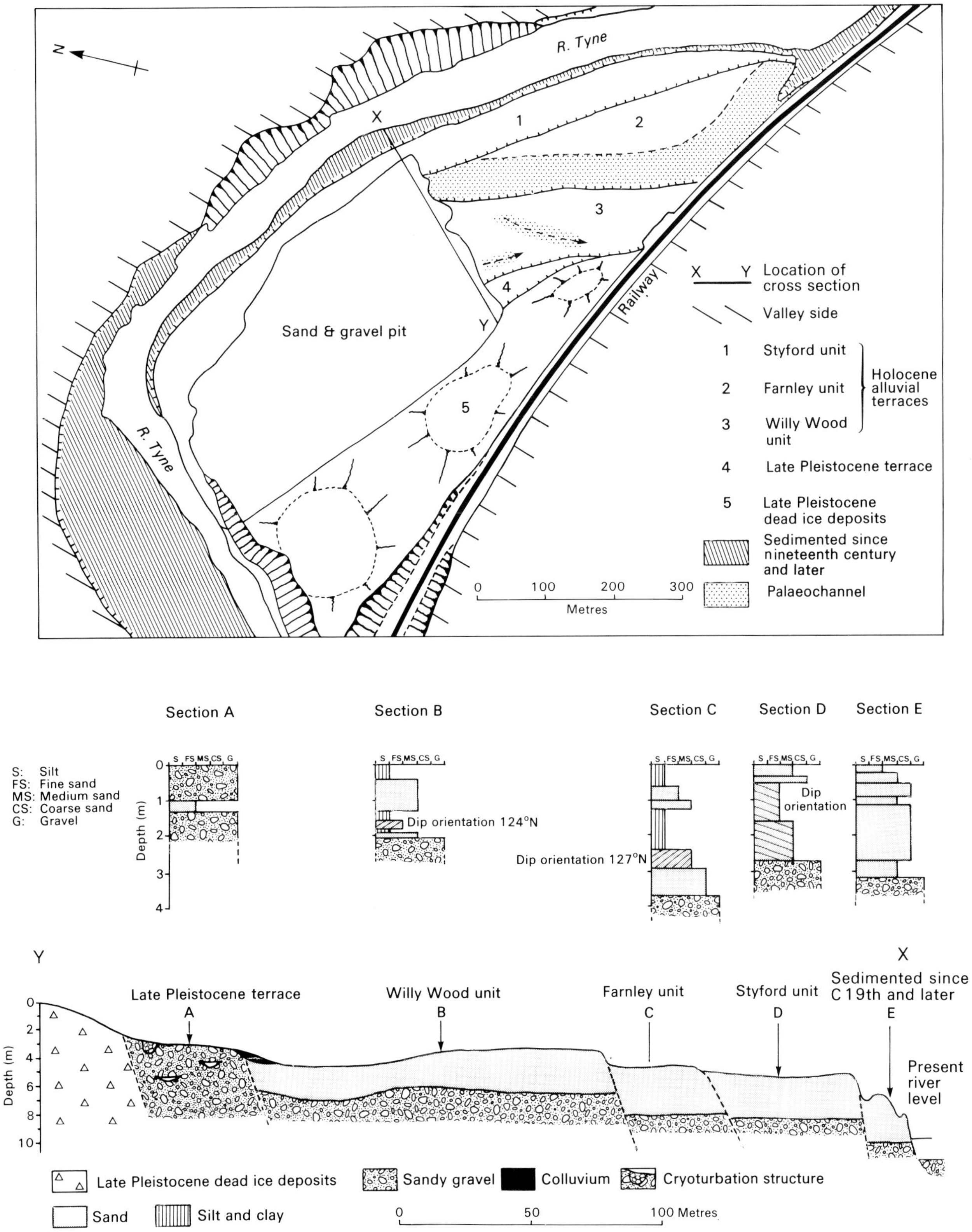

***Fig. 12.3.** Late Pleistocene and Holocene river terrace morphologies and sedimentary sequences at Farnley Haughs, Middle Tyne valley.*

concentrations in the former unit (Pb, Zn, Cu, Fig. 12.4) below 180 cm do not exceed maximum background levels (as recorded in the Middle Holocene Willy Wood unit) and would appear to have been deposited during a period when there was no metal mining. At 180 cm and, more notably, at around 45 cm metal concentrations rise sharply. Metal contamination above 180 cm would appear to be associated with large-scale mining activity which in the Tyne catchment began at the beginning of the seventeenth century (Raistrick and Jennings 1965). Lead levels in the top 50 cm of the section, however, are similar to those in fine-grained alluvium laid down elsewhere in the Middle Tyne valley during a period of mining in the middle and late eighteenth century (Rumsby and Macklin, unpublished). Corroboration for significant Post-medieval alluviation in piedmont reaches of the Tyne has recently come from investigations at Lambley (Fig. 12.1, NY 672605), in the South Tyne valley, where 3 m of fine-grained sediments were deposited after 1450–1660 cal AD but before the early nineteenth century (Macklin and Passmore, unpublished).

Deposition of the Styford unit was followed by a period of major channel downcutting with river bed levels falling by nearly 2 m (Fig. 12.3). This phase of channel entrenchment and widening, which is evident throughout the Middle Tyne valley, has been dated to the late eighteenth century at Broomhaugh (NY 945643, 6 km east of Farnley Haughs) using cartographic sources and trace metal analysis (Rumsby 1991). This phase of channel incision coincided with renewed fine-grained sedimentation on the surface of the Styford fill-terrace, witnessed by a major textural discontinuity in the upper part of the section in the form of a prominent and laterally- extensive layer of coarse sand between 50–30 cm (Fig. 12.4). This reverses the overall up-section fining sequence in the Styford unit and, following Knox (1987) and Macklin, Rumsby and Newson (1992), is considered

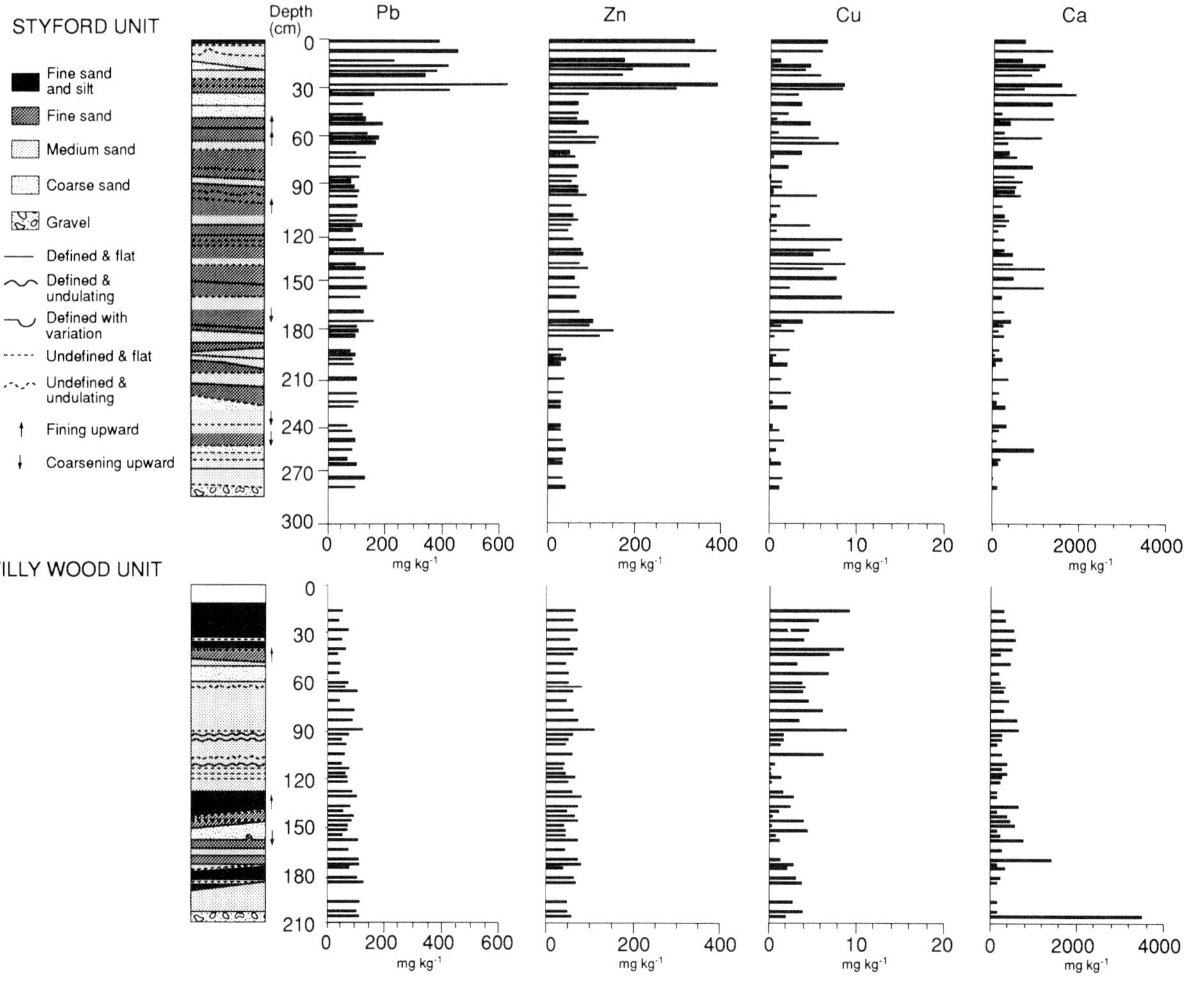

Fig. 12.4. Sedimentary logs and geochemical composition of Willy Wood and Styford units (upper fine members), Farnley Haughs.

to represent a single overbank unit deposited by an exceptionally large flood, possibly that of 1771 (the largest documented flood in the Tyne basin).

Up until the end of the nineteenth century a period of net sediment storage in the Middle Tyne resulted in the partial refilling (predominantly by fine-grained sediment) of the valley floor entrenched in the eighteenth century. Since the turn of the century, however, the river Tyne has become increasingly vertically unstable with periods of rapid floodplain accretion (1898–1939; 1966–1990) alternating with closely-spaced intervals of erosion (1880–1898; 1940–1965). The timing of the latter has been shown to be associated with an increase in magnitude and frequency of flooding in the Tyne catchment (Macklin *et al* 1992a; Rumsby 1991).

The geochemical compositions of the four Holocene alluvial units at Farnley Haughs have been compared using multivariate ordination methods (Ter Braak 1988). Based on geochemical analysis of 150 fine-grained inorganic sediment samples, element envelopes (delimiting geochemical variation within a unit) are presented for each alluvial fill (Fig. 12.5). In an attempt to identify provenance of Holocene alluvium, the geochemical characteristics of contemporary fine-grained channel sediments and Pleistocene tills in both the North and South Tyne basins were also analysed. South Tyne sediments are relatively rich in lead and zinc (Fig. 12.5, associated with base-metal mineralisation in the catchment) and calcium (reflecting a significant limestone outcrop in the basin). In contrast, North Tyne sediments, including those collected from its principal tributary, the river Rede, have low trace metal and calcium concentrations.

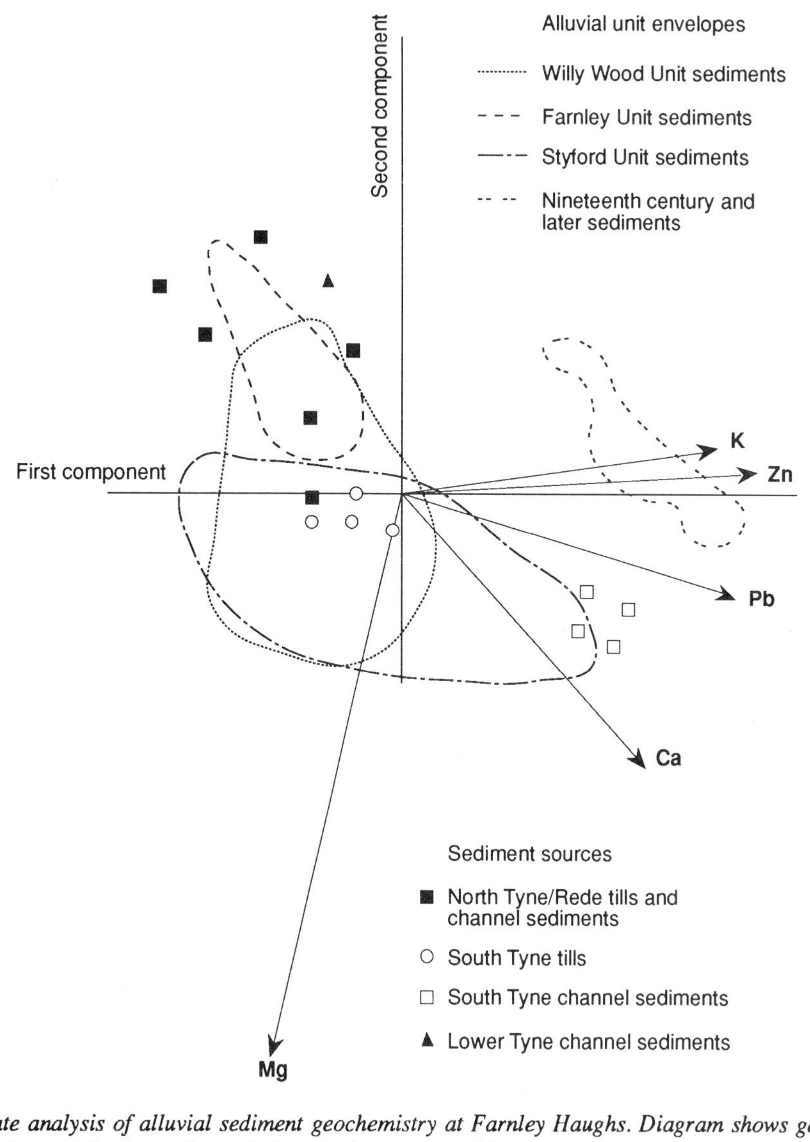

Fig. 12.5. *Multivariate analysis of alluvial sediment geochemistry at Farnley Haughs. Diagram shows geochemical envelopes for Willy Wood, Farnley and Styford units and nineteenth century and later sediments, together with sediment source samples (tills and contemporary channel fines), plotted on their first and second principal component axes. Also shown are vectors of Mg, K, Ca, Zn and Pb.*

For the Styford and later alluvial units, that were deposited (at least in part) while lead, and later zinc, were being mined in the South Tyne catchment, heavy metals provide a most reliable means of assessing sediment sources and (as discussed earlier) dating fluvial sediments. In alluvium that pre-dates metal mining, geochemical analysis is still useful for identifying the relative contribution of erosion within the North and South Tyne catchments to alluviation in the Middle Tyne valley for particular intervals over the last 6500 years. The chemical signature of the Willy Wood unit indicates sediment delivery from both the North and South Tyne during the Middle Holocene, although the proportion of sediment derived from the latter catchment appears to decrease towards the end of the period. In contrast, sands and silts comprising much of the Farnley unit appear to consist almost entirely of sediments originating from the North Tyne basin, and suggest substantial erosion beginning in the Iron Age and continuing possibly into Middle Saxon times. This phase of soil erosion coincides with, and was probably caused by, larger-scale and more continuous agricultural clearance during the Late Iron Age and Romano- British periods in the Rede and North Tyne valleys (Jobey 1977; 1978; Higham 1986).

Thinhope Burn: Northern Pennine uplands

Thinhope Burn is a steep and deeply-incised stream that drains a small moorland catchment on the north-west flank of the Northern Pennines (NY 680550), and runs into the river South Tyne 9 km north-west of Alston (Fig. 12.1). Recent investigations (Macklin, Rumsby and Heap 1992) have revealed that over the last 1600 years Thinhope Burn has experienced unprecedented (viewed in the context of earlier Postglacial valley floor development) rates of channel incision (locally up to 8 m), punctuated by periods of relative stability and limited valley floor sedimentation. This sequence is especially clear in the upper reaches of the basin where three cut-and-fill river terraces are well developed at 8, 6 and 4 m above the present stream bed (Fig. 12.6). The highest fill-terrace consists of gravels overlain by 2 m of sands and silts which in the upper part of the unit, and towards the valley side, interdigitate with peat. Radiocarbon dates of 7060–6500 cal BC and 240–530 cal AD, from wood within the basal gravels and at the interface between silt and peat 0.5 m below the surface of the terrace indicate that Thinhope Burn was characterised by relatively slow valley floor sedimentation until early historic times. The detailed history and nature of alluviation over this period, and its relation to catchment environmental change, is the subject of ongoing research and will be published elsewhere.

Sometime after 240–530 cal AD the first phase of major Holocene valley floor entrenchment occurred in Thinhope and resulted in nearly 4 m of channel bed erosion. Before 660–980 cal AD, however, channel incision had ceased and was followed by valley floor sedimentation. A second period of accelerated river incision occurred after 660–980 cal AD. The valley floor was subsequently refilled (predominantly with fine-grained alluvium) to a level ca. 4 m above the present stream bed. This fill-terrace is the most extensive Holocene fluvial unit in Thinhope and has a

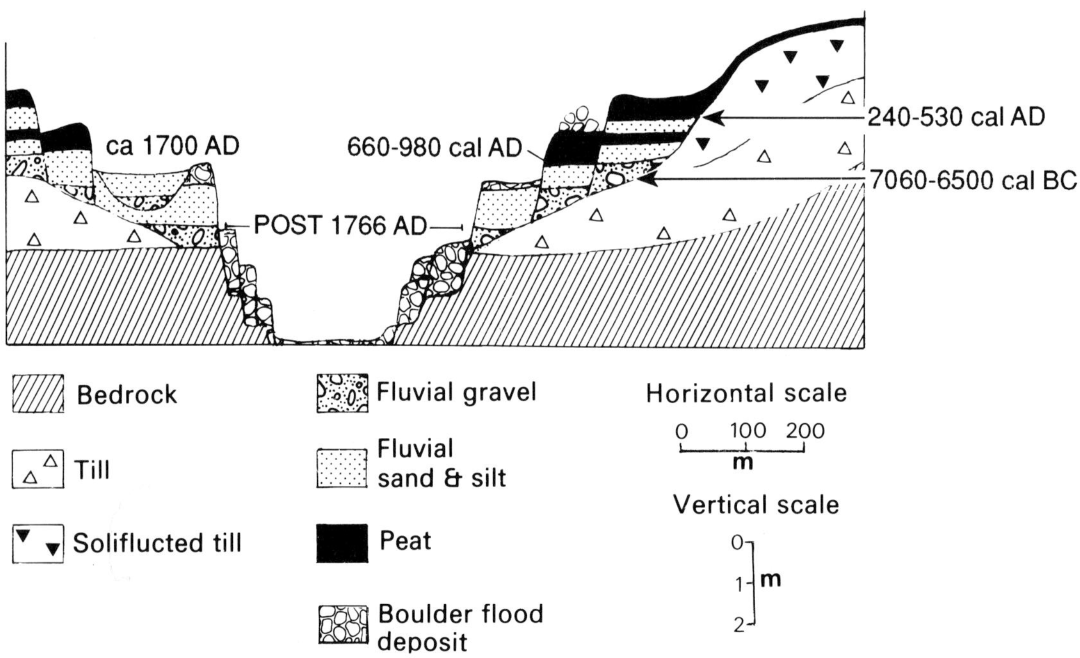

Fig. 12.6. Schematic valley floor cross-section of Thinhope Burn, Northern Pennines, showing terrace morphology, sedimentary characteristics and ages of Holocene alluvium.

number of well- developed palaeochannels on its surface that generally have higher sinuosities and lower width/depth ratios than the present channel. Pollen analysis of organic-rich silts that infill one of these palaeochannels shows a vegetation sequence very similar to that recorded by Roberts *et al* (1973) in nearby Upper Weardale, and radiocarbon dated to 1700–1780 cal AD. This would suggest that this channel was cut off sometime at the turn of the seventeenth-eighteenth centuries and thus pollen evidence provides a tentative *terminus ante quem* for this terrace unit.

Lichenometric dating of extensive coarse flood sediments in Thinhope Burn (Macklin, Rumsby and Heap 1992), resulting from a series of large floods in the periods 1780–1820, 1840–1880 and 1920–1950 show a third major phase of Late Holocene channel incision began in the early part of the eighteenth century. Dated and levelled flood deposits indicate more than 4 m of stream erosion (locally through bedrock) since ca. 1766, with most downcutting accomplished in the period up to 1780–1815. From the mid-nineteenth century there was limited channel change until 1940 when incision was renewed that continues, on a more limited scale, up to the present day.

To summarise, the Late Holocene (post 240–530 cal AD) witnessed a striking transformation of channel and floodplain sedimentation styles in Thinhope. It saw the replacement of a relatively stable meandering channel, and a floodplain accreting fine-grained sediment, by a vertically and laterally active, low-sinuosity, boulder-bed stream with a high coarse sediment load. Channel and floodplain metamorphosis towards the end of the Roman period, and most notably during the eighteenth century, were both associated with pronounced valley floor erosion.

TOWARDS A HOLOCENE ALLUVIAL HISTORY OF THE TYNE BASIN

Nature of climate, land-use and alluvial records

In Fig. 12.7 periods of known (i.e. dated) Holocene valley floor sedimentation and episodes of river erosion and incision (time bracketed by alluviation events) in the Tyne drainage basin are summarised. The time-span, thickness and height (relative to present river bed level in upland and lowland reaches, and to Ordnance Datum in the Lower Tyne) of all alluvial units are plotted together with probable dates and depths of valley floor incision. In addition, to assess both the temporal and spatial patterns of valley floor erosion and deposition in the region, Holocene alluvial sequences at the upland, piedmont and lowland sites described above are juxtaposed. Although some details of our current scheme of Holocene river development will undoubtedly require some modification in the light of ongoing and future investigations, we believe that we have probably identified most of the important episodes of alluviation in the basin. Nevertheless, until more dating control is available, prominent gaps in some alluvial sequences (e.g. between 890–1420 cal AD at upland and piedmont sites) should not be interpreted simply as indicating prolonged periods of river erosion.

In a preliminary attempt to relate fluvial regimen changes in the Tyne to their ultimate causative processes, alluvial chronologies are compared with selected independent records of Holocene climate in northern Britain, north-west Europe and Scandinavia. Three types of proxy data have been used;

1. Holocene glacier fluctuations in Scandinavia (Karlen 1991). Glaciers respond to changes in temperature and/or precipitation. Thus glacial advances plotted in Fig. 12.7 most probably reflect a decrease in temperature and/or an increase in precipitation. In most cases the maximum extent of glacier advance has been dated and therefore dates on moraines post-date climatic change.
2. Deuterium/hydrogen ratios from dated tree stumps in the Cairngorms, Scotland (Dubois and Ferguson 1985). Stable isotope analysis opens the prospect of being able to realise the "cooler and/or wetter" conundrum with excessively low deuterium values related to periods with very heavy rainfall.
3. Recurrence surfaces in raised mires in north-west Europe (Haslam 1988) and Bolton Fell Moss, Cumbria (Barber 1981). These are identified in bog stratigraphies by a change from darker humified peat to a lighter brown unhumified peat that is believed to reflect an increase in surface wetness controlled by changes in the precipitation/evaporation ratio. If the Little Ice Age is taken to be a reasonable synoptic analogue to earlier neoglacial phases in the Holocene, times of glacial advance in Scandinavia, and accelerated peat growth in raised mires in Britain and Europe, would most likely have been periods of increased run-off in the Tyne basin.

Episodes of extensive woodland clearance in north-east England (after Turner and her co-workers) are also indicated on Fig. 12.7. Forest removal in the Tyne followed by cultivation and grazing, as has been shown in studies of many other river basins (e.g. Knox 1972; 1977), would have resulted in a significant increase in catchment run-off, flood flashiness and slope erosion. Agriculturally-induced soil erosion would have also augmented delivery of fine sediment to valley floors. However, what proportion of this material reached local rivers and accelerated floodplain sedimentation, and how much went into long-term storage on the valley floor or lower slopes is a matter for conjecture. Although the progressive conversion of the Tyne basin from natural

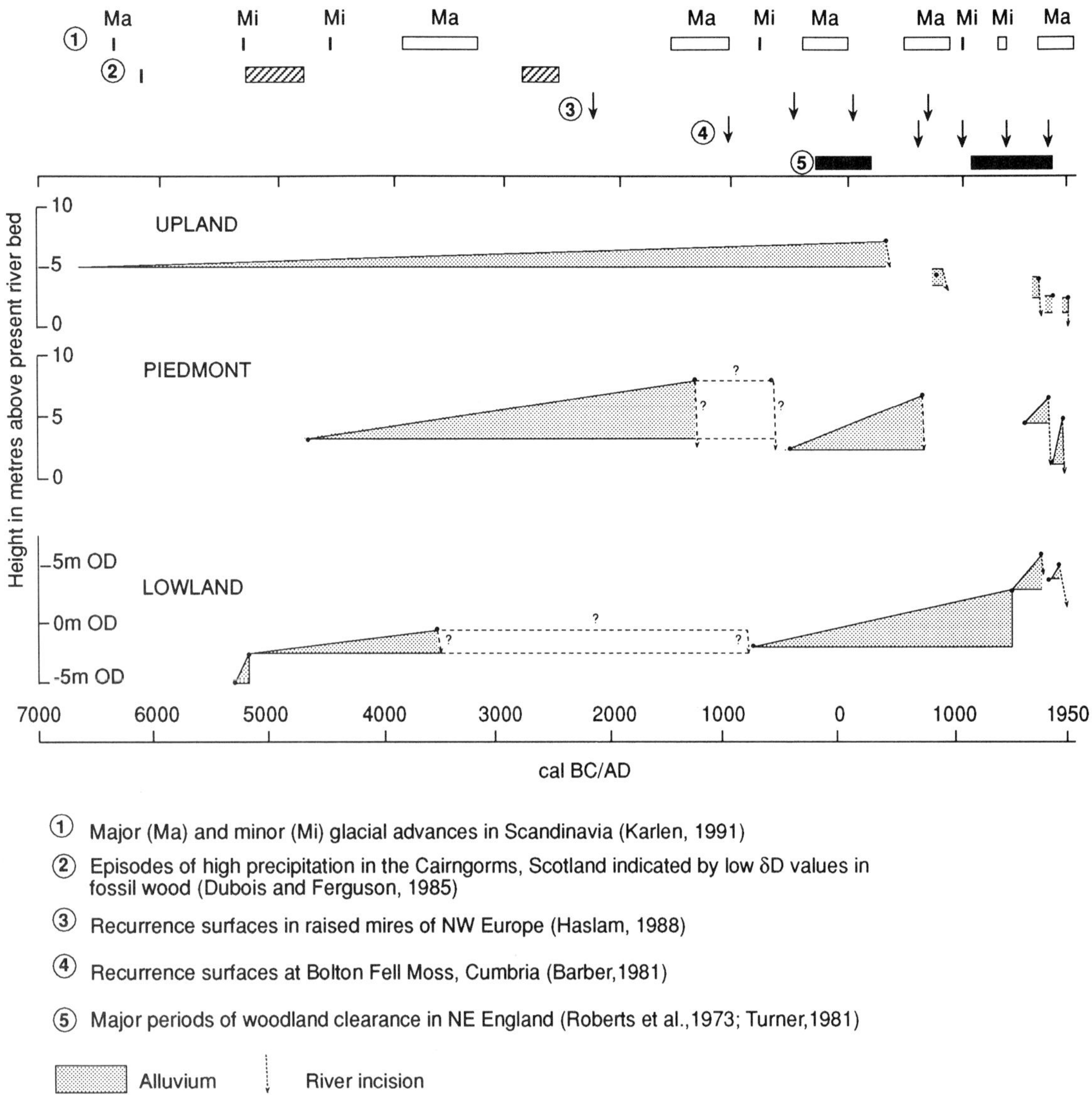

Fig. 12.7. Time-level diagram for Holocene alluvial units in upland, piedmont and lowland reaches of the Tyne. Selected independent proxy records of Holocene climatic change in Britain, Scandinavia and Europe, together with episodes of extensive woodland clearance in north-east England, are also presented.

woodland to agricultural land use would have increased the severity of flooding (decreased interception, infiltration and evaporation, and increased overland flow and flood peaks), recent work by Higgs (1987) in the Upper Severn has shown that land-use change has had negligible effects on larger magnitude flow events. Recent studies have demonstrated, however, that it is exceptional floods that control large-scale patterns of lateral and vertical channel development in most northern English rivers, including the Tyne (Macklin and Lewin 1989; Newson and Macklin 1990).

To summarise, while later prehistoric and historic agricultural activities have undoubtedly made the Tyne and its tributaries more responsive to relatively small changes in precipitation (principally through vegetation destruction and soil disturbance which are important regulators of run-off and sediment supply) the timing of floods would have been unaffected and must therefore directly relate to Holocene climatic fluctuations.

River response to Holocene climate and land-use change

The Early and Middle Holocene (ca. 6000–1000 cal BC) would appear to have been a period of valley floor sedimentation throughout the Tyne basin (Fig. 12.7). In the Lower Tyne, alluviation shortly before 5000 cal BC was associated with increasing base levels related to sea-level rise, though it also coincided with a period of increased precipitation and probably flooding in northern

Ballantyne, C.K. 1991. Late Holocene erosion in upland Britain: climatic deterioration or human influence? *The Holocene*, 1 (1), 81–85.

Barber, K.E. 1981. *Peat stratigraphy and climatic change*. Rotterdam: Balkema.

Bell, M. 1982. The effects of land-use and climate on valley sedimentation. In A.F. Harding (ed.), *Climatic Change in Later Prehistory*, 127–142. Edinburgh: University Press.

Blackford, J.J. and Chambers, F.M. 1991. Proxy records of climate from blanket mires: evidence for a Dark Age (1400 BP) climatic deterioration in the British Isles. *The Holocene*, 1 (1), 63–67.

Brakenridge, G.R. 1980. Widespread episodes of stream erosion during the Holocene and their climatic cause. *Nature*, 283, 655–656.

Brown, A.G. 1990. Holocene floodplain diachronism and inherited downstream variations in fluvial processes: a study of the river Perry, Shropshire, England. *Journal of Quaternary Science*, 5 (1), 39–51.

Brown, A.G. and Barber, K.E. 1985. Late Holocene palaeoecology and sedimentary history of a small lowland catchment in central England. *Quaternary Research*, 24, 87–102.

Burrin, P.J. and Scaife, R.G. 1988. Environmental Thresholds, Catastrophe Theory and Landscape Sensitivity: Their Relevance to the Impact of Man on Valley Alluviations. In J.L. Bintliff, D.A. Davidson and E.G. Grant (eds.), *Conceptual Issues in Environmental Archaeology*, 211–232. Edinburgh: University Press.

Davies, G. and Turner, J. 1979. Pollen Diagrams from Northumberland. *New Phytologist*, 82, 783–804.

Dubois, A.D. and Ferguson, D.K. 1985. The climate history of pine in the Cairngorms based on radiocarbon dates and stable isotope analysis, with an account of the events leading up to its colonisation. *Review of Palaeobotany and Palynology*, 46, 55–80.

Dunham, K.C. 1990. *Geology of the Northern Pennine Orefield, Volume 1, Tyne to Stainmore* (2nd edition). Economic memoir of the British Geological Survey, sheets 19 and 25 and parts of 13, 24, 26, 31, and 32 (England and Wales). London: H.M.S.O.

Gates, T. 1981. Farming on the Frontier: Romano-British fields in Northumberland. In P.Clack and S.Haselgrove (eds.), *Rural settlement in the Roman North*, 21–42. Durham: Council for British Archaeology Group 3.

Gates, T. 1983. Unenclosed settlements in Northumberland. In J.C. Chapman and H.C. Mytum (eds.), *Settlement in North Britain, 1000 BC – AD 1000*, 103–48. Oxford: British Archaeological Reports, British Series, 118.

Harbottle, B. and Newman, T.G. 1973. Excavation and survey on the Starsley Burn, North Tynedale, 1972. *Archaeologia Aeliana*, 5 (1), 137–173.

Harbottle, B. and Newman, T.G. 1977. Excavation and survey in North Tynedale, 1973–1975. *Archaeologia Aeliana*, 5 (5), 121–150.

Harvey, A.M., Oldfield, F., Baron, A.F. and Pearson, G.W. 1981. Dating of Post-Glacial Landforms in the Central Howgills. *Earth Surface Processes and Landforms*, 6, 401–12.

Haslam, C.J. 1988. Late Holocene peat stratigraphy and climatic change – a macrofossil investigation from the raised mires of north-west Europe. University of Southampton, unpublished PhD thesis.

Higgs, G. 1987. Environmental change and hydrological response: flooding in the upper Severn catchment. In K.J. Gregory, J. Lewin and J.B. Thornes (eds.), *Palaeohydrology in Practice*, 131–159. Chichester: Wiley.

Higham, N. 1986. *The Northern Counties to AD 1000*. London: Longman.

Houghton, J.T., Jenkins, G.J. and Ephraums, J.J. (eds.) 1990. *Climate change: The IPCC Scientific Assessment*. Report prepared for IPCC by Working Group 1. Cambridge University Press.

Hunt, C.J. 1984. *The lead miners of the northern Pennines in the eighteenth and nineteenth centuries*. Newcastle upon Tyne: Davis.

Jackson, R.G. 1978. Preliminary evaluation of lithofacies models for meandering alluvial streams. In A.D. Miall (ed.), *Fluvial Sedimentology*, 543–577. Calgary: Canadian Society of Petroleum Geologists Memoir 5.

Jobey, G. 1961. Further notes on rectilinear earthworks in Northumberland: some medieval and later settlement. *Archaeologia Aeliana*, 4, 87–102.

Jobey, G. 1977. Iron Age and later farmsteads on Belling Law, Northumberland. *Archaeologia Aeliana*, 5 (5), 1–38.

Jobey, G. 1978. Iron Age and Romano-British settlements on Kennel Hall Knowe, Northumberland. *Archaeologia Aeliana*, 5 (6), 1–28.

Johnson, R.W. 1895. *The Making of the Tyne*. London: Walter Scott.

Karlen, W. 1991. Glacier fluctuations in Scandinavia during the last 9000 years. In L. Starkel, K.J. Gregory and J.B. Thornes (eds.), *Temperate Palaeohydrology*, 395–412. Chichester: John Wiley and Sons.

Knox, J.C. 1972. Valley alluviation in southwestern Wisconsin. *Annals of the Association of American Geographers*, 62, 401–410.

Knox, J.C. 1977. Human impact on Wisconsin stream channels. *Annals of the Association of American Geographers*, 67 (3), 323–342.

Knox, J.C. 1983. Responses of river systems to Holocene climates. In H.E. Wright Jnr (ed.), *Late-Quaternary Environments of the United States, The Holocene, Vol.2*, 26–41. Minneapolis: University of Minnesota Press.

Knox, J.C. 1987. Stratigraphic evidence of large floods in the Upper Mississippi Valley. In L. Mayer and D. Nash (eds.), *Catastrophic flooding*, 155–180. Boston: George Allen and Unwin.

Kozarski, S. 1983. River channel adjustment to climatic change in west central Poland. In K.J. Gregory (ed.), *Background to Palaeohydrology*, 355–374. Chichester: John Wiley and Sons.

Lamb, H.H. 1977. *Climate: Past, Present and Future. Volume 2: Climate History and the Future*. London: Methuen.

Lamb, H.H. 1982. *Climate, History and the Modern World*. London: Methuen.

Lewin, J. and Macklin, M.G. 1987. Metal Mining and Floodplain Sedimentation in Britain. In Gardner (ed.), *International Geomorphology 1986, Part 1*, 1009–1027. London: John Wiley and Sons.

Macklin, M.G. 1985. Floodplain sedimentation in the upper Axe valley, Mendip, England. *Transactions of the Institute of British Geographers*, (new series) 10, 235–244.

Macklin, M.G. (in press). Holocene river environments in north-east England. In K.J. Gregory (ed.), *British Holocene River Environments*. English Nature.

Macklin, M.G. and Dowsett, R.B. 1989. The chemical and physical speciation of trace metals in fine grained overbank flood sediments in the Tyne basin, North-East England. *Catena*, 16, 135–151.

Macklin, M.G. and Lewin, J. 1986. Terraced fills of Pleistocene and Holocene age in the Rheidol Valley, Wales. *Journal of Quaternary Science*, 1 (1), 21–34.

Macklin, M.G. and Lewin, J. 1989. Sediment transfer and transformation of an alluvial valley floor: The River South Tyne, Northumbria, U.K. *Earth Surface Processes and Landforms*, 14, 233–246.

Macklin, M.G. and Lewin, J. (in press). Holocene river alluviation in Britain. In I. Douglas and J. Hagedorn (eds.), *Geomorphology and Geoecology, Fluvial Geomorphology*. Zeitschrift fur Geomorphologie (supplement) 85.

Macklin, M.G. and Passmore, D.G. 1988. *Late Quaternary Sedimentation in the Lower Tyne Valley, North East England*. University of Newcastle upon Tyne: Dept. of Geography Seminar Paper No.55.

Macklin, M.G., Passmore, D.G., Stevenson, A.C., Cowley, D.C., Edwards, D.N. and O'Brien, C.F. 1991. Holocene alluviation and land-use change on Callaly Moor, Northumberland, U.K. *Journal of Quaternary Science*, 6 (3), 225–232.

Macklin, M.G., Rumsby, B.T. and Heap, T. 1992a. Flood alluviation and entrenchment: Holocene valley floor development and transformation in the British uplands. *Geological Society of America Bulletin* 104, 631–643.

Macklin, M.G., Rumsby, B.T. and Newson, M.D. 1992b. Historic floods and vertical accretion of fine-grained alluvium in the Lower Tyne Valley, North East England. In P. Billi *et al* (eds.), *Dynamics of Gravel-Bed Rivers; Proceedings of the Third International Workshop on Gravel-Bed Rivers*, 564–580. Chichester: John Wiley and Sons.

Miall, A.D. 1985. Architectural-element analysis: A new method of facies analysis applied to fluvial deposits. *Earth Science Reviews*, 22, 261–308.

Newson, M.D. and Lewin, J. 1991. Climatic change, river flow extremes and fluvial erosion – scenarios for England and Wales. *Progress in Physical Geography*, 15 (1), 1–17.

Newson, M.D. and Macklin, M.G. 1990. The geomorphologically-effective flood and vertical instability in river channels – a feedback mechanism in the flood series for gravel-bed rivers. In W.R. White (ed.), *International Conference on River Flood Hydraulics*, 123–140. Chichester: John Wiley and Sons.

Passmore, D.G., Macklin, M.G., Heap, T. and Anderson, J. 1991. Geoarchaeological investigations of Late Holocene alluvial fills and valley side sediments at Habitancum Roman Fort, West Woodburn, Northumberland: A re-evaluation of "The Piercebridge Formula". *Universities of Durham and Newcastle upon Tyne Archaeological Reports for 1990*, 14, 29–32.

Passmore, D.G., Macklin, M.G., Stevenson, A.C., O'Brien, C.F. and Davies, B.A.S. 1992. A Holocene alluvial sequence in the lower Tyne valley, northern Britain: A record of river response to environmental change. *The Holocene* 2(2), 138–147.

Patton, P.C. and Boison, P.J. 1986. Processes and rates of formation of Holocene alluvial terraces in Harris Wash, Escalante River basin, south-central Utah. *Geological Society of America Bulletin*, 97, 369– 378.

Pearson, G.W., Pilcher, J.R., Baille, M.G., Corbett, D.M. and Qua, F. 1986. High precision ^{14}C measurement of Irish oak to show the natural ^{14}C variations from AD 1840–5210 BC. *Radiocarbon*, 28, 911–934.

Piggott, S. 1949. A wheel of Iron Age type from Co.Durham. *Proceedings of the Prehistoric Society*, 15, 191.

Raistrick A. and Jennings, B. 1965. *A History of Lead Mining in the Pennines*. London: Longmans.

Roberts, B.K., Turner, J. and Ward, P.F. 1973. Recent forest history and land-use in Weardale, Northern England. In H.J.B. Birks and R.G. West (eds.), *Quaternary plant ecology*, 207–221. Oxford: Blackwell Scientific Publications.

Robinson, M.A. and Lambrick, G.H. 1984. Holocene alluviation and hydrology in the upper Thames basin. *Nature*, 308, 809–14.

Rowell, T.K. and Turner, J. 1985. Litho-, Humic- and Pollen Stratigraphy at Quick Moss, Northumberland. *Journal of Ecology*, 73, 11–25.

Rumsby, B.T. 1991. Flood frequency and magnitude estimates based on valley floor morphology and floodplain sedimentary sequences: The Tyne Basin, N.E.England. University of Newcastle upon Tyne, unpublished PhD thesis.

Schirmer, W. 1983. Criteria for the differentiation of late Quaternary river terraces. *Quaternary Studies in Poland*, 4, 199–204.

Shotton, F.W. 1978. Archaeological inferences from the study of alluvium in the Lower Severn-Avon valleys. In S. Limbrey and J.G. Evans (eds.), *Man's Effect on the Landscape: The Lowland Zone*, 27–32. London: Council for British Archaeology, Research Report No.21.

Starkel, L. 1983. The reflection of hydrologic changes in the fluvial environment of the temperate zone during the last 15000 years. In K.J. Gregory (ed.), *Background to Palaeohydrology*, 213–235. Chichester: John Wiley and Sons.

Starkel, L. 1985. Late glacial and Postglacial history of river valleys in Europe as a reflection of climatic changes. *Zeitschrift für Gtetscherkunde und Glazialgeologie*, 21, 159–164.

Stuiver, M. and Becker, B. 1986. High-precision calibration of the radiocarbon timescale, AD 1950–500 BC. *Radiocarbon*, 28, 863–910.

Ter Braak, C.J.F. 1988. CANOCO – A FORTRAN program for canonical community ordination by partial detrended canonical correspondence analysis, principal components analysis and redundancy analysis (version 2.1). *Technical report: LWA-88-02*. GLW, Postbus 100, 6700 AC Wageningen.

Thomas, R.G., Smith, D.G., Wood, J.M., Vissier, J., Claverley-Range, E.A. and Koster, E.H. 1987. Inclined heterolithic stratification - terminology, description, interpretation and significance. *Sedimentary Geology*, 53, 123–179.

Topping, P. 1989. Early Cultivation in Northumberland and The Borders. *Proceedings of the Prehistoric Society*, 55, 161–179.

Turner, J. 1979. The environment of north east England during Roman times as shown by pollen analysis. *Journal of Archaeological Science*, 6, 285–290.

Turner, J. 1983. Some pollen evidence for the environment of North Britain 1000 BC to AD 1000. In J.C. Chapman and H.C. Mytum (eds.), *Settlement in North Britain, 1000 BC – AD 1000*, 3–28. Oxford: British Archaeological Reports, British Series, 118.

Wendland, W.M. 1982. Geomorphic responses to climatic forcing during the Holocene. In C.E. Thorn (ed.), *Space and Time in Geomorphology*, 355– 371. Boston: Allen and Unwin.

Wendland, W.M. and Bryson, R.A. 1974. Dating climatic episodes of the Holocene. *Quaternary Research*, 4, 9–24.

Weyman, J. 1984. The Mesolithic in North-East England. In C.B. Burgess and R. Miket (eds.), *Between and Beyond the Walls: Essays on the Prehistory and History of North Britain in Honour of George Jobey*, 38–51. Edinburgh: John Donald.

13

Alluvium and archaeology in the Herefordshire valleys

James Dinn and Rebecca Roseff

Summary: *The contributions of two associated projects to an understanding of the archaeology of the alluviated river valleys of Herefordshire are described. The Herefordshire Valleys Survey tested a number of techniques and methodologies for the discovery and assessment of archaeological sites and deposits under alluvium, and some of these are introduced here. On the basis of this and of a database of river valley sites in the county, recommendations have been made for the management of suballuvial archaeological landscapes, especially in the face of threats from sand and gravel quarrying. A draft report on this survey was completed in early 1991. A SERC studentship at Birmingham University took as its starting point an alluviated site at Wellington, Herefordshire, which was not identified prior to the granting of planning permission for gravel extraction. It has used evidence from this site to illuminate the hydrological and land-use history of the Lugg valley. Work on this project continues.*

INTRODUCTION

The two projects described in this paper focus on the interrelationship of archaeology and alluvium in the Herefordshire valleys, and in particular in the Lugg valley of north Herefordshire. Although the two projects are linked, their points of origin and their aims are rather different.

The Herefordshire Valleys Survey was a one year pilot study, run by the Archaeology Section of Hereford and Worcester County Council, and funded by English Heritage. Its brief took in a number of themes, including:

i) reviewing the existing evidence for prehistoric and Roman lowland settlement, and assessing the degree of preservation of environmental evidence, buried soils and other archaeological deposits,
ii) evaluating techniques and methodologies for prospection for and survey of deeply buried archaeology, in the face of threats from sand and gravel extraction,
iii) producing a policy for the management of well-preserved damp or wetland archaeology in this area, much of which may not be visible on the surface.

A broad approach was thus followed, involving the collection of data from a wide area, which encompassed all of the major river valleys in Herefordshire, and the application of evaluation techniques to a small number of sites across the county. A summary of the results is offered here. The full report is currently in draft (Dinn 1991), and is to be published as a report of the Hereford and Worcester County Council Archaeology Section.

The SERC Research Studentship studied a large river valley in this area in detail, using as a starting point data from a gravel quarry at Wellington in the lower Lugg valley, 8 km north of the city of Hereford. At this site, sections of more than 150 m in length and 2 to 3 m in depth have been exposed for several months at a time. Extensive Roman archaeological deposits are present over much of the quarry area at ca. 0.5 m below the modern ground surface.

GEOMORPHOLOGY, GEOLOGY AND ALLUVIUM

Herefordshire, now the western part of Hereford and Worcester county, is bowl-shaped, with a rim of hills and a central lowland. The Wye is the major river, and nearly all the rivers in the county, including the Lugg, drain into it; all of these rivers run through valleys which are alluviated to a variable extent. In addition there are a number of infilled pro-glacial lake beds, such as Lake Wigmore, which also contain alluvial sediments.

About half of the drainage basin of the Lugg (the north-west) is upland (over 150 m OD) with an underlying geology of mudstones and limestones of Silurian age, the south-eastern half being generally lower-lying ground on Devonian rocks (Fig. 13.1). The Devonian here is divided into the St Maughan's and Raglan formations. Rock types in both are similar mudstones, siltstones and sandstones with occasional calcareous bands — but the St Maughan's formation, largely to the east of the Lugg, has a higher percentage of sandstone, the more resistant rock type, and a greater frequency of outcrops of calcareous bands (Brandon 1989). Much of the lower-lying land is overlain by deposits of the last glaciation, consisting for the most part of locally derived (i.e. Silurian and Devonian) material. East of the Lugg there are small patches of till from an earlier glacial episode with rocks of provenance from the north and west (Clarke 1934; Luckman 1970; Brandon 1989; Hey 1991).

The Lugg at Wellington drains an area of approximately 750 km^2 and has an average daily mean flow of 11.29 m^3s^{-1}; for comparison, the Wye at Hereford drains 1900 km^2, with a daily mean flow of 45.75 m^3s^{-1} (National Rivers Authority, Monmouth). The floodplain at Wellington is 2 km wide and near-level (at ca. 55 m OD). There has been some discussion as to why a river with a mean flow one-quarter that of the Wye should have a floodplain in places twice its width. Merewether (1877) suggested that this plain was once a lake, while it has also been interpreted as the course of a larger river, either in the Miocene (Clarke 1934), or in the Pleistocene (the Teme, Lugg and Onny; Brandon 1989). Today this plain is frequently flooded during times of heavy flow as water backs up upstream from the confluence of the Lugg, Frome and Wye, 12 km to the south. In this part of the valley there are substantial

Fig. 13.1. Geology of the Lugg catchment and location of fieldwork sites.

deposits of fluvio-glacial gravels, which are up to 8 m deep at Wellington, deriving from the last glacial episode (Brandon 1989; Hey 1991). Over much of the width of the valley these are covered by up to 3 m of alluvium.

ARCHAEOLOGICAL AND PALAEOENVIRONMENTAL WORK IN HEREFORDSHIRE

Archaeology in Herefordshire has long suffered from a lack of systematic research. The limited work which has been done has tended to focus on obvious monuments on higher ground (for instance the hillforts), to the detriment of lowland archaeology. The most recent detailed overviews of prehistoric and Roman archaeology in the county are nearly 40 years old (Gavin-Robinson 1954; Dudley 1954), though Stanford (1980; 1991) has provided short summaries by period. Subsequent work, including aerial survey, some field survey, and the field element of the Herefordshire Valleys Survey, has begun to change our understanding of the archaeology of the area (Dinn 1991). The air photographic evidence for Shropshire and the surrounding areas (from SO 70 northwards) has recently been collated and published by Whimster (1989). There are many clear parallels between Herefordshire and the area to the north; similar types of cropmark are present (especially enclosures and ring-ditches), and in both areas they only appear at very irregular intervals.

Palaeoenvironmental work in the immediate area has been limited to a single pollen analysis study, from Rhosgoch Common on the south-western edge of the Lugg basin (Bartley 1960). This analysis of 1 m of undated upland peat shows a sequence of pine, followed by mixed deciduous woodland, and by a synchronous elm decline and rise in moss and heathers. One peat core shows in its upper layers a sharp increase in grasses which may be interpreted as anthropogenic.

THE HEREFORDSHIRE VALLEYS SURVEY: EXTENSIVE FIELDWORK

Fieldwork on sites in the county other than Wellington included auger transects across valleys, fieldwalking and geophysical survey (Geophysical Surveys 1990). Locations were selected with reference to the known presence of sites close to alluvium. Sample transects were set out in alluviated areas in all of the larger valleys in central and north Herefordshire: the Lugg, Wye and Frome, as well as the Worm Brook.

Although this work was necessarily limited in extent, it was sufficient to demonstrate the presence of archaeological deposits, partly or wholly covered by alluvium, in a number of areas. These include Roman settlement deposits at Stretton Grandison, a ditched enclosure at Castle Frome, and enclosures at Kingsland. The cross-valley transects also provide a context into which more detailed work can be fitted.

FIELDWORK AT WELLINGTON

Introduction

As a background to the two projects, it is necessary to examine in some detail the history of fieldwork at Wellington (Fig. 13.2). The Wellington Quarry area was not evaluated prior to the granting of permission for extraction (in 1985), for there was no known archaeological interest. The discovery in 1986 of the stone footings of a Roman villa, by the City of Hereford Archaeology Unit, led to an evaluation in 1987, by the Archaeology Section of Hereford and Worcester County Council (Clarke *et al* 1988). This showed the site to be both extensive and complex. Over most of the area examined Roman soil layers were preserved under alluvium of varying thickness. As a result, an area of over 4 ha was designated as a core area of archaeological importance, with the intention that this should be preserved *in situ* if at all possible. The quarry operators, Redland plc, have substantially altered their programme to accommodate this, and have funded salvage recording work during topsoil stripping since 1988.

The results of work at Wellington since 1988 have expanded the chronological sequence considerably, as well as enlarging on the environmental potential of the site. Three areas were salvage-recorded, to the south and west of the core area, during 1989 and 1990 (Edwards 1989; Shelley 1989; Edwards 1990). The salvage recording revealed Roman and other deposits and features over a wide area, two ring-ditches, and a number of infilled river channels. The Lugg today flows about 500 m to the east of the site.

Survey techniques

The core area, to the south of the villa building, has been examined using a number of techniques, to supplement the test pits dug as part of the 1987 evaluation. These included magnetometer survey, and an extensive augering programme (auger holes at 10 m intervals) which provided sufficient data to reconstruct the Roman topography, as well as samples (taken at 0.2 m vertical intervals) for phosphate and magnetic susceptibility analysis. The long sections mentioned earlier have been recorded in detail (Fig. 13.4), and samples from the alluvial sequence have been taken for micromorphological and chemical analysis, and for radiocarbon and infrared-stimulated luminescence dating. A number of dates are now available (see Appendix). Environmental investigation of the site has included the assessment of samples for pollen (by James Greig, University of Birmingham), plant macrofossils and snails.

Magnetometer survey by John Gater (Gater and Gaffney 1989) revealed the presence of an enclosure, possibly double-ditched, to the south of the villa. Although these enclosures are now known to be common in the region, this example is likely to be significantly better preserved than most, due to the overlying layer of alluvium. To the west of the enclosure was a stream channel. Both the channel and the interior of the enclosure showed as areas of magnetic 'noise'.

The ditches and channel were also recorded in the auger survey. Through this survey it has been possible to determine the varying level of the Roman ground surface and other layers, while a computer surface modelling program has allowed the rapid production of

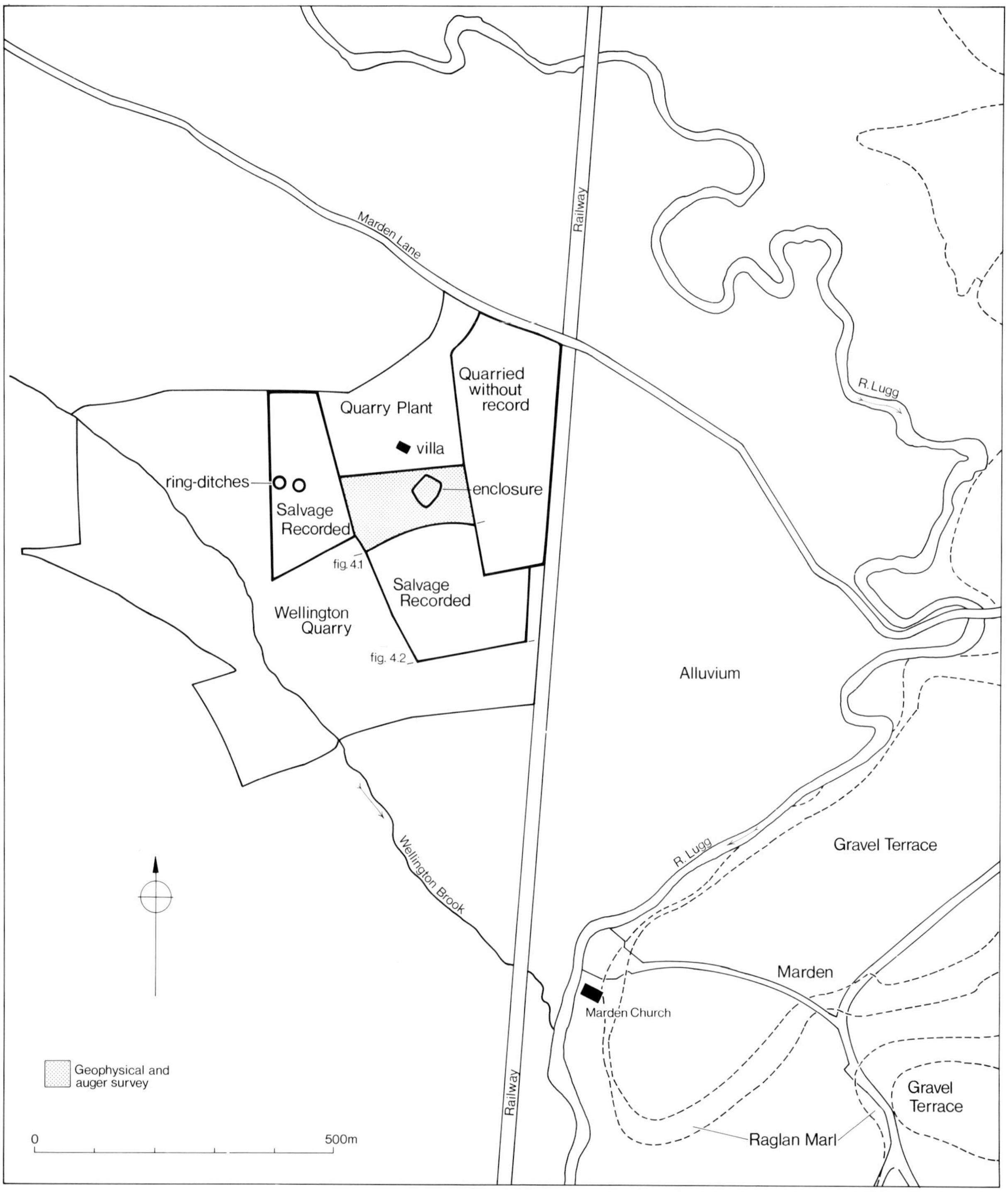

Fig. 13.2. Wellington: fieldwork areas and major recorded features.

isometric views and contour plots of buried deposits (Fig. 13.3), for visualisation and interpretation. It is also possible, using this program, to calculate or plot the volume of material between the modern ground surface and early deposits, or the volume of surviving occupation deposits, which can help in planning excavation work.

Geophysical survey and augering may be used together, to provide a plan view and a three-dimensional ground surface on which to place that plan. More work is required on phosphate and magnetic susceptibility testing to assess their utility in alluviated conditions; in particular, the results should be checked against data from any subsequent excavations. Trial trenches may also be necessary to assess the character of the deposits seen through augering. It will only be if the core area of the site cannot be preserved *in situ* that it will be possible to achieve a full assessment of the techniques, by testing them to destruction.

Study of sections at Wellington

The description below uses Soil Survey nomenclature (Hodgson 1976) and Munsell soil colour codes. Fig. 13.4 is a simplified synthesis of two east-west sections; these were drawn at 1:20, and thus only the most remarkable features can be shown here. The sections are about 200 m apart. About 100 m to the west, the alluvium thins to ca. 1 m, of the same material as the uppermost unit of the north section.

Sediments can be broadly divided into two types, backwater channel fills containing much shell and organic matter, and non-channel fills. Non-channel fills are stone-free, clearly laminated in places, remarkably uniform in texture (silty clays and silty clay loams), and either reddish brown (5YR) or brown (7.5YR) in colour. The sediment colour may derive either from the parent material (the reddish brown from Devonian areas, the brown from Silurian), or from post-depositional changes in iron oxide content and form induced by wetting and drying cycles. Future work will establish this. Channel fills are also stone-free silty clay loams, composed of different coloured sediments, but none are directly comparable to non-channel fills in colour, structure or texture. They contain large amounts of broken mollusc shell and organic matter, and in places bands of tufaceous material, heavier concentrations of shell and patches of gley. In section they are prismatic in structure and not laminated, with many reddish yellow (7.5YR) and grey mottles.

Description of north section *(Fig. 13.4a)*

At the base are glacial gravels from the last glaciation (Brandon 1989; Hey 1991). These are red, well sorted, rounded, and banded in places. They are not level, but rise and dip, and this is reflected to some extent in overlying layers and surface contours which follow the same pattern. They are overlain by an unsorted red (2.5YR) clay and tufaceous gravel mix of ca. 0.5 m in depth. Above this is ca. 0.75 m of yellowish red (5YR) stone-free silty clay loam (27% clay), clearly laminated in places. This is overlain by 0.10 m of heavily gleyed

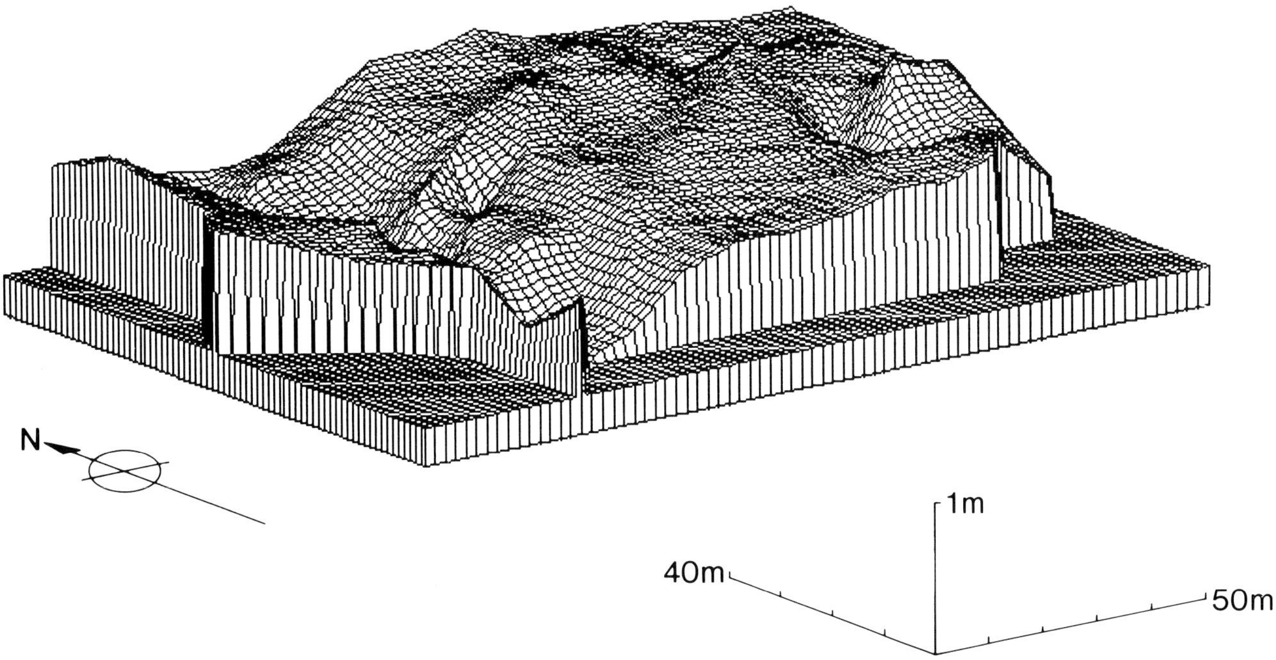

***Fig. 13.3.** Wellington: isometric model of Roman ground surface, interpreted from auger samples.*

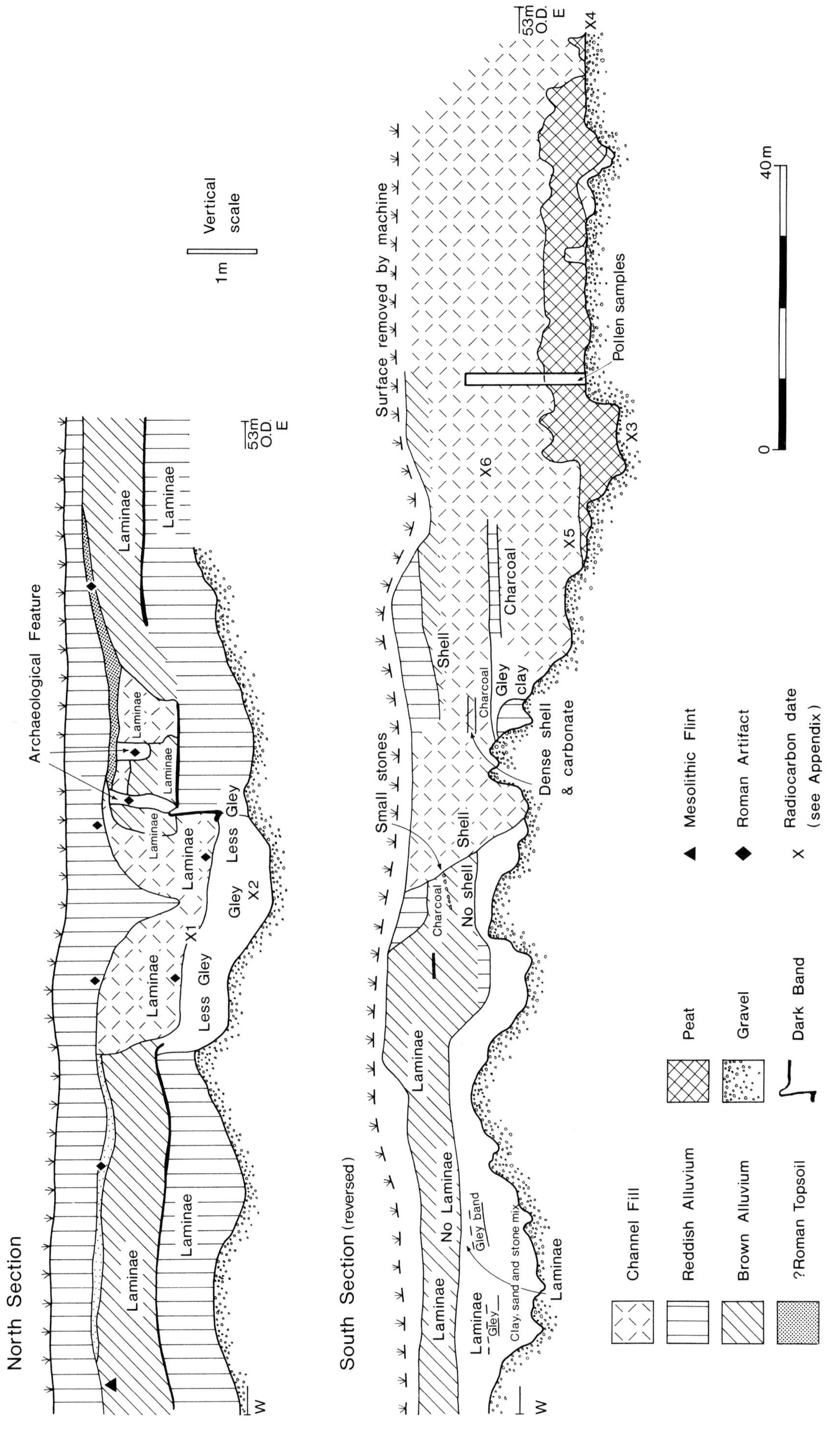

Fig. 13.4. Wellington: east-west sections (vertical scale exaggerated).

silty clay (38% clay) and a 0.10 m thick dark band of clay and humic matter at 54.10 m OD. Above this there is a clear boundary and about 1 m of brown (7.5YR) silty clay loam (30% clay), in places laminated and mottled with reduced and oxidised iron, increasingly more so with depth. Overlying this is 0.20 m of dark soil. To the east of the channel this soil layer is very dark and contains many Roman artifacts; to the west it is not immediately apparent, and there are fewer Roman artifacts, but soil thin sections taken from this level show faunal activity and clay cutans indicating this to be a soil. Overlying this and over the entire site is 0.5 m of non-laminated, stone-free, red brown (5YR) silty clay (37% clay).

Several Mesolithic or Early Neolithic flints have been found in or very near Roman levels. There is one large channel feature, about 40 m in width, cutting through this stratigraphy. Two radiocarbon dates were obtained from this, at the base 478–127 cal BC , and midway up cal AD 680–990. A programme of infrared-stimulated luminescence dating has produced dates for two of the deposits; three dates were obtained for a sample from a layer stratified within the Roman sequence, probably on the edge of the channel (AD 170, 200 and 280, all ± 200; lab ref 784b), and there was a single date of AD 1030 ± 250 (lab ref 784a) from the alluvium immediately above the Roman topsoil (Spooner *et al* 1990: 86; Martin Aitken, pers. comm.).

Description of south section *(Fig. 13.4b)*

In the west the gravels rise steeply and are overlain by an unsorted and compacted clay, sand and stone mix. This is overlain by the brown (7.5YR) silty clay loam comparable to that of the north section and the 0.50 m of red silty clay which is uniform over the site. There is no evidence of a Roman soil. In the east the gravels dip and peat has developed in the declivity. Analysis of plant macrofossils from this indicate that it formed from an alder carr with rather disturbed conditions nearby, the sort of vegetation to be found for example on an unstable river bank: stinging nettle, elder, thistle, blackthorn etc. Preliminary pollen results from a 1.5 m column, taken from the base of the peat into the overlying clay with humic matter, shows the presence of oak/elm/lime and indicate that much of the peat deposit is Atlantic. Towards the top of the column (at 52.91 m OD), from samples taken in gleyed alluvium, there are signs of forest clearance, and at 53.50 m OD grass and cereal type pollen are present (James Greig, pers. comm.). Radiocarbon dates substantiate this, with dates of 5900–5630 cal BC from the base of the peat and 3980–3640 cal BC above its surface. A date of 1510–1129 cal BC was obtained 10 m to the west of the pollen core at the same level as the cereal pollen sample (± 0.2 m). This was derived from charcoal found within a 0.1 m thick band of fine gravelly tufaceous material which overlay a gleyed clay layer with organic matter similar to that of the pollen sample. They are interpreted as being broadly contemporary.

The peat has formed in a channel which cuts through brown sediment of the same texture and colour as that in the north section. Overlying the peat are channel fills that are increasingly gleyed and humic towards the base and grade into the peat.

Interpretation of sections

The structure and particle size of the non-channel fill deposits suggest they have been formed by overbank sedimentation (Allen 1965; Selley 1970; Miall 1987). This alluvium has been deposited quickly, for the laminae which are clearly observable in section have not been disturbed by faunal and floral activity as would occur in a period of stasis. Channel fill deposits, with their high percentage of clay and silt, have accrued from slow moving water that has already deposited its heavier particles and represent vertical accretion in the depressions resulting from former watercourses, where flood water might naturally be expected to flow and remain.

North Section

The gravels and unsorted clay and gravel mix were deposited unequally across the site, in a fluvio-glacial environment. These were overlain by flood deposits of vertical accretion type which quickly built up, perhaps in the early Postglacial before vegetation was established. The dark band is interpreted as a humic gley soil, and represents a period of stasis. Another period of aggradation by overbank sedimentation of a brown silty clay loam followed. In the south, brown sediment of the same texture and colour as that of the north section is cut by a large migrating channel with a radiocarbon date at its base of 5900–5630 cal BC It is reasonable to assume that this sediment was deposited at the same time and that it therefore pre-dated 5900–5630 cal BC .

Mesolithic or Early Neolithic flints in or very near Roman levels in the north section indicate that soils have formed and been eroded between the Mesolithic and Roman periods and soil thin-sections do show evidence of this.

The dark soil layer east of the channel has been interpreted as a Roman soil. The presence of Roman finds at the base and surface of the channel fill shows the stream was flowing during the Roman period and for some time after. There is a radiocarbon date of cal AD 680–990 within the channel. Roman features are found within 1–2 m of this contemporary stream, and it seems likely that the stream was utilised as a resource in some way; finds of iron slag and hammer scale indicate ironworking, and the dark colour of the soil to the east of the stream might be the result of burning and incorporation of waste from industry. The main Roman occupation was to the north of this section, and it seems that the south-eastern area at this time was lower-lying and subject to frequent flooding, and therefore not used in the Roman period.

The dates from the channel and the dark soil layer suggest that the channel in the north section was open for between 950 and 1800 years; even at the minimum period this makes it quite a stable stream. At about 40 m wide, it is about as large as the present day Lugg. However, there is no sign of it at all to the south, even 100 m away in a section temporarily exposed during gravel extraction, though smaller channels were exposed, with Roman finds in their fills.

This stream may represent a position of the Lugg from the mid to late second century BC (latest date for inception) to the later first millennium AD (the radiocarbon date at the base of the channel fill), which was managed during the Roman period in this stable form, perhaps for use in industry. To the south it divided into a braided channel in a lower-lying and aggrading area.

South Section

To the west the gravels rise steeply, and are overlain by an unsorted glacial deposit. There is a clear boundary marking an erosion episode and then a similar sequence to that of the north section (see above). To the east is a large area of channel fill deposits, more than 120 m in width. Part or all of this lower lying area must have at one time been a watercourse, in the early sixth millennium BC or before, which created the cut observed to the west. It occupies a depression in the gravels, with a width several times greater than the modern Lugg, and must have been created by a laterally migrating stream in glacial, early Postglacial or Flandrian times.

A period of stasis followed, of some 2000 years, represented by the peat which formed from alder carr.

At some time in the early fourth millennium BC this channel was filled by vertically accreting deposits. The average annual rate of aggradation within the channel was about 0.5 mm. For the uppermost layers representing aggradation from about AD 1000 to the present day, the average annual rate is between 0.4 mm (minimum) and 0.7 mm (maximum). Overbank deposits observed at this site after floods in January 1990 were in places 2–10 mm in depth, with several 1mm laminae in the form of darker and lighter bands (perhaps reflecting organic matter content) observed from the one event. It would be wrong therefore to say that one lamina equals one flooding event, and it is not suggested here that deposits were laid down regularly. However, the figures are comparable to rates observed by authors elsewhere (Alexander and Prior 1971; Shotton 1978; Chester and James 1990) and the dates do indicate an overall steady accretion.

Overall

A parish boundary 500 m east of the site follows the present day twisted course of the Lugg (see Sheppard 1979). Parish boundaries are generally held to have been established at least by 1180 AD, and in most cases a good deal earlier (Rackham 1986: 19). The radiocarbon date of 680–990 cal AD at the base of the channel fill in the northern section may represent a *terminus post quem* for river course change of the Lugg into its present day course. Alternatively this channel represents a tributary of the Lugg when it was more in the form of a braided river. The change to the incised meandering form of today must have taken place at least by the time of the establishment of the parish boundary.

In post-Roman times 0.5 m of red alluvium has built up. The radiocarbon date of 680–990 cal AD in the channel fill underlying this alluvium, and the infrared-stimulated luminescence date of AD 1030 ± 250 at its base give *termini post quem* for its deposition. Ridge and furrow survives as earthworks in the fields to the east of the quarry; there appears to have been little alluviation since its abandonment, and thus this can be used as a *terminus ante quem*. Estate and tithe maps of the early 19th century indicate that the area was used as meadow rather than open-field arable at this period. This seems to bracket the deposition of the red alluvium to the period between Late Saxon and the 19th century AD.

Summary of interpretation

Gravels and unsorted clay and gravel were deposited unequally over the site in a fluvio-glacial environment, and were overlain quickly by a reddish brown silty clay. Before the early sixth millennium BC, a period of stasis represented by a humic gley soil followed, and there was then a further period of aggradation by overbank sedimentation. Subsequently, a laterally migrating river cut a channel in the south. A period of stasis, represented by the peat, occurred between the sixth and early fourth millennia BC in this lower lying southern area. This channel filled with flood-water deposits at an average rate of 0.5 mm per annum. In the higher northern part of the site, where Mesolithic and Roman artifacts are found in close association, a stable period is also indicated, representing a period of 5000–7000 years, though erosion of surface soil also took place. In the Roman period the northern higher ground was occupied by a villa, a ditched enclosure, and associated small-scale ironworking. At this time a river flowed through the site, breaking into at least two smaller streams to the south. This northern channel filled some time after the later first millennium AD, since when 0.5 m of alluvium has been deposited over the site. The Lugg today is confined to a meander river 500 m to the east, and parish boundaries indicate that it has occupied this course for at least 1000 years. Three main periods of aggradation are represented, one in the early Postglacial, one in the early fourth millennium BC or later, and one in the Late Saxon period.

THE MANAGEMENT OF ARCHAEOLOGY IN THE HEREFORDSHIRE VALLEYS

The problems of archaeological prospection in the Herefordshire river valleys are compounded by the lack of survey in these areas in the past, due to their low perceived potential. The continued development of prospection techniques, in particular non-destructive techniques, remains a major priority. Alluviated sites have a particular potential to yield information, which extends beyond the physical boundaries of the site itself; this potential has been summarised by Needham (1989: 29–32), and the results of the work at Wellington show what may be learned from a single site in Herefordshire. These sites are very vulnerable, especially as effective protection may depend on known archaeological significance.

A number of counties, in particular Berkshire and Cambridgeshire, have pioneered the concept of site and landscape ranking. Standard criteria are used to divide landscapes into a number of categories for minerals planning purposes, ranging from 'preferred' areas to 'areas of maximum objection'. Although the rankings produced have no statutory force, they make it possible to define, on archaeological grounds, areas where mineral extraction would be least damaging. This ranking system should enable an appropriate archaeological response to be made to any application for extraction. The ranking categories suggested here (Table 13.1) are based on those in operation in Cambridgeshire and Berkshire (French and Wait 1988), which pay particular regard to the potential of alluviated areas.

In areas where development is proposed, a staged approach to evaluation should be followed. This should attempt to define, in increasing detail, areas of both high and low archaeological potential. A first stage is to identify the presence and location of archaeological deposits. Subsequent stages should aim to define the extent (both horizontal and vertical) of deposits, and to assess their character, and to test apparently 'blank' areas.

The recommended approach combines the strategic grading of sites and areas, and a requirement for staged evaluation of proposed development areas before the consideration of a planning application. In this way the possibility of the preservation of significant deposits *in situ* is kept open, and the special potential of alluviated sites is acknowledged.

Acknowledgements

Thanks are due to the funding bodies, English Heritage and SERC, for their support for the two projects, and to Anthony Streeten and Helen Keeley of English Heritage, and Susan Limbrey of the University of Birmingham, for their advice, interest and encouragement throughout. The authors are grateful to the quarry operators, Redland plc, for funding the continuing salvage recording programme, and for allowing virtually unrestricted access for other fieldwork and recording.

The following staff of the Archaeology Section provided valuable help and advice: Adrian Tindall, Rachel Edwards, Simon Woodiwiss and Malcolm Cooper. The illustrations were drawn by Laura Templeton (Figs. 13.1 and 13.2) and Nigel Dodds (Fig. 13.4). Augering and other fieldwork was carried out by John Hemingway, Rodney Cottrill, Dave Cox, Ruth Bruniges, and Kath Evans.

Specialist services and reports were supplied by John Gater and Geophysical Surveys; James Greig, University of Birmingham; Martin Aitken and Rupert Housley, University of Oxford. The very large number of phosphate and magnetic susceptibility samples from Wellington were processed by Nuala Moran, who deserves especial thanks; our thanks also go to the Departments of Archaeological Science and

Table 13.1. *Proposed ranking categories for archaeological sites and landscapes in Hereford and Worcester*

Category	Description	Archaeological response	Planning designation
1	Scheduled ancient monuments	Preservation *in situ*	Maximum objection
2	Areas of high potential (often alluviated)	Preservation *in situ*, evaluation if development proposed	Maximum objection
3	Recognised significant archaeological sites	Evaluation	Restricted
4	Unknown archaeological potential	Evaluation	Prospect
5	Known low archaeological significance	Watching brief	Preferred
6	Archaeological deposits known to be absent	None	Preferred

Environmental Science at Bradford University for their generosity with the use of their facilities.

References

Alexander, C.S. and Prior, J.C. 1971. Holocene sedimentation rates in overbank deposits in the Black Bottom and the Lower Ohio River, Southern Illinois. *American Journal of Science*, 270, 361–72.

Allen, J.R.L. 1965. A review of the origin and characteristics of recent alluvial sediments. *Sedimentology*, 5, 89–191.

Bartley, D.D. 1960. Rhosgoch Common, Radnorshire. Stratigraphy and pollen analysis. *New Phytologist*, 59, 238–62.

Brandon, A. 1989. *Geology of the country between Hereford and Leominster*. London: Her Majesty's Stationery Office.

British Geological Survey. 1990. *Mid-Wales and Marches. 1:250 000 map series. Solid geology*. Keyworth.

Chester, D.K. and James, P.A. 1990. Holocene alluviation in the Algarve, Southern Portugal: the case for an anthropogenic cause. *Journal of Archaeological Science*, 18, 73–89.

Clarke, A.T., Taylor, G.J. and Woodiwiss, S.G. 1988. Evaluation excavation at Marden Quarry, Wellington (HWCM 5522). Internal report, Archaeology Section, Hereford and Worcester County Council.

Clarke, B.B. 1934. Geomorphology of the Lower Wye Valley. Unpublished dissertation. University of Birmingham.

Darvill, T.C. 1987. *Ancient monuments in the countryside: an archaeological management review*. London: Historic Buildings and Monuments Commission.

Dinn, J.L. 1991. The Herefordshire Valleys: an assessment of survey techniques and archaeological potential. Draft report. Worcester: Archaeology Section, Hereford and Worcester County Council.

Dudley, D.R. 1954. The Herefordshire area in the Roman period. In F. Morgan (ed.), *Herefordshire: its natural history, archaeology and history*, 120–9. Gloucester.

Earp, J.R. and Hains, B.A. 1971. British regional geology. *The Welsh borderland*. London: Her Majesty's Stationery Office.

Edwards, R.E. 1989. Salvage recording at Wellington Quarry, Marden Lane (HWCM 5522). Internal report 27. Worcester: Archaeology Section, Hereford and Worcester County Council.

Edwards, R.E. 1990. Salvage recording III at Wellington Quarry, Marden Lane (HWCM 5522). Internal report 49. Worcester: Archaeology Section, Hereford and Worcester County Council.

French, C.A.I. and Wait, G.A. 1988. *An archaeological survey of the Cambridgeshire river gravels*. Cambridge: Cambridgeshire County Council.

Gater, J.A. and Gaffney. C. 1989. Report on geophysical survey, Wellington Gravel Quarry, Herefordshire. Unpublished typescript report. Bradford: Geophysical Surveys.

Gavin-Robinson, R.S. 1954. Prehistoric man in Herefordshire. In F.C. Morgan (ed.), *Herefordshire: its natural history, archaeology and history*, 107–119. Gloucester.

Geophysical Surveys. 1990. Report on geophysical survey: Hyde Farm, Castle Frome, Lower Kinsham and Kingsland. Unpublished typescript reports 90/21–24. Bradford: Geophysical Surveys.

Hey, R.W. 1991. Pleistocene gravels in the Lower Wye Valley. *Geological Journal*, 26, 123–36.

Hodgson, J.M. 1976. *Soil Survey field handbook*. Soil Survey technical monograph 5. Harpenden.

Luckman, B.H. 1966. Some aspects of the geomorphology of the Lugg and Arrow valleys. Unpublished MA thesis. University of Manchester.

Luckman, B.H. 1970. The Hereford Basin. In C.A.Lewis (ed.), *The glaciations of Wales and adjoining regions*. London: Longman.

Merewether, F. 1877. Drift at Pentelow Brook. *Transactions of the Woolhope Naturalist and Historical Field Club*, 9, 18.

Miall, A.D. 1987. Recent developments in the study of fluvial facies models. In F.G. Ethridge, R.M. Flores and M.D.Harvey (eds.), *Recent developments in fluvial sedimentology*. Society of Economic Palaeontologists and Mineralogists Special Publication 39. 1–13.

Needham, S. 1989. River valleys as wetlands: the archaeological prospects. In J.M. and B.J. Coles (eds.), *The archaeology of rural wetlands in England*. 29–34.

Rackham, O. 1986. *The history of the countryside*. London: Dent.

Rowley, R.T. 1986. *The landscape of the Welsh Marches*. London: Michael Joseph.

Selley, R.C. 1970. *Ancient sedimentary environments*. London: Chapman and Hall.

Shelley, D.C. 1989. Salvage recording at Wellington Quarry, Wellington. Internal report 32. Worcester: Archaeology Section, Hereford and Worcester County Council.

Sheppard, J.A. 1979. *The origins and evolution of field and settlement patterns in the Herefordshire manor of Marden*. Department of Geography, Queen Mary College, University of London. Occasional Paper 15.

Shotton, F.W. 1978. Archaeological inferences from the study of alluvium in the Lower Severn-Avon valleys. In S. Limbrey and J.G. Evans (eds.), *The effect of man on the landscape: the Lowland Zone*, 27–32. London: Council for British Archaeology Research Report 21.

Spooner, N.A., Aitken, M.J., Smith, B.W., Franks, M. and McElroy, C. 1990. Archaeological dating by infrared-stimulated luminescence using a diode array. *Radiation protection dosimetry*, 14 no 1/4, 83–6.

Stanford, S.C. 1980. *The archaeology of the Welsh Marches*. London: Collins.

Stanford, S.C. 1991. *The archaeology of the Welsh Marches*. 2nd edition. Leinthall Starkes (privately published).

Whimster, R. 1989. *The emerging past*. London: Royal Commission on Historical Monuments.

APPENDIX
Dates from Wellington

Radiocarbon dates

Site ref.	Lab ref.	Material	δ13C	Age BP	Cal date range (2 sigma)
1	OxA-2878	red deer bone	-22.8%	1180 ± 60	AD 680–990
2	OxA-2879	oak wood	-29.7%	2260 ± 70	BC 478–474,410–170,139–127
3	OxA-2880	willow wood	-29.3%	6790 ± 80	BC 5820–5500
4	OxA-2881	willow wood	-29.5%	6930 ± 80	BC 5900–5630
5	OxA-2882	oak charcoal	-24.2%	4990 ± 70	BC 3980–3640
6	OxA-2883	indet.charcoal	-26.7%	3070 ± 70	BC 1510–1150,1148–1129

All the uncalibrated dates are given in radiocarbon years BP, using the half-life of 5568 years and the measured δ13C values quoted (to within ± 1.0% only). The results have been normalised to -25% in accordance with the usual convention.

Calibration has been carried out using the CALIB program, and the following calibration curves:

Samples 1, 2 and 6: Stuiver, M. and Becker, B. 1986. High-precision decadal calibration of the radiocarbon time scale, AD 1950–2500 BC. *Radiocarbon,* 28, 863–910.

Sample 5: Pearson, G.W., Pilcher, J.R., Baillie, M.G., Corbett, D.M. and Qua, F. 1986. High-precision ^{14}C measurement of Irish oaks to show the natural ^{14}C variations from AD 1840–2500 BC. *Radiocarbon,* 28, 911–34.

Samples 3 and 4: Bidecadal weighted average of data from:

Linick, T.W., Suess, H.E. and Becker, B. 1985. La Jolla measurements of radiocarbon in South German oak tree-ring chronologies. *Radiocarbon,* 27, 20–32. [for the interval 5219–7199 BC].

Stuiver, M., Kromer, B., Becker, B. and Ferguson, C.W. 1986. Radiocarbon age calibration back to 13,300 years BP and the ^{14}C age matching of the German oak and US bristlecone pine chronologies. *Radiocarbon,* 28, 969–79.

Kromer, B., Rhein, M., Bruns, M., Schoch-Fischer, H., Munnich, K.O., Stuiver, M. and Becker, B. 1986. Radiocarbon calibration data for the 6th and 8th millennia BC. *Radiocarbon,* 28, 954–60. [for the interval 5229–7207 BC].

Linick, T.W., Long, A., Damon, P.E. and Ferguson, C.W. 1986. High-precision radiocarbon dating of bristlecone pine from 6554 to 5350 BC. *Radiocarbon,* 28, 943–53.

Infrared-stimulated luminescence dates

Lab reference	Age BP	Date range	Location
784a	920 ± 200	1030 ± 250 AD	Above Roman topsoil layer
784b	1670 ± 200	280 ± 200 AD	Edge of channel?
784b	1750 ± 200	200 ± 200 AD	Edge of channel?
784b	1780 ± 200	170 ± 200 AD	Edge of channel?

SECTION 4

Archaeology and alluvial landscapes in the English Midlands: lowland river systems

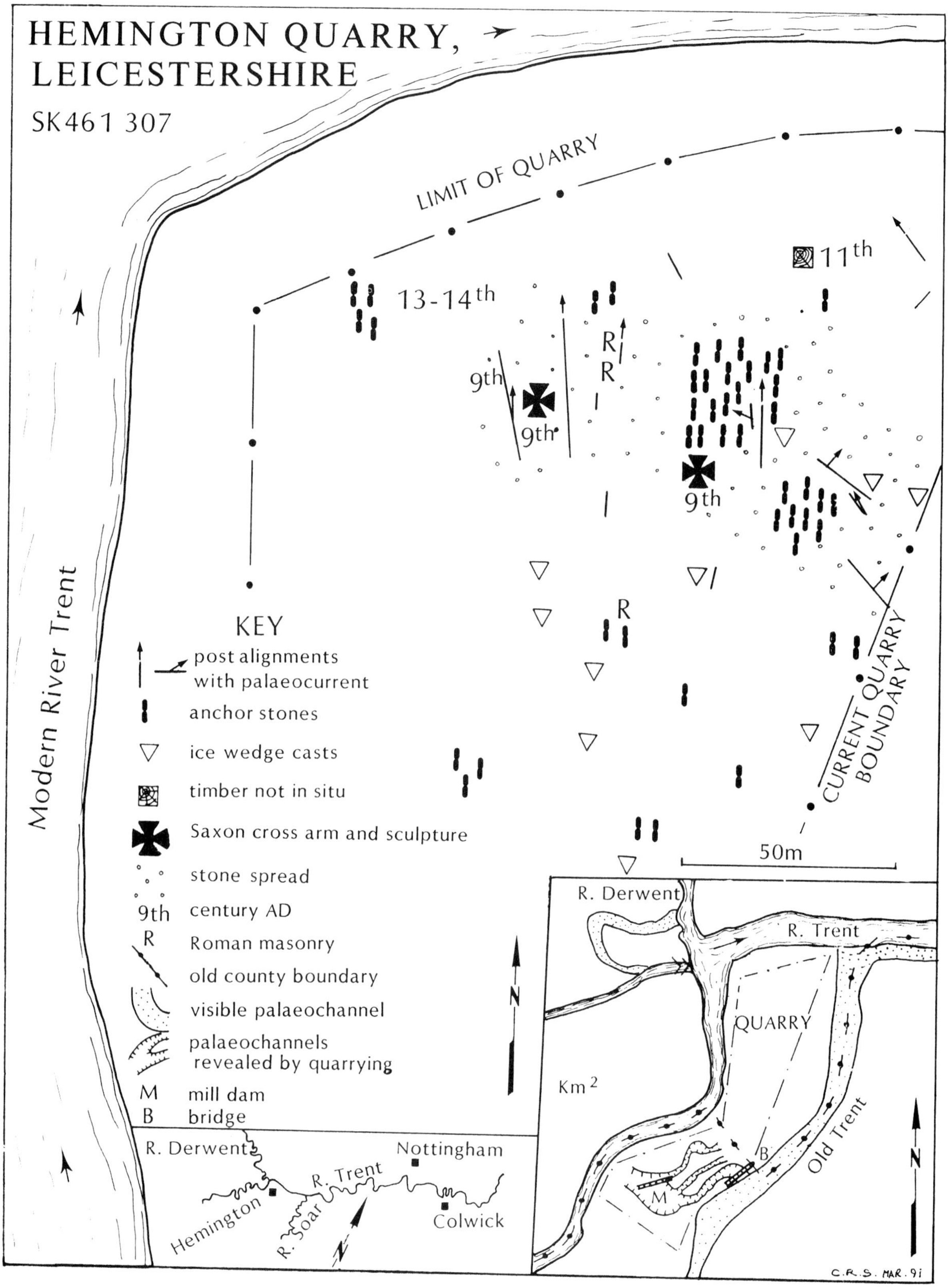

Fig. 14.2. Hemington Quarry. The southern area associated with the mill dam and bridge has coherent palaeochannels. The northern area suggests a continuous migration of the river (Trent or Derwent?) to the east. The post alignments are probably mainly fishweirs although some may be revetments, as may be the stone spread.

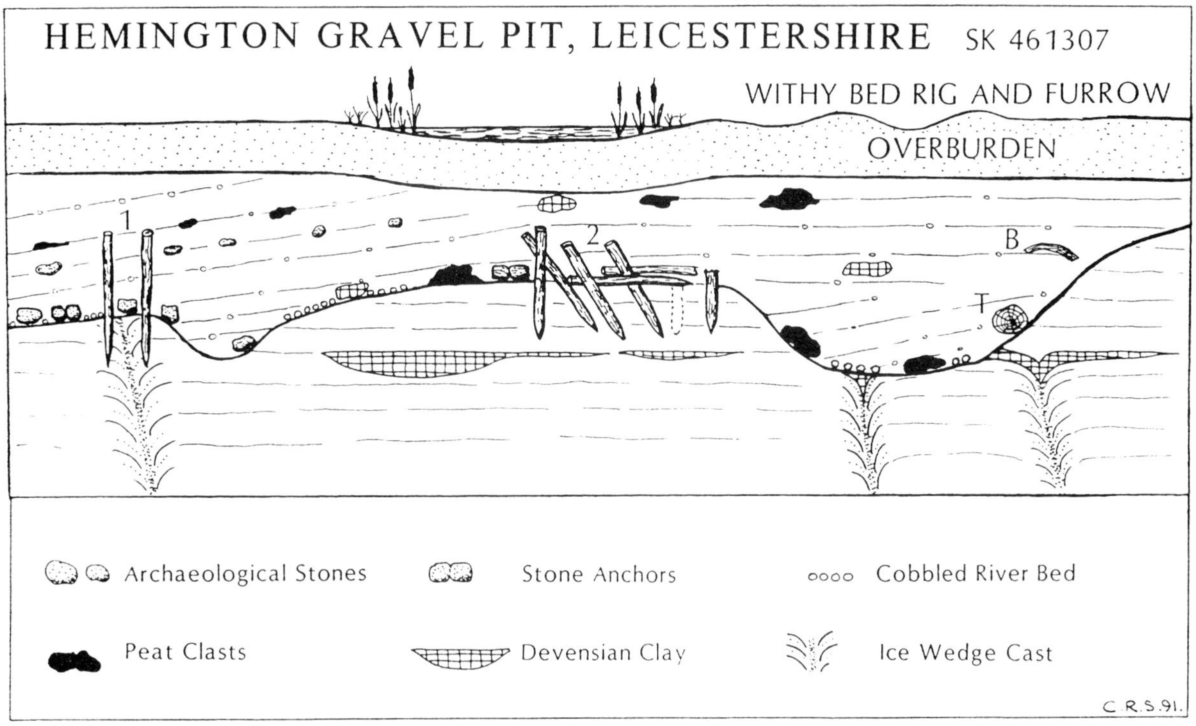

Fig. 14.3. *A composite section of the northern area of the gravel pit at Hemington. (1: a post alignment (fishweir?) driven into an ice-wedge cast; 2: a damaged alignment indicating current direction; T: tree trunk; B: boat timber).*

(intercept, 2–sigma; CAM QF3–1985). The first was found in the same channel as two anchor stones and the last lay beneath the double row of massive squared posts of oak, each 18–22 cms wide and up to 5 m long, which formed a mill dam. This dam has been dendrochronologically dated by Robert Howard of the Nottingham University Tree Ring Dating Laboratory, giving a felling date with a 98% confidence level of AD 1127–1158. A series of radiocarbon and dendrochronology dates for the river structures in the north of the Hemington pit is currently being obtained with the help of funding from English Heritage and the Kiln Trust. This will indicate the rate of lateral movement of the river channel. The first result is from a post alignment with a 9th century AD date (Fig. 14.2).

An unusual feature of the gravel pit at Hemington is a spread of large stones of Triassic and Carboniferous sandstone over the Medieval river bed. These have an average weight of 19 kilograms, although there is great variation in size, with the largest approaching 100 kilograms. The total weight is expected to reach 200 tonnes when gravel quarrying ceases in about two years. They comprise a mixture of ancient stone quarry waste, millstones, querns and unfinished or re-used building stones, some of which are Roman (Blagg, T F C, pers. comm.), and include the arm of a ninth century Saxon cross and another carved stone of the same period (Gaimster *et al* 1990: 199). Known quarries upstream of Hemington and close to the Trent occur at Castle Donington, of Triassic sandstone, and at Melbourne, of Carboniferous millstone grit. Carboniferous sandstone is also found in the catchment area of the Derwent. In Saxo-Norman times hardcore could have been shipped to Hemington both from these quarries and from abandoned buildings in former Roman towns such as Derby and Leicester. At Hemington the hardcore was used to form an armoured bed or as bank revetments to stabilise a very mobile river.

Some of these stones have been re-used yet again as anchor stones by the cutting of grooves to hold twisted willow withy bands (Figs. 14.2, 14.4). One of these withies has been given a radiocarbon date in the 13th-14th centuries: 1175–1410 cal AD (Groningen 2–sigma; OxA-2289). Seventy-seven anchor stones have been found in Hemington pit and seem to be a unique artifact type in British inland waters. They were almost certainly used to hold fishing tackle and occur most frequently in the vicinity of post alignments (fishweirs). The ratio of Triassic to Carboniferous anchors is 48% to 52%, very close to that of the stone spread which is 52% to 48%, suggesting that the stones have been chosen randomly to make anchors.

At Hemington the lower suite of gravel is an undisturbed fluvial glacial deposit probably of Devensian age. Its interface with the upper suite of disturbed meander belt gravel is usually marked by the truncation of ice wedge casts, of which 33 have been found. In some areas this interface rises to within 1.5

Fig. 14.4. Stone anchors for Medieval fishing tackle grooved to hold a twisted withy rope. Centre top is a re-used Roman block, complete with lead clamp grooves. Scale in centimetres.

Fig. 14.5. Quarry face at Hemington gravel pit showing a post alignment set into the Saxo-Norman river bed above an ice wedge cast (post 1 in Fig. 14.3). The channel fill is partly obscured by quarry scree.

metres of the top of the gravel, and it is then most likely that channels cut into the Devensian gravel are either avulsions or man-made. One such channel, probably man-made, was spanned by a double row of clustered piles with an approximate felling date of AD 1214 ± 10 established by dendrochronology (Fig. 14.2). This is probably a bridge spanning the tail race of the Norman mill, of which the dam was found in the gravel 200 metres upstream (Clay 1986, Clay and Salisbury 1990). Two of the 18 post alignments have been built directly into ice wedge casts (Fig. 14.5, Fig. 14.3, post 1) suggesting that the cast may have offered a firmer and slightly raised area of river bed which attracted the weir builders. A discontinuous stratum of very pure clay lies in the Devensian gravel and contains microcrystalline fissures typical of tundra conditions (A. Brown pers. comm., Fig. 14.3). It was deposited at a time of very low discharge in the river, and when present, caps the ice wedge casts and forms part of their contents (Fig. 14.3). Reworked clasts of this clay are frequently found in the upper suite of gravel.

Hemington, at the confluence of the rivers Trent and Derwent, and only some two kilometres upstream of the Soar Trent confluence was well positioned for an inland port serving the towns of Derby and Leicester (Fig 14.2). This suggestion is strengthened by the presence of a Medieval road to the south called "the Portway" which leads to the bridge postulated above (pers.comm. F. Hartley, Leicestershire Museums).

Sub-fossil oak trunks are frequently uncovered during quarrying. A total of 124 such trunks were studied at Colwick; 64 examples had enough rings to be worth counting, of which 38 were correlated to Bronze Age and Neolithic floating chronologies (Salisbury 1984: 192, 202). It is postulated that whole stands of trees would be uprooted by a flood. These would have been carried down the ancient course, crown first, roots dragging behind, eventually becoming waterlogged and sinking to the river bed. Each weighing up to 16 tonnes, they would barely have been moved by subsequent river meandering and so, when uncovered by quarrying, they still lay in a broad sinuous band indicating the ancient river course. Seventy-five per cent of the trunks had their roots upstream, confirming the direction of flow as established by observation of the gravel deposits. The Neolithic group of trunks has recently formed part of the complex of chronologies that has enabled Michael Baillie to date more accurately the Bronze Age Sweet Track, a wooden path found beneath peat deposits in the Somerset Levels (Hillam *et al* 1990: 210–9).

A full account of the results of surveillance at Colwick can be found in Salisbury (1984) and the fishweirs and anchor stones recorded at Hemington to date are described in Salisbury (1991a).

FUTURE RESEARCH

It may be that ground probing radar or other ground geophysical techniques may help to reveal these river structures (Allsop 1990), but their flimsy nature and depth of 3–4 metres beneath the surface makes this very unlikely. Rapidly filled channels contain gravel of the same type as that surrounding them and are unlikely to present anomalies detectable by geophysical techniques. For the foreseeable future, regular surveillance of gravel quarrying is the only feasible method of recording these ancient riverscapes – an ideal project for amateurs.

Acknowledgements

I wish to thank Ennemix Aggregates Limited for allowing me access to the Hemington Gravel Pit and especially their employee, Mr Harry Gasson, who has been a constant source of help. Leicestershire Archaeological Unit with whom I work as an independent agent, has given me continuing support and encouragement throughout the six years surveillance at Hemington and sharing direction with Pat Clay has always been a special pleasure. Archaeological and geological advice has been given by the following: Jenny Alexander, Jenny Allsop, Tony Brown, Tim Charsley, Robert Howard, Peter Jones, Geof Petts, Harold Potter, Keith Reedman, Gavin Simpson, Mick Stanley and Howard Usher. Generous grants from the Kiln Trust and English Heritage have provided an extensive and continuing programme of dendrochronological and radiocarbon dating. Finally I am particularly grateful to my wife Hazel (nee Wheeler) for doing all the typing and editing as well as ensuring that my archaeology never strayed into heresy or fancy.

References

Allsop, J. 1990. Recent Applications of Geoprospection Techniques to Archaeology. In E. Prenicka, and B. Wagner (eds.), *Archaeometry 1990*, 667–76. Basel: Birkhauser Verlag.

Clay, P. 1986. Castle Donington. *Current Archaeology*, 102, 208–11.

Clay, P. and Salisbury, C.R. 1990. A Norman Mill Dam, and other sites at Hemington Fields, Castle Donington, Leicestershire. *The Archaeological Journal*, 147, 276–307.

Gaimster, D.R.M., Margeson, S. and Hurley, M. 1990. Medieval Britain and Ireland in 1989. *Medieval Archaeology*, 34, 162–252.

Garmonsway, G.N., 1954. *The Anglo-Saxon Chronicle*. London: J.M. Dent & Sons Ltd.

Hillam, J., Groves, C.M., Brown, D.M., Baillie, M.G.L., Coles, J.M. and Coles, B.J. 1990. Dendrochronology of the English Neolithic. *Antiquity*, 64, 210–219.

MacCormick, A.G. 1968. Three Dug Out Canoes and a Wheel from Holme Pierrepoint, Nottinghamshire. *Transactions of the Thoroton Society of Nottinghamshire*, 72, 14–28.

Salisbury, C.R. 1984. Flandrian Courses of the River Trent at Colwick, Nottinghamshire. *The Mercian Geologist*, 9(4), 189–207.

Salisbury, C.R. 1985. The Taming of the Trent. *East Midlands Archaeology*, 1, 5–12. Council for British Archaeology, Group 14.

Salisbury, C.R. 1988. A Saxon and Norman Fish Weir at Colwick, Nottinghamshire. In M. Aston (ed.), *Medieval Fish, Fisheries and Fishponds in England*, 329–51. Oxford: British Archaeological Reports, British Series, 182.

Salisbury, C.R. 1991a. Primitive British Fishweirs. In G.L. Good, R.H. Jones and M.W. Ponsford (eds.), *Waterfront Archaeology. Proceedings of the Third International Conference, 1988*, 76–87. London: Council for British Archaeology Research Report 74.

Salisbury, C.R. 1991b (forthcoming). A Watching Brief at Hemington Fields, Castle Donington, SK 461 307. *Transactions of the Leicestershire Archaeological and Historical Society*, 91.

Straw, A. 1963. The Quaternary Evolution of the Lower and Middle Trent. *East Midland Geographer*, 3, 171–89.

15

A Norman mill dam at Hemington Fields, Castle Donington, Leicestershire

Patrick Clay

Summary: *The discovery of a 12th century dam and probable millrace at Hemington Fields, Castle Donington illustrates the problems and potential for archaeology within the floodplain of the river Trent. Buried beneath two metres of silt and gravel alluvium the site was only discovered during a watching brief as gravel extraction was taking place. Salvage excavation, however, has recovered vital information on the technology of Medieval vertical wheeled mills and of water management on major rivers (Clay and Salisbury 1990).*

INTRODUCTION

Hemington Fields lies near the centre of a floodplain, 4 km (two and a half miles) wide, built up at the confluence of the rivers Trent, Derwent and Soar (Fig 15.1). The plain consists of ca. 5 metres of gravel and sand underlain by the Mercian mudstone group (formerly Keuper Marl) and capped by over a metre of silts and clays which have accumulated since Flandrian times. The lower levels of the gravel were probably deposited after the maximum of the last glaciation (Devensian). Below these a discontinuous pure clay horizon probably represents a time of maximum braiding across the Late Devensian floodplain when clay deposits were able to settle out.

With the final disappearance of the ice and the onset of Flandrian times about 10,000 BP, the Trent shrank and settled into its present meander belt. During this time, it migrated across the plain, often braiding or forming "ox-bow" cutoffs.

As this volume emphasises, one of the major challenges now facing archaeologists is in the detection and recording of archaeological remains buried beneath deep alluvial deposits such as these (Needham 1989). This was one of the considerations when an agreement was reached in Leicestershire in line with the CBI Code of Practice for Mineral Operators whereby planning permission for extractive industries includes a constraint allowing a watching brief even where no sites have been previously located. Although this is not an ideal situation, as it only permits salvage recording of any discoveries made, it does allow some cover of these less "visible" areas.

Following the granting of planning permission for gravel extraction within the floodplain (Fig 15.1) a watching brief was arranged with the kind co-operation of the gravel company involved, Pontylue Sand and Gravel Company. The watching brief was undertaken by the Burleigh Archaeological Fieldwork Group, one of Leicestershire Museums' Community Archaeology groups (Liddle 1985), under the direction of Ian Hind. The extraction work commenced in October 1984 and in April 1985 the Group observed that the gravel company's dragline was uncovering large timber posts and several stone blocks. Recording was undertaken by the Group and C.R. Salisbury who had experience of working on similar sites uncovered by gravel quarrying elsewhere in the Trent valley (Salisbury *et al* 1984; Salisbury this volume). The exposed timbers were recorded in the quarry face and measured drawings were made of the disturbed timbers which had been deposited on the quarry floor. A salvage excavation was undertaken by the Leicestershire Archaeological Unit in conjunction with C.R. Salisbury and the Burleigh Archaeological Fieldwork Group in June 1985. Following this, an evaluation was undertaken to the east to ascertain whether further excavation was required and a watching brief of the gravel extraction is continuing.

THE EXCAVATION

Within the time available an area of ca. 120 m^2 was examined by excavation (Figs. 15.2 & 15.3). The metre thick silt and clay overburden overlying the gravel deposits had already been removed by the quarry

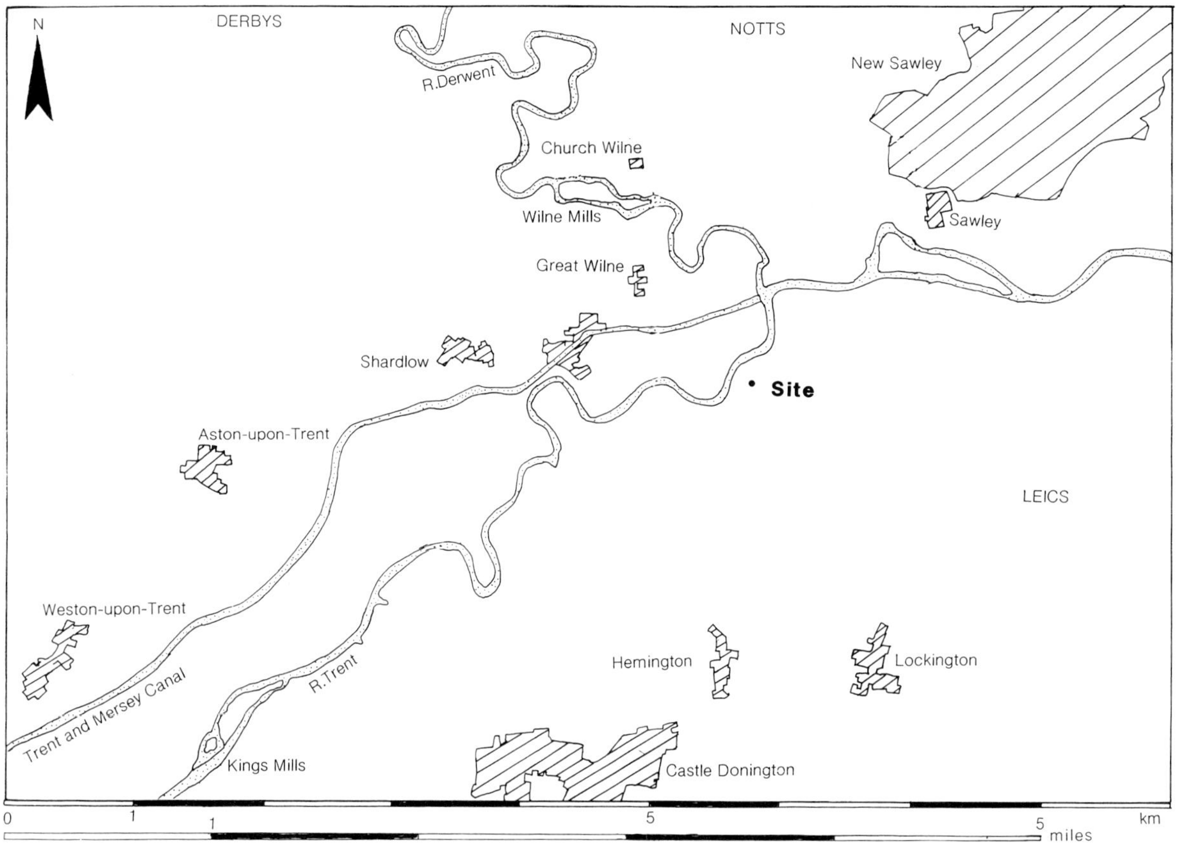

Fig. 15.1. Castle Donington: site location.

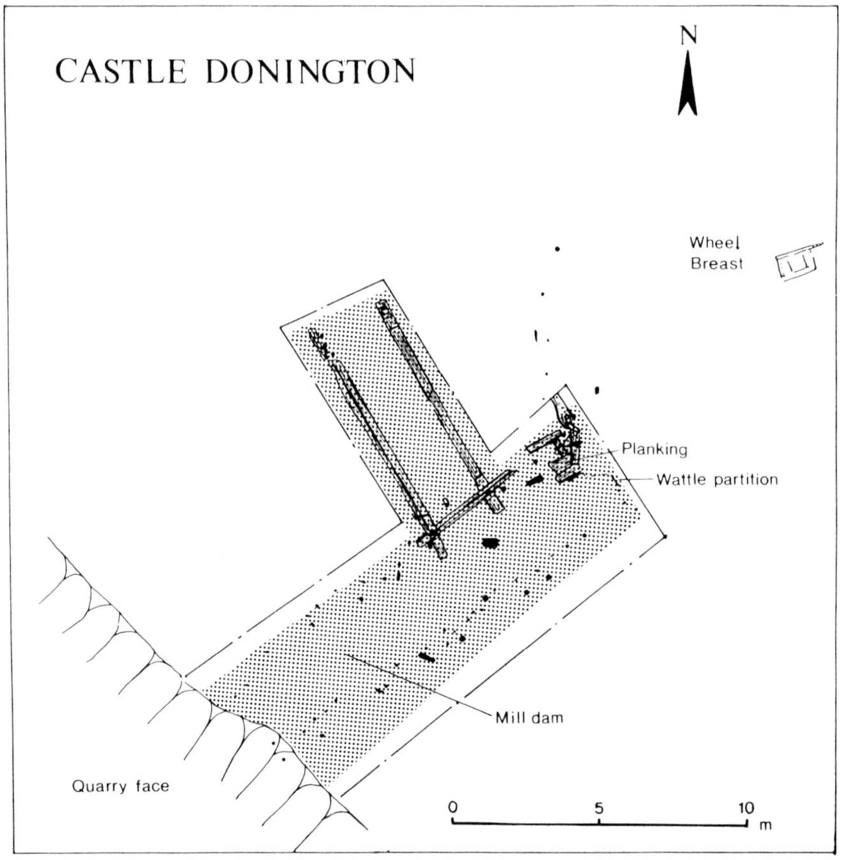

Fig. 15.2. Castle Donington: plan of excavation.

Fig. 15.3. Castle Donington: view of excavation from the north-west.

company and a JCB 3C with clawed bucket was then employed to remove the gravel until the tops of the timbers were exposed. This occurred at a depth of 0.25 m from the top of the gravel. In view of the limited time available a further metre depth of gravel surrounding the timbers was removed with the machine until stones and horizontal timbers were exposed. The quarry company's dragline was later used to help expose timbers first located by hand digging. The total time available before the site was destroyed by further quarrying was eight days with a labour force of eight.

Two phases of timber structure within the mill dam were revealed during the excavation. The earliest of these consisted of two rows of posts and an associated wattle partition. This may originally have been a fish weir (Losco-Bradley and Salisbury 1979) which having silted up was incorporated within the construction of the later main mill dam. Overlying this was another timber structure consisting of two rows of wattled oak posts supporting a core of stones, gravel and brushwood. From the angle of the dam and the slope of the upright posts it can be suggested that the river had been flowing in an east-south-easterly direction with the dam aligned north-east/south-west across the flow to direct water towards a mill race (Fig. 15.5). Traces of a head race may be evident from a silt-filled channel to the north of the dam. A secondary dam at right angles to the main dam was located to the north consisting of oak posts braced by horizontal timbers with a stone, gravel and brushwood core. This secondary dam was perhaps constructed to create a head of water which could then be controlled through a sluice before entering the mill race. Dendrochronological dates were obtained from the oak timbers used in the second phase of the structure which suggested a felling date in the second quarter of the 12th century (Howard 1990). Radiocarbon dates obtained from samples in the first phase timber structure suggested use between the 9th and 12th centuries.

Perhaps the most significant discovery made during the work on this site was of a recessed oak structure (Fig. 15.4) which appeared to be part of a wheel breasting within which a vertical water wheel would have turned. Similar stone structures are known from various sites, for example Nettleton, Wiltshire, of possible Roman date (Spain 1984: 118–9 figs 8 & 9), and, still extant, at Kings Mills, Castle Donington (discussed below). Timber parallels are less common; however a late 17th century timber wheel breasting has been excavated at Chingley, Kent (Cleere and Crossley 1985: 233 fig.49). That example had similar side panels to those present at Hemington and was located at the head of a long timber-lined tail race. (For a full discussion of this structure see Clay and Salisbury 1990).

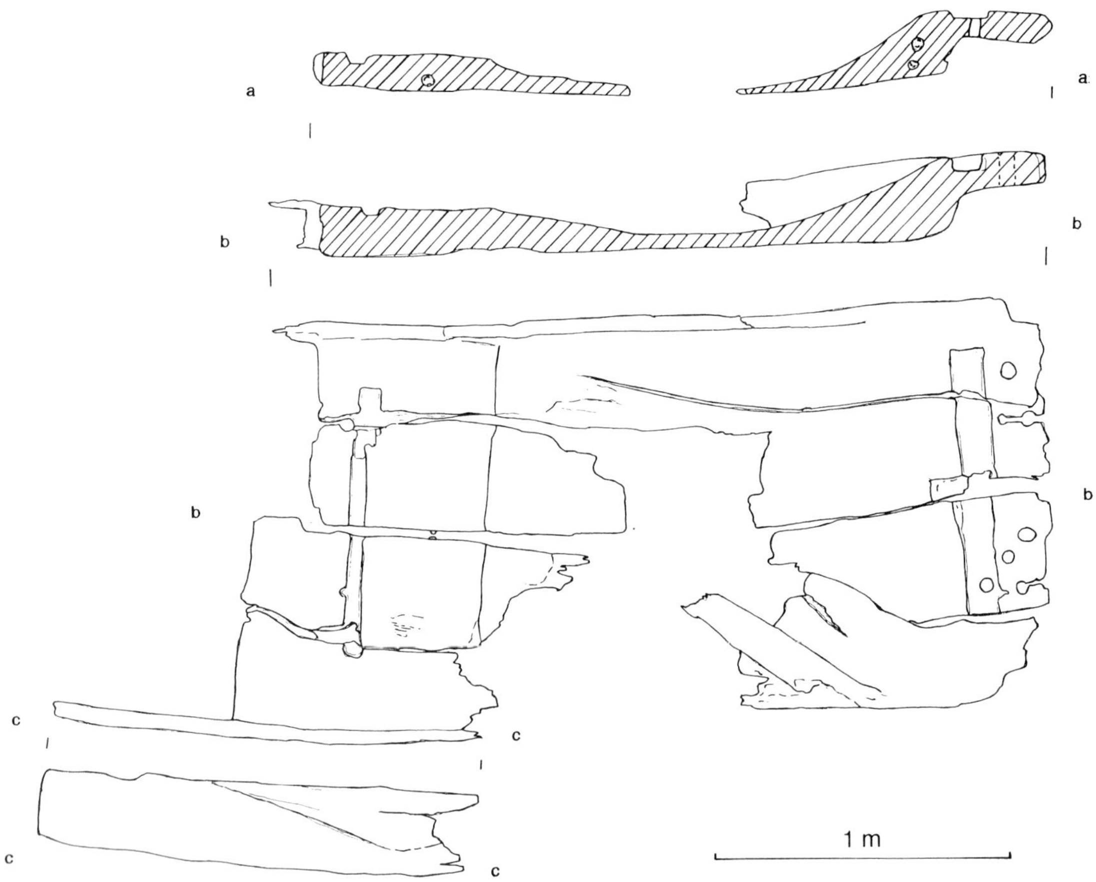

Fig. 15.4. Castle Donington: oak wheel breasting.

At Castle Donington the wheel breasting would probably have been secured within the mill race at an angle of ca.10° to facilitate the outflow of the water as it passed beneath the wheel. The wheel breasting was located during the watching brief following the excavation and may not have been *in situ*. However, its size and orientation when discovered suggests that it was close to its original position. If this is a correct assumption, the level at which the structure was discovered may be of significance. The level of ca.100.5 m O.D., which is ca.2 m below the surviving height of the secondary dam, is consistent with the fall expected in a mill race so that a head of water could be maintained by the dams and controlled by sluices. Silt observed during the evaluation 3 m to the east of the wheel breasting may denote the position of the mill race.

Broken and unfinished millstones located on the site suggest that the mill was used for all or part of its life to grind corn. However, some of the unfinished stones are small enough to have been roughouts for hand querns so their association with a powered mill is unclear. They may have been waste from quarries near Melbourne 12.5 km (8 miles) upstream along the Trent (M. Stanley – pers. comm.).

DISCUSSION

Medieval water mills are incompletely understood, very few having been excavated (Holt 1988: 117). Their location next to water courses has often led either to destruction through erosion or burial through deposition of alluvium (Spain 1984: 101). Both processes are evident at Castle Donington where much of the site has been destroyed by river movement although the high water-table has meant that some elements have survived.

The building of mill dams across major rivers is well documented and references to the methods of construction show the use of a timber framework as early as ca. AD 500 (Wikander 1985: 151). 7 km (4.5 miles) upstream from the Castle Donington site a similar mill dam arrangement across the Trent is still extant at Kings Mills, Castle Donington (Fig. 15.1). Here a dam 350 m long crosses the river at an angle to divert the waters into the mill race. The date of this dam is uncertain, but as the mill is mentioned in the Domesday Book (Morgan 1979: 12) it is not inconceivable that a similar structure was in existence at this time. These structures appear to have become numerous enough to cause problems by the late 14th century when a Royal Commission was appointed to investigate the "many weirs, mill dams, pales and kiddles fixed or rased in the water of the Trent impeding the passage of ships" (Stevenson 1882).

From the angle of the surviving upright timbers the original direction of flow of the river appears to be west-north-west/east-south-east with the mill dam positioned at an oblique angle across the flow. This suggests that the mill may have been located on the north bank of the river with a mill stream located to the east (Fig. 15.5). If the mill was to the north of the river it may have been associated with the settlement of Great Wilne, now within the parish of Shardlow and Wilne, Derbyshire. This settlement, situated in the angle between the confluence of the Trent and the Derwent, had been known as South Wilne (Glover and Riden 1985: 58), although from the 12th century it was commonly known as Wilne by Shardlow, or Wilne and Shardlow (Cameron 1959). The movement of the river may be indicated by fields to the south of the present course of the Trent still bearing the names Shardlow and Wilne pasture presumably relating to the village on the north bank. From the 11th-16th century the area was part of an estate also including the settlements of Shardlow and Aston, centred at Weston. In the Domesday Book reference is made to a mill at Weston (value 19s 4d), the estate's centre, but no mills are mentioned for Shardlow or Aston (Morgan 1978: 273). Although mills were often situated a long way from the settlement it seems unlikely that the mill mentioned would be located 6.5 km (4 miles) from the estate's centre. The dendrochronological dates obtained from the excavated timbers suggest that most elements of this mill were constructed sometime after 1086, probably ca. 1140.

There is no evidence to establish how long the mill was in operation as no dating evidence other than that from the timbers was obtained. The site did however undergo some erosion and the upper layers of the mill dam have been washed away. The site was then covered by alluvial deposits of gravel, clay and silts – a process which continued until its destruction by gravel extraction. The site may have been abandoned for many reasons – the propensity of the Trent to flood may have resulted in the mill structures being damaged and constantly needing repair. If other mill dams were positioned downstream of the site this may also have had a detrimental effect. If the level of the water retained by the lower dam was too high this would affect the fall of the water at the higher dam. This was evident with the Bazacle dams on the Garonne, near Toulouse, where law suits were often taken out because the owners of downstream dams illegally raised the height of their dams to increase the waterpower driving their mills. This practice eventually led to the

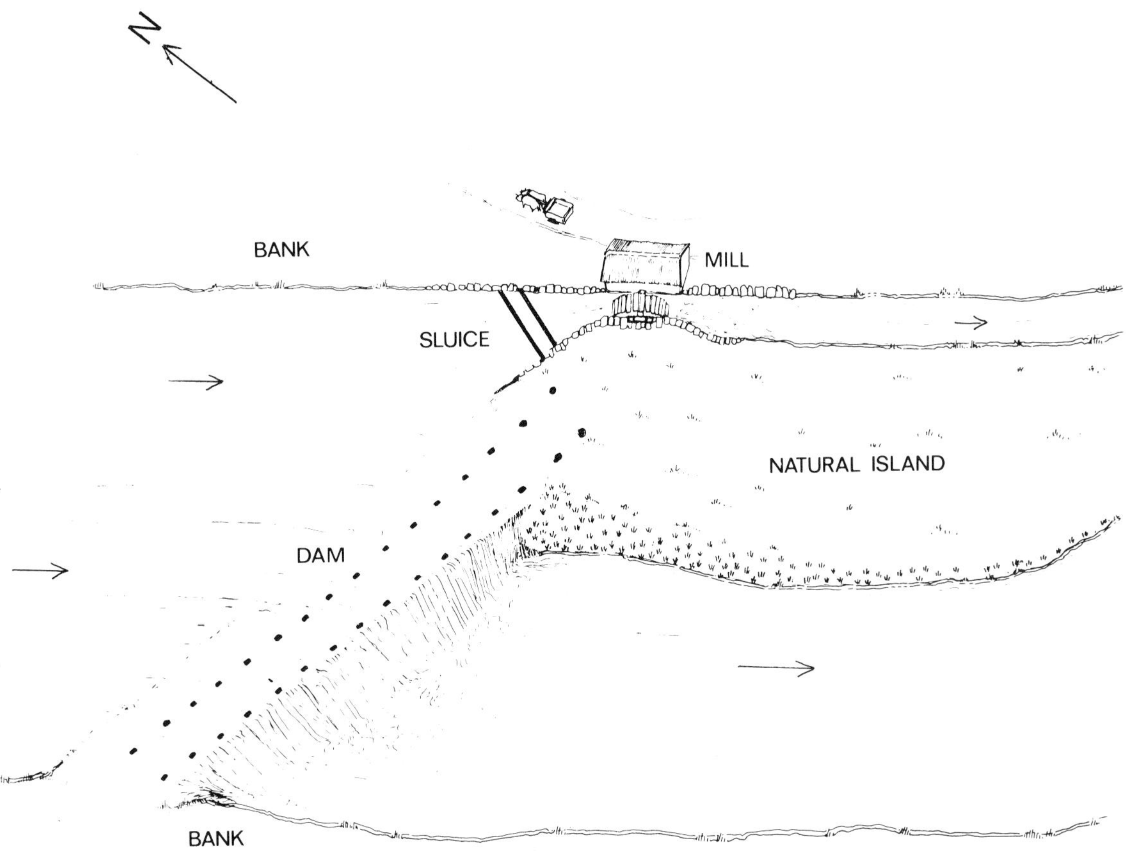

Fig. 15.5. Castle Donington: suggested arrangement of mill dam.

bankruptcy of some of the mills upstream (Gimpel 1977: 18–20). Alternatively, a major hydrological change such as an avulsion of the Trent from one course to another may also have left the site without any water power.

CONCLUSION

The movement of the river Trent within its floodplain and the quantity of silt and gravel alluvium deposited creates special problems for archaeological investigations. Geophysical prospecting and infra red aerial photography may be a way forward (Allsop and Greenbaum, this conference; Cornwell 1985; Zalesiewicz *et al* 1985). However the palaeochannels observed during watching briefs in this area (Salisbury, this volume) often consist of gravel deposits within gravel contexts which may remain elusive for geophysical survey. The need for arrangements for watching briefs to be maintained, with contingency time and funding agreed before extraction takes place, remains paramount.

References

Cameron, K. 1959. *The Place Names of Derbyshire.* English Place Names Society, 28.

Clay, P. and Salisbury, C.R. 1990. A Norman Mill Dam and other sites at Hemington Fields, Castle Donington, Leicestershire. *Archaeological Journal*, 147, 276–307.

Cleere, H. and Crossley, D. 1985. *The Iron Industry of the Weald.* Leicester: Leicester University Press.

Cornwell, J.D. 1985. Applications of Geophysical Methods to Mapping Unconsolidated sediments in East Anglia. *Modern Geology*, 9, 187–205.

Gimpel, J. 1977. *The Medieval Machine.* London: Book Club Associates.

Glover, C. and Riden, P.(eds). 1981. *William Wolley's History of Derbyshire.* Derbyshire Records Series, 7.

Holt, R. 1988. *The Mills of Medieval England.* Oxford & New York: Blackwells.

Howard, R. 1990. Dendrochronological Analysis. *In* Clay and Salisbury 1990, 298–300.

Liddle, P. 1985. *Community Archaeology.* Leicestershire Museums, Arts and Records Service.

Losco-Bradley, P.M. and Salisbury, C.R. 1979. A Medieval Weir at Colwick, Nottinghamshire. *Transactions of the Thoroton Society of Nottinghamshire*, 83, 15–22.

Morgan, P. 1978. *Domesday Book, Derbyshire.* Chichester: Phillimore.

Morgan, P. 1979. *Domesday Book, Leicestershire.* Chichester: Phillimore.

Needham, S. 1989. River valleys as wetlands: the archaeological prospects. In J.M. Coles and B.J. Coles (eds.), *The Archaeology of Rural Wetlands in England*, 29–34. Exeter & London: Wetland Archaeological Research Project and English Heritage.

Salisbury, C.R., Whittley, P.J., Litton, C.D. and Fox, J.L. 1984. Flandrian Courses of the River Trent at Colwick, Nottinghamshire. *Mercian Geologist*, 9, No.4, 185–207.

Spain, R.J. 1984. Romano-British Watermills. *Archaeologia Cantiana*, 100, 101–128.

Stevenson, W.H. (ed.) 1882. *Records of the Borough of Nottingham (vol 1).*

Wikander, 0. 1985. Mill Channels, weirs and ponds — The Environment of Ancient Water-mills. *Opuscula Romana*, XV (13), 149–154.

Zalasciewicz, J.A., Mathers, S.J. and Cornwell, J.D. 1985. The Application of ground conductivity measurements to geological mapping. *Quarterly Journal of Engineering Geology*, London, 18, 139–148.

16

Archaeology and palaeochannels in the Lower Welland and Nene valleys: alluvial archaeology at the fen-edge, Eastern England

C. A. I. French, M. G. Macklin and D. G. Passmore

Summary: *Preliminary results of palaeoenvironmental investigations in the Lower Welland and Nene valleys, north-west Cambridgeshire, have begun to elucidate regional Holocene alluvial histories and the environmental context of archaeological sites on the valley floor. Former channel systems have been mapped (from aerial photographs), and valley floor sediments exposed in excavations and machine trenches subjected to sedimentological, geochemical, mineral magnetic, soil micromorphological and palynological analyses. Lateglacial to Early Holocene river development in the Maxey-Etton area of the Lower Welland valley saw replacement of a low-sinuosity, braided channel system by a higher sinuosity meandering river. By the Neolithic period an anastomosed channel system had developed that persisted until recent times. Fine-grained alluvial aggradation of the valley floor during and after the Neolithic period resulted in a shift of settlement foci to locally higher "islands" and the burial of former field systems, settlements and monuments. This process is also demonstrated in the Nene valley at Fengate, where Bronze Age fen settlement and field systems were buried beneath encroaching reedswamp during later Bronze Age and Iron Age times, and subsequently by inorganic silts and clays probably during the later Roman and Early Medieval periods. Base-level changes (associated with rising Postglacial sea-levels) and accelerated influx of fine-grained sediments (from cleared and cultivated catchment slopes) appear to have been particularly significant influences on regional patterns of valley floor development and settlement strategies.*

INTRODUCTION

Despite the extensive amount of archaeological, palaeobotanical and soil micromorphological work carried out over the past decade in the Peterborough area (French 1983; 1988a; 1988b; 1990), the Holocene alluvial histories of the neighbouring Lower Welland and Nene valleys have received relatively minor attention. Accordingly, a new project has been instigated whose principal objectives are to elucidate Holocene valley floor development with particular attention to channel morphologies, patterns of alluviation and colluviation, and the interaction of human settlement with valley floor environments. The project has four components:

1. Geomorphological mapping of former channel systems and sedimentological, geochemical and mineral magnetic analyses of alluvium to determine sediment provenance and depositional environments (M.G. Macklin and D.G. Passmore, University of Newcastle-upon-Tyne).
2. Micromorphological studies of the channel sediments and adjacent buried soils (C. French, University of Cambridge).
3. Palynological analyses of basal peats in palaeochannels and archaeological deposits (R.G. Scaife, Fenland Archaeological Trust/University of Southampton).
4. Radiocarbon dating (where possible) of major sedimentary units, (East Kilbride laboratory) supplemented by a programme of optically-stimulated luminescence (R. Parish, University of Durham) and archaeomagnetic dating (C. Batt, University of Durham).

The landscape under investigation comprises two main areas: first, the Lower Welland valley between Stamford and the fen-edge around Peakirk, centred on the Maxey-Etton-Northborough area (Fig. 16.1); and second, Fengate at the fen-edge on the eastern side of Peterborough (Fig. 16.3). Both river valleys are underlain by extensive Late Devensian gravel deposits, with Peterborough marking the south-eastern extremity

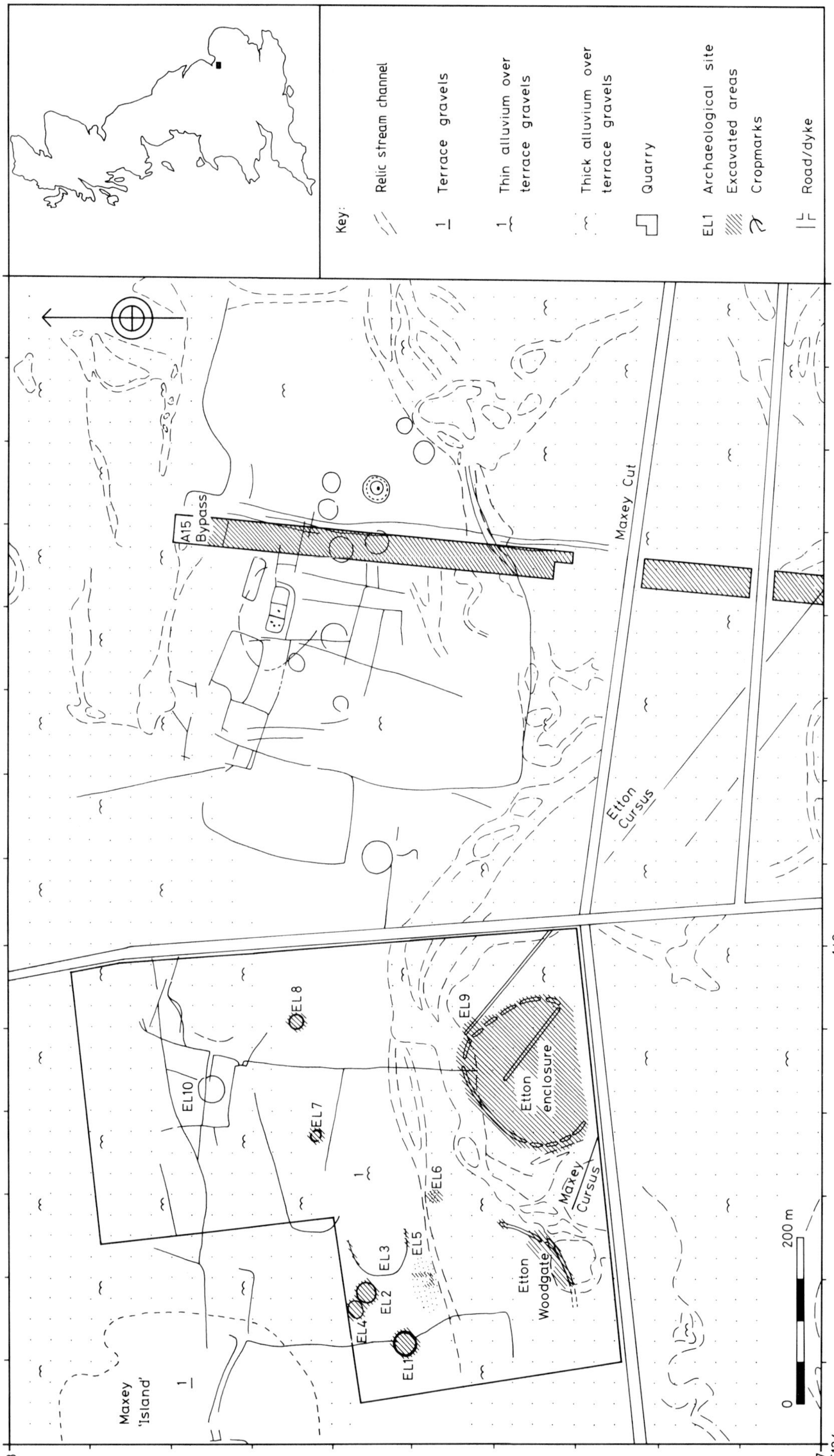

***Fig. 16.1.** Location of study area (inset) and detail of archaeological features and palaeochannels between Maxey and Etton.*

of the limestone upland which rises to the north-west. The Nene and Welland rivers coalesce with the fen basin to the east, which during the Postglacial period has been infilled with various freshwater and marine deposits (Hall 1988).

As the first season of field and laboratory work has just been completed, results reported here are of a preliminary nature. This paper summarizes work in progress in the two main areas: first, an extensive study of the Lower Welland valley around Maxey and Etton and, second, an intensive, site-specific investigation in the Lower Nene valley/fen-edge interface at Fengate, on the eastern side of Peterborough.

THE LOWER WELLAND VALLEY

Much of the Holocene alluvial valley floor and Pleistocene gravel terraces of the Lower Welland valley between the villages of Maxey, Etton and Northborough has been subject to archaeological investigation over the past decade (Pryor and French 1985; Pryor *et al* 1985; French and Pryor, forthcoming). The principal sites include the Middle Neolithic causewayed enclosure site of Etton (Fig. 16.1), the Etton-Woodgate Middle Neolithic ditched settlement site (Fig. 16.1), several Late Neolithic and Early Bronze Age midden deposits (Fig. 16.1: A15 Bypass and Etton Woodgate), numerous Late Neolithic henge monuments (Fig. 16.1: sites EL2, 4 and 7) and Bronze Age barrows (Fig. 16.1: sites EL 1 and 8, A15 Bypass).

The lower parts of the First Terrace mapped by the British Soil Survey are covered by silty clay alluvium, varying in thickness from ca. 30–120 cm (Booth 1981; Burton 1981). This alluvium surrounds the villages of Maxey and Northborough, leaving them as 'islands' in an extensive alluvial flat which broadens and coalesces with the fen-edge 2 km to the east. Buried soils are relatively well preserved beneath both the alluvial overburden and the prehistoric monuments and earthworks that have been investigated in the area (Fig. 16.1) (Pryor and French 1985; Pryor *et al* 1985; French 1988a and 1990). These have been extensively examined using micromorphological techniques.

Soil record

To date some 25 soil profiles have been examined using micromorphological techniques (French 1985) ranging from those less well preserved relic examples which were preserved beneath prehistoric earthworks on the high ground (ca. 9 m OD) of Maxey 'island', to the extensive areas of well preserved buried soils on the lower ground of the Holocene alluvial valley floor. Soils buried by Holocene river sediments, or located above the level of Holocene alluviation, are either argillic brown earths or brown earths. Argillic brown earths are created by the process of clay translocation under stable forested conditions in which clay particles are moved down the profile (leaving an eluvial or Eb horizon) and redeposited (illuviated) in a textural B (an argillic or Bt) horizon (Avery 1980; Fisher 1982). This is considered to be a modal soil type under the climax (Atlantic/Sub-Boreal) forests of southern England. Brown earths are essentially less well developed argillic brown earths, and consist of an A horizon overlying a Bw or cambic horizon in which there has been some alteration of the parent material and minor illuviation of fine material (Avery 1980; Limbrey 1975). In general, the argillic brown earths are better developed on the higher, better drained ground of Maxey 'island' and less well developed on the edges of the former floodplain (French 1985; French 1988a; French 1990).

A sequence of major events is observable in the soil record. First, the development of argillic brown earths testifies to widespread woodland development by the Neolithic period, at least on the higher parts of the terrace just above the Holocene alluvial flat in this part of the river valley. Second, much of this woodland element was substantially cleared by the latter half of the Neolithic period, prior to the construction of the Etton causewayed enclosure and numerous adjacent and later henge monuments. Third, shortly after deforestation, several coincident and probably concomitant processes began to affect the lower parts of the floodplain terrace (ca. 7 m OD), including seasonal flooding episodes, alluviation of silts and clays, and localised erosion and colluvial deposition on the lower/downstream edges of Maxey 'island'. Fourth, throughout later prehistory, alluviation continued and gradually encroached onto higher parts (ca. 7–8 m OD) of the adjacent First Terrace. All these events were associated with channel migration across the valley floor and widespread, low-velocity freshwater flooding and alluviation of fine-grained sediment (discussed below). In addition, from the later Neolithic period on there was a seasonally higher and fluctuating ground water-table that resulted in gleying of the lower half of soil profiles and, locally, minor deposition of calcium carbonate derived from the underlying base-rich terrace subsoil. Finally, widespread alluviation of the area began in the post-Roman period and continued to dominate the landscape until recent times (French 1983).

Alluvial record

The Lower Welland valley is characterised by a series of interconnected alluvial belts (separated by upstanding Pleistocene gravel terraces) that are between 200–350 m wide and are underlain by fine-grained sediment (silts and clays overlying gravelly sands). Palaeochannels (evident on aerial photographs as cropmarks) are clearly evident in all alluvial belts, although in the field they usually have barely discernible topographic expression. Interpretation of palaeochannel histories and pattern, however, is frequently problematic with the morphology of earlier channels often being masked by subsequent alluviation, and/or truncation by later channels.

A preliminary Lateglacial and Holocene channel sequence has been established in the Lower Welland valley, between Northborough (TF 5150 3080) and the Maxey Cut 1 km north-west of Etton (TF 5134 3072), from aerial photographs, field survey and levelling, and excavation of selected palaeochannels (Figs. 16.1 and 16.2). Within this 3 km reach, the morphology and stratigraphy of palaeochannels in four alluvial belts have been examined in some detail.

The most northerly alluvial belt lies immediately south of Deeping Gate (Fig. 16.2) and from aerial photographic interpretation has a meandering, single-thread palaeochannel system with prominent point bars on which ridge and swale accretionary features are well-developed. There is no evidence of subsequent erosion and/or superimposition by later channels which suggests only one phase of channel development before the alluvial belt was dismembered from the main river. The subsurface stratigraphy and dimensions of these channels have not yet been investigated, and their age and precise relation to the archaeological record is presently unknown. They do, however, appear to pre-date most archaeological features, although an enclosure (probably of prehistoric age) at the apex of one meander loop seems to respect the channel margin.

The second alluvial belt is located immediately west of Northborough (Fig. 16.2). This contains a series of divided palaeochannels with a meandering pattern that has both point and mid-channel bars together with larger islands around which lower-sinuosity anabranches

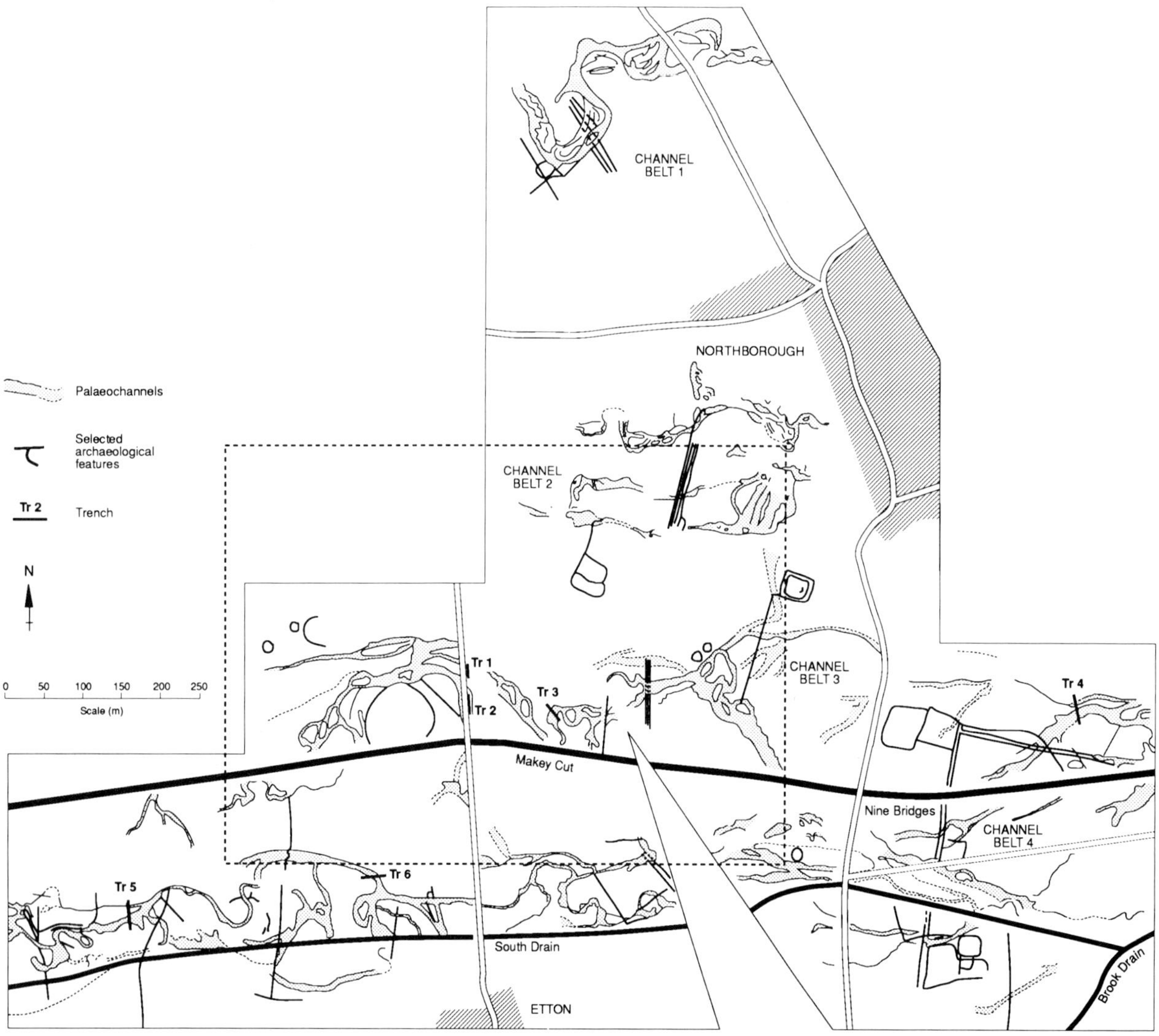

Fig. 16.2. Palaeochannel belts, modern drainage networks, selected archaeological features and trench locations in the Maxey-Etton-Northborough area, lower Welland valley.

divide. This channel system has an anastomosing-like morphology closely resembling some Canadian rivers described by Smith (1983). The sedimentology, stratigraphy and archaeology of this channel system await investigation.

The third alluvial belt is aligned approximately parallel to the Maxey Cut (Fig. 16.2) and contains at least two palaeochannel systems which differ in morphology and age. On the basis of morphological relations, the oldest channels are those evident 0.5 km north-west of Nine Bridges (Fig. 16.2) and run in a north-easterly direction towards Northborough. These have the lowest sinuosities of all channels investigated (with the exception of recent drainage channels) and are truncated by a channel system that was active during the construction and occupation of the Etton Neolithic enclosure (see below, Figs. 16.1 and 16.2). Lozenge-shaped bars within these earlier palaeochannels are more common in laterally-mobile braided rivers than in stable anastomosed systems. Excavations (Trench 4, Fig. 16.2) showed a relatively wide (20 m) and shallow (0.7 m) channel with numerous sloughs formed in gravelly sands. An argillic brown earth palaeosol that had developed on the surface of these gravelly sands indicates that this earlier channel system had probably been inactive since the Early Holocene, or perhaps even earlier. There is therefore a significant hiatus between soil formation and the deposition of overlying clays which infill these channels.

The second channel system in this alluvial belt runs around and, in places, abuts against the Middle Neolithic causewayed enclosure site at Etton. These are anastomosed palaeochannels with a relatively high-sinuosity which were active in the Neolithic period, and caused part of the enclosure ditch at Etton to be abandoned (Pryor *et al*, forthcoming) (Fig. 16.1). Trenches 1 and 2 (Fig. 16.2) across this system showed channels about 1.3 m deep and between 29–45 m wide, developed in silts and clays; in places they truncate the underlying Early Holocene palaeosol. A channel (1 m deep and ca. 18–20 m wide) 100 m east of the Etton enclosure overlies Late Neolithic midden deposits containing Grooved Ware and is infilled with gravelly sand and interbedded organic material (radiocarbon dated to 3875 ± 50 BP (Q-3149) V. Switsur, pers. comm.) overlain by silty clay (French and Pryor, forthcoming). At Trench 3 organic material interbedded with fluvial gravelly sands recovered from below the Neolithic channel system (at a depth of 1.38 m) has been radiocarbon dated to between 10,070 ± 190 years BP (GU-5124) and 10,900 ± 120 years BP (GU-5123).

The fourth alluvial belt runs parallel and adjacent to the South Drain (Fig. 16.2) and has an extremely complex sequence of meandering – anastomosed palaeochannels with evidence of several phases of meander migration and channel avulsion. It is noticeable, however, that east of Etton (TF 5140 3065) channels become less sinuous, narrower and more diffuse in form. Trenches 5 and 6 (Fig. 16.2) showed a similar stratigraphy to the Maxey Cut alluvial belt with gravelly sands, which display inclined heterolithic stratification (Thomas *et al* 1987), overlain by 0.75 m of structureless silts and clays. A radiocarbon date of 7350 ± 90 BP (GU-5126) from organic material interbedded within the lower sand unit (at a depth of 1.2 m) in Trench 5 (Fig. 16.2) indicates actively migrating meandering channels, transporting relatively coarse bedload, existed in the Lower Welland valley throughout Early Holocene times.

Although several studies (e.g. Brown 1987; Burrin and Scaife 1984; Seale 1979) have alluded to the probable importance of anastomosing rivers to the development of Holocene alluvial valley floors in lowland southern and eastern England, this study is the first to map such channel systems in detail and relate them, in a provisional way, to the archaeological record. Lateglacial-Early Holocene river development in the Lower Welland appears to have been quite similar to that evident in the Gipping valley, Suffolk, documented by Rose *et al* (1980), with a low-sinuosity, braided channel system being replaced in the Early Holocene by a higher-sinuosity meandering river that nevertheless still transported a significant component of sand and fine gravel as bedload. By the Neolithic, however, an anastomosing-like river morphology had developed which persisted until recent times, when streams in the area were canalised (the Maxey Cut was enlarged in 1953) and modified by land drainage and farming.

The development of an anastomosing channel system in the Lower Welland valley during the Holocene, and perhaps in other similar areas in eastern and southern England, is likely to reflect the interaction between two extrinsic factors. First, increasing local base levels associated with Holocene sea-level rise and, second, the influx of silts and clays into the basin which provided cohesive stream bank sediment and promoted lateral stability of channels. This latter event was linked, at least in part, with accelerated soil erosion in the Welland catchment resulting from Neolithic and later vegetation clearance and cultivation (French 1983; Scaife and Robinson, in Pryor *et al*, forthcoming).

THE FENGATE AREA

The Fengate area, on the fen-edge east of Peterborough, has revealed a huge complex of sites over the past two decades. Initial discoveries identified an extensive rectilinear field system (developed on the upstanding First Terrace), defined by ditches and banks and divided by major double-ditched droveways, all aligned at right angles to the fen-edge of the day (Pryor 1978; 1980)

(Fig. 16.3). This 2nd millennium BC field system is believed to have been flood-free and to have enclosed winter pasture, mainly for cattle.

Recently, it was possible to trace this field system eastwards towards the fen-edge and beneath the fen 'skirt' deposits. The ditches of one droveway were observed to narrow and become fence lines, ending at about the 3 m contour, and in another example both turned outwards through a right angle, so that the droveway in between opened out onto the fen to the east. Pollen analysis (R.G. Scaife, pers.comm.) suggests these fields were fringed by flood meadows by the later Bronze Age (circa 1200–1000 BC) which were subject to variable levels of seasonal freshwater flooding. It is therefore possible to envisage this land as open fen-edge pasture in the spring, summer and autumn, with a shallow reed fen to the south-east, and permanent flood-free enclosed winter pasture immediately inland to the west.

This area of flood meadow and drier pasture was linked with the Bronze Age Flag Fen wooden platform (Pryor *et al* 1986), located some 150 m into the fen reedswamp, by a timber avenue constructed at ca. 2850 BP (V.R. Switsur, pers.comm.) (Fig. 16.3). Excavation revealed this avenue to be flanked on both sides by low parallel banks of sediment ca. 50 m long and 1.5 m wide. These comprised 0.15 m of fine-medium sand overlain by 0.2 m of silty fine sand that becomes finely laminated in the upper part of the deposit. These waterlain sediments suggest the wooden avenue crossed a shallow stream channel ca. 20 m wide that followed the fen-edge contour and the course of an earlier, probably Lateglacial, channel located at the site by geotechnical investigations (by Ove Arup). This river crossing location may go some way to explaining the position and concentration of votive metal objects deposited to either side of the avenue.

Sections through the fen sediments exposed during the 1989 Fengate excavation (Fig. 16.3) have allowed a palaeoenvironmental investigation of local fen development from the later Bronze Age onwards, the preliminary results of which are summarised as follows. Encroachment of reedswamp and fine-grained alluviation during later Bronze Age and Iron Age times

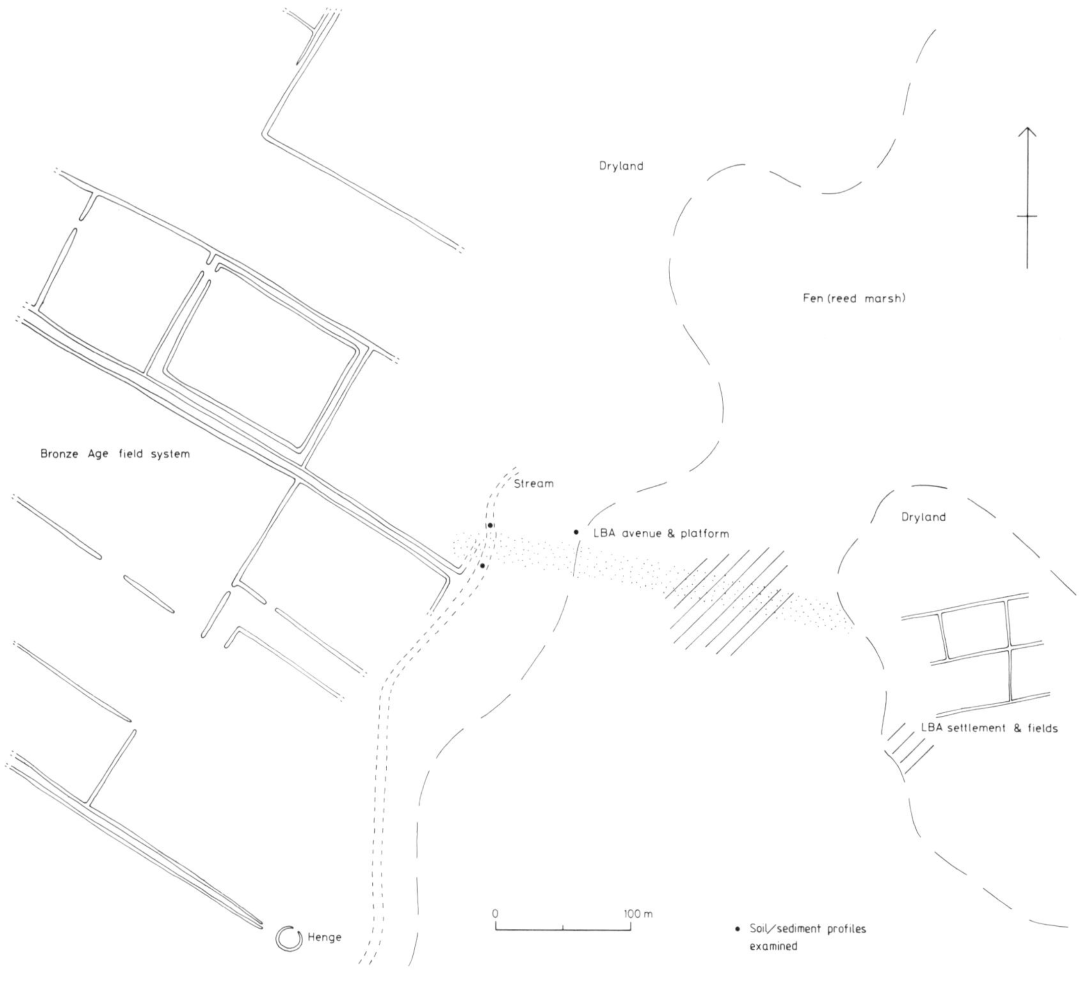

Fig. 16.3. *Bronze Age field systems, avenue and platform at Fengate, lower Nene valley, Peterborough.*

is reflected in aggradation of 0.4 m of silty peat, overlain by 0.5 m of peaty sandy silt interbedded with thin peat horizons. This period sees the abandonment and burial of the Flag Fen platform and timber avenue, as well as the lowermost extremities of the Bronze Age Fengate field system. Overlying these organic sediments, and extending over the entire fen-edge area, are inorganic silty clays (ca. 0.25 m thick) deposited throughout the later Roman and possibly the Early Medieval periods. Markedly higher values of frequency-dependent ("quadrature") magnetic susceptibility within this unit suggests these sediments are derived, at least in part, from eroded catchment topsoils (Oldfield 1983; Dearing *et al* 1985), and thus are likely to reflect regional intensification of deforestation and cultivation during the mid-late first millennium AD.

CONCLUSIONS

Preliminary results of geoarchaeological investigations in the Lower Welland and Nene valleys have begun to elucidate regional Holocene alluvial histories and the environmental context of archaeological sites on the valley floors. Lateglacial-Early Holocene river development in the Maxey-Etton area of the Lower Welland valley saw replacement of a low-sinuosity, braided channel system by a higher-sinuosity meandering river. By the Neolithic period, however, a combination of base-level change associated with the rising Postglacial sea, and increased delivery of fine-grained sediment to valley floors, following anthropogenic catchment disturbance, had promoted development of an anastomosed channel system. Neolithic and Bronze Age ceremonial monuments appear to have been sited close to the floodplain and from the later Neolithic period were affected by river sedimentation. Valley floors in the region continued to receive sediment throughout the later prehistoric and historic periods, forcing settlements onto higher parts of the valley floor. By Early Medieval times this had left areas of higher ground in the valley, such as the site of Maxey village, as 'islands' surrounded by several sets of stream channels.

Investigations at Fengate, on the fen-edge in the Nene valley, demonstrate the abandonment and burial of Bronze Age fen settlement and field systems beneath encroaching reedswamp during later Bronze Age and Iron Age times, and subsequently by river silts and clays during the later Roman and Early Medieval periods.

Future research will centre on a comprehensive dating programme of alluvial belt palaeochannels (particularly through direct-sediment dating techniques including OSL and palaeomagnetic dating) to determine more precisely the chronology of Postglacial channel system development in the study area. In addition, the provenance of fine-grained alluvium will be examined to investigate possible links with deforestation and agricultural activity. Through these exercises it is hoped to further elucidate the interaction of climatic, land-use and sea-level related controls on Holocene alluvial histories in this part of the Lower Welland and Nene valleys.

Acknowledgements

The authors would like to thank R.G. Scaife, R.I. Macphail, F. Pryor, R. Parish and C. Batt, as well as V.R. Switsur for supplying radiocarbon dates in advance of publication.

References

Avery, B.W. 1980. *Soil Classification for England and Wales*. Harpenden: Soil Survey Technical Monograph 14.

Booth, S.J. 1981. *The sand and gravel resources of the county between Stamford, Lincolnshire, and Maxey, Cambridgeshire. Description of 1:25,000 sheets TF 00 and TF 10*. Mines Assessment Report, Institute of Geological Science No.80.

Brown, A.G. 1987. Long-term sediment storage in the Severn and Wye catchments. In K.J. Gregory, J. Lewin and J.B. Thornes (eds.), *Palaeohydrology in Practice*, 307–332. Chichester: John Wiley and Sons.

Burrin, P.J. and Scaife, R.G. 1984. Aspects of Holocene valley sedimentation and floodplain development in southern England. *Proceedings of the Geological Society*, 95, 1, 81–96.

Burton, R.G.O. 1981. *Soils in Cambridgeshire (2)*. Harpenden: Soil Survey Record No.69.

Dearing, J.A., Maher, B.A. and Oldfield, F. 1985. Geomorphological linkages between soils and sediments: the role of magnetic measurements. In K. Richards (ed.), *Geomorphology and Soils*, 245–266. London: George Allen and Unwin.

Fisher, P.F. 1982. A review of lessivage and Neolithic cultivation in southern England. *Journal of Archaeological Science*, 9, 299–304.

French, C.A.I. 1983. An environmental study of the soil, sediments and molluscan evidence associated with prehistoric monuments on river terrace gravels in northwest Cambridgeshire. University of London: unpublished PhD Thesis.

French, C.A.I. 1985. Soil, sediment and molluscan analyses of excavated features. In F.M.M. Pryor and C.A.I. French 1985, 205–216.

French, C.A.I. 1988a. Aspects of buried prehistoric soils in the lower Welland valley and the fen margin north of Peterborough, Cambridgeshire. In W. Groenman-van Waateringe and M. Robinson (eds.), *Man-made Soils*, 115–128. Oxford: British Archaeological Reports, International Series, 410.

French, C.A.I. 1988b. Further aspects of the buried prehistoric soils in the fen margin northeast of Peterborough, Cambridgeshire. In P. Murphy and C.A.I. French (eds.), *The Exploitation of Wetlands*, 193–211. Oxford: British Archaeological Reports, British Series, 186.

French, C.A.I. 1990. Neolithic soils, middens and alluvium in the lower Welland valley. *Oxford Journal of Archaeology*, 9, 305–311.

French, C.A.I. and Pryor, F.M.M. (forthcoming). *Archaeology and Environment of the Etton Landscape*. Fenland Archaeological Trust Monograph.

Hall, D.N. 1988. *The Fenland Project, No.2: Fenland Landscapes and Settlement between Peterborough and March*. East Anglian Archaeology Monograph No.35.

Limbrey, S. 1975. *Soil Science and Archaeology*. London: Academic Press.

Oldfield, F. 1983. The role of magnetic studies in palaeohydrology. In K.J. Gregory (ed.), *Background to Palaeohydrology*, 141–165. Chichester: John Wiley and Sons.

Pryor, F.M.M. 1978. *Excavation at Fengate, Peterborough, England: the Second Report*. Toronto: Royal Ontario Museum Archaeology Monograph No.5.

Pryor, F.M.M. 1980. *Excavation at Fengate, Peterborough, England: the Third Report*. Northampton and Toronto: Northamptonshire Archaeological Society Monograph, 1/Royal Ontario Museum Archaeology Monograph No.6.

Pryor, F.M.M. and French, C.A.I. 1985. *Archaeology and Environment in the Lower Welland Valley*. East Anglian Archaeology Monograph, 27.

Pryor, F.M.M., French, C.A.I. and Taylor, M. 1985. An Interim Report on Excavation at Etton, Maxey, Cambridgeshire, 1982–1984. *Antiquaries Journal*, 65, 275–311.

Pryor, F.M.M., French, C.A.I. and Taylor, M. 1986. Flag Fen, Fengate, Peterborough I: discovery, reconnaissance and initial excavation (1982–85). *Proceedings of the Prehistoric Society*, 52, 1–24.

Pryor, F.M.M. *et al* (forthcoming). *Excavation at Etton, Maxey, Cambridgeshire, 1982–1988*. Historic Buildings and Monuments Commission Monograph.

Rose, J., Turner, C., Coope, G.R. and Bryan, M.D. 1980. Channel changes in a lowland river catchment over the last 13,000 years. In R.A. Cullingford, D.A. Davidson and J. Lewin (eds.), *Timescales in Geomorphology*, 159–175. Chichester: John Wiley and Sons.

Seale, R.S. 1979. Ancient courses of the Great and Little Ouse in Fenland. *Proceedings of the Cambridgeshire Antiquarian Society*, 69, 1–19.

Smith, D.G. 1983. Anastomosed fluvial deposits: modern examples from Western Canada. In J.D. Collinson and J. Lewin (eds.), *Modern and Ancient Fluvial Systems*, 155–168. Special Publication No.6, International Association of Sedimentologists. Oxford: Blackwell Scientific Publications.

Thomas, R.G., Smith, D.G., Wood, J.M., Vissier, J., Claverley-Range, E.A. and Koster, E.H. 1987. Inclined heterolithic stratification – terminology, description, interpretation and significance. *Sedimentary Geology*, 53, 123–179.

17

Life on the edge: archaeology and alluvium at Redlands Farm, Stanwick, Northants.

Graham D. Keevill

Summary: *Elsewhere in this volume, Dr Mark Robinson provides an overview and comparison of the alluvial sequences of the Nene, Ouse and Thames valleys. This paper focuses on a part of the Upper Nene valley which has been studied by the Oxford Archaeological Unit since 1989. A number of prehistoric ritual monuments are described, after which the development of the late prehistoric, Roman and post-Roman landscape is discussed. A summary of the relationship between the archaeological remains and the local alluvial sequence is presented within the wider context of the Raunds Area Project (Fig. 17.1).*

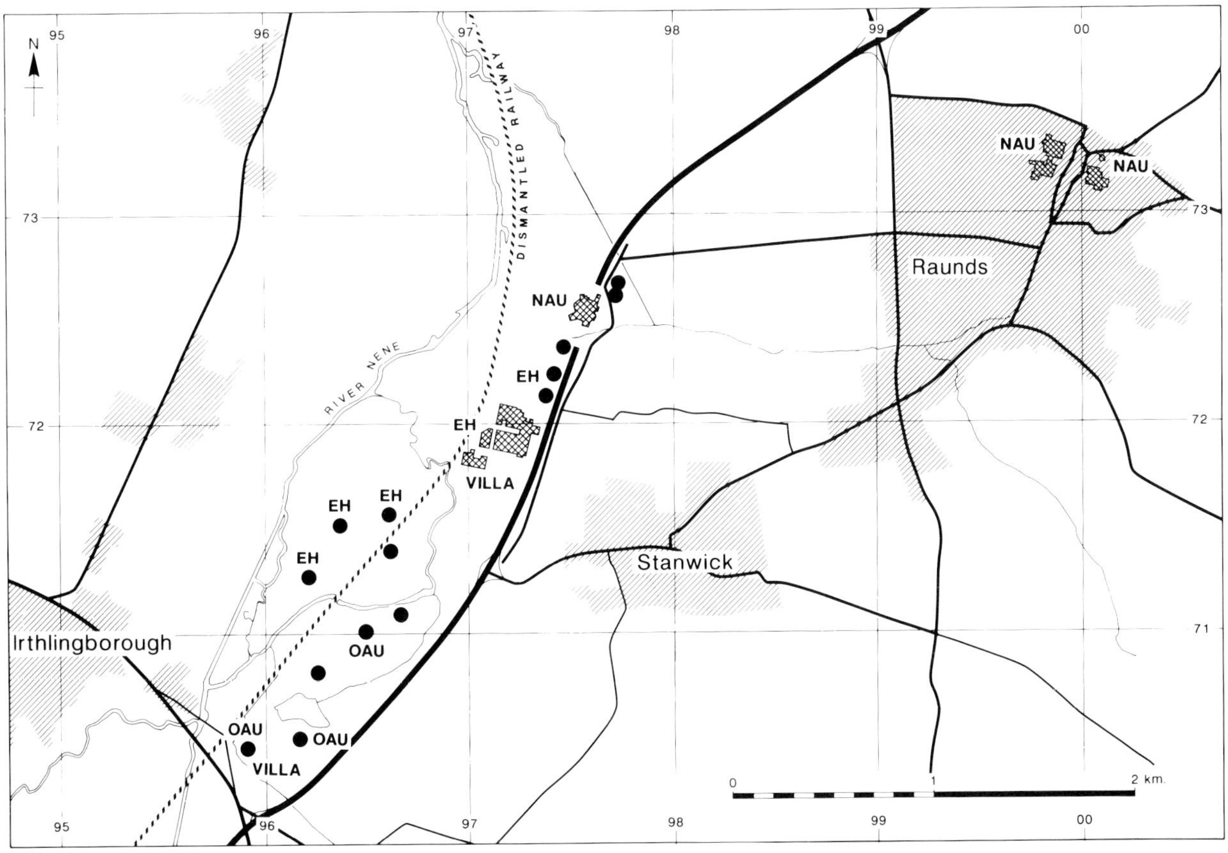

Fig. 17.1. *Redlands Farm and the Raunds Area Project: sites excavated by the Oxford Archaeological Unit, English Heritage, and the Northamptonshire Archaeology Unit.*

INTRODUCTION

The Upper Nene valley is a landscape rich in archaeological remains, and consequently it has been the subject of intensive fieldwork, most recently under the aegis of the Raunds Area Project. The floodplain of the river and the first gravel terrace on either bank have been particularly productive. In 1989, the Oxford Archaeological Unit conducted an assessment of ca. 86 hectares of arable land at Redlands Farm (centred on SU 964708), predominantly within the floodplain, south-west of Stanwick village for ARC Central Ltd. This was followed by major excavations in 1989 and 1990, with watching briefs carried out in 1990–2. All of the fieldwork has been wholly funded by ARC.

Three principal sites have been excavated (see Figs. 17.1 and 17.4): a long barrow (1989), a round barrow (1990), and a Romano-British settlement (1990). Other sites identified by the assessment lie within designated reserve areas and are thus preserved from destruction. The major sites and their relationship to alluvial deposits are briefly described here.

PREHISTORIC SITES

Evidence for activity in the Neolithic period and the Bronze Age is largely confined to ritual monuments, although flint scatters have been found on the lighter soils along the valley side (Parry 1990) and a timber platform, probably Neolithic, has been excavated at West Cotton (Windell *et al* 1990: 13). Tree-throw holes suggest that clearance in the floodplain had started during the Late Mesolithic/Early Neolithic; the valley bottom seems to have supported alder woodland, with some hazel, and stands of lime and oak on drier ground (Robinson 1990).

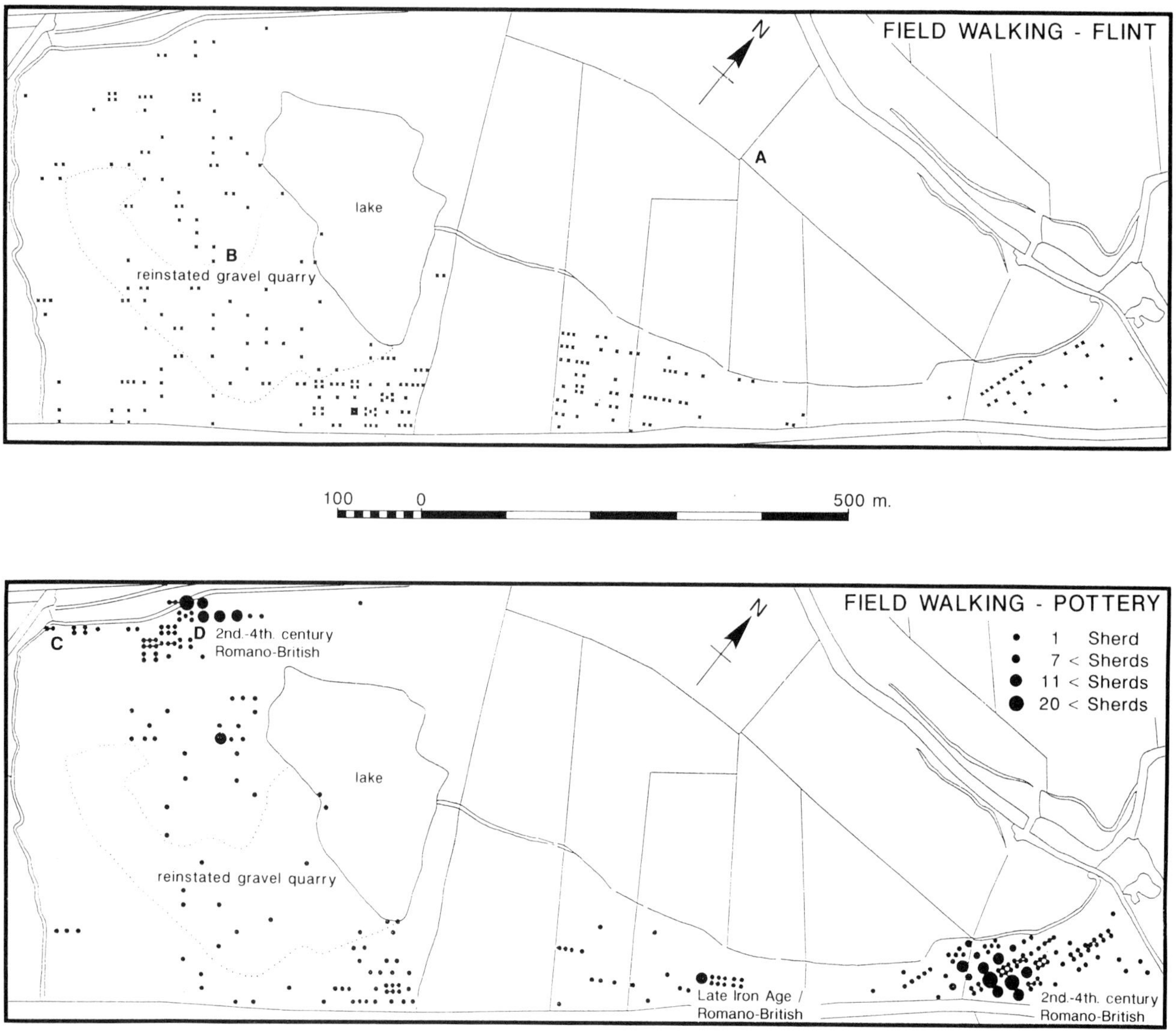

Fig. 17.2. Fieldwalking results at Redlands Farm. A = Long Barrow; B = Round Barrow; C = Iron Age site located during trenching; D = Villa.

Fieldwalking at Redlands Farm produced only a background scatter of flintwork, with no concentrations even in the vicinity of the known monuments described below (see Fig. 17.2). This does not reflect the depth of the alluvium, as flints were found even in the area of thickest deposits to the west of the infilled quarry (Fig. 17.2); similarly during the 1990 excavations flints were recovered from the Roman villa site, but not from the round barrow. The location of the villa, however, is clearly identifiable in the pottery distribution, although this is partly because of its raised position. It should also be noted that flints were recovered from the topsoil over the quarry itself. The fieldwalking data, therefore, reflect later disturbance more than any archaeological trends.

The long barrow at Redlands Farm consisted of two large quarry ditches flanking a mound 50 m long and approximately 10 m wide, occupying a low gravel island in the floodplain (Fig. 17.3); plough damage had reduced the mound to a maximum height of 0.6 m at its north-east end. Orientation was NE-SW, and the mound construction was of turf and tipped gravel. There was no evidence for separation of the mound into bays. The mound was, however, constrained on at least three sides by a timber palisade. Plough damage meant that the palisade could not be traced around the south-west end of the barrow. Thin timber planks were preserved in waterlogged deposits in the quarry ditches; the planking probably derived from the palisade.

Two other features were contemporary with the mound: a stone cist at the south-west end, and a central pit which might also have been a cist but which had been deliberately destroyed. Each feature contained two fragments of bone and no other finds. Study of environmental samples from the barrow ditches suggests

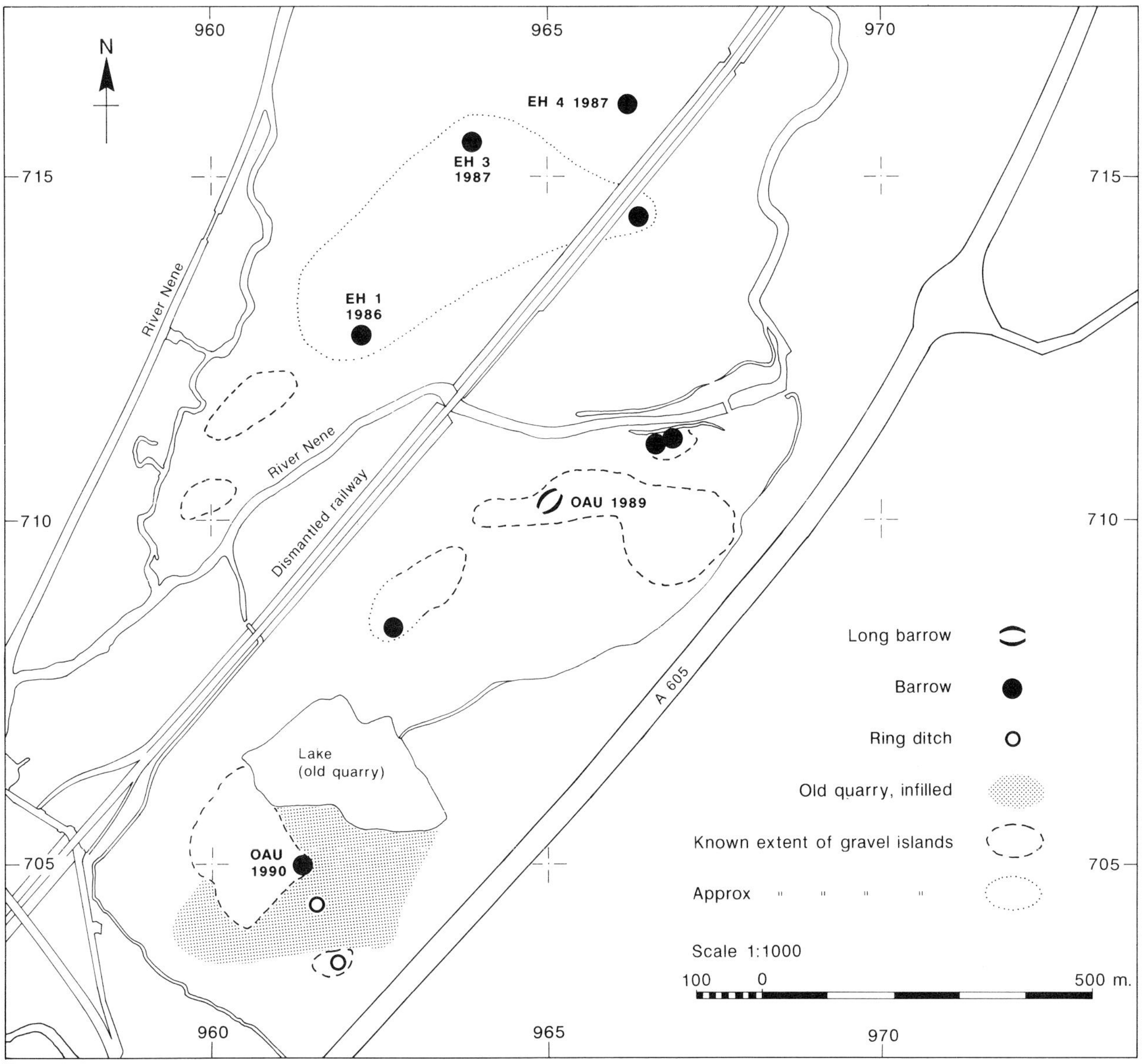

Fig. 17.3. Prehistoric ritual monuments at Stanwick.

that the monument lay in open country (Moore and Jackson 1990). This may represent locally-cleared ground rather than widespread clearance, as evidence elsewhere points to a wooded environment (see above).

During the Early Bronze Age ritual activity continued on and around the long barrow. Three Beaker inhumations were inserted into the mound along its central axis, while 16 cremations were located in front of the north-east end (Moore and Jackson 1990). Furthermore, ten round barrows/ring-ditches were loosely clustered around the long barrow (see Fig. 17.4); the long enclosure at West Cotton, 2 km to the north, similarly became the focus for the siting of later monuments, though in a more formal, linear arrangement (Windell *et al* 1990: 7–15).

In common with the long barrow, most of the round barrows/ring-ditches lie on low gravel islands in the floodplain. Stony ploughsoils sealed by alluvium and characteristically different from the stone-free soils buried by the mounds have been found around some of the barrows (Robinson 1990). Possible evidence for agricultural land use during the Bronze Age has been found at Stanwick villa (Neal 1989: 152) and at Redlands Farm, where a series of parallel N-S gullies at the base of the sequence of field systems may date to the Bronze Age. Some of these gullies are likely to be field boundaries, although others appear to mark droveways (Neal 1989: 152).

The siting of the funerary monuments close to a major confluence of the Nene and several subsidiary palaeochannels is noteworthy; similar relationships have been noted at West Cotton (Windell *et al* 1990: 15) and Grendon, Northants (Gibson and McCormick 1985: 65). Barrows/ring ditches are indeed commonly to be found on low-lying ground close to the river throughout the upper and middle reaches of the Nene (RCHM 1975: Figs 1–8). For instance, the Stone Pit Lodge cropmark complex straddling the boundary between the parishes of Fotheringhay and Southwick contains at least six ring ditches on the west bank of the Nene at its confluence with the Willow Brook, below the 30 m (100 ft) contour (RCHM 1975: fig. 53). The Fotheringhay Lodge complex contains 22 ring ditches, again below the 30 m (100 ft) contour and close to the Nene (RCHM 1975: fig. 50). Only one group of four ring ditches in the parish of Collyweston differs markedly from this norm; all four occupy a hilltop above the 91 m (300 ft) contour (RCHM 1975: 30, and fig. 40).

Some of the round barrows/ring-ditches are to be preserved, but where this has not been possible, excavation has been necessary. Several have been excavated to the north of the long barrow by English

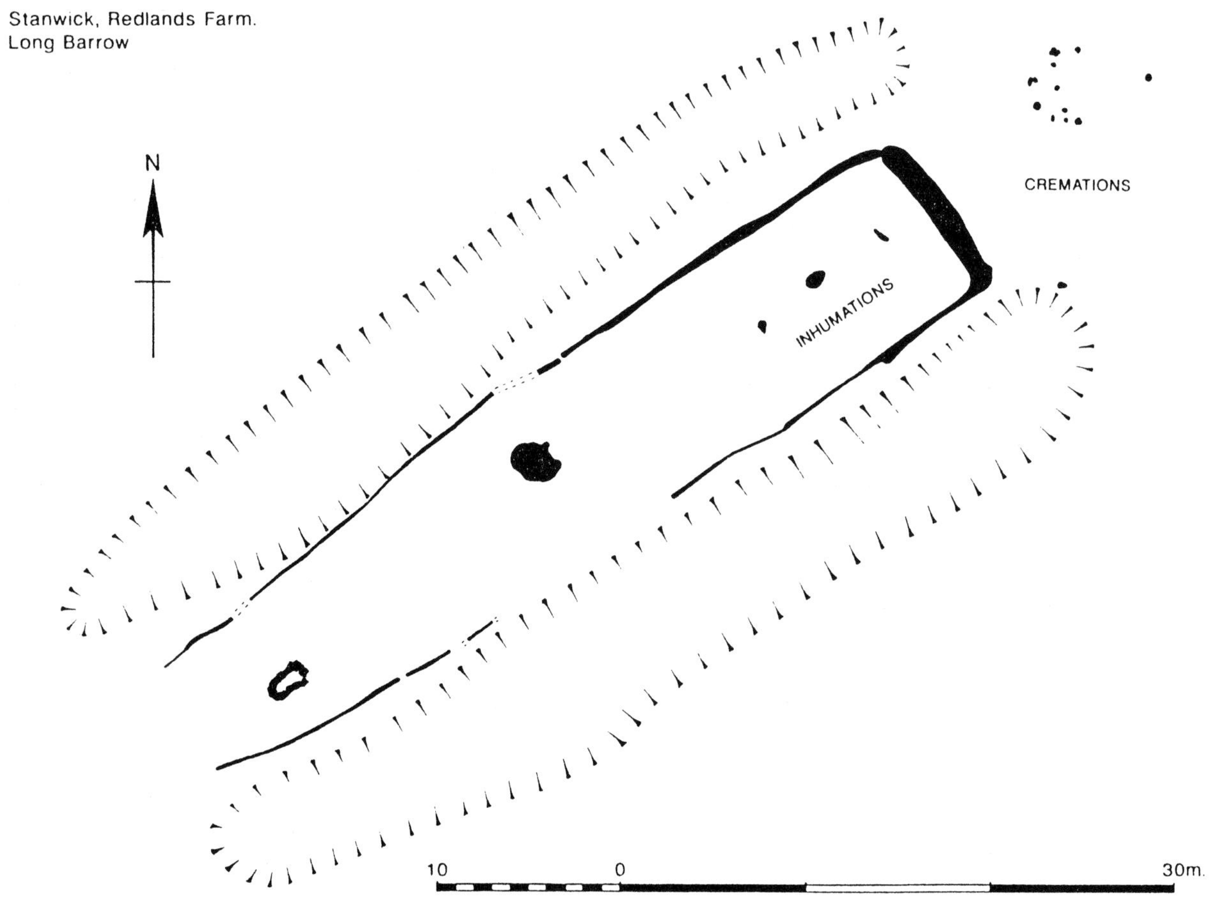

Fig. 17.4. Plan of the Redlands Farm Neolithic long barrow.

Heritage (Halpin 1987: 5–8), and one ploughed-out example was examined by the OAU in 1990 (Fig. 17.4). The barrow was defined by two concentic ring-ditches; the extent of plough damage removed any stratigraphic relationship between the ditches. Four graves, containing five inhumations, were sited within the circuit of the inner ditch. One of these – not central within the barrow – was accompanied by a Beaker, but the burials were otherwise devoid of grave-goods. Worked flints were notable by their absence, although numerous flakes were recovered from features and layers associated with the villa complex described below.

Most of the Neolithic and Bronze Age monuments were partially or wholly covered by alluvium, as at West Cotton (Windell *et al* 1990: 7–9), Irthlingborough (Halpin 1987: 5) and Redlands Farm. In the former and latter cases no trace of the monuments survived on the ground; their presence was only revealed by excavation. It is clear in most of these cases, however, that the alluvial deposits are of a later, probably Medieval date. Ploughing, probably during the later Iron Age and Roman periods, had caused varying degrees of damage culminating in total destruction of the mound of the round barrow excavated in 1990. During these periods at least the barrows must have been visible, recognisable monuments.

THE LATE IRON AGE AND ROMANO-BRITISH PERIODS

Exploitation of the gravel terraces and floodplain increased in the later Iron Age and achieved its zenith during the Roman period. David Neal's dissection of the Stanwick villa complex has revealed a substantial zone of nucleated, Late Iron Age settlement underlying the Roman villa complex, with some evidence for continuity between the two periods (Neal 1989: 156). Environmental material points to mixed agricultural land use, with hay meadow management of at least part of the floodplain; the major cereal crop was spelt wheat, which appears to have been planted on heavy calcareous clay soils (Campbell 1990).

Fieldwalking (Fig. 17.2) and trial trenching at Redlands Farm produced evidence for Late Iron Age settlements in the south-west corner and along the east side of the assessment area. These sites have been preserved within the extraction plan. A substantial increase in the level of activity is indicated throughout the Roman period (see Fig. 17.2). This was most graphically illustrated by the excavation of a small settlement complex in 1990 (see Figs. 17.2 and 17.5), in the heart of the floodplain.

Phase I refers to the Early Bronze Age barrow and the possible field system noted above. The first Roman phase (IIa) consisted of a mill and two rectangular barns (Fig. 17.5) set within a system of ditched enclosures and field boundaries; the latter appear to originate in the Late Iron Age, although no structures of this date could be positively identified. The mill lay on a low sand island in the floodplain, and was fed from the south and west by two leats, each of which drained a subsidiary palaeochannel of the Nene (Figs. 17.5 and 17.6). The east end of the mill building was cellared, and a barn door in its south wall opened out onto the surrounding ground level. The reason for the settlement developing so far into the floodplain is now apparent: the complex was initially designed around the exploitation of water power.

Phase IIb was characterised by the conversion of the mill into a winged villa with rear corridor, south-facing facade, an hypocaust with painted wall plaster, and a tesselated pavement (Figs. 17.5 and 17.6). This necessitated the destruction of the west leat and the northern end of the south one, although the remainder of the latter may have been retained as a landscape feature. A marked dip in the ground to the south of the villa was noted, and it is possible that the leat was diverted into this to create an artificial pond; this could not be confirmed, however, as work on this part of the site was curtailed following agreement on its preservation.

The field systems continued in use, but were now defined by stone walls, at least one of which merged into a boundary wall around the villa and its associated buildings. One of the Phase IIa barns remained in use.

Phase IIc saw further development of the villa, with the provision of a verandah/corridor to the facade, division of the wing rooms, insertion of a mosaic pavement, and the enclosure of the hypocaust furnace (Figs. 17.5 and 17.6). Elsewhere two stone-built round houses superseded the Phase IIb boundary wall (Fig. 17.5). The field systems continued in use.

The villa declined in Phase IId (Fig. 17.5). The wings were abandoned and demolished. The north gable-ends of each wing were found where they had been pulled over; that of the east wing would have stood at least 6 m high, suggesting that this part of the building had had an upper storey (Fig. 17.6). This may have been a later addition, perhaps in Phase IIc, as the character of the wall fabric changed from herringbone masonry at its base to *tabula* courses in the upper storey. Certainly the demolition appears to have been caused by the obvious structural defect of building the wing over the old mill-leat; the latter had only been infilled with limestone fragments and chippings. A substantial buttress had been added to the east face of the north-eastern corner of the wing in an apparently unsuccessful attempt to shore up the structure.

Elsewhere in the building, the hypocaust was disused and the tesselated pavement was partially removed and re-used to seal a group of three infant burials in the abandoned west wing. A further four burials, of three

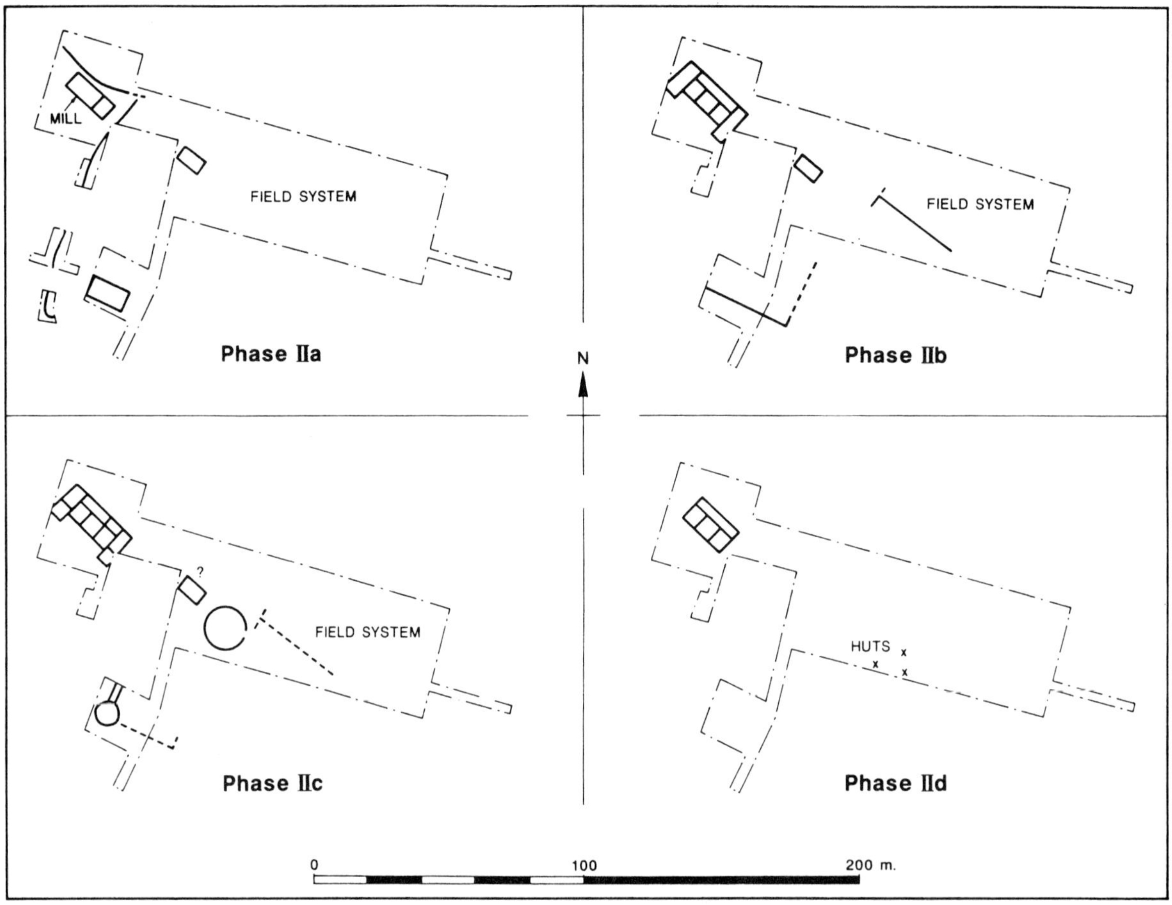

Fig. 17.5. The Romano-British sequence, Redlands Farm.

Fig. 17.6. The villa viewed from the north; note the leat approaching from top left.

infants and an adult, were found in and around the villa; two infant burials were also located in the field system. Three sunken-featured huts were built to the east of the villa, probably at the same time as the latter was declining (Fig. 17.5). Thus a small bi-focal settlement was formed, as occurred at West Cotton (Windell *et al* 1990: 16); the siting of both complexes astride river channels is interesting, but it is not clear that the Redlands Farm channel would have been flowing at this time. Limestone slabs had been laid across the channel during the Roman period, presumably because the ground was still damp and subject to occasional flooding. At the very least, however, the course of the channel would still have been obvious as a depression in the landscape.

Following the final abandonment of the villa, it is notable that little robbing of walls occurred. Most of the rooms were covered with thick layers of rubble, most of which were clearly derived from the gradual decay of the remaining structure. Thereafter over-bank flooding began to cover the ruinous structure with thick layers of alluvium. At least two separate, major episodes of deposition were evident, and it seems that the earlier of these occurred after the sixth century AD and extended 50 m east of the villa. Elsewhere alluviation appears to have become widespread in the Late Saxon period and to have continued into the Medieval period (Campbell 1990).

After the initial alluviation, a ridge-and-furrow field system was established; the early Saxon huts had been abandoned, as two of the furrows cut into their back-filled hollows. Thus the ridge-and-furrow was probably established in the Mid Saxon period. Its physical extent is significant, as the system terminated to the west at a ditch and hedge line. This coincided with the maximum extent of the first alluviation. Only the larger round house lay to the east of this boundary (see Fig. 17.5), and this was the only Roman structure to suffer Medieval and later damage.

In the Medieval period, the ridge-and-furrow was sealed by a thick deposit of alluvium, as at West Cotton (Windell *et al* 1990: 41). This must have been caused by massive and persistent over-bank flooding. Thereafter the fields were left to pasture. The land was only brought back into cultivation during the nineteenth century, when field drains were cut through the alluvium along the length of each furrow.

CONCLUSION

The occupational sequence at Redlands Farm can be broadly defined as follows: after at least partial tree clearance, the floodplain saw the development of a ritual landscape during the Neolithic and Bronze Ages, with some evidence for agricultural land use, but not for settlement, in the later period. During the later Iron Age activity began to intensify, culminating in the most intense use during the Roman period. A rising water level and over-bank alluviation forced a retrenchment in the Saxon period, and the land was eventually abandoned to pasture in the Medieval period.

The pattern of alluviation at Redlands Farm not only affected human exploitation of the land; it also had a direct effect on the survival of the archaeology. The villa and most of its associated buildings survived in remarkably good condition because they were sited at the limit of usable land in the Roman period. The increasing over-bank flooding and attendant deposition of alluvium meant that the buildings eventually became hidden from anyone wanting to quarry them for stone. Similarly they did not suffer from ridge-and-furrow ploughing. The round house which lay just above the contemporary high-water point was the only Roman building to suffer extensive damage.

Acknowledgements

I am grateful to John Moore, Mark Robinson and Dave Windell for discussing various aspects of this paper, both while the excavations were in progress and subsequently. The drawings are by Paul Hughes, Danyon Ray and Simon Chew of the Oxford Archaeological Unit. The high-level photograph was commissioned from Charles Ward Photography, Northampton.

References

Campbell, G. 1990. Aspects of the Roman to medieval environment and agricultural economy. The Raunds Area Project: The Archaeology of a Northamptonshire Landscape. (unpublished abstracts).

Gibson, A. M. and McCormick, A. 1985. Archaeology at Grendon Quarry, Northamptonshire, Part 1: Neolithic and Bronze Age sites excavated in 1974–5. *Northamptonshire Archaeology*, 20, 23–66.

Halpin, C. 1987. Irthlingborough Bronze Age Barrow Excavation. In B.Dix (ed.), The Raunds Area Project: second interim report. *Northamptonshire Archaeology*, 21, 3–30.

Moore, J. and Jackson, D. 1990. Stanwick, Redlands Farm. *South Midlands Archaeology*, 20, 55–6.

Neal, D. S. 1989. The Stanwick Villa, Northants: An Interim Report on the Excavations of 1984–8. *Britannia*, 20, 149–68.

Parry, S. 1990. Field Survey; the prehistoric and Roman periods. The Raunds Area Project: The Archaeology of a Northamptonshire Landscape. (unpublished abstracts).

RCHM 1975. *An Inventory of Historic Monuments in the County of Northampton Volume 1: Archaeological Sites in North-East Northamptonshire*. London: HMSO.

Robinson, M. 1990. Aspects of the earlier prehistoric environment of the valley bottom. The Raunds Area Project: The Archaeology of a Northamptonshire Landscape. (unpublished abstracts).

Windell, D., Chapman, A. and Woodiwiss, J. 1990. *From Barrow to Bypass: Excavations at West Cotton, Raunds, Northamptonshire, 1985–1989*. Northamptonshire County Council.

geoarchaeological implications. These implications will be discussed later in the paper in relation to a model of floodplain evolution. The result is that we have a floodplain stratigraphy made up of "parcels" of different ages and with different sedimentary characteristics (Needham 1989). The geochronology of low-energy floodplains, may, however, be rather more difficult to establish than in upland contexts. The most important dating method is usually radiocarbon assay and both the geomorphological implications and the information content of radiocarbon dates varies as does their precision and reliability (Pilcher 1991; Brown and Keough, in press). Rates of sedimentation vary both spatially and temporally (Brown 1989) and they will affect artifact density as has been shown by Ferring (1986). Although there is a great potential for the use of a variety of geophysical techniques including ground penetrating radar, seismicity, conductivity/resistivity and magnetic mapping, in the archaeometry of floodplains the detailed investigation of floodplain evolution and palaeoenvironments requires excavation, as described from the sites discussed here. Archaeological excavations can provide excellent exposures for lithostratigraphic analysis, especially if augmented by boreholes and augering.

THE MIDDLE NENE

The Nene valley, which was unglaciated during the Devensian, is characterised by a very low regional slope. The last ice limit during the Devensian is generally believed to have reached about 80 km north of the sites in the Raunds area (Institute of Geological Sciences 1977). The orientation of the Nene valley across the boulder clay plateau from Northampton to Peterborough partly follows a drift-filled buried channel which is almost certainly pre-Devensian in age. The catchment is geologically rather mixed with substantial areas of Jurassic limestone and ironstone, and Lias clay, below the boulder clay of the plateau. It also has a relatively unflashy hydrological regime which is today significantly altered by abstraction and river management. The Middle Nene valley has a relatively shallow Holocene floodplain fill of 0–4 m resting upon Devensian periglacial gravels. The area investigated was from Northampton to Peterborough (Fig. 18.1), but the sites were clustered towards Northampton as this has been the main area for aggregate extraction in recent years. The sites used here are Little Houghton, Wollaston, Higham Ferrers and Raunds, although work has also been conducted at Ditchford.

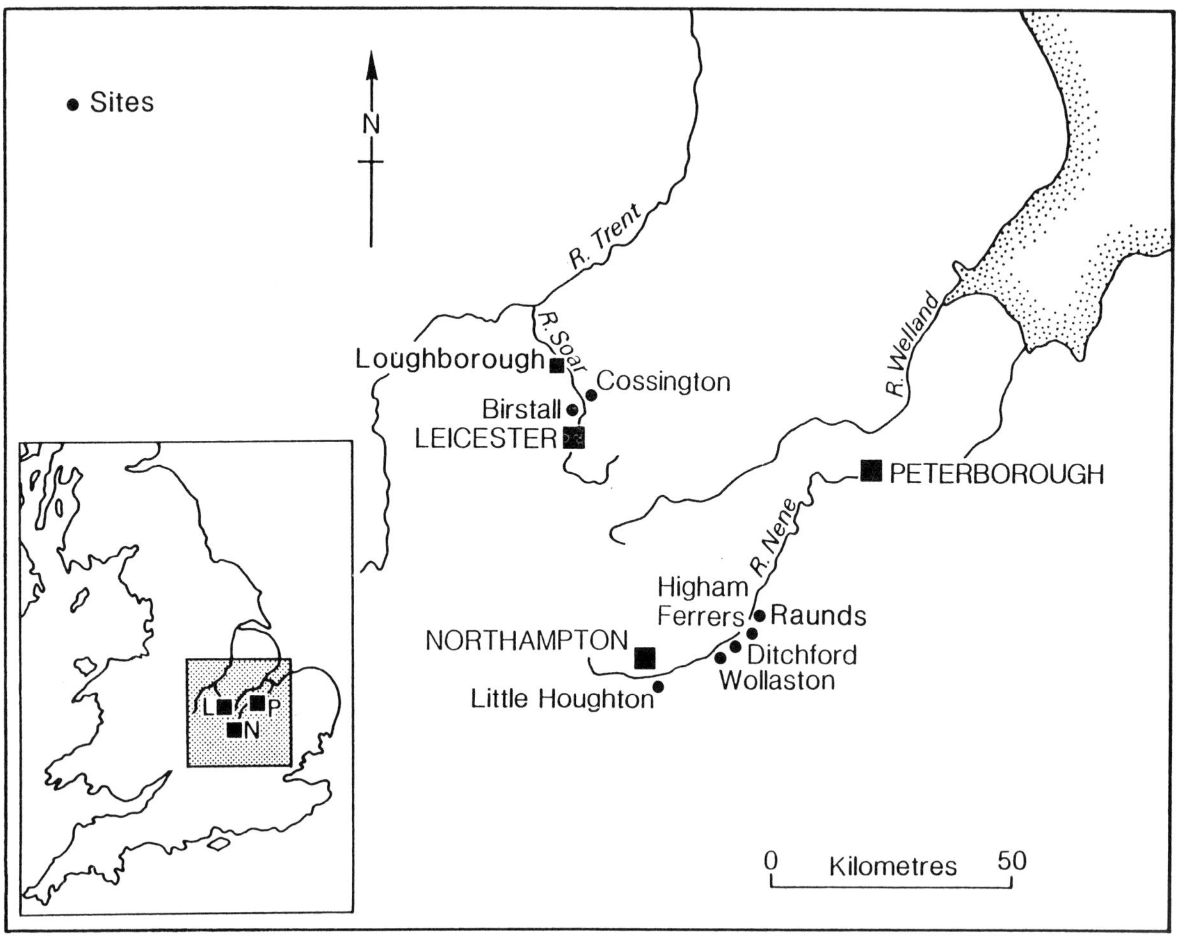

Fig. 18.1. Location map of the Nene valley and Soar valley sites described in the text.

ALLUVIAL ISOPACH AND PALAEOCHANNEL MAPPING

The generalised stratigraphy is overbank and channel-fill silts, clays and organics sitting on top of a very variable mixed gravelly loam which itself rests upon the basal periglacial sand and gravel. This relatively simply stratigraphy facilitated plotting of the isopach of the fine Holocene alluvium (Fig. 18.2). This has been done for the Raunds reach from aggregate exploration boreholes and records provided by ARC Ltd. and C. Halpen. The surface has been smoothed (by fitting a fifth order polynomial) in order to link up the deeper belts of alluvium which sit in depressions in the underlying gravels. These are not palaeochannels but the palaeochannel belts which can be regarded as the preferred location of the channel or channels during the Holocene. We would therefore expect most palaeochannels to occur in the belt, and islands with archaeological sites outside the belt. As can be seen on Figure 18.2 the palaeochannels range in age from approximately 12,400 BP to 2,000 BP. With the exception of those palaeochannels at the edge of the floodplain, all were relatively straight in planform. This is because they were created by avulsion rather than by neck cutoff processes, which tend to preferentially preserve meander bends. All the palaeochannels were invisible to ground survey as they have little or no topographic expression and little difference in soil type. This is the result of the deposition of fine overbank alluvium across the entire floodplain.

PALAEOCHANNEL POLLEN DIAGRAMS

These palaeochannels not only provide dating for abandonments and therefore channel change but also provide a local vegetation record from pollen analysis. The value of this data source is optimised by the small size of the sites and the proximity of the sites to the floodplain edge. These sites are also often very close to occupation sites on terrace islands, as is the case at the multi-period site at West Cotton in the Raunds reach.

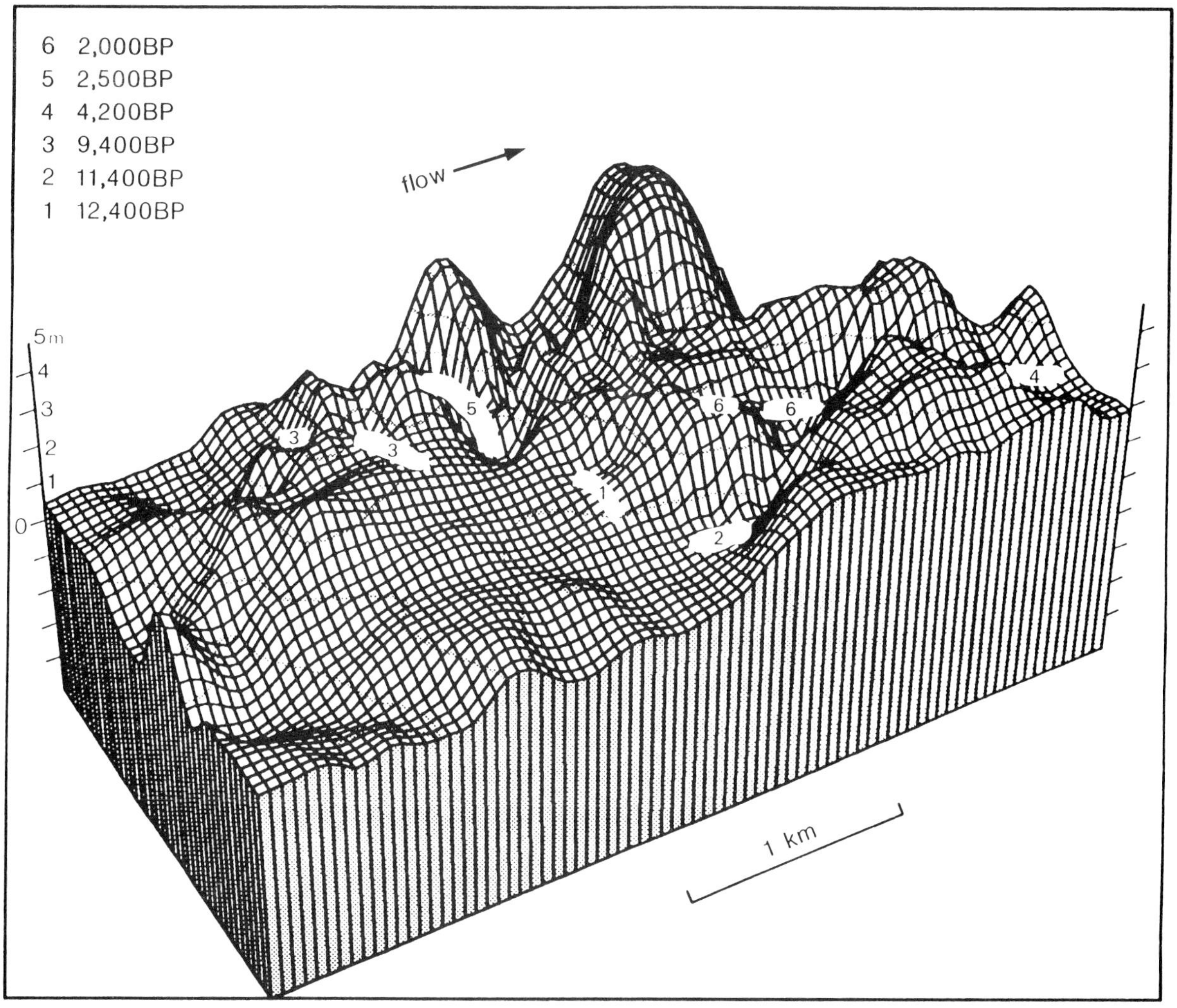

Fig. 18.2. Three-dimensional plot of the fine alluvial sediments in the Raunds reach. The surface was generated from the inverse of the overburden (silt and clay) giving an approximate relative gravel-surface topography. See text for further details. Data supplied by C Halpen and ARC Ltd.

the wood and peat samples used for dating plotted against radiocarbon years BP. The dated wood was either the only wood present at that level or the same species as the most common wood species found at that level. This allows Figure 18.4 to be read as a proxy-wood occurrence/age diagram. In the Early Holocene all the dates come from non-woody material, largely *in situ* mixed (more than one component) peat (Fig. 18.4a). After 6,000 years BP wood is more common and initially dominated in the Middle Holocene by oak (*Quercus* sp.) and *in situ* alder (*Alnus glutinosa*). The first occurrence of beech (*Fagus sylvatica*) is dated at 3270 ± 40 BP (SRR-3599) which confirms its presence either on free-draining parts of the floodplain, such as gravel islands, or at the floodplain edge during the Middle Bronze Age. Beech wood has been recovered from the Somerset Levels trackways as early as 4,500 BP (Coles and Orme 1980), and its expansion to form a significant component of woodlands in later prehistory has been linked to anthropogenic disturbance (Godwin, 1975). Between 3,000 BP and 2,000 BP the most common wood types are blackthorn/cherry (*Prunus* sp.), elder (*Sambucus nigra*) and hazel (*Corylus avellana*). What is suggested is a change in floodplain trees from those associated with both wetlands and tall deciduous woodlands to dry-land and scrub-woodland species. These species were probably derived from the understorey and clearings in the Atlantic woodland and

Fig. 18.5. Little Houghton root (alder) penetrating into gravels at the South Site. Grid squares 0.1 m and ranging rod divisions at 0.5 m.

Fig. 18.6. Wollaston. Younger palaeochannel scar (foreground) with peat and woody macrofossils buried by silty clay alluvium.

are typical of cleared and managed floodplains. They probably grew in patches of scrub-woodland, in the riparian zone, on natural levees and as hedges.

PALAEOLANDSURFACES AND PALAEOWOODS

Much of the *in situ* wood comes from intact and truncated palaeolandsurfaces which are relatively common in both the Nene and Soar floodplains. These provide potential links between pollen evidence and macrofossil evidence, with the possibility of linking woodland pattern to the pollen signal. Very often, as at the site of Cossington on the Soar, there is a sequence of superimposed palaeolandsurfaces which have to be 'unpacked' to be related to macrofossil remains (Keough, Brown and Rice, unpublished). At Little Houghton two sections revealed palaeolandsurfaces (Fig. 18.5). One contained a complex sedimentary feature associated with alder roots, dated to around 3,400 BP. The feature has all the characteristics of a tree-throw pit or pits. These include, asymmetric dish-shaped cross-profiles, stratified infill suggesting inwashes, root material and the interdigitation of poorly sorted mound-sediment with overbank sediments. An alder root penetrating through the mixed gravelly loam into the underlying sand and gravel indicates that at this site deforestation of the alder woodland post-dated 3,400 BP.

At Wollaston the situation is more complex with a depression in the basal gravels containing two palaeochannel scars separated by a sandy bar. All of this is buried by overbank sediments (Fig. 18.6). The depression was 75 m wide and the inset palaeochannel scars approximately 22 m and 10 m wide, separated by 20 m of bedded sand. The depression extended to a depth of just over 3 m below the ground surface and the palaeochannels reached a depth of 2.5 m. Both palaeochannels were filled with organic rich muds and peat. The infill of the older and wider palaeochannel dates to the very early Holocene, while the younger palaeochannel dates to around 3,000 BP. Under the younger palaeochannel lies a diachronous sandy unit containing transported oak wood. The convex bedding of the sand bar which interdigitates with overbank sediments suggests that it accumulated subaerially. This inference is strengthened by the fact that all the wood from the feature comes from dry-land tree types. The sand bar was probably deposited as a levee and may have had some scrub species growing on or near it. A land surface existed across the older palaeochannel (A) around 3,600 BP as indicated by *in situ* alder roots.

The site at Higham Ferrers also displays a complex structured unit between the periglacial gravels and the superficial overbank unit (Fig. 18.7). The stratigraphy consisted of a basal sandy gravel, a gravelly loam and a superficial clayey silt. The gravelly loam was 0.25 m to 0.5 m thick with a poorly defined lower boundary grading into the basal planar bedded gravel unit, which was 3.0 m to 3.3 m thick. Although the gravelly loam had an extremely complex and variable internal structure, graded bedding was common at its base, while clay drapes and interdigitation with overbank silts was more common towards the top. The evidence suggests this unit accumulated by overbank, biotic and channel processes. The resulting uneven landsurface supported willows around 3,200 years ago while an adjacent palaeochannel scar was infilling with organic sediments.

At Raunds three dated sections have been investigated. One is Lateglacial in age while the other two are Holocene. The upstream section shows a Lateglacial paleaochannel scar and a Middle Holocene landsurface which was covered by alder woodland. It was possible to produce a plan of the alder woodland for this site (Fig. 18.8), although it is not known if the pattern represents one generation (the last) or a combination of several generations. The alder site at Raunds is only 30 m from a Neolithic landsurface with numerous tree-throw pits but which was slightly higher and may have been covered by oak rather than alder. Charcoal, soil micromorphology and artifacts in the pits have been interpreted as evidence that the clearing may have been created by Neolithic peoples (Mcphail and Goldberg 1990), although it could equally well result

Fig. 18.7. Organic clay with wood (elder, 2385 ± 50 BP) on reworked sand and gravel (middle unit) at Higham Ferrers.

a

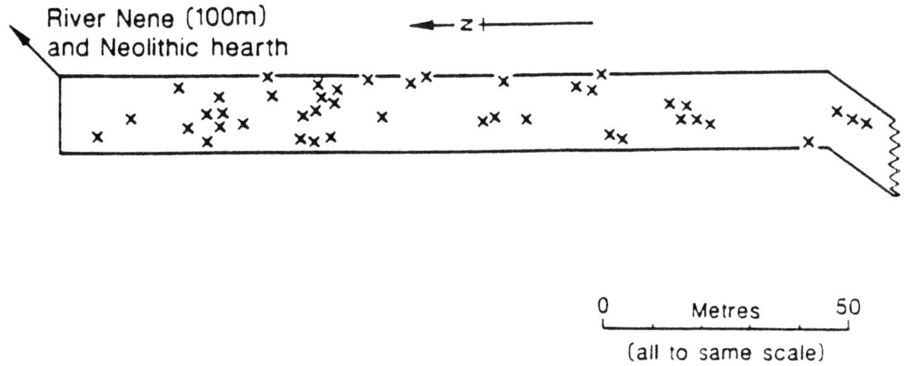

Fig. 18.8. Plan of the alder woodland at Raunds.

from the opportunistic exploitation of a natural clearing (Brown, unpublished). The alder roots at all the sites mentioned (and all lie beneath 2 ± 0.1 m of alluvium) are associated with disturbance of the underlying sand and gravel producing a texture and structure typical of the gravelly loam that exists at all the sites above the periglacial gravels and below the fine overbank unit. It is suggested that this was the floodplain subsoil prior to the metamorphosis of the floodplain and channel system that was caused by a change in sediment load (Brown and Keough, in press). The downstream site at Raunds shows a disturbance feature also probably associated with tree-throw, but upon which an overbank sand mound developed after 2,100 BP. The formation of overbank mounds caused by floodplain roughness elements, such as old grass tussocks and trees, has been observed on contemporary floodplains (Brown 1983).

This evidence of palaeolandsurfaces comes from islands of floodplain preserved between palaeochannels and on top of the oldest palaeochannels. All these palaeolandsurfaces have been buried by the post-3,000 BP silty clay unit. However, from the dating of this contact it is apparent that the base of the overbank unit is diachronous, partly because of the uneven Middle Holocene floodplain surface, with deposition occurring earlier in the depressions and later on the mounds. This means that the sandy and gravelly islands must have become progressively smaller as fine sedimentation proceeded. The present size of surviving islands may be considerably smaller than their size in the Bronze Age. This may have relevance to the interpretation of burial mound location and settlement sites, as flooding was restricted to smaller floodplain belts and large areas of the floodplain were covered by permanently dry land. This situation changed relatively rapidly during the first millennium BC. The burial of landsurfaces with *in situ* tree roots allows the location of past floodplain woodlands, adding a spatial dimension to any inferences from pollen analysis. However, this is only possible for brief time-windows providing a few snap-shots of changing vegetation and land use.

RADIOCARBON DATES AND ALLUVIAL CHRONOLOGY

In the Nene and the Soar the authors attempted to date every site which produced organic remains which was exposed over the three years of the study. In all over 50 radiocarbon dates were used in the construction of a cumulative frequency graph for the floodplains (Fig. 18.9). The overall graph has two steep sections, one in the Early Holocene 10,000–9,000 BP and the other in the Late Holocene, 4,000–2,000 BP. The material forming these peaks also differs, with the Early Holocene (and Lateglacial) samples coming from *in situ* organics, largely channel fills, and the Late Holocene samples coming from both *in situ* and allochthonous organics. Similarly the palaeoecological evidence comes from these two time periods since it is exclusively derived from organic channel infills and palaeolandsurfaces. This leaves a geoarchaeological gap, which is a function of the evolutionary history of these reaches of the valleys. During the Middle Holocene there was relative channel stability with few avulsion events and as a result few palaeochannels were created. The lack of palaeolandsurfaces, which interestingly is less marked, is more difficult to explain, but there are two possible mechanisms. One explanation is widespread floodplain erosion, in the form of topsoil-stripping during the second and third millennia BC. Although some topsoil may have been lost there is little recognisable stratigraphic evidence of this. The alternative explanation is little or no overbank deposition of fines, but continuing soil development and bioturbation which eventually produced the gravelly loam unit. The Late Holocene age of all the roots is best explained by a preservation 'event' caused by a rise in floodplain water-tables prior to burial associated with increased overbank sedimentation. Robinson and Lambrick (1984) have proposed a similar time-lag between hydrological change and increased sedimentation for the Upper Thames basin.

This geoarchgaeological gap illustrates a general problem which is that our ability to see hydrological

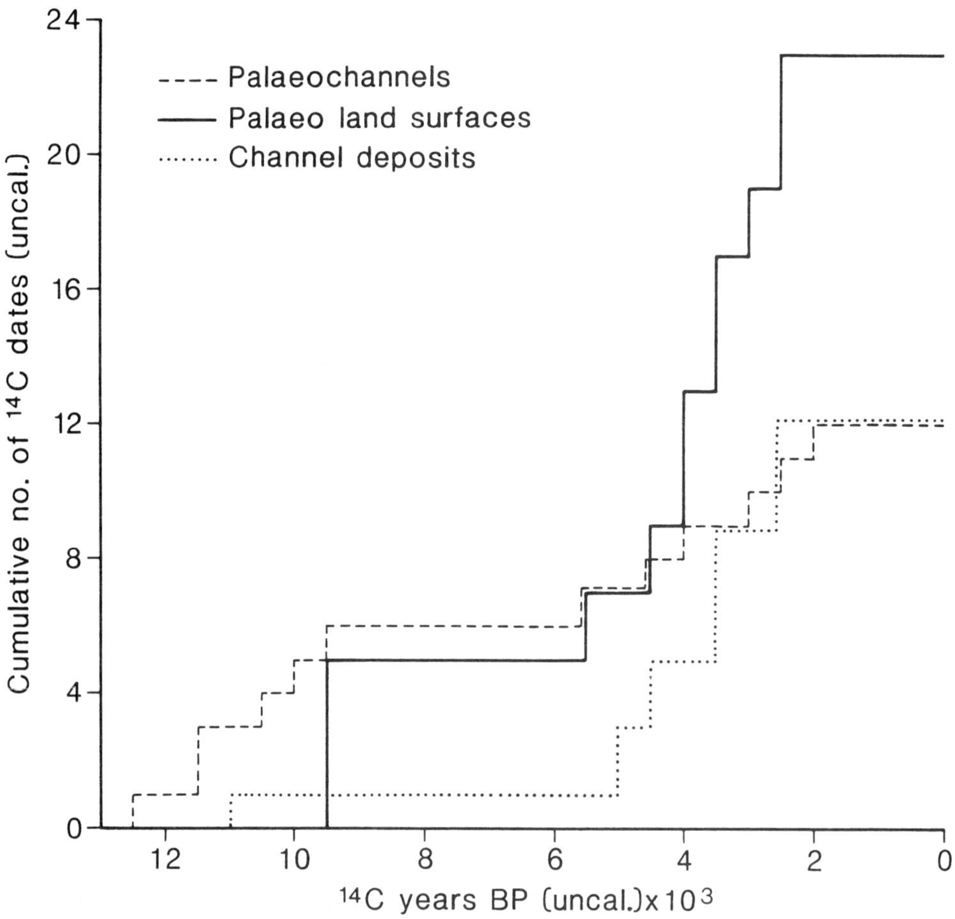

Fig. 18.9. Cumulative graph of all the radiocarbon dates (uncalibrated) from the floodplains of both the Nene and the Soar.

change, in detail, is partially dependent upon sediment availability, and sediment availability can easily be affected by human activity. This is one reason why we cannot easily separate anthropogenic from climatic effects during the Holocene. However, this gap, although widespread in lowland alluvial systems is not universal. In particular it is not a problem where lake systems survived in floodplains into the Middle Holocene, as was the case in the Cheshire-Shropshire lowlands (Brown 1987a) or where valley or raised mires formed. It is, however, typical of southern Britain, and an excellent example is the floodplain of the river Gipping in Suffolk, as investigated at Sproughton by Rose *et al* (1980). Many lowland floodplains have undergone a particular pattern of floodplain evolution characterised by relative channel stability and an aggrading floodplain (i.e. a rise in floodplain elevation with a constant bed height) and occasional channel avulsion (Brown 1987b).

THE STABLE-BED AGGRADING-BANKS (S.B.A.B.) MODEL AND FLOODPLAIN ENVIRONMENTS

Channels can become deeper and narrower through floodplain deposition rather than bed erosion, and change location through avulsion rather than by lateral migration and meander neck-cutoffs. This model of floodplain and channel evolution has been termed the stable-bed aggrading-banks model by Brown, Keough and Rice (unpublished). The probable cause is a disequilibrium between bed aggradation and floodplain aggradation, itself the result of an increase in the fine sediment load in a situation where banks are resistant to erosion due, at least partly, to dense floodplain vegetation. In effect channels became deeper by the addition of a silt/clay bank and levee associated with the deposition of the superficial fine unit found at all the sites. This increase in the transport of fine sediment through the channel system changed channel conditions, as work on Coleoptera has verified. Osborne (1988) has used fossil Coleoptera to show how the bed of the river Avon (West Midlands) changed from a clean stoney bed in the Late Bronze Age to the blanket of mud and silt which it retains today. In some systems channel siltation may have been the process responsible for a reduction in the number of channels and the creation of palaeochannels by the plugging of the neck with fine sediments. It is quite likely that there were commonly two or more Early and Middle Holocene channels, between which floodplain developed. This anastomosing pattern is associated with low gradient systems (Smith 1983) and it may well have been partially inherited from a braided pattern that existed in the valleys during the Lateglacial (Brown, Keough and Rice, unpublished) as was probably the case in the Upper Thames (Limbrey and Robinson 1988). Some contemporary floodplains still retain this pattern, the best known being parts of the river Lee in Eire (The Gearagh), and there are many records of the siltation of secondary channels during the historical period (for example the Severn in Gloucester). Indeed the Nene itself has an anastomosing pattern in many locations (generally two channels), some components of which may be natural rather than anthropogenic in origin.

CONCLUDING REMARKS

Within a period of approximately 2,500 years covering the Bronze and Iron Ages the Middle Nene floodplain was transformed from a stable, probably multiple channel system covered by dense woodland, to a cleared agricultural landscape with managed channels. This period was one of tremendous change in the floodplain environment and resources, and much of the change must have taken place on human timescales. Communities must have had to adapt to the changing channel locations, cutoff and secondary channel siltation and periods of severe flooding. In addition floodplain land-use changes reveal an increasing exploitation of the floodplain as part of the agricultural resource.

The work described here also highlights two aspects of the geoarchaeological potential of floodplains. Firstly, even what at first might seem to be stratigraphically simple and monotonous floodplain stratigraphies can contain a wealth of palaeoenvironmental and geoarchaeological evidence, especially associated with buried palaeochannels and palaeolandsurfaces. Secondly, however thorough the analysis of floodplain sections, boreholes etc., the geoarchaeological potential of floodplain environments is constrained by the sedimentary evolution of the systems. This can bias the geoarchaeological record and will vary from valley to valley or, in complex terrains, even from reach to reach. The use of inappropriate models of floodplain evolution will inhibit the search for, and recognition of, sedimentary facies capable of producing valuable geoarchaeological evidence. Archaeological excavations in valleys in lowland Britain have rarely produced stratigraphic sequences comparable to those expected from typical meandering systems (compare Needham 1985 with Allen 1965 or Walker 1984). The increased awareness of the variety of floodplain sedimentary and post-depositional processes such as colluviation (Brown, in press), aeolian deposition and biological activity that may produce floodplain stratigraphies will improve the interpretation of the palaeoenvironments associated with valley settlements and land use. The climatic and catchment controls on river and sediment discharge along with inherited factors not only condition floodplain evolution but also the geoarchaeological potential of alluvial valley floors.

Acknowledgements

Most of the radiocarbon dates mentioned in this chapter were provided by the Natural Environmental Research Council along with a studentship to one of us (MKK) as part of a study on floodplain evolution in the East Midlands. Four radiocarbon dates at Raunds were provided by the Ancient Monuments Laboratory, English Heritage and we would like to thank all those involved in the Raunds Area Projects for their help with many aspects of this work.

References

Allen, J.R.L. 1965. A review of the origin and characteristics of recent alluvial sediments. *Sedimentology*, 6, 89–191.

Behre, K.E. 1981. The interpretation of anthropogenic indicators in pollen diagrams. *Pollen et Spores*, 23, 225–245.

Brown, A.G. 1983. An analysis of overbank deposits of a flood at Blandford-Forum, Dorset, England. *Revue de Geomorphologie Dynamique*, 32, 95–99.

Brown, A.G. 1987a. Long-term sediment storage in the Severn and Wye catchments. In K.J. Gregory, J. Lewin and J.B. Thornes (eds.), *Palaeohydrology in Practice*, 307–332. Chichester: Wiley.

Brown, A.G. 1987b. Holocene floodplain sedimentation and channel response of the lower river Severn, United Kingdom. *Zeitschrift für Geomorphologie*, (Neue Folia) 31, 293–310.

Brown, A.G. 1989. Holocene floodplain diachronism and inherited downstream variations in fluvial processes: a study of the river Perry, Shropshire, England. *Journal of Quaternary Science*, 5, 39–51.

Brown, A.G. in press. Slope erosion and colluviation at the floodplain edge. In M. Bell and J. Boardman (eds.), *Past and Present Soil Erosion*. Oxford: Oxbow.

Brown, A.G. and Keough, M.K. in press. Holocene floodplain metamorphosis in the Midlands, United Kingdom. *Geomorphology*, 4.

Castleden, R. 1976. The floodplain gravels of the river Nene. *Mercian Geologist*, 6, 33–47.

Coles, J.M. and Orme, B.J. 1980. *The Prehistory of the Somerset Levels*. Somerset Levels Project.

Ferguson, R. 1981. Channel form and channel changes. In J. Lewin (ed.) *British Rivers*, 90–125. London: Allen and Unwin.

Ferring, C.R. 1986. Rates of fluvial sedimentation: Implications from archaeological variability. *Geoarchaeology*, 3, 259–274.

Hooke, J.M., Harvey, A.M., Miller, S.Y. and Redmond, C.E. 1990. The chronology and stratigraphy of the alluvial terraces of the river Dane valley, Cheshire, N.W. England. *Earth Surface Processes and Landforms*, 15, 717–738.

Institute of Geological Sciences, 1977. *Quaternary Map of the U.K.: South.*

Limbrey, S. and Robinson, S. 1988. Dry land to wet land: soil resources in the upper Thames valley. In P. Murphy and C. French (eds.), *The Exploitation of Wetlands*, 129–144. Oxford: British Archaeological Reports, British Series, 186.

Macklin, M.G. and Lewin, J. 1986. Terraced fills of Pleistocene and Holocene age in the Rheidol valley, Wales. *Journal of Quaternary Science*, 1, 21–34.

Macphail, R.I. and Goldberg, P. 1990. The micromorphology of tree subsoil hollows: Their significance to soil science and archaeology. In L.A. Douglas (ed.), *Soil Micromorphology*, 425–429. Amsterdam: Elsevier.

Needham, S. 1985. Neolithic and Bronze Age settlement on the buried floodplains of Runnymede. *Oxford Journal of Archaeology*, 4, 125–137.

Needham, S. 1989. River valleys as wetlands: the archaeological prospects. In J.M. Coles and B.J. Coles (eds.), *The Archaeology of Rural Wetlands in Britain*, 29–34. Exeter: Wetland Archaeologicsal Research Project/London: English Heritage.

Osborne, P.J. 1988. A late Bronze Age insect fauna from the river Avon, Warwickshire, England: Its implications for the terrestrial and fluvial environment and for climate. *Journal of Archaeological Science*, 15, 715–727.

Pilcher, J.R. 1991. Radiocarbon dating. In P.L. Smart and P.D. Frances, *Quaternary Dating Methods — A Users Guide*. Quaternary Research Association Technical Guide 4, 16–44. Cambridge.

Robinson, M.A. and Lambrick, G.H. 1984. Holocene alluviation and hydrology in the upper Thames basin. *Nature*, 308, 809–814.

Rose, J., Turner, C., Coope, G.R. and Bryan, M.D. 1980. Channel changes in a lowland river catchment over the last 13,000 years. In R.A. Cullingford, D.A. Davidson and J. Lewin, (eds.), *Timescales in Geomorphology*, 159–176. Chichester: Wiley.

Smith, D.G. 1983. Anastomosed fluvial deposits: Modern examples from Western Canada. In J.D. Collinson and J. Lewin (eds.), *Modern and Ancient Fluvial Systems*, 155–168. Oxford: International Association of Sedimentologsits, Special Publication, Blackwell.

Walker, R.G. 1984. *Facies Models*. Geoscience Canada, Reprint Series 1, Toronto, Canada.

19

Environment, archaeology and alluvium on the river gravels of the South Midlands

Mark Robinson

Summary: *Within less than 20 years, many of the preconceptions about the Holocene geomorphology and archaeology of the floodplains of the major river systems of the south Midlands/central southern England have been overturned. Large-scale gravel extraction combined with the rise of rescue archaeology have resulted in many quarry exposures being available for study and much detailed excavation in the Upper Thames, Nene and Ouse valleys. Their channels have been shown to be much more stable than previously thought, with little migration and reworking of the Pleistocene gravels during the past 10,000 years. Overbank alluviation with fine sediments was not a feature of these rivers until comparatively recently, indeed little or no flooding was taking place during much of the Holocene. The broad expanses of alluvium which comprise the modern floodplains to these rivers are not blank areas but cover rich archaeological landscapes with a high potential for environmental studies.*

INTRODUCTION

The Upper Thames, Nene and Ouse drain the northern half of the belt of Jurassic and Cretaceous sedimentary rocks which strike from Dorset north-east to the Wash (Fig. 19.1). The region under consideration extends from the headwaters of the Thames towards the eastern boundaries of Northamptonshire and Bedfordshire, beyond which the Nene and Ouse come under the influence of the Fens and the effects of sea level. The Jurassic limestone hills of the Cotswolds and Northampton Uplands comprise the watershed to the north-west while the Chalk scarp of the Berkshire Downs and Chilterns, through which the Thames exits at the Goring Gap, delimit the south-easterly extent of the region. The rivers have shaped the solid geology to create a landscape of clay vales and plains with limestone, or to a lesser extent, sandstone ridges. Except in the valley bottoms, there is an extensive covering of chalky boulder clay in the north-eastern half of the region. Where the rivers run through the clay vales, broad flat expanses of Pleistocene gravel terraces step down to the Holocene floodplains. Such terraces are most extensive in the Oxford Clay vale of the Thames above Oxford and the Kimmeridge/Gault Clay vale of the Thames between Abingdon and Dorchester, but there are also substantial gravel deposits on the Oxford Clay of the Ouse Valley below Bedford, and on the Upper Lias Clay of the Nene Valley below Northampton. Throughout most of the region, the river terraces have a high limestone content, but where they have been derived from the chalk, flint is more important. The modern floodplains, which in the case of the Thames can be up to 2 km across, occupy the lowest part of the most recent or first terrace, the gravels being covered with up to 2 m of alluvial clays to clay loams. The entire "terrace", both that covered by, and that free of, Holocene alluvium is sometimes referred to as the floodplain terrace.

The region is in the heart of lowland England and arable agriculture extends to the watersheds. Its soils are mostly calcareous or circumneutral and only in a few areas of acid sands and plateau gravels are there serious problems of soil fertility. The limestone and gravel terraces have easily-worked soils although some of the soils on the limestone are shallow and brashy. The Jurassic and Cretaceous clays are relatively intractable and suffer from impeded drainage but cultivation is possible and most are now ploughed. The chalky boulder clay has a higher silt and sand content in its soils so does not present such serious problems. The floodplains of the major rivers are still regularly inundated where there is no flood protection, but modern drainage and river management has enabled their cultivation, mostly for grass leys.

The Holocene sequence of the floodplains is

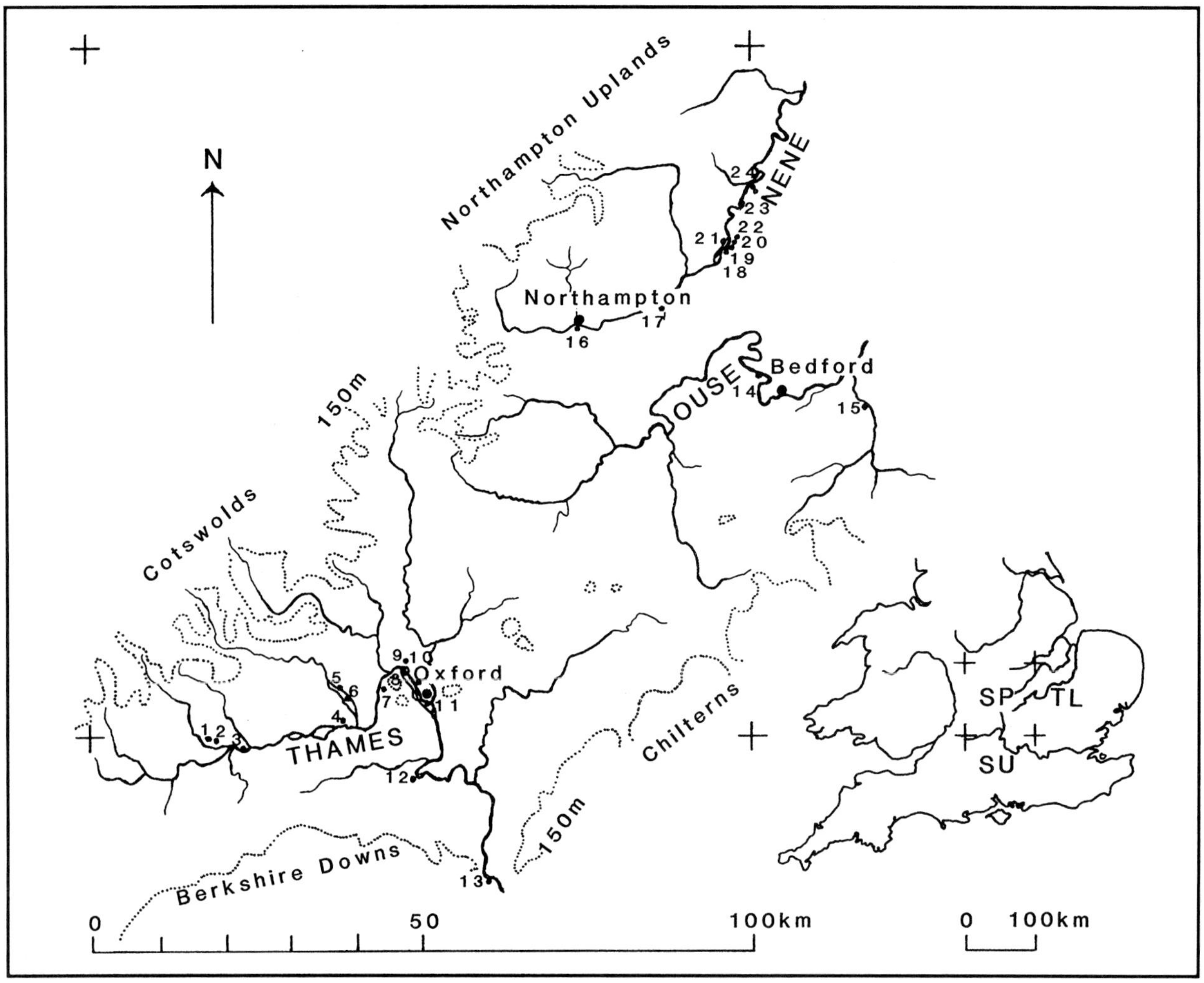

Fig. 19.1. The Upper Thames, Nene and Ouse basins. Main sites mentioned in the text: 1. Thornhill Farm, 2. Claydon Pike, 3. Buscot Lock, 4. Old Shifford, 5. Gill Mill, 6. Mingies Ditch, 7. Farmoor, 8. King's Weir, 9. Yarnton, 10. Port Meadow, 11. Grandpont/St. Aldates, 12. Drayton, 13. Wallingford, 14. Bromham, 15. Warren Villas, 16. St. James' Square, 17. Earls Barton, 18. South Stanwick Villa, 19. South Stanwick Barrow, 20. Stanwick, 21. Irthlingborough, 22. West Cotton, 23. Titchmarsh, 24. Aldwincle.

considered under four headings: channels, palaeohydrology and alluviation, vegetation including agriculture, and archaeology. The basic outline has already been established for the Upper Thames basin (Robinson and Lambrick 1984) and it is now possible to give provisional sequences for the other two valleys.

CHANNELS

The Holocene regime of all three rivers has been one of channel stability and simplification of multiple channel systems by silting. The transition from a system of minor braided channels, which were reworking and probably lowering part of the floodplain terraces, to multiple, broad incised channels occurred towards the end of the Late Devensian. A radiocarbon date of 10,860 ± 130 BP (HAR-8356) was obtained on waterlogged twigs from a minor channel within the cross-bedded gravels of the lower Windrush at Mingies Ditch (Allen and Robinson, forthcoming), while organic sediments at the base of a major incised channel of the Thames at Farmoor which carried at least a seasonal flow in the Iron Age were dated to 10,600 ± 250 BP (BIRM-590) (Lambrick and Robinson 1979: 141–2). Both deposits produced fruits of *Betula nana* (dwarf birch), an arctic-alpine shrub and fragments of *Helophorus glacialis*, a beetle of snow meltwater pools. A temperate biological assemblage from a channel cutting the gravels at Mingies Ditch was dated to 9,380 ± 110 BP (HAR-8366). Organic material from a palaeochannel of the Nene near Raunds, which was filled with fine sediments, gave a date of 9370 ± 170 BP (HAR-9243).

A palaeochannel of the Nene adjacent to the

Medieval village of West Cotton (Windell *et al* 1990: 13–14) illustrates Holocene channel stability and silting well. The section (Fig. 19.2) was through a bend in the channel which had been working its way southwards into the undisturbed gravels of the first terrace. On the inside of the bend at the northern edge of the channel, lenses of fine sediment were interleaved with gravel, representing the transition to more stable conditions in the Late Devensian. *Helophorus glacialis* was identified from organic material in a sandy lens on the inside edge of another section through the channel (projected onto Fig. 19.2'A'.) confirming the early origin of this channel. Little sedimentation took place during the first half of the Holocene and organic sediments of Late Neolithic date covered the Pleistocene gravels on the bed of the channel on either side. Within these sediments was a platform of cut alder branches and brushwood which has yielded radiocarbon dates of 3990 ± 54 BP (UB-3319) and 4062 ± 54 BP (UB-3321). There followed a period when clay sedimentation reduced the width of the channel by half. The next dated period of activity was Middle Saxon when wooden pegs dated to 1264 ± 52 BP (UB-3323) and 1297 ± 49 BP (UB-3328) were driven through silts which incorporated the Neolithic structure and gravel was deposited over it. A series of three Saxon mills was constructed alongside the channel on the south-eastern bank; radiocarbon dates of 941 ± 53 BP (UB-3325) and 1014 ± 51 BP (UB-3327) being obtained on oak timbers from the latest mill. Gravel from the mill leats had been dumped into the channel and the contemporary channel bed itself was used for the retting of flax. The sediments with the flax also contained remains of flowing water invertebrates. Thereafter, the channel filled rapidly with clay. The remains of the Saxon mills were sealed by a Medieval flood embankment which elsewhere carried a leat, presumably to maintain the head of water for another mill. Eventually Medieval overbank alluvium covered both the floodplain and this channel, lapping up against the flood embankment. The topography of the floodplain had been so changed by the alluviation that the former leat was diverted to flow along the silted channel to the extant channel of the Nene in the opposite direction to that in which the palaeochannel had flowed.

The palaeochannel at West Cotton was typical in that although it contained sediments of various dates, episodes of deposition were relatively short and there were no long organic sequences. Sedimentation often tends to correspond to periods of activity on the bank close to the site, which can be very useful to archaeologists. For example the West Cotton channel contained both Late Neolithic and Saxon organic deposits, both periods of significant activity nearby. A channel of the Thames at Wallingford into which a Late Bronze Age timber waterfront structure had been built

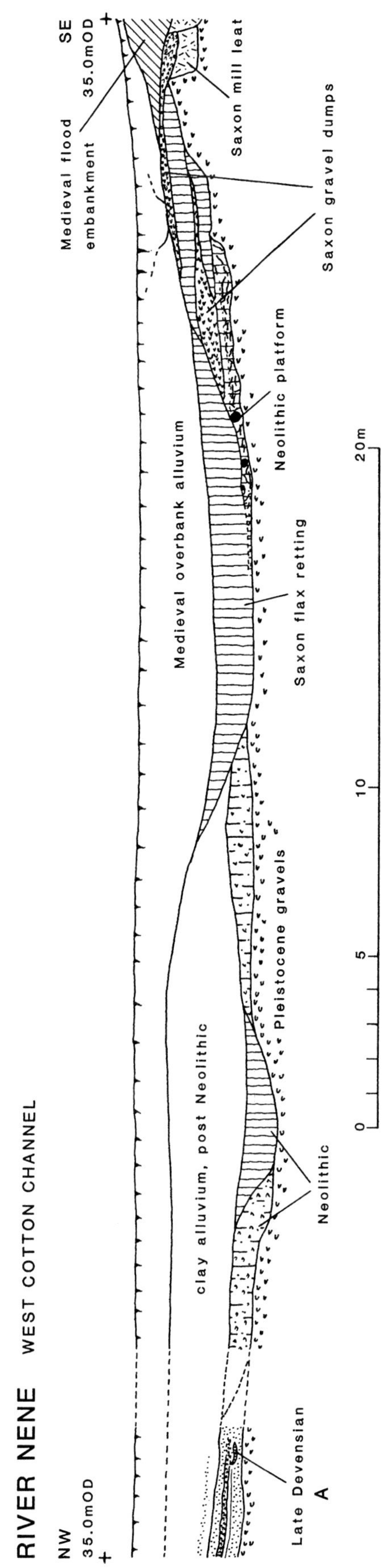

Fig. 19.2. Section through a palaeochannel of the River Nene at West Cotton.

likewise contained contemporaneous organic sediments (Lambrick 1987). Channels became redundant at various dates but the Early Medieval period probably saw the greatest simplification of all three river systems. Channel flow was being altered by the construction of mills, large fish weirs and river crossings. There was also a considerable fine sediment load being carried by the rivers at this period. Numerous channels of the Thames at Oxford fell out of use where the causeway and bridges of the Grandpont carried a road southwards from the town across the floodplain (Durham 1984: 87–90).

Those channels which survived were narrow and deeper than they had been in the earlier Holocene. This was achieved almost entirely as a result of sedimentation, overbank alluviation building up the height of the banks and encroaching into former channel edges. The palaeochannel at West Cotton only showed slight evidence for post-Neolithic incision into its gravel bed.

PALAEOHYDROLOGY AND ALLUVIATION

During the Early and Mid Holocene, pedological processes predominated over alluvial accretion on floodplains, and although limited flooding could have occurred, the water-tables were seasonally low. All three basins, however, showed a Late Holocene rise in water-table followed by extensive flooding and, eventually, alluviation, although these events were not synchronous between them (Fig. 19.3).

The action of Late Devensian braided streams left an undulating surface to the parts of the first terraces that are now covered with fine alluvium and in places gravel islands project through the clay. The horizontally bedded gravels of the higher parts never covered by alluvium give flatter surfaces. The lowest areas of floodplain tend to have a clay covering directly above the gravel, perhaps the result of alluviation shortly after the transition to channel flow. The higher parts of the floodplains have soils more similar to those on the still exposed gravel terrace, being sandy loams, silty clay loams and silty clays. There is often evidence for soil decalcification. These palaeosols have been studied in most detail on sites in the Upper Thames valley, where they have been found beneath prehistoric monuments that were subsequently covered with alluvium (Limbrey and Robinson 1988). Their high silt and chlorite content showed a substantial loess component.

The Mid Holocene water-table was at least seasonally low on the floodplains of all three drainage basins. The fills of earlier prehistoric features are oxidised and do not contain preserved organic remains whereas adjacent features of similar depth, but more recent date, have gleyed fills with waterlogged preservation. This rise in water-table was already underway between the Late Bronze Age and Middle Iron Age in the Upper Thames valley and continued until the onset of flooding (Robinson and Lambrick 1984). This was best illustrated at Mingies Ditch, where a radiocarbon date of 2,800 ± 90 BP (HAR-4489) was obtaind on *Alnus* charcoal from a non-waterlogged layer in a tree-throw pit, while an Iron Age ditch of similar depth contained waterlogged twigs at the bottom which gave a date of 2170 ± 90 BP (HAR-4485; Robinson and Lambrick 1984). A "recurrence surface" for peat preservation in the upper fills of the Iron Age ditches suggested a continuing rise in water-table after the abandonment of the site.

The rise in water-table occurred at a later date in the Ouse basin. A group of Iron Age and Roman ditches were excavated by Mr. M. Dawson on the floodplain of

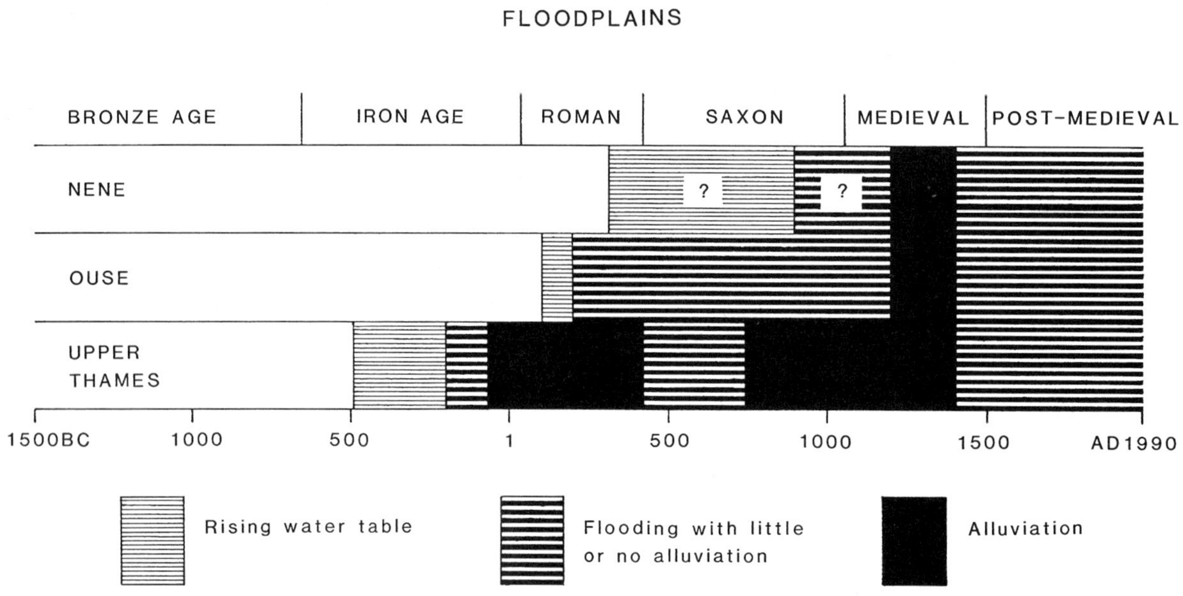

Fig. 19.3. Generalised floodplain sequences for the Upper Thames, Nene and Ouse basins.

the River Ivel at Warren Villas. On one part of the site the fills of Iron Age ditches were fully oxidised whereas adjacent Roman ditches of the 2nd/3rd century AD had gleyed lower fills and some badly preserved organic material. A layer of sedge peat, also of 2nd/3rd century date covered a ploughsoil on the floodplain and filled the tops of the ditches (Robinson, unpublished).

Evidence for the date of any rise in water-table in the Nene Valley is at present less clear. However, the level of organic preservation in a series of Roman ditches running along the edge of the floodplain at the Stanwick Roman Villa, excavated by Mr. D. Neal, was highest in the latest (Robinson, unpublished).

Rising water-tables lead to flooding. At Farmoor, a Middle Iron Age settlement on a low-lying part of the Thames floodplain adjacent to a former channel experienced flooding which deposited molluscs of flowing water, such as *Bithynia tentaculata*, both on the ground surface and in ditches dated to 2060 ± 70 BP (HAR-1926) (Lambrick and Robinson 1979: 111). However, little sedimentation was taking place. Flooding was probably occurring in the Ouse basin at Warren Villas from the Mid Roman period onwards, again initially with little sedimentation, enabling peat formation to occur on the general surface of the floodplain. It is possible that Saxon flooding with little alluviation was taking place in the Nene Valley because somewhat organic sediments which pre-dated the main period of alluviation covered demolition rubble of a late 4th century building on a low-lying site at South Stanwick (Keevill, this volume). General alluviation of the main body of the floodplain began first in the Upper Thames valley. At Drayton, the gravel upcast from a Neolithic cursus ditch sealed a well oxidised stony calcareous brown earth which was only 0.08 m thick (Limbrey and Robinson 1988: 138; Robinson and Lambrick 1984). The upcast bank was subsequently covered with 0.64 m of clay alluvium, the lower half of which was gleyed, and within which a Roman ditch was interstratified. The alluvium of the Thames floodplain has not proved very amenable to magnetic dating, perhaps because seasonal drying causes deep cracking of the clay. However, successful results were obtained by Dr. A. Clark where the sediments were at their greatest thickness above the largely silted cursus ditch. He showed that alluviation was already underway by 50 BC. By tracing the layers of sediment from above the ditch to the alluvium beyond its edges, it is possible to extrapolate the rate of sedimentation to the general surface of the floodplain. This gives an average rate of accumulation of about 0.5 mm per annum from about AD 1 to AD 400. This slowed down to a rate of 0.2 mm or less between about AD 400 and AD 800, but it was not possible to show whether alluviation ceased entirely during the Early Saxon period. The sedimentation rate returned to an average of 0.5 mm per annum between about AD 800 and AD 1100. It is uncertain when alluviation ceased, but Post-medieval artefacts were not found below the modern "A" horizon and, if sedimentation continued at the rate of 0.5 mm per annum, deposition would have been complete sometime between about 1400 and 1500. These dates confirm the sequence advanced by Robinson and Lambrick (1984) based on the stratigraphic relationship between dated archaeological sites and alluvium. Further sites have also been excavated which fall into a similar pattern. An exposure by Ms. G. Hey at Yarnton, near Oxford, had a humic horizon within the alluvium which possibly represented an Early Saxon slow-down of sedimentation.

Alluvium containing Medieval sherds covered the peat on the floodplain at Warren Villas whilst at Bromham, on the floodplain of the Ouse itself, alluvium sealed a Roman settlement of the late 3rd century AD (Tilson 1973).

Whereas alluviation in the Upper Thames valley was a relatively gradual process and floodwaters need never have been very deep, slowly levelling up the floodplain, it seems to have been overwhelming in parts of the Nene valley. Late Saxon or Early Medieval ridge-and-furrow at West Cotton was sealed by up to 2 m of clay alluvium. In the Upper Thames, the depth of alluvium on the main body of the floodplain rarely exceeds 1 m and was deposited over a longer period. Also, floodwaters seem to have been deep in the West Cotton area because true alluvium slopes a little way up the valley sides, being at a higher level than the top of the alluvium in the valley centre. Alluvium sealed a ploughsoil which truncated Roman features at Titchmarsh (Windell 1985), whilst at South Stanwick, alluvium covered ridge-and-furrow, again suggesting a late date for the onset of alluviation in the Nene valley.

In all these drainage systems alluviation largely ceased before the end of the Medieval period, although flooding has continued to the present day. Post-medieval artefacts or features are not found sealed beneath alluvium and near West Cotton a complete but crushed 14th century pot was found below the topsoil on the surface of the alluvium (S. Parry, pers. comm.).

The sequences described above and in Fig. 19.3 illustrate the general floodplain trends in the three valleys. Until substantial silting had taken place in the channels, hydrological changes in the catchment are not necessarily reflected in the sediments other than in palaeochannels. For example, a Late Bronze Age soil with a dry grassland molluscan fauna developed above Neolithic or Bronze Age channel sediments of the Thames at Wallingford only to be buried by the general alluvial deposit which covers the floodplain (Thomas *et al* 1986: 178–84). The channel at West Cotton also contained a large volume of clay sediments which pre-dated general overbank alluvium, and *Alnus* root clusters *in situ* within it suggest that the surface of this deposit was at least seasonally exposed for a period, thus allowing tree colonization (probably prior to the Iron Age).

The uneven surface of the floodplain would have meant that the initial overflow was restricted to the lowest part of the floodplain and the first overbank alluvium would have been deposited in hollows left by palaeochannels. Frequent inspection of the gravel pit at West Cotton by Mr. D. Windell (pers. comm.) while excavating at West Cotton did not reveal deposits other than the main body of Late Saxon/Early medieval alluvium. The earlier-dated bodies (e.g. Brown, this volume) are presumably silting in hollows. Flooding in the Upper Thames valley did not reach its full extent until the Medieval period. At Claydon Pike for example, alluvium filled the tops of Late Roman ditches on islands of gravel terrace (Robinson and Lambrick 1984).

VEGETATION AND AGRICULTURAL PRACTICES

The three floodplains probably experienced similar vegetational succession in the earlier Holocene, resulting in dense alder woodland on the lower areas and oak/alder woodland on the higher areas (G. Campbell, pers. comm.; Robinson and Wilson 1987: 28–31). Initially, post-clearance vegetation probably resembled lightly-grazed grassland on the gravel terraces. However, just as agricultural intensification was resulting in heavier grazing and ploughing extending onto the floodplains, so the hydrological conditions were changing. There followed the development of wet, overgrazed pasture and the eventual cessation of cultivation. Increasingly, floodplain grassland was managed for hay production and alluvial floodmeadow became the major category of vegetation until recent times.

Without long organic sequences, vegetational succession for the first half of the Holocene can only be established by putting together the results from radiocarbon-dated samples from a number of channel sections. Such a study of pollen and macroscopic plant remains from Mingies Ditch showed that the arctic vegetation of the Late Devensian floodplain had, by 9380 ± 100 BP (HAR-8366), been replaced by *Betula pendula* or *pubescens*, *Salix* sp. and *Populus tremula*, but tree cover was by no means complete (Allen and Robinson, forthcoming). *Quercus* sp., *Corylus avellana* and *Cornus sanguinea* had joined these species in open woodland by 7590 ± 80 BP (HAR-8354). A sample dated to 6540 ± 80 BP (HAR-8355) was dominated by remains indicative of *Alnus glutinosa* including much pollen, fruits, catkins, leaf mite galls and the now extinct beetle *Agelastica alni*. Dense alder woodland seems to have shaded out some of the trees and shrubs that were formerly present, including all the *Betula* sp. and most of the *Salix* sp., but *C. avellana* and *Quercus* sp. remained important. The rise in water-table came too late for much true alder carr to develop on any of the these floodplains.

Evidence for two episodes of Neolithic clearance at Drayton came from tree-throw pits showing signs of burning after the trees had fallen (Lambrick and Moore 1987). The interpretation put on them is that clearance was achieved by killing the trees (mostly *Quercus* on the basis of abundant charcoal), either allowing them to fall or pulling them down once their roots had decayed and burning the fallen trunks. The radiocarbon dates on them either fell within the date range for the construction of the Drayton cursus in the Early Neolithic (combined date of 4750 ± 42 BP, OXA-2073, OXA-2074, OXA-2071, HAR-6478) when due allowance is made for the age of the trees at death, or corresponded to other archaeological evidence for Late Neolithic habitation on the site (4220 ± 80 BP, OXA-2076; 3880 ± 70 BP, OXA-2078). Similar, but as yet undated tree-throw pits have been found in the West Cotton/Stanwick area of the Nene valley.

The ditches of a long barrow in the same general area of the Nene Valley at South Stanwick were unusually deep for Neolithic features and contained preserved organic material (Moore and Jackson 1990). A preliminary investigation of macroscopic plant remains showed a light-demanding flora of sparsely vegetated weedy ground to have been present with species such as *Aphanes arvensis*, *Potentilla reptans* and *Polygonum persicaria*. Interestingly, seeds of *Verbena officinalis*, not an obvious member of such a community, were quite well represented. Its presence was perhaps a reflection that earlier Neolithic open country floras were not fully developed.

The Neolithic timber platform in the channel at West Cotton mentioned above was perhaps part of an alignment of Neolithic monuments which included the South Stanwick barrow. Organic deposits associated with the platform and also from another channel exposure dated to 4300 ± 150 BP (HAR-9241) gave strong evidence that alder woodland predominated on the floodplain. There was, however, an open country element in the biota identified by Mrs. G. Campbell including seeds of *Plantago major* and the mollusc *Vallonia pulchella*. Seeds of *Verbena officinalis* were also present. All the channel deposits of this date from West Cotton contained significant amounts of charcoal, perhaps the result of clearance activities. The contrast between the evidence from Stanwick with that from West Cotton, which being from channel deposits would have had a much larger catchment, might suggest that the Neolithic monuments were set along a linear clearance through a largely wooded landscape. Alternatively, woodland regeneration occurred amongst them.

Although there was some clearance of the Thames floodplain in the Neolithic, waterlogged channel sediments from Buscot Lock presented a similar picture to those from the West Cotton channel, with strong evidence of alder woodland and only limited evidence of open habitats (Robinson and Wilson 1987: 31–2).

It is possible that much clearance took place on the floodplains during the Early and Middle Bronze Age, but evidence from waterlogged deposits is lacking. However, round barrows at West Cotton/Irthlingborough (Windell *et al* 1990: 11–13; Halpin 1987) sealed stone-free sandy loam whereas the surrounding alluvium covered a stony ploughsoil. Some of the cremations in the barrows contained charred tubers of *Arrhenatherum elatius*, a grass which will not tolerate heavy grazing. Taken together, these results would be consistent with the barrows having been built on a floodplain of lightly grazed grassland.

The development of floodplain grassland in the Upper Thames valley from the Iron Age onwards has already been considered in detail (Lambrick and Robinson 1988). At Mingies Ditch, where clearance of alder woodland did not occur until the Late Bronze Age (see above), the Middle Iron Age floodplain before the onset of flooding supported damp mesotrophic pastureland but with some mixed scrub present (Allen and Robinson, forthcoming; Robinson 1981: 256–9). The impact of waterlogging and inundation of the Iron Age pasture was combined with increasing grazing pressure. The proportion of *scarabaeoid* dung beetles, particularly members of the genus Aphodius, which feed on the droppings of domestic animals on grassland, from these sites can exceed 25% of the terrestrial Coleoptera. These changes seem to have caused the development of overgrazed marshy pasture with tussocks of the *Juncus effusus* type of rushes (mostly probably *J. inflexus*). Numerous seeds of *Rumex maritimus* and *Chenopodium rubrum* suggest that livestock created an expanse of nutrient-rich mud around each settlement. The Middle Iron Age sites with evidence of such conditions at Farmoor and Port Meadow (Lambrick and Robinson 1988: 65–71) have now been joined by a series of Late Iron Age sites at Claydon Pike/Thornhill Farm (Miles and Palmer 1990).

In the later Iron Age, cultivation was extending from the gravel terraces onto the edge of the floodplain of the Upper Thames just as conditions were becoming less suitable for arable. The uneven surface of an early Roman field, showing the pattern of cross-ploughing, was preserved by the upcast of a Roman ditch on the edge of the floodplain at Drayton (Lambrick and Moore 1987). This in turn had been covered with a thin layer of alluvium that had been disturbed by a single episode of uni-directional ploughing which inverted some of the sediments. Finally, further alluviation took place.

In the Ouse valley, a rising water-table was also causing problems for earlier Roman agriculture. Upcast from the 2nd/3rd century ditches at Warren Villas (see above) sealed a plough soil in which organic matter was not preserved. There were marks of cross-ploughing in the underlying gravel. The soil which lapped over this upcast, however, had a significant organic content and was associated with uni-directional asymmetrical plough marks. The waterlogged seeds from samples of four of these plough marks which together made up 1 kg are listed in Table 19.1.

Table 19.1

Waterlogged seeds from Roman ploughmarks at Warren Villas

Taxon	Number
Ranunculus cf. *repens* L.	6
R. sceleratus L.	1
Brassica rapa L. ssp. *sylvestris* (L.) Jan.	2
Reseda luteola L.	1
Hypericum sp.	10
Lychnis flos-cuculi L.	4
Cerastium cf. *fontanum* Baug.	1
Sagina sp.	3
Spergula arvensis L. — large seeded	2
Chenopodium album L.	4
Atriplex sp.	4
Potentilla anserina L.	1
Epilobium sp.	1
Polygonum persicaria L.	66
Rumex sp.	2
Urtica dioica L.	1
Plantago major L.	5
Sambucus nigra L.	1
Sonchus asper (L.) Hill	11
Juncus effusus tp.	25
J. bufonius gp.	42
J. articulatus tp.	143
Juncus spp.	36
Isolepis setacea (L.) R.Br.	161
Carex spp.	6
Gramineae indet.	3
Total	562

1 kg sample sieved to 0.2 mm

The discovery of plant remains in the ancient ploughsoil in which they had been growing is extremely unusual. They suggest that the field was still being cultivated when conditions were very wet indeed. The most abundant seeds were from *Isolepis setacea*, an annual plant more normally associated with such habitats as wet sand along stream margins and marshy meadows. The weed flora shows some affinities to the *Chenopodietalia* communities of spring sown crops, with numerous seeds of *Polygonum persicaria* present. Interestingly, seeds of *Eleocharis palustris*, a marshland plant, which are frequently found in carbonised cereal assemblages and are assumed to come from wet field margins, were absent. No seeds of certain crop plants were identified. *Reseda luteola* was cultivated as a dye plant but also grows on field margins. However, a large-seeded variety of *Spergula arvensis*, which is associated with flax cultivation (Pals and Dierendonck 1988: 249), was present. The other British archaeobotanical records of large-seeded *S. arvensis* are from Saxon flax-retting deposits (West Cotton channel, see below; Westbury, Milton Keynes – J. Letts, pers. comm.).

The higher parts of the Nene floodplain at West Cotton presented no problems for Roman and Early Medieval ploughing. Some of the prehistoric monuments were damaged by ploughing. The onset of heavy alluviation that sealed the ridge-and-furrow seems to have been sudden and there was no evidence of ploughing under wet conditions.

The earliest evidence for hay meadow in the region comes from the Thames floodplain at Claydon Pike, where the over-grazed marshy pasture had been replaced by grassland with such herbs as *Pedicularis palustris*, *Leucanthemum vulgare* and *Rhinanthus* sp. in the 2nd century AD. The main expansion of hay meadow in the Thames valley was Late Saxon or Early Medieval (Lambrick and Robinson 1988: 72–3). The full hay meadow flora was well illustrated by waterlogged deposits from a Late Medieval well on a gravel island at Claydon Pike. The waterlogged deposits contained a coin of 1473–77. The well seems to have been surrounded by floristically diverse hay meadow belonging to either *Alopecurus pratensis – Sanguisorba officinalis* flood meadow or *Cynosurus cristatus – Centaurea nigra* meadow. The species identified included:

Ranunculus cf. *acris*	*Plantago* cf. *lanceolata*
Rumex acetosa	*Leucanthemum vulgare*
Primula cf. *veris*	*Centaurea nigra*
Rhinanthus sp.	*Leontodon* sp.

Such grassland only retains its species composition if it is mown in June or July and the aftermath grazed. There is molluscan evidence from alluvial deposits on many sites in the Upper Thames valley for meadowland (Robinson 1988). Closely grazed flood-pasture only supports a resticted fauna of amphibious species such as *Anisus leucostoma* and *Lymnaea truncatula* while fully terrestrial species are absent. On floodmeadows, they are joined by a typical damp grassland fauna of such species as *Succinea* or *Oxyloma* sp., *Vallonia pulchella*, *Arion* sp. and *Trichia hispida* gp.

There is documentary evidence for Medieval hay meadow on the floodplains of the Ouse and Nene as well as the Thames. The palaeochannel at West Cotton formed the eastern boundary of West Cotton Meadow, which was first documented in 1349 and extended westwards up to the channel of the Nene which is extant today (Hall *et al* 1988: 35, 63). The archaeobotanical remains from the Late Saxon deposits in the channel included a similar range of species to those from the Thames valley hay meadow assemblages. At this date, the higher part of the floodplain to the south-east of the palaeochannel would still have been arable.

ARCHAEOLOGY

During the Neolithic and earlier Bronze Age, the wooded landscape of the floodplains was gradually transformed into a landscape organised around monuments. There was probably little distinction made between the gravel terraces and all but the lowest areas of floodplain. Thus the early Neolithic cursus at Drayton in the Thames valley (see above) ran from the first terrace onto the floodplain, then its line was picked up on the second terrace beyond. Bronze Age round barrows occur under alluvium, for example at King's Weir in the Thames valley (Bowler and Robinson 1980) and at Earls Barton in the Nene valley (Jackson 1984). Many more sites are probably entirely obscured under alluvium. However, islands of gravel rising above the general level of the floodplain surface were particularly favoured for monuments. The long barrows at South Stanwick occupied almost all of an elongate gravel island and the Neolithic complex at West Cotton was aligned on it. Several round barrows in the adjacent area at Irthlingborough gained enhanced prominence because they had been situated on gravel islands.

Traces of Neolithic and earlier Bronze Age settlement on the floodplains are slight, but no less so than on the gravel terraces. Flint scatters are present and there are several examples of Late Neolithic pits containing domestic debris such as at Drayton in the Thames valley (Lambrick and Moore 1987) and at Stanwick in the Nene valley.

Little is known about the Late Bronze Age archaeology of the region at present, but sites are being discovered on the floodplain. A possible Late Bonze Age system of parallel ditches at Stanwick extends from the gravel terrace to under the alluvium. The Late Bronze Age site with timber waterfront structures on a former island of the Thames at Wallingford (Lambrick 1987) is a reminder of the importance of the rivers as lines of communication. This site showed a great similarity to the Late Bronze Age site further downstream in the Middle Thames at Runnymede Bridge (Needham, this volume).

Prior to the onset of inundation, later prehistoric settlements were present, but the floodplains seem not to have been as intensively occupied as the gravel terraces. The rise in water-table in the Upper Thames influenced the type of settlement. For example, Mingies Ditch was a small, enclosed pastoral site which made use of the stream system of the floodplain for the corralling and watering of animals (Allen and Robinson forthcoming). The settlement was integrated into the economy of the remainder of the valley and obtained its cereals from dry ground. At Stanwick, in the Nene Valley, however, a larger Iron Age settlement which extended a little way under the alluvium probably undertook mixed farming.

There were two responses in the Iron Age to the onset of extensive flooding in the Upper Thames valley. Some settlements were occupied seasonally, for example the Middle Iron Age pastoral enclosures at Farmoor (Lambrick and Robinson 1979: 125) but other areas were exploited from specialised pastoral settlements on islands of higher ground as with the Middle Iron Age sites at Port Meadow (Lambrick and McDonald 1985: 98–100) and the Late Iron Age phase at Claydon Pike (Miles and Palmer 1990). In both cases, animals would have been overwintered on sites elsewhere. There is some evidence from the bone assemblages that during the Iron Age, the open grassland of the Upper Thames floodplain was used for the raising of horses for higher status sites, perhaps the hillforts of the region (Allen and Robinson, forthcoming).

From the Roman period onwards, the floodplain was managed as part of larger agricultural units which included tracts of the valley side. In the Nene and Ouse valleys, some occupation sites remained prior to flooding, for example at Bromham on the Ouse floodplain (Tilson 1973). Roman engineering on the floodplains included bridge building and water management. The abutment and parts of the oak timberwork of a series of Roman bridges across a channel of the Nene were discovered at Aldwincle, the earliest being 1st century AD (Jackson and Ambrose 1976).

An exceptionally well-preserved Late Roman villa on the Nene floodplain at South Stanwick was built on the site of an earlier water mill (Keevill, this volume). At Mingies Ditch, the stream adjacent to the Iron Age site was diverted in the Roman period. Also in the Thames valley, Early Roman drainage ditches on the floodplain at Claydon Pike were probably part of the improved management, as hay meadow, of what had been marshy pasture. These ditches related to an Early Roman "official" site on an island of higher ground which may have functioned as a prata supplying fodder to the civil administration in Cirencester (Miles and Palmer 1990; Robinson, unpublished). Extensive winter floods clearly presented problems of communications in the Thames valley. Alongside a Roman road crossing the floodplain of the river Windrush at Gill Mill was a double row of oak posts which probably supported a plank walkway (Wallis and Lambrick 1989). Similar structures still exist in the Thames valley.

Very little is known about the earlier Saxon archaeology of the floodplains. Later Saxon and Medieval structures again tend to be related to water management or river crossings. The mills and flood embankment at West Cotton have already been discussed in the section on channels (Windell *et al* 1990). The suburbs of Bedford, Northampton and Oxford spread onto their floodplains not long after the Saxon foundation of these towns. Habitation was initially on reclaimed land alongside roads leading to river crossings and the entire tenements were gradually raised above flood levels by dumping (Durham 1977: 182–5; 1984: 95–7). An activity that was occurring in the Late Saxon suburbs on the floodplains alongside all of these towns was flax retting, in ditches and wattle-lined pits extending below the water-table (e.g. Brown 1977; Robinson 1983). Flax retting was also an important floodplain activity on rural sites. It was taking place in the West Cotton channel during the Late Saxon period. Wattle-lined pits on the floodplain of the river Ivel at Warren Villas, one of which gave a dendrochronological date of AD 1125/6, contained flax remains and a wicker fish trap. There is evidence of Roman flax retting in the Upper Thames valley. A Late Roman pit beneath a thin layer of alluvium at Old Shifford (Hey 1990) contained numerous flax capsules, while a beet (twisted bundle) of flax stems with capsules still attached was found in channel sediments of possible Roman or Saxon date at Yarnton.

DISCUSSION

It has been argued elsewhere that the channels of the Upper Thames reached their greatest size in the very early Holocene when the river would have been fed by the meltwaters from a thick snow cover to its catchment each spring (Robinson and Lambrick 1984). Thereafter the river never had sufficient flow for significant channel migration. This explanation seems equally applicable to the Ouse and Nene.

It was also possible to correlate hydrological changes and periods of alluviation on the Upper Thames floodplain with agricultural activity in the catchment rather than any supposed climatic change (Robinson and Lambrick 1984). The rise in water-table and onset of extensive flooding were possibly the result of clearance in the Cotswolds from the Late Bronze Age onwards primarily for pastoralism. Extensive ploughing in the Cotswolds perhaps began in the Late Iron Age, and certainly continued throughout the Roman period, when there were many villa estates. Roman settlement also expanded onto the clays. This would have exposed slopes formerly under grass to erosion on an unprecedented scale, and thus provided the necessary sediment for the alluviation of this period. Important new evidence suggesting a slowing of alluviation in the Early Saxon period was provided by the Drayton archaeomagnetic dating sequence. This was probably a consequence of the agricultural collapse at the end of the Roman period. During the Late Saxon and Early Medieval periods, cultivation of the slopes and alluviation of the floodplain were both intensive. The cessation of alluviation in the Late Medieval period corresponded to a general shift to grassland on the slopes due to the catastrophic decline in population with the Black Death. Flooding, however, continued because the channel capacity had been reduced and there was no general reversion of the catchment to woodland.

It is harder to relate the onset of waterlogging and flooding in the other two valleys to human activity. The boulder clay region of both the Ouse and the Nene Valleys was extensively settled during the Iron Age (Simco 1982: 10). Fieldwalking in the Raunds area of the Nene valley has shown even more evidence for Roman activity on the boulder clay, with traces of settlement even in those areas which were avoided in the Iron Age (S. Parry, pers. comm.). This, however, was not reflected by any major changes on the Nene floodplain. In the Ouse catchment, there was much Roman settlement on the chalky boulder clay but not the Oxford Clay or Gault Clay (Simco 1985: 20–3). The waterlogging of the Ouse floodplain during the Roman period would be consistent with Roman agricultural expansion, but it is uncertain why it was not accompanied by alluviation. Possibly there was not such a high proportion of arable as in the Thames valley or perhaps it was related to a low degree of slope of the land under cultivation and the availability of minor sediment sinks (topographic features in which colluvial sediments were trapped before they reached the stream system).

The Medieval pattern of alluviation by the Ouse and the Nene was, however, more similar to that of the Upper Thames. This might imply a greater area under cultivation in their catchments during the Early Medieval period than in the Roman period; this does seem to have been the case in the Nene valley. There was wholesale clearance and ploughing up of grassland resulting from increased pressure on land that was already quite densely settled. This only reached its maximum during the 13th and early 14th centuries (Steane 1974: 95–109). The evidence from the Ouse basin is at present less clear. The cessation of alluviation by the Ouse and Nene was synchronous with this event in the Upper Thames basin and the Black Death again provides the only plausible explanation. Much ridge-and-furrow on the claylands remained under grass until this century.

The diachronicity of some of the hydrological events between the drainage basins tends to support the view that the causes were not climatic change. Indeed, major climatic change would be required to inundate regularly the entire floodplains of the three rivers when previously peak flow had been restricted within their banks. To confine the most serious modern floods of the Thames to the channel would require it to have three times its actual capacity (information from Lt.-Col. D. Williams, formerly of the Thames Conservancy). On the other hand, it would probably take major changes in agricultural activity in the catchment to become reflected in the floodplain sequences. However, minor changes in the catchment prior to the Iron Age, both those induced by human activity and those due to climatic variation, could have influenced the palaeohydrology and sedimentation of the channels. For example, the Late Bronze Age site at Wallingford was on dry ground that had developed from channel deposits (Thomas *et al* 1986: 179). Neolithic and Bronze Age clearance activities could easily have been responsible for some of the earlier sediments in the palaeochannels.

The human usage of the floodplains until the early years of the 20th century can be divided into three phases. In the first phase, which lasted until the Middle Iron Age in the Upper Thames valley, the Early Roman period in the Ouse valley and into the Late Saxon or even the Early Medieval period in the Nene valley, there was little difference between much of the floodplains and the lower gravel terraces. Only on the very lowest areas of floodplain, where the proximity of the water-table, the greater likelihood of clay soils and perhaps even some flooding could have favoured lush growth of grass was a distinction likely to have been made.

During the second phase, flooding presented serious problems and the floodplains were not managed to their best potential. There was damage to pastureland caused by overstocking under wet conditions in the Upper Thames valley during the Middle and Late Iron Age. Early Roman ploughing in the Upper Thames and Ouse valleys was occurring in waterlogged soils which had become most unsuitable for cultivation. In the Nene valley, a mill leat had to be put on the top of a flood embankment to maintain a head of water, and alluviation resulted in the village protected by the embankment being at a lower level than the general surface of the surrounding floodplain.

The third phase saw a coming to terms with the changed conditions of the floodplain and its management as a valuable part of an organised agricultural system. With the rise of towns in the Roman period came an increasing demand for fodder. Once some alluvium had been deposited on the floodplains, they became ideal for growing hay. Fertility is maintained by the winter flooding with its input of silts and deposition of decaying organic matter. The land is waterlogged in winter, but the free flow of water through the underlying gravels enables sufficient of the alluvial profile to drain in summer for a good rooting depth without giving problems of drought. At least some hay meadow in the Upper Thames valley had a Roman origin, but by the Medieval period meadowland was the major use of the floodplains in all three valleys. Floodmeadow became some of the most valuable agricultural land in the region. In the early 19th century, Thames meadows were being rented for 60s per acre when the average annual rent for farmland in the county was 22s 10d per acre (Young 1813,36–9). It is unlikely that it could have commanded such a high rental if there had been no alluviation and flooding was not taking place.

Acknowledgements

I am grateful to all the excavators listed in Table 19.2 whose sites I visited. In particular I must thank Mr. D. Windell, formerly of the Northamptonshire Archaeology Unit, for much patience with stratigraphy of West Cotton, Mr. M. Dawson and Ms. H. Duncan of the Bedfordshire Archaeological Unit for making details of Warren Villas available so rapidly and Mr. G. Lambrick of the Oxford Archaeological Unit for 17 years of discussions on the Thames floodplain. Mrs. G. Campbell kindly let me use her archaeobotanical results from the West Cotton channel. The laboratory work was all undertaken in the University Museum, Oxford, much of it funded by the Ancient Monuments Laboratory of HBMC (England).

***Table 19.2** Site locations and excavators*

Site	River basin	Grid Ref.	Excavator
Aldwincle	Nene	SP999801	D. Jackson
Bromham	Ouse	TL017523	P. Tilson
Buscot Lock	Thames	SU230980	M. Robinson & D. Wilson
Claydon Pike	Thames	SU190996	D. Miles & S. Palmer
Drayton	Thames	SU490943	G. Lambrick
Earls Barton	Nene	SP870627	D. Jackson
Farmoor	Thames	SP443057	G. Lambrick & M. Robinson
Gill Mill	Thames	SP380070	J. Wallis & G. Lambrick
Grandpont/St. Aldates	Thames	SP514055	B. Durham
Irthlingborough	Nene	SP964713	C. Halpin
King's Weir	Thames	SP476102	D. Bowler & M. Robinson
Mingies Ditch	Thames	SP391059	T. Allen & M. Robinson
Old Shifford	Thames	SP382023	G. Hey
Port Meadow	Thames	SP492085	G. Lambrick
St. James' Square	Nene	SP752602	J. Williams & D. Jackson
South Stanwick Barrow	Nene	SP965710	J. Moore & D. Jackson
South Stanwick Villa	Nene	SP960704	G. Keevill
Stanwick	Nene	SP970718	D. Neal
Thornhill Farm	Thames	SU181997	S. Palmer
Titchmarsh	Nene	TL006797	D. Windell
Wallingford	Thames	SU607882	G. Lambrick
Warren Villas	Ouse	TL180473	M. Dawson
West Cotton	Nene	SP976725	D. Windell
Yarnton	Thames	SP479109	G. Hey

All sites were visited and sampled by the author with the exception of Aldwincle, Bromham and Earls Barton.

References

Allen, T. and Robinson, M. forthcoming. *The prehistoric landscape and the Iron Age enclosed settlement at Mingies Ditch, Hardwick with Yelford, Oxon.* Oxford: University Committee for Archaeology.

Bowler, D. and Robinson, M. 1980. Three round barrows at King's Weir, Wytham, Oxon. *Oxoniensia*, 45, 1–8.

Brown, A. 1977. Plant remains. In B. Durham, Archaeological investigations in St. Aldates, Oxford. *Oxoniensia*, 42, 169–72.

Durham, B. 1977. Archaeological investigations in St. Aldates, Oxford. *Oxoniensia*, 42, 83–203.

Durham, B. 1984. The Thames crossing at Oxford: archaeological studies 1979–82. *Oxoniensia*, 49, 57–100.

Hall, D., Harding, R. and Putt, C. 1988. *Raunds; picturing the past.* Raunds: F.W. March with Buscott Publications.

Halpin, C. 1987. Irthlingborough Bronze Age barrow excavation. In B. Dix (ed.), The Raunds Area Project: first interim report. *Northamptonshire Archaeology*, 21, 5–8.

Hey, G. 1990. Aston, Bampton and Shifford, Old Shifford Farm. *South Midlands Archaeology*, 20, 86–8.

Jackson, D.A. 1984. The excavation of a Bronze Age barrow at Earls Barton, Northants. *Northamptonshire Archaeology*, 19, 3–30.

Jackson, D.A. and Ambrose, T.M. 1976. A Roman timber bridge at Aldwincle, Northamptonshire. *Britannia*, 7, 39–72.

Lambrick, G.H. 1987. Wallingford By-pass. *South Midlands Archaeology*, 17, 99–100.

Lambrick, G.H. and McDonald, A. 1985. The archaeology and ecology of Port Meadow and Wolvercote Common, Oxford. In G.H. Lambrick (ed.), *Archaeology and nature conservation*, 95–109. Oxford: University Department for External Studies.

Lambrick, G.H. and Moore, J. 1987. Drayton: Cursus. *South Midlands Archaeology*, 17, 85–6.

Lambrick, G.H. and Robinson, M.A. 1979. *Iron Age and Roman riverside settlements at Farmoor, Oxfordshire.* London: Council for British Archaeology Research Report, 32.

Lambrick, G.H. and Robinson, M.A. 1988. The development of floodplain grassland in the Upper Thames Valley. In M. Jones (ed.), *Archaeology and the flora of the British Isles*, 55–75. Oxford: University Committee for Archaeology Monograph, 14.

Limbrey, S. and Robinson, S. 1988. Dry land to wet land: soil resources in the Upper Thames Valley. In P. Murphy and C. French (eds.), *The exploitation of wetlands*, 129–44. Oxford: British Archaeological Reports, British Series, 186.

Miles, D. and Palmer, S. 1990. Claydon Pike and Thornhill Farm. *Current Archaeology*, 11, 19–23.

Moore, J. and Jackson, D.A. 1990. Stanwick, Redlands Farm. *South Midlands Archaeology*, 20, 55–6.

Pals, J.P. and Dierendonck, M.C. van 1988. Between flax and fabric: cultivation and processing of flax in a medieval peat reclamation settlement near Midwould (Prov. Noord Holland). *Journal of Archaeological Science*, 15, 237–51.

Robinson, M.A. 1981. The Iron Age to early Saxon environment of the Upper Thames terraces. In M. Jones and G. Dimbleby (eds.), *The environment of man: the Iron Age to the Anglo-Saxon period*, 251–86. Oxford: British Archaeological Reports, British Series, 87.

Robinson, M.A. 1983. The late Saxon environment. In J.H. Williams and D. Farwell, Excavations on a Saxon site in St. James' Square, Northampton. *Northamptonshire Archaeology*, 18, 150 and fiche 30–38.

Robinson, M.A. 1988. Molluscan evidence for pasture and meadowland on the floodplain of the Upper Thames basin. In P. Murphy and C. French (eds.), *The exploitation of wetlands*, 101–12. Oxford: British Archaeological Reports, British Series, 186.

Robinson, M.A. and Lambrick, G.H. 1984. Holocene alluviation and hydrology in the Upper Thames basin. *Nature*, 308, 809–14.

Robinson, M.A. and Wilson, R. 1987. A survey of environmental archaeology in the South Midlands. In H.C.M. Keeley (ed.), *Environmental Archaeology: a regional review*, 2, 16–100. London: Historic Buildings and Monuments Commission for England, Occasional Paper 1.

Simco, A. 1982. Roman Bedfordshire: 100 years on. In *Programme of the summer meeting of the Royal Archaeological Institute at Bedford*, 8–14. London: Royal Archaeological Institute.

Simco, A. 1985. *Survey of Bedfordshire in the Roman period.* Bedford: Bedfordshire County Concil.

Steane, J. 1974. *The Northamptonshire landscape.* London: Hodder and Stoughton.

Thomas, R., Robinson, M., Barrett, J. and Wilson, R. 1986. A late Bronze Age settlement site at Wallingford, Oxon. *Archaeological Journal*, 143, 174–200.

Tilson, P. 1973. A Belgic and Romano-British site at Bromham. *Bedfordshire Archaeological Journal*, 8, 23–66.

Wallis, J. and Lambrick, G.H. 1989. Ducklington: Gill Mill Farm. *South Midlands Archaeology*, 19, 49–50.

Windell, D. 1985. Titchmarsh. *Northamptonshire Archaeology*, 20, 151.

Windell, D., Chapman, A. and Woodiwiss, J. 1990. *From barrows to bypass: excavations at West Cotton, Raunds, Northamptonshire 1985–1989.* Northampton: Northamptonshire County Council.

Young, A. 1813. *General view of agriculture of Oxfordshire.* London: Sherwood, Neely and Jones.

20

Alluvial archaeology of the Holocene in the Upper Thames basin 1971–1991: a review

George Lambrick

Summary: *Investigations on the Upper Thames floodplain since 1971 have been of considerable significance in developing our understanding of the socio-economic context and the impact of past human land use on the environment. In the last twenty years the exceptional value and potential of archaeological research on alluvial Holocene floodplains has come to be recognized. This paper uses the example of the Upper Thames river basin to illustrate the wide range of research issues which have emerged from this recognition. A number of topics are identified where further research is most needed.*

Archaeological sites in or under alluvium tend to be particularly well preserved, but are difficult to locate through superficial survey techniques. Current planning policies have targeted apparently archaeologically 'blank' areas on the floodplain as being suitable for mineral extraction. It is already clear that these areas are far from 'blank' and the appropriateness of such policies is questioned in the light of the quality of preservation and the problems of dealing with indirect impacts resulting from dewatering.

The pattern of hydrological and alluvial change in the Holocene on the Upper Thames floodplain is reviewed with reference to results of recent excavations. In interpreting the causes of change it is suggested that purely natural (i.e. climatic) explanations cannot be put forward without exploring the possibility that the observed phenomena are explained by human interference with the natural hydrological cycle. In the case of the Upper Thames valley the evidence of changing human settlement and land use in the catchment appears to offer an adequate explanation of the changes in hydrology and alluviation on the floodplain in the last 6000 years or more.

INTRODUCTION

Almost all the work discussed in this paper (Fig. 20.1) has been carried out by the Oxford Archaeological Unit, mostly through individual projects funded by English Heritage, and more recently by mineral operators, with some general input from a British Academy funded research project in 1980–81. A major contribution to these studies has been made by Dr Mark Robinson whose paper in this volume complements this one.

Prior to 1971 very little excavation had been conducted on the floodplain of the Thames. Exceptions, such as some of the early investigations of the Oxford Blackfriars (reported in Lambrick and Woods 1976), at Port Meadow just north of the city (Atkinson 1942), at Burroway near Clanfield (Sutton 1966) and at Wallingford (Wymer 1960), did not effectively address many of the issues of stratigraphy, hydrology and sedimentary processes which are now fundamental to the subject. The floodplain was not ignored, at least in that it was considered as an important topographical factor in relation, for example, to the siting of Oxford (Jope 1956) or the prehistoric sites on Port Meadow (Atkinson 1942; Case 1963). However, there was no concept of the the floodplain having fundamentally changed its physical character in the last 3000 years, and there was little idea that early socio-economic systems might have had very different relationships to the floodplain in terms of settlement and land use from those of more recent historical periods: the Medieval and Post-medieval pattern was simply adopted as a model and extrapolated backwards.

Three excavations were particularly important in establishing the value of research into the archaeology of the Upper Thames floodplain. The potential of urban sites on the floodplain was revealed most clearly in Oxford by work at 79–80 St Aldates in 1971–2 (Durham 1977) and at nearby Blackfriars in 1972–4 (Lambrick and Woods 1976). The potential for rural archaeology, and more importantly for studying human

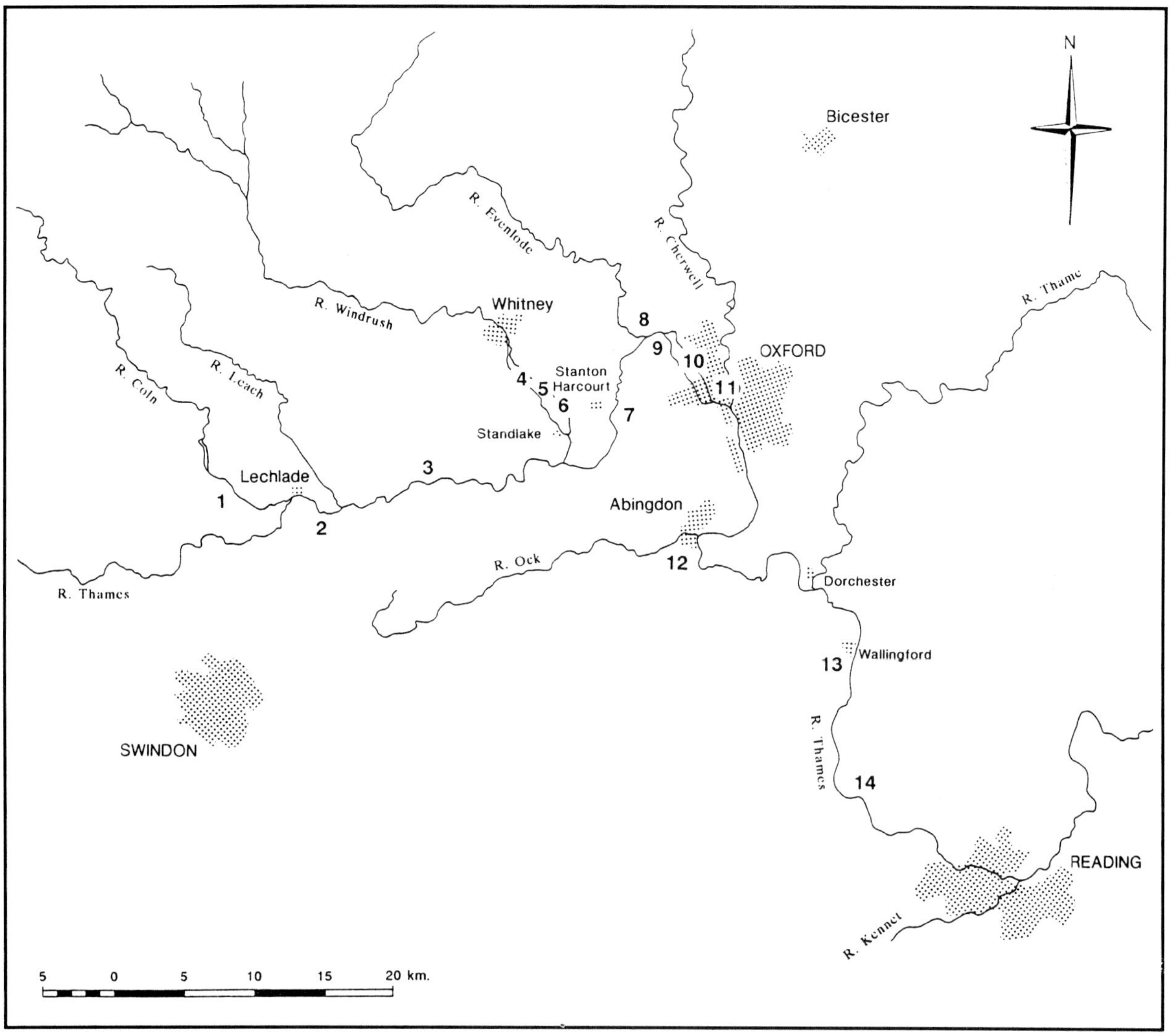

Fig. 20.1. Map of the Upper Thames valley showing the location of sites mentioned in the text: 1. Claydon Pike and Thornhill Farm; 2. Buscot; 3. Burroway; 4. Gill Mill; 5. Mingies Ditch; 6. Gravelly Guy; 7. Farmoor; 8. Cassington and Yarnton; 9. King's Weir; 10. Port Meadow; 11. Oxford suburbs (The Hamel, Blackfriars, St.Aldates); 12. Drayton; 13. Wallingford; 14. Goring.

impact on the environment in general, emerged from excavations at Farmoor in 1974–6 (Lambrick and Robinson 1979).

Other projects followed and a number of general papers synthesising the results have been published, each providing an integrated approach to the palaeoenvironmental and archaeological evidence. These have dealt in turn with the changes in hydrology and patterns of alluviation during the Holocene (Robinson and Lambrick 1984), the development of grassland resources on the floodplain (Lambrick and Robinson 1988), and, developing ideas from this, consideration of the floodplain as a resource within a wider consideration of the development of late prehistoric and Roman farming on the Thames gravels (Lambrick in press). Numerous more detailed studies and site reports are quoted in this paper and in Mark Robinson's complementary paper which compares the Thames sequence with those of the Nene and Ouse.

This review of the progress and opportunities in alluvial archaeology as exemplified by the Upper Thames basin is divided into four broad themes:

i) Discovery of sites and deposits of interest
ii) Preservation of sites, stratigraphy, deposits and objects
iii) Identification of the stages of development in hydrology and alluviation, their chronology and causation
iv) Aspects of past human settlement and economy in relation to floodplain land use and resources.

DISCOVERY

Patterns of discovery

Ever since the days of Major Allen in the 1930s, lowland river valleys such as the Thames have been known archaeologically from their cropmarks. To a substantial extent this remains the single most productive means of discovering sites, particularly in dry summers such as 1990, despite 60 years' worth of intensive coverage. However, there is now also a much clearer appreciation that the limits of the technique in relation to subsoil conditions do not define the limits of the archaeology itself. The major turning point in this respect was the chance discovery at Farmoor of Iron Age farmsteads sealed beneath alluvium on the floodplain adjacent to some minor cropmarks on the first gravel terrace. Such discoveries, often related to adjacent cropmark sites, are now a normal expectation, exemplified recently by a Roman field system at Drayton extending into the alluvium beyond the limits of cropmarks (Fig. 20.2).

The regular requirement now placed on mineral operators and other developers to evaluate the archaeological potential of proposed development sites has also demonstrated the regular occurrence of sites hidden beneath alluvium or otherwise not revealed by air photography. A good example of this is the Roman settlement at Gill Mill on the Windrush floodplain south of Witney (Fig. 20.3), and other sites at Northmoor, Cassington and Goring.

Recently the Oxford Archaeological Unit in conjunction with The Trust for Wessex Archaeology carried out a review of the results of evaluation by trenching sites on various geologies. This suggested that on the Upper Thames river gravels (almost all alluvial or low first terrace sites) about 40% of evaluations have produced either substantially new evidence or entirely new sites (Brereton *et al* forthcoming).

These studies have also shown how surface finds scatters like cropmarks may not indicate the full extent of sites which are partly sealed by alluvium thicker than the depth of the ploughsoil, as was particularly clear at Gill Mill (Fig. 20.3). This is especially important in discovering the extent and quality of sites since there is clearly an inverse relationship between the clarity of superficial evidence and the quality of preservation of subsoil archaeology (see below). There is some evidence of a preference for settlements to occupy slightly higher, drier parts of the floodplain, as at Claydon Pike (Miles and Palmer 1990) or Port Meadow (Lambrick and McDonald 1985), but this is not a universal pattern, as is evident from the discovery at Farmoor of one of the Iron Age farmsteads (Enclosure Group 3) under some of the deepest alluvium on the site in the top of an old channel.

It is thus arguably almost as dangerous to limit detailed evaluation to such 'gravel island' areas, as it is to assume that cropmarks provide a full picture of the archaeology. A recent assessment by Gill Hey at Yarnton, just north of Oxford, included 9 trenches across old river channels, 7 of which produced archaeological evidence, 4 of these representing significant remains such as timber platforms and stakes, occupation evidence and an earth causeway.

Future trends

The means of discovering sites beneath alluvium and of assessing their character, extent and quality need careful development. Improved geophysics and phosphate sampling may provide useful means of detecting areas of enhanced activity, and multi-spectral remote sensing has been suggested as being of value. However, inspection of imaging of the floodplain in the Cassington and Yarnton areas, made available by the Institute of Hydrology, suggests only limited value for identifying archaeological features, although the pattern of former channels is clarified. To be reliable, any non-interventionist evaluation technique must be capable of indicating genuine negative areas as well as positive traces of archaeology. For the present, trenching is likely to remain the most reliable single means of establishing the presence or absence of archaeology and is perhaps the only means of clarifying the detailed character of buried sites. Even then very careful consideration needs to be given to sampling strategy in relation to the types of deposit which need to be identified. The real challenge is to find a balance of complementary techniques which are reliable and cost-effective.

The trends in mineral planning policies in the Upper Thames valley reflect an attempt to avoid known (i.e. cropmark) sites. This has reinforced in the current position in which all current workings are situated on the floodplain or low first terrace areas, where the problems of archaeological detection are greatest and potential preservation is highest (see below). The long term implications of this trend have yet to be established, but it is already clear from the discovery of well-preserved sites in these areas, through detailed field evaluation prior to gravel extraction, that serious problems are likely to occur. There is an increasingly obvious need for reviewing the wisdom of current Minerals and Structure Plan policies in this respect.

PRESERVATION

Stratigraphy

Generally speaking, archaeological sites on the Thames floodplain are well preserved. More sites exist than had once been appreciated and they tend to preserve better

Fig. 20.2. Roman arable fields sealed by alluvium at Drayton: upper, banks ditches and ard marks; lower, distribution of Roman pottery.

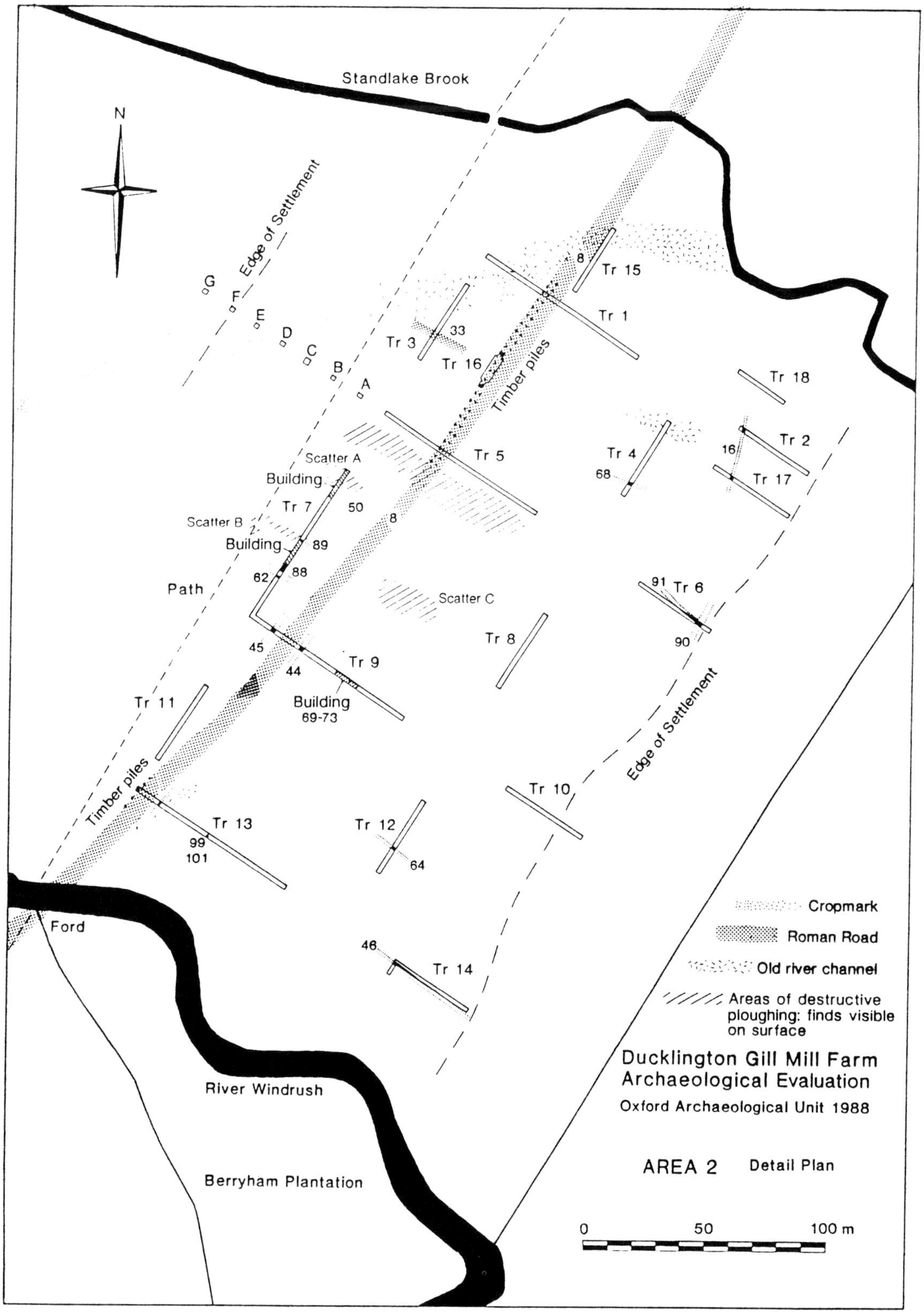

Fig. 20.3. Roman road and settlement at Gill Mill, Ducklington: the results of evaluation trenches, identifying a well preserved site with in situ timber structures, road metalling, buildings with floor and occupation levels and ditch and bank boundaries. Surface scatters of finds show where ploughing has cut through the overlying alluvium and disturbed the archaeology.

stratigraphic sequences and more extensive organic deposits than sites on the higher terraces. On the second gravel terrace, Bronze Age barrows normally exist only as subsoil ring ditches, whereas on the floodplain there are several examples of upstanding earthworks, as at Buscot, Wytham (Bowler and Robinson 1980), Port Meadow (Lambrick and McDonald 1985) and Goring (Allen forthcoming). At Burroway (Lambrick 1984) an Iron Age fort survives with its internal occupation surfaces and stake and post holes sealed by a protective blanket of alluvium beneath the modern ploughsoil. Its ploughed down but still visible burnt ramparts seal a corduroy of charred timbers.

The ancient common grazing land of Port Meadow and Wolvercote Common just outside Oxford preserves a remarkable range of sites, including at least six Bronze Age barrows, three Iron Age farmsteads with attached paddocks, a seventeenth century civil war entrenchment, a later seventeenth century racecourse, early Post-medieval gravel quarries and a First World War army camp (Lambrick and McDonald 1985). Preserved organic remains from the waterlogged Iron Age ditches have revealed a closely similar flora to that of the present day. Not far upstream are the well preserved earthworks of the deserted Medieval hamlet of Summerford, one of a number of such sites representing the extension of settlement onto the floodplain at the height of Medieval expansion in the thirteenth century.

A survey in 1981 indicated that the area of floodplain which has never been in cultivation is diminishing rapidly towards a core of areas which are of ecological interest for their unimproved grassland. Surviving earthworks in arable areas are clearly still under threat (Lambrick 1982).

Both earthwork and non-earthwork sites on the floodplain tend to have exceptionally well preserved subsoil stratigraphy. The pre-alluvial ground surface at Drayton preserved an *in situ* scatter of flintwork and burnt spreads, pits and the *in situ* bank of the Neolithic Drayton cursus (Ainslie and Wallis 1987; Lambrick and Moore 1987). At Wallingford 0.6 m of alluvium seals *in situ* Late Bronze Age occupation and midden deposits on a former eyot in the river, with an adjacent shallow river channel (Thomas *et al* 1986; Lambrick 1987; Figs. 20.4 & 20.5).

Iron Age occupation levels at Mingies Ditch on the Lower Windrush floodplain revealed gravel paths (as at Farmoor), details of the floors of houses and structural evidence of turf walling and hearths (Allen and Robinson in press). Recent work at Gill Mill (Fig. 20.3) a little way upstream has revealed parts of a Roman road crossing the floodplain with *in situ* occupation levels associated with buildings indicated by fragile traces of shallow sleeper walls (Wallis and Lambrick 1989).

Some rural sites have provided unusually well-preserved traces of early cultivation, which very rarely survives on the higher terraces. Possible Bronze Age plough marks were discovered under alluvium at the Hamel in Oxford (Palmer 1980). At Gravelly Guy Iron Age and Roman cultivation of the adjacent gravel terrace was revealed by colluvial ploughsoils interdigitated with the alluvial covering of the floodplain (Lambrick 1985b). Stratified within the alluvium at Drayton, the first fully intact physical evidence for Roman fields on the gravels has been recovered (Fig. 20.2). The full 0.1 m depth of the ploughsoil survived, containing pottery from manure scattering and sealing cross-ploughed ard-marks. It was also stratigraphically directly related to the upcast banks and ditches surrounding the fields (Lambrick and Moore 1987).

In the urban situation too, the best stratigraphy in Oxford has been encountered in the Late Saxon and Medieval floodplain suburbs of the city in St Aldates and St Thomas's (Durham 1977; 1984; 1990; Palmer 1980). Because of the lack of later cellarage, these sites have provided Oxford's least disturbed build-up of successive structural phases. As a result, they have provided the key pottery sequences for dating the region's Medieval archaeology, supported by good documentary evidence for the well-preserved buildings.

The Medieval suburban floodplain at Oxford was also significant as the home of the Dominican and Franciscan friars, and the Blackfriars in particular was well preserved because of the need both before and after the Dissolution to build up the surface of the floodplain as a means of flood defence, a problem specifically documented by the Blackfriars. The gradual sampling of this major site prior to various redevelopment schemes resulted in one of the most complete pictures of a Medieval friary yet achieved (Lambrick and Woods 1976; Lambrick 1985a).

Another very impressive, perhaps unique feature of Oxford's Medieval floodplain was the extraordinary Norman Grandpont, a massive stone bridge replacing the 'Oxen Ford', and extending intermittently for over a kilometer (Durham 1984; 1990).

Organic remains

The preservation of organic deposits is another aspect of preservation which is of special importance on floodplain sites, both for artifacts and biological remains. Neolithic and Bronze Age organic deposits are rarely preserved except in channel deposits, as at Buscot Wick (Robinson and Wilson 1987) and Mingies Ditch (Allen and Robinson, in press), but two successive shallow gullies on the pre-alluvial ground surface on the Windrush floodplain at Gravelly Guy, with pollen and other organic remains, have been radiocarbon dated to 2274–2044 and 844–803 cal BC (GU-5097, GU-5098;

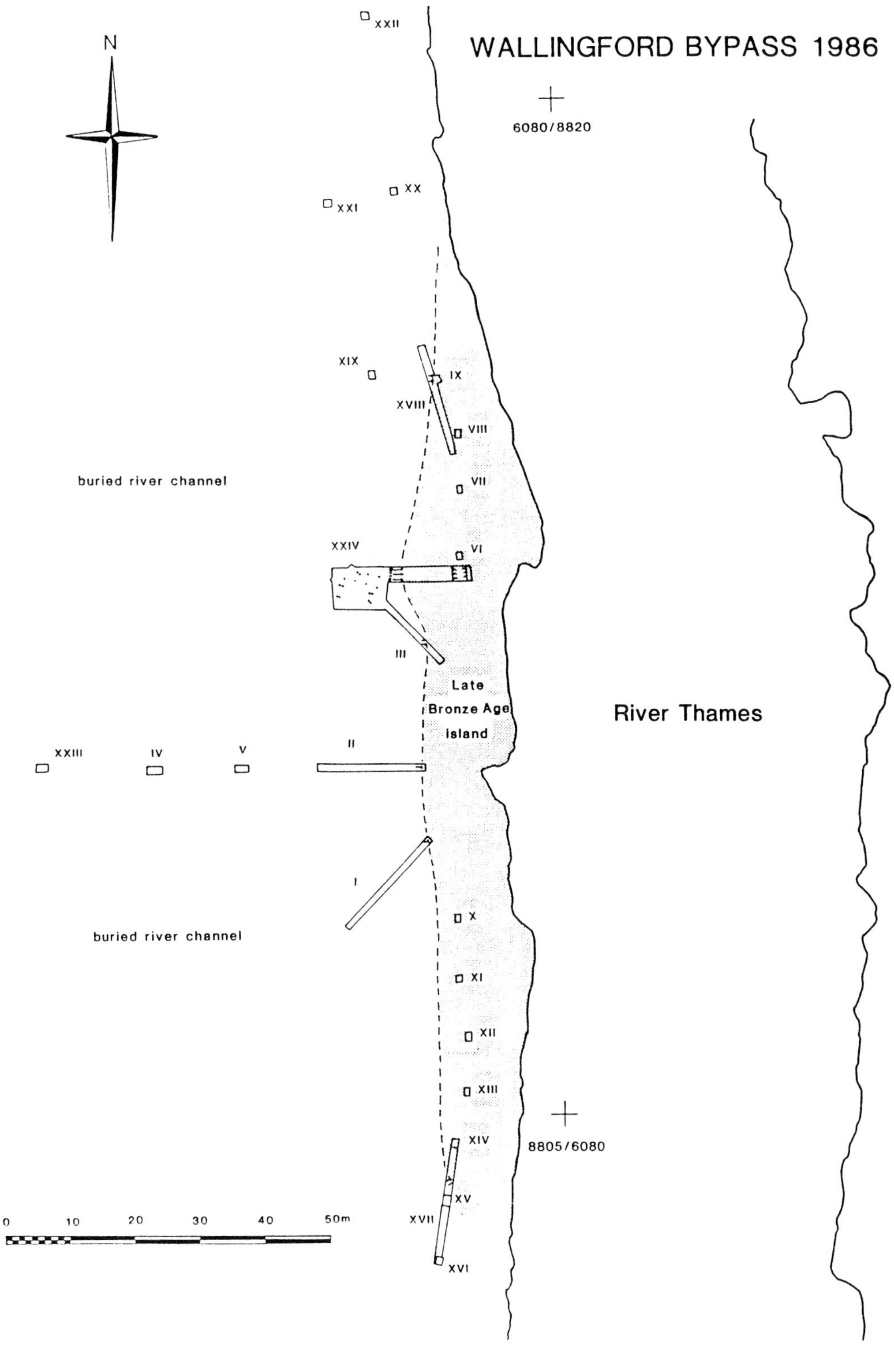

Fig. 20.4. Wallingford Late Bronze Age island settlement.

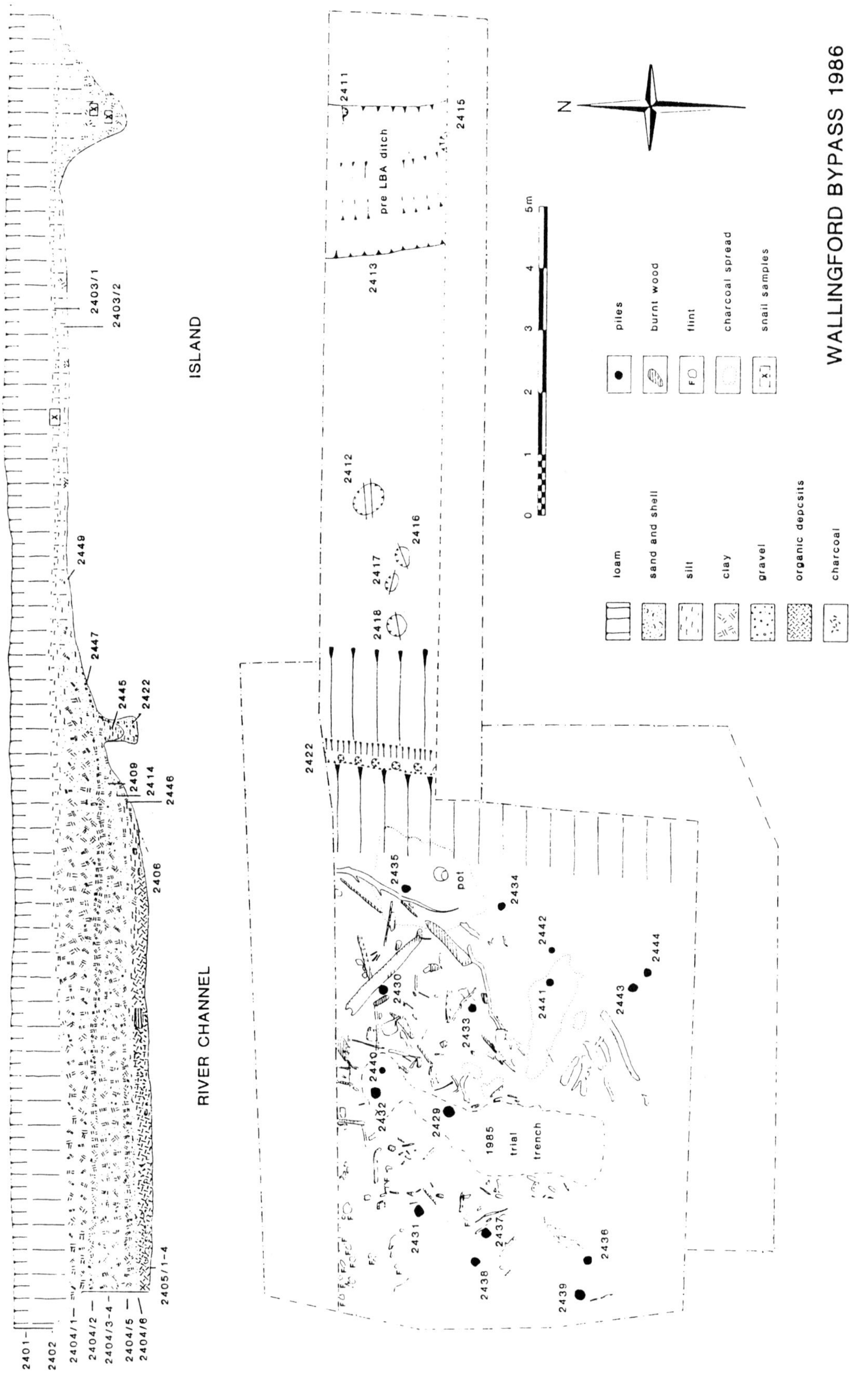

Fig. 20.5. Wallingford Late Bronze Age settlement: transect across island and channel deposits with preserved wood.

intercept, 1–sigma). The relative lack of early prehistoric organic remains in man-made features, although disappointing compared with the Ouse or Nene catchments, is of significance in indicating a later rise in the permanent water-table on the Upper Thames floodplain (see Robinson, this volume).

For later periods both artifacts and biological remains have provided invaluable evidence. For example, at Wallingford excavation of a buried river channel next to the Late Bronze Age eyot produced numerous timbers reflecting virtually all the different ways of cutting and splitting wood with bronze implements. They also illustrate the careful selection of oak for split structural timbers and hazel for straight coppiced rods to form a structure which had been burnt down before being dumped in the river (Lambrick 1987; Fig. 20.5).

At Farmoor the quality of the organic biological remains was such that it was possible to infer not only the broad environmental context of the settlements on the floodplain (grazed grassland) but also that they were subjected to seasonal flooding and that they were only occupied for perhaps 6–10 years (suggested by the absence of perennial weed seeds).

Comparable evidence is available for the Roman period when the first indications of hay making are evident at Claydon Pike and Farmoor in the first to third century AD. Artifacts such as shoes and bowls are also common finds on floodplain sites.

Urban Medieval deposits on the floodplain have likewise provided valuable evidence of living conditions and diet. For example, the superior diet of the inhabitants of the Blackfriars friary compared with houses in nearby Saint Aldates is evident from seeds and pips of the fruit they ate (Robinson in Lambrick 1985a).

One of the most serious worries for the future is what effect dewatering for gravel extraction may have on neighbouring sites. At Drayton this was dramatically demonstrated when a Roman ditch, shown in 1981 to have good organic preservation at the start of gravel digging, was found to have been completely desiccated up to at least 100 m away when extraction moved into the adjacent field five years after dewatering began in 1981. French and Taylor (1985) suggested that discernible deterioration in the preservation of wood occurred at Etton within two to three months of the initial dewatering.

Currently the Roman site at Gill Mill, much of which is to be preserved, is being protected by provision of clay bunding. While sufficient clay was being exposed beneath the gravel to achieve this, water was discharged into a cleared out field ditch running between the pit and the site to help keep a flow of water through the upper levels of gravel and thereby diminish the speed of desiccation. The water-table is being monitored, and it should be possible to sample the site to determine whether organic preservation has started to be affected. Another, more extensive and elaborate study of the effects of dewatering on groundwater and the benefits of clay bunding and recharge drains is being carried out at Yarnton, adjacent to the rich hay-meadows.

Both studies are being funded by the gravel companies concerned. They are likely to provide some further insights into the problem, and possible solutions. Despite the valuable observations made by French and Taylor (1985), there is still a need for more detailed research since the relationship between height of water-table and organic preservation is not likely to be a simple one, at least in the short term: very little is known of the rate of decay, whether it accelerates with time (as is likely), or how the rate of decay varies between different organic materials, or how far it is affected by different soil conditions.

This is a key planning issue, not least because dewatering can have severe impacts on archaeological deposits *outside* a development site. Given the limitations of existing knowledge of the archaeological resource sealed beneath alluvium already discussed, this raises the thorny question of whether evaluations of development sites required under Planning Policy Guidance Note 16 should extend into neighbouring areas to assess whether significant waterlogged deposits exist whose value could indirectly be destroyed through hydrological drawdown.

THE DEVELOPMENT OF HYDROLOGY AND ALLUVIATION ON THE THAMES FLOODPLAIN, CHRONOLOGY AND CAUSES OF CHANGE

The stages of development

For both the Pleistocene and Holocene sequences repeated observation of sites on the gravels have allowed the development of a model of sedimentary and hydrological processes in the Upper Thames basin. The interpretation of the Holocene sequence was published in 1984 (Robinson and Lambrick 1984) and is described in some detail elsewhere in this volume by Mark Robinson. Only the main stages need briefly be summarised here:

1. Early Holocene: stabilisation of channels, formation of non-alluvial soils over gravel.
2. Mesolithic to Bronze Age: relatively low water-tables, minimal flooding and alluviation.
3. Late Bronze Age to Early Iron Age: rise in water-table.
4. Early to Middle Iron Age: increased flooding with minimal alluviation.
5. Late Iron Age and Roman: extensive flooding and substantial alluviation.
6. Early to Middle Saxon: continued flooding, probable slow down of alluviation.

7 Late Saxon to earlier Middle Ages: extensive flooding and substantial alluviation.
8 Late Middle Ages and Post-medieval: extensive flooding and limited alluviation.

Since the initial publication of this sequence at least nine sites have been observed which either replicate or refine one or more of these stages. These observations have been made throughout the Upper Thames, at Lechlade, Clanfield, Ducklington, Stanton Harcourt, Northmoor, Cassington, Oxford, Drayton and Goring. Several of these sites are briefly described in other sections of this paper or in Mark Robinson's. Of particular value has been the sequence at Drayton where Tony Clark has successfully applied magnetic dating to clarify the Early to Middle Saxon slow-down in alluviation and date the Late Saxon or Early Medieval transition from pasture to hay meadow evident from differences in the snail faunas. These observations are discussed in more detail by Mark Robinson (this volume) and are summarised here in Figure 20.6.

At Gravelly Guy, in the Windrush valley near Stanton Harcourt, an unusual sequence has been investigated. This consisted of a pre-Iron Age palaeosol associated with successive Late Neolithic and Late Bronze Age waterlogged gullies, sealed by Iron Age colluvial ploughsoil, followed by Roman and later interdigitated colluvium and alluvium (Fig. 20.7). This is entirely consistent with the general sequence, but the pre-Iron Age waterlogged preservation close to the contemporary ground surface is anomalous. It is worth noting that organic deposits of these periods were also recovered at Mingies Ditch just upstream. This may suggest that the Lower Windrush tributary floodplain had a generally higher water-table in the earlier Holocene than the main Thames floodplain, though the Mingies Ditch sequence still indicates an Iron Age rise in the water-table (Allen and Robinson in press).

Soils and sediments on the floodplain

Work on the soils and sediments of the floodplain has been carried out by Simon Robinson and Richard Macphail in an attempt to test observations made through standard archaeological recording and sampling for biological evidence. These analyses have been undertaken both to improve understanding of the pedological and sedimentary processes involved, and to investigate the possible source material of the alluvium.

Particle size analysis has been undertaken by Simon Robinson on deposits at Burroway, Gravelly Guy, Drayton and Wallingford (Limbrey and Robinson 1988), and by Richard Macphail at Drayton, as part of a more detailed micromorphological study (Macphail 1990). These studies have confirmed the observable difference between the pre-alluvial palaeosol and the alluvial sediments. The analyses have also confirmed the transition to coarser deposits in the upper part of the alluvial sequence noted in 1984 (Robinson and Lambrick 1984).

Micromorphological analysis of the pre-alluvial soils beneath the Neolithic cursus bank at Drayton has revealed evidence for the very immature and often disturbed character of the pre-alluvial floodplain soils. This fits well with archaeological observation that there is often very little evidence of a distinct soil horizon between the alluvium and the subsoil horizon of sandy clay and gravel overlying the gravel itself. This varies considerably, however, not only from one site to another, but also within small areas, such as at Drayton. On this site deeper, though still poorly developed soils filled hollows in the gravel, while elsewhere there was virtually no pre-alluvial soil cover.

An important aspect of the work at Drayton was to provide an archaeological explanation for the commonly disturbed character of these deposits. Approximately 30% of the pre-alluvial ground surface on this site was occupied by tree-throw holes. Carcful archaeological excavation of the stratigraphy of the infilled holes has revealed the detailed formation processes of these features and suggested (from the regular occurrence of charcoal and scorched soil) some degree of human interference, presumably associated with clearance of standing or fallen trees, which may well have been deliberately killed. These features are noticeably more common on the floodplain than the higher gravel terraces, possibly because of differences in the compactness of the gravel or the height of the water-table affecting the efficiency of the trees' root anchorage.

Attempts to use mineralogical and other techniques to establish the original sources of the alluvium have so far had limited success. The most extensive study has been by Simon Robinson, who compared heavy mineral components of the alluvium with possible parent materials in the surrounding catchment (Limbrey and Robinson 1988). Unfortunately this was inconclusive, largely because of the insufficiently clear and reliable distinctions in the heavy mineral composition of the rocks involved. In his work on the Drayton sequence Richard Macphail has observed chalk fragments in the upper layers of alluvium, suggesting a significant input of sediment from the chalk, or perhaps more likely chalk head deposits, which occur ca. 7–8 km south-west of the site. This material would have been transported via the river Ock, a tributary of the Thames draining the Vale of White Horse.

Causes of hydrological and alluvial change

The problems of elucidating the causes of hydrological and alluvial change are very different from those of establishing the basic pattern of development. The

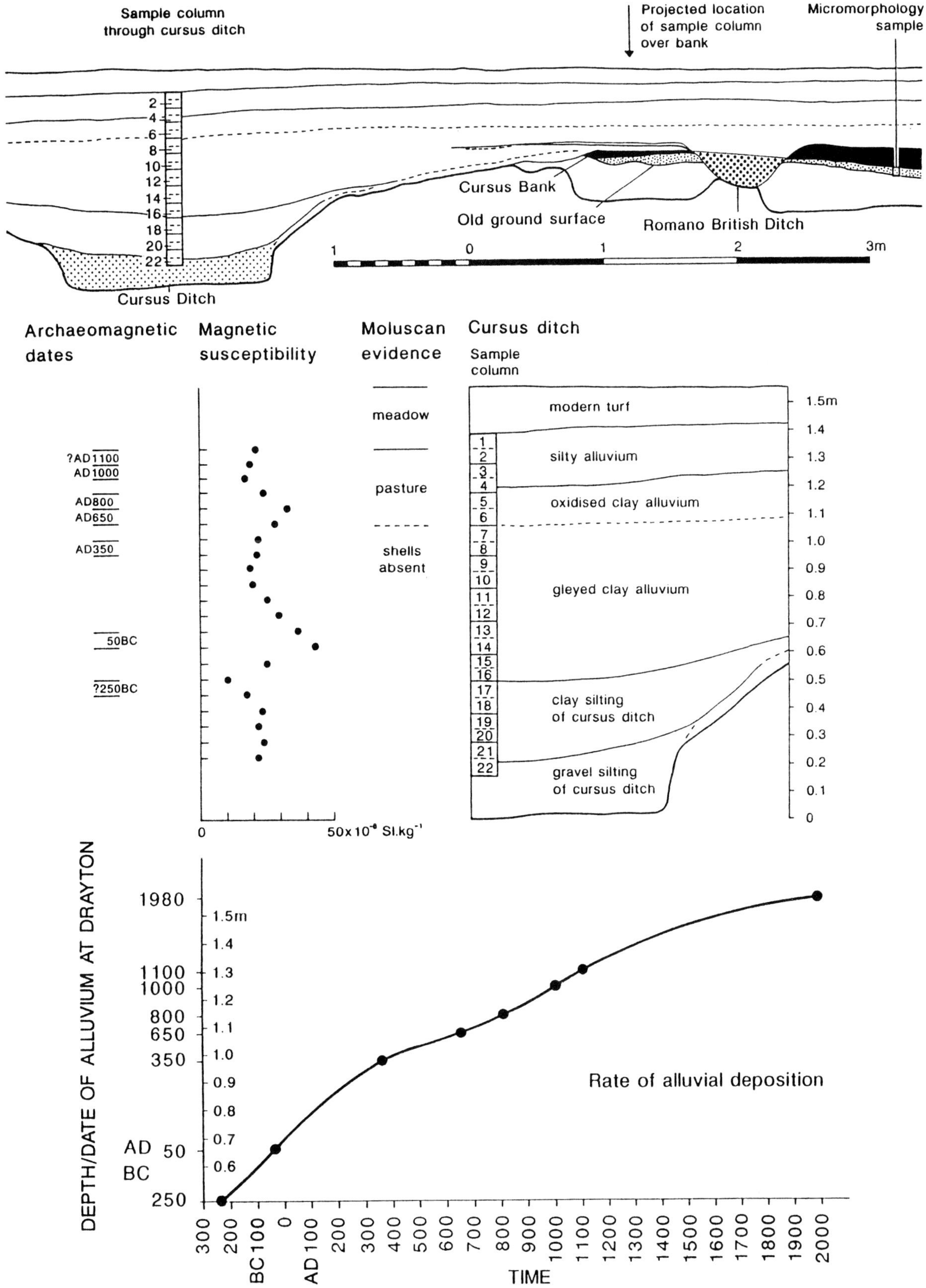

Fig. 20.6. Drayton: alluvial stratigraphy and summary of dating and environmental sequence.

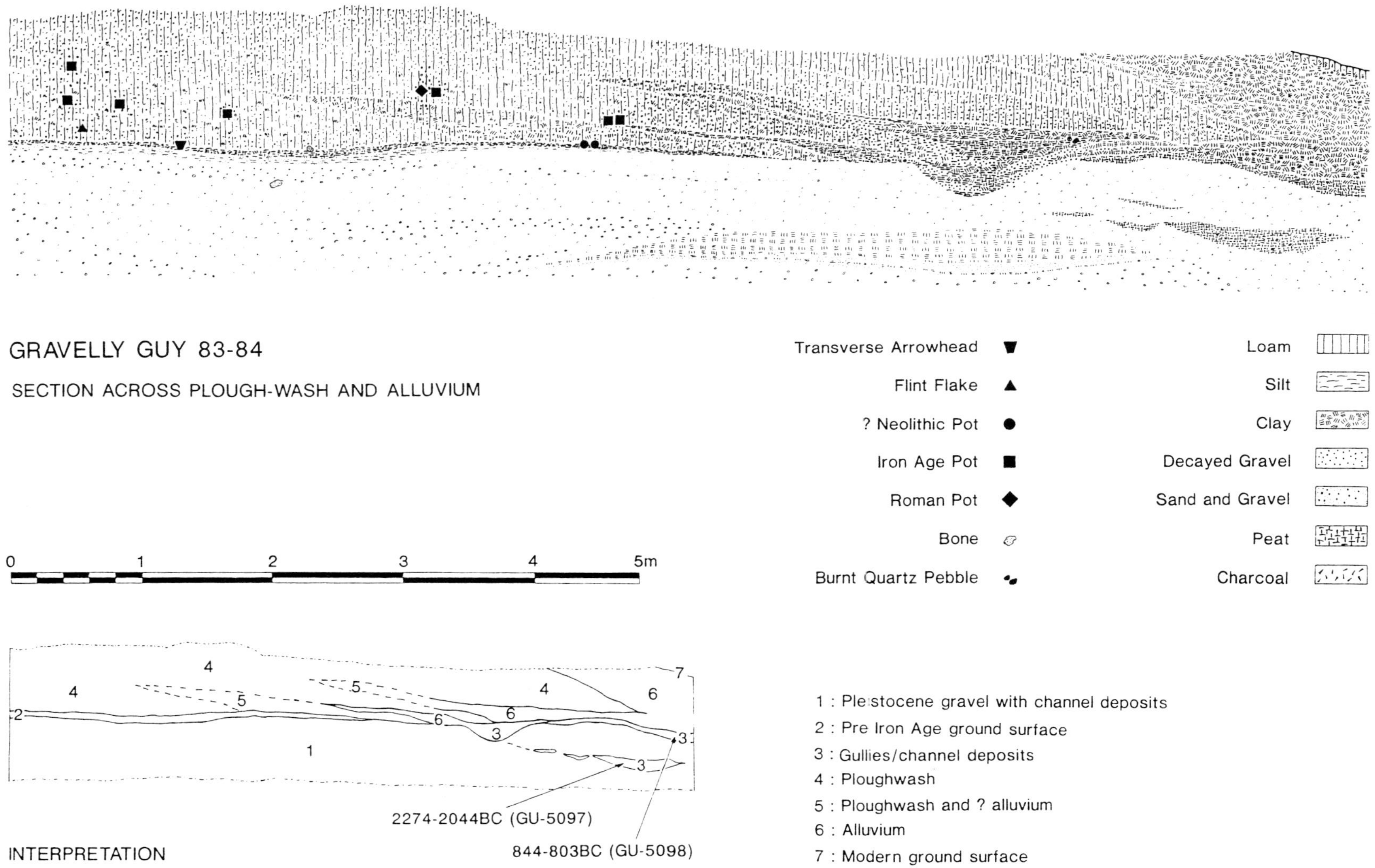

Fig. 20.7. *Gravelly Guy: sequence of interdigitated colluvium and alluvium overlying pre-Iron Age floodplain surface and organic deposits.*

intrinsic nature of river systems is such that hydrological conditions and alluvial sequences on their floodplains provide a genuine generalisation about changing conditions within their upstream catchments. It is therefore possible to establish the basic pattern of change from a small sample of individual sites on the floodplain. However, this does not apply in reverse: it is much more difficult to establish the possible causes of change from a very small sample of sites in the wider catchment.

Two principal causes of hydrological and alluvial change can be put forward: climatic change and the impact of human interference on the environment. Both are complex factors to investigate. Although climate represents an even more generalised phenomenon than any individual river system, its character can only be established through interpretation of indirect evidence, including, most notably, biological sub-fossils. But these also reflect human interference with the environment. Rather like any scientific study investigating causal relationships between naturally occurring phenomena, it is necessary to consider whether any observed patterns could be accounted for by human interference in the system rather than by purely natural relationships. As such, the effects of climate as a cause of hydrological and alluvial change on river floodplains cannot be determined independently of studying the impact of human activity on the environment, particularly for periods after the introduction of agriculture.

It is also difficult, however, to investigate in any detail the scale, spatial distribution and permanence of past human interference in the environment: a very large number of sites needs to be considered to provide an adequate sample to generalise about broad patterns of change. The distribution of rescue-led archaeological research, concentrated along the river valleys because of the nature of threats, has been essential to the elucidation of the archaeology of the floodplain, but has not been particularly conducive to establishing an adequate body of suitable data in the rest of the catchment. Nevertheless, since this issue was briefly discussed in 1984 (Robinson and Lambrick 1984) there has been a notable flowering of general reviews and other discussions of the archaeology and environmental history of the region (Briggs *et al* 1986; Saville 1984a; Darvill 1987; Holgate 1988; Hingley 1984; Miles 1989; Robinson and Wilson 1987).

In the early post-Devensian period climate was the essential determinant of hydrological change, but once the temperate climate was fully established neither climatic change nor human clearance and settlement in the Upper Thames catchment was initially sufficient to create detectable changes in general hydrology or alluviation on the floodplain. However, as Mark Robinson suggests elsewhere in this volume, there may have been some effect on sedimentation in channels. Recent research has indicated something of the pre-Iron Age pattern of activity through the combined study of earlier prehistoric monuments and flint scatters (Marshall 1985; Holgate 1988; Gaffney and Tingle 1989; Lambrick 1988).

More generally, there is growing environmental evidence for the survival of significant woodland resources into the Early Bronze Age or later. This seems apparent in the occurrence of wild, collected plant foods in carbonised seed assemblages (Moffett *et al* 1989) and the predominance in animal bones of cattle and pigs which are capable of browsing and rooting in woodland, or at least a not fully cleared environment (Robinson and Wilson 1987). Some monuments may have been built in woodland or only recently cleared land or areas reverted to woodland (Saville 1983; Saville 1990; Lambrick 1988; Evans 1972). The virtual absence so far of any evidence of sedentary farmsteads in the region before the Late Bronze Age may reflect the long survival of a shifting pattern of domestic activity (Holgate 1988), in which some regeneration of woodland may have been typical.

This is supported by results of pollen analysis at Sydlings Copse 4 km north-east of Oxford, where there is evidence of minimal, sporadic interference from burning, with virtually no effect on the tree canopy until a period of clearance of woodland on well drained soils around 3800 ± 100 BP (OxA-2049, OxA-2050), followed by a major increase in bracken, suggesting conversion to rough heathland grazing. It is not until the early Roman period that grasses and herbs clearly oustripped the proportion of bracken and fern spores in the pollen record (Day 1991).

There is growing evidence of Late Bronze Age activity in the Upper Thames valley, and in particular the emergence of land boundaries and field or paddock systems on the gravels (Thomas 1980; Lambrick 1979; Miles and Palmer n.d.; Lambrick, in press) and to some extent on the surrounding hills (Ford 1982; Lambrick 1988; Darvill 1987). This is coupled with the construction of major defensive enclosures, such as Rams Hill and Ivinghoe on the chalk (Bradley and Ellison 1975; Cotton and Frere 1968) and probably Nottingham Hill, Chastleton and other sites on the Cotswolds (Saville 1984b). Although arable agriculture was clearly increasingly important on the gravels from the Early Iron Age, there is some indication from the admittedly sparse evidence on the Cotswolds and Berkshire Downs, that crop growing was not equally important across the catchment. None of the excavated hillforts on the Cotswolds has yet produced the sort of density of storage pits evident at some Wessex hillforts (Cunliffe 1984) or on the gravels (e.g. Williams 1951; Parrington 1978; Lambrick 1979, 1985b), or even on non-hillfort Middle Iron Age settlements on the Cotswolds (Saville 1979; Lambrick 1988).

Differences in the Middle Iron Age settlement pattern between the Cotswolds and the Thames gravels (Lambrick 1983; Hingley 1984; Lambrick 1988) may well reflect considerable differences in the character and intensity of land use. It can at least plausibly be argued that the Early to Middle Iron Age saw widespread woodland clearance to provide improved pastoral resources, the economic importance of which may well be reflected in the labour investment required to construct the hillforts. This argument at the very least suggests that the Iron Age rise in water-table and flooding on the floodplain could have been induced by the impact of later prehistoric farming and, more specifically, clearance for pastoralism, which could have induced much greater run-off without releasing much sediment into the drainage system.

The expansion of Roman settlement and large mixed farming establishments in the region is well known (Young 1986; Miles 1989), but their origins are poorly understood. It is quite clear that Roman settlement expanded onto clay soils in the catchment, and it is tempting to suggest that the very clayey character of Roman alluvium might particularly reflect expansion of arable onto the relatively impermeable clay slopes, coupled with the provision of ditched drainage. As yet, however, there has been insufficient detailed work on the settlement pattern and land use on the clays at this period to substantiate such ideas. The dating of both the onset of alluviation and the origins of Roman settlements off the gravels is too inaccurate to clarify the issue.

However, Petra Day's work at Sydlings Copse (Day 1991) is again valuable, illustrating the creation of a fully cleared agricultural landscape at ca. 1820 ± 80 BP (OxA-2047), with the regular occurrence of cereals, herbs associated with ground disturbance and grasses, and a substantial decline of bracken. Although there was no obvious acceleration of sedimentation in this small valley, there is an increase in magnetic susceptibility values suggesting increased human interference with the soils in the area. It is thus again the case that changes in human settlement pattern and land use may well provide an adequate explanation of the observed changes in alluviation on the floodplain.

Likewise, the disruption of the economy and settlement pattern, possible disease and general de-intensification of land use and infrastructure during the Early Saxon period (Hawkes 1986) may well explain the decline in the rate of alluviation which has now been demonstrated for the first time from the magnetic dating of the Drayton profile. Again, there is a problem with the relative dearth of good archaeological and environmental data from the catchment, especially off the gravels. One point worth noting, however, is that carbonised crop remains tend to be less common on Saxon settlements on the gravels than their Iron Age or Roman predecessors. This is perhaps most obvious in a comparison of a series of sites at Abingdon: Iron Age Ashville, Late Iron Age to Saxon Barton Court and the immediately adjacent, more substantial Barrow Hills Saxon settlement (Jones in Parrington 1978; Jones in Miles 1986; Moffett pers. comm.). This could reflect some reversion to pastoralism.

The rapid expansion of settlement in the Late Saxon and Early Medieval period is well attested, and Domesday provides the first clearly documented picture of land use in the catchment. The greatest density of plough teams occurred on the Cotswolds, in the Vale of Aylesbury and parts of the Vale of White Horse (Jope and Terret 1962). Physical evidence of the very extensive cultivation of the clays and other soils is provided by the occurrence of ridge-and-furrow, much of it still surviving as earthworks (Sutton 1965). At the height of the Medieval expansion of arable cultivation, ridge-and-furrow expanded onto the floodplain in some places. In the Cassington to Oxford reach of the river there are several examples, typically aligned at right angles to the river and lacking a headland at the end nearest the river, presumably to allow floodwater to drain away rather than remaining trapped in the furrows. With the expansion of arable it became increasingly important to maximise grassland resources by creating very extensive common meadows on the floodplain to provide fodder to overwinter stock (Lambrick and Robinson 1988).

Once again the pattern of human settlement and land use provides an entirely adequate explanation of the observed acceleration of alluviation on the floodplain at this period. The intensity of this expansion may well have surpassed that of the Late Iron Age to Roman period.

The Late Medieval and Post-medieval slow-down in alluviation, despite continued flooding, is explicable in terms of changes in population, settlement and land use, reflected in the numerous deserted and shrunken villages in the area. Bond (1986) has concluded that about 150 Medieval settlements, roughly 24% of the total known to have existed in Oxfordshire, have become deserted, and another 20% show evidence of contraction. Of the abandoned sites, only 9% were deserted before 1350, compared with 30% between 1350 and 1450, and 45% between 1450 and 1700, but only 3% since 1700 (the remaining 13% being undatable). There is some indication that desertion was commonest in the clay vales, but this is only really clear-cut in the Vale of Aylesbury around Thame.

As part of the process of economic readjustment in the 14th to 16th centuries, particularly after the Black Death, much arable land reverted to grassland – as is still evident from extensive ridge-and-furrow under permanent pasture. Bond suggests that desertion particularly affected minor settlements relying on marginal arable with only limited pastoral resources. In the fifteenth century enforced evictions, to allow manorial lords to enclose land for more profitable sheep pasture, reached their height. The Cotswolds became famous for the Late Medieval development of the wool industry. Early enclosure was especially common on the Chilterns and in the clay vales (Emery 1974).

The absence of any obvious Post-medieval acceleration of alluviation, despite the notorious flooding which continued into the days of photography (Emery 1974: Pl. 20), is perhaps surprising given the re-expansion of arable agriculture. However, the clays generally came back into cultivation late on, when improved piped drainage became widespread after ca. 1840 (Emery 1974). Enclosure and the abandonment of ridge-and-furrow cultivation may well have been further factors tending to reduce the scale of sediment transport off the surface of arable fields, despite the substantial run-off.

This attempt to explain the causes of changes in hydrology and alluviation on the Upper Thames floodplain has concentrated on anthropogenic factors. It is by no means clear how much effect climate has had in changing the region's Holocene environment in general (Robinson 1984; Robinson and Wilson 1987). The differences between the Upper Thames floodplain sequence and those of neighbouring catchments (Robinson this volume) suggest that more locally variable factors to do with human exploitation of the environment provide, on the evidence outlined above for the Thames, an entirely adequate explanation of the observed changes.

One cautionary note, however, is that these neighbouring sequences do not appear to reflect directly similar correlations with the settlement archaeology. Underlying differences in geology, topography and the innumerable subtle factors affecting whether significant sediment transport takes place (Lambrick and Robinson 1984) may be among the reasons why superficially similar patterns of settlement may not result in the same effects on different floodplains. If so, these same factors could also have affected how far climatic factors would be reflected in floodplain sequences.

The causes of hydrological and alluvial change still require further investigation, preferably using the same techniques for each catchment. One promising line of enquiry for the Upper Thames is to try a standard technique used for establishing the environmental sequence of upland catchments by investigating pollen cores from the surprisingly numerous and well distributed small peat deposits recently identified in the catchment (Day 1991; Adrian Parker pers. comm.). Much more extensive surface collection surveys in the wider catchment would also be very valuable to elucidate settlement and land-use patterns, as would specific excavations aimed at establishing the economic basis of settlements at critical periods, especially Iron Age and Saxon.

THE VALUE OF ALLUVIAL ARCHAEOLOGY FOR INVESTIGATION OF SETTLEMENTS AND SOCIAL AND ECONOMIC RELATIONSHIPS

An aspect of the archaeology of the Upper Thames floodplain which has gradually emerged as being of particular value is the light it has cast on a variety of social and economic issues. An example of this at its simplest level is the indication of differences in living standards between the Blackfriars and their ordinary urban neighbours, already mentioned above.

More complex, and certainly more significant in revealing whole new aspects of Middle Iron Age settlement in the region, was the evidence of the highly specialised pastoral floodplain settlement at Farmoor (Lambrick and Robinson 1979). This site is still unique for the social and economic implications of its exceptional waterlogged biological evidence, indicating specialised pastoralism, probable transhumance and an exceptionally short period of occupation. What remains unclear is the precise social context of these farmsteads: whether they represent specialised summer shielings associated with more permanent year-round settlements on higher ground, or an independent, almost nomadic element in local Iron Age society.

One of the original conclusions of the Farmoor work was that grassland resources were clearly much more significant than had previously been appreciated. In this case grassland was sufficiently important to determine the location of settlement, in an otherwise unsuitable environment, through social arrangements geared to the specialist exploitation of this resource. Further work, particularly at Mingies Ditch, Northmoor, Lechlade, and Drayton, has continued to address this issue and consider the development of grassland land use on the floodplain together with its economic and social implications (Lambrick and Robinson 1988; Lambrick, in press; Robinson this volume). In the Middle Iron Age, Farmoor and other pastoral sites appear to represent a social rather than a technical development of pastoralism, based on the special physical organization of settlement. Hay-making was a relatively late (Roman) development, a labour intensive, technical solution to the problem of providing sufficient fodder to overwinter large herds when grass growth is limited. The evidence for the expansion of very extensive communal hay meadows on the floodplain in the early Middle Ages, exemplified by the Drayton snails and the well documented and still floristically rich Picksey and West Meads (McDonald 1988), reflects further developments in economic and social arrangements in a period of expanding population.

As a direct result of these considerations of the exploitation of floodplain grassland, it has been possible to cast new light on the intensification of farming in the late prehistoric and Roman periods, emphasising the central importance of pastoralism as a key area of development (Lambrick in press), and the floodplain as a key resource in the wider pattern of settlement and land use in the valley.

At Stanton Harcourt the sequence of Iron Age colluvium sealed by alluvium on the Windrush floodplain provides a rather different key to the interpretation of the local settlement pattern. This sequence indicates that the adjacent settlement, and by implication a series of others also set back from the edge of the gravel terrace, each had a strip of arable along the terrace margin. These linear settlements surrounded the core of the gravel terrace where the remains of numerous Bronze Age barrows and the Devils Quoits henge circle still survived. Enclosures opened into this area from the linear settlements, on the opposite side from the arable fields, and there is some indication that soils away from the settlements in these areas remained relatively stone-free (i.e. not mixed with the underlying gravel) into the Roman period, suggesting that they had not been ploughed. It is possible that the settlements surrounding this core area may have shared it as a large communal area of grazing.

This admittedly tentative interpretation (not fully elaborated here), together with detailed evidence from excavation of the settlement on the gravel terrace, has interesting implications for recognizing a hierarchy of how land use may have been related to a series of socio-economic entities of household, farmstead and community amongst these Iron Age settlements. Identifying the probable location of the arable fields from the colluvial deposits spreading onto the floodplain was a crucial step in developing this interpretation.

CONCLUSIONS

Alluvial archaeology can contribute enormously to numerous broad themes concerning the relationships of past human societies, the social and economic context of their exploitation of natural resources and their impact on the environment. Much work to date has been the result of making the most of relatively fortuitous opportunities for appropriate research. In the future it will increasingly be necessary to undertake projects specifically to address outstanding issues concerning both the identification and preservation of this outstanding source of evidence and its interpretation. Those which most obviously warrant attention are as follows:

1. Careful study of the feasibility and cost-effectiveness of different discovery techniques for sites sealed by alluvium.
2. Review of planning policies encouraging gravel extraction in alluvial areas.
3. Research on processes of desiccation and methods to avoid hydrological draw-down.
4. Greater use of magnetic dating to establish the chronology and rate of deposition of alluvium.
5. Further detailed study of pre-alluvial floodplain soils in relation to micro-topography of floodplain.
6. More detailed study of channel formation/infill pattern and relationship to human activity.
7. Identification of sources of alluvium and especially changes in sources through time.
8. Investigation of changing patterns of land use in the catchment by surface collection, sampling of key sites (by geophysics/excavation) and recovery where possible of pollen sequences, for correlation with hydrological and alluvial change.
9. Further research on patterns of clearance in relation to grazing and browsing requirements of domestic stock.
10. Research to establish more generally the date and extent of the conversion of pasture to hay meadow in the early Middle Ages, by use of magnetic dating and snail sequences.
11. Investigation of specialised uses of floodplain/river resources.
12. Investigation of social relationships between floodplain settlements/land use and adjacent settlements.
13. Generally improved structuring of research to reflect catchment and sub-catchment divisions.

Acknowledgements

This paper would have been impossible to write without the co-operation of many colleagues over several years, much of whose work is drawn upon or quoted here. Particular thanks go to Tim Allen, Brian Durham, Gill Hey, David Miles, John Moore and Simon Palmer. Above all I would like to acknowledge the contribution of Dr Mark Robinson, not only for the results of his innumerable analyses of such a wide range of biological remains, but also for many hours of fruitful discussion. For most of the illustrations I am grateful to Danyon Rey.

Most of the work reported here was funded by English Heritage, and I would also like to acknowledge the financial support of ARC, G. Curtis and Sons, and J. Smith and Sons of Bletchington on various of the sites discussed. I am grateful to the British Academy for financing a study of the Upper Thames floodplain in 1981–2.

References

Ainslie, R. and Wallis, J. 1987. Excavations on the Cursus at Drayton, Oxon. *Oxoniensia*, 52, 1–10.

Allen, T. G. and Robinson, M. in press. *The Prehistoric Landscape and the Iron Age Enclosed Settlement at Mingies Ditch, Hardwick with Yelford, Oxon.* Oxford: Oxford University Committee for Archaeology.

Allen, T. G. forthcoming. Report on excavations at Goring, Gatehampton Farm.
Atkinson, R. J. 1942. Archaeological Sites on Port Meadow, Oxford. *Oxoniensia*, 7, 24–35.
Bond, J. 1986. The Oxford Region in the Middle Ages. In G. Briggs, J. Cook and T. Rowley (eds.), *The Archaeology of the Oxford Region*, 135–59. Oxford: Oxford University Department for External Studies.
Bowler, D. and Robinson, M. 1980. Three Round Barrows at King's Weir, Wytham, Oxon. *Oxoniensia*, 45, 1–8.
Bradley, R. J. and Ellison, A. 1975. *Rams Hill*. Oxford: British Archaeological Reports, British Series, 19.
Brereton, S. and Lambrick, G. H. forthcoming. A Survey of the Results of Field Evaluations undertaken by the Oxford Archaeological Unit and Wessex Archaeology, 1988–90.
Briggs, G., Cook, J. and Rowley, T. (eds.) 1986. *The Archaeology of the Oxford Region*. Oxford: Oxford University Department for External Studies.
Case, H. J. 1963. Notes on Ring Ditches in the Oxford Region. *Oxoniensia*, 28, 19–53.
Cotton, M. and Frere, S. S. 1968. Ivinghoe Beacon Excavations 1963–5. *Records of Buckinghamshire*, 18 pt 3, 187–260.
Cunliffe, B. W. 1984. *Danebury, an Iron Age Hillfort in Hampshire*. London: Council for British Archaeology Research Report 52.
Darvill, T. C. 1987. *Prehistoric Gloucestershire*. Gloucester: Alan Sutton and Gloucestershire County Library.
Day, P. 1991. Post-glacial Vegetational History of the Oxford Region. *New Phytologist*, 119, 445–470.
Durham, B. G. 1977. Archaeological Investigations in St Aldates, Oxford. *Oxoniensia*, 42, 83–203.
Durham, B. G. 1984. The Thames Crossing at Oxford: Archaeological Studies 1979–82. *Oxoniensia*, 49, 57–100.
Durham, B. G. 1990. The City of Oxford. *Current Archaeology*, 121, 28–33.
Emery, F. 1974. *The Oxfordshire Landscape*. London: Hodder and Stoughton.
Evans, J. G. 1972. *Land Snails in Archaeology*. London: Seminar Press.
Ford, S. 1982. Excavation and Fieldwork on the Berkshire Grim's Ditch. *Oxoniensia*, 47, 13–36.
Gaffney, V. and Tingle, M. 1989. *The Maddle Farm Project*. Oxford: British Archaeological Reports, British Series, 200.
Hawkes, S. 1986. The Early Saxon Period. In G. Briggs, J. Cook and T. Rowley (eds.), *The Archaeology of the Oxford Region*, 64–108. Oxford: Oxford University Department for External Studies.
Hingley, R. 1984. Towards Social Analysis in Archaeology: Celtic Society in the Iron Age of the Upper Thames Valley (400–0 BC). In B. Cunliffe and D. Miles (eds.), *Aspects of the Iron Age in Central Southern England*, 72–88. Oxford: Oxford University Committee for Archaeology.
Holgate, R. 1988. *Neolithic Settlement of the Thames Basin*. Oxford: British Archaeological Reports, British Series, 194.
Jope, E. M. 1956. Saxon Oxford and its Region. In D. B. Harden (ed.), *Dark Age Britain*.
Jope, E. M. and Terret, I. B. 1962. Oxfordshire. In H. C. Darby and E. M. J. Campbell (eds.), *The Domesday Geography of South East England*, 186–238. Cambridge: The University Press.
Lambrick, G. H. 1979. Berinsfield, Mount Farm. *CBA Group 9 Newsletter*, 9, 113–5.
Lambrick, G. H. 1982. Thames Floodplain Survey. *South Midlands Archaeology*, 12, 129–34.
Lambrick, G. H. 1983. *The Rollright Stones*. Oxford: Oxford Archaeological Unit.
Lambrick, G. H. 1984. Clanfield: Burroway. *South Midlands Archaeology*, 14, 104–5.
Lambrick, G. H. 1985a. Further Excavations on the Second Site of the Dominican Priory, Oxford. *Oxoniensia*, 50, 131–208.
Lambrick, G. H. 1985b. Stanton Harcourt, Gravelly Guy. *South Midlands Archaeology*, 15, 107–11.
Lambrick, G. H. 1987. Wallingford By-pass. *South Midlands Archaeology*, 17, 99–100.
Lambrick, G. H. 1988. *The Rollright Stones, Megaliths, Monuments and Settlement in the Prehistoric Landscape*. London: English Heritage Archaeological Report 6.
Lambrick, G. H. in press. The Development of late Prehistoric and Roman Farming on the Thames Gravels.
Lambrick, G. H. and McDonald, A. 1985. The Archaeology and Ecology of Port Meadow and Wolvercote Common, Oxford. In G. H. Lambrick (ed.), *Archaeology and Nature Conservation*, 95–109. Oxford: Oxford University Department for External Studies.
Lambrick, G. H. and Moore, J. P. 1987. Drayton Cursus. *South Midlands Archaeology*, 17, 85–6.
Lambrick, G. H. and Robinson, M. A. 1979. *Iron Age and Roman Riverside Settlements at Farmoor, Oxfordshire*. London: Council for British Archaeology Research Report, 32.
Lambrick, G. H. and Robinson, M. A. 1988. The Development of Floodplain Grassland in the Upper Thames Valley. In M. Jones (ed.), *Archaeology and the Flora of the British Isles*, 55–75. Oxford: Oxford University Committee for Archaeology.
Lambrick, G. H. and Woods, H. 1976. Excavations on the Second Site of the Dominican Priory, Oxford. *Oxoniensia*, 41, 168–231.
Limbrey, S. and Robinson, S. 1988. Dry Land to Wet Land: Soil Resources in the Upper Thames Valley. In P. Murphy and C. French (eds.), *The Exploitation of Wetlands*, 129–44. Oxford: British Archaeological Reports, British Series 186.
Macphail, R. 1990. Soil Report on Drayton Cursus near Abingdon, Oxfordshire. English Heritage Ancient Monuments Laboratory Report 57/90.
Marshall, A. J. 1985. Neolithic and Earlier Bronze Age Settlement in the Northern Cotswolds: a preliminary outline based on the distribution of surface scatters and funerary areas. *Transactions of the Bristol and Gloucestershire Archaeological Society*, 103, 23–54.
McDonald, A. 1988. Changes in the Flora of Port Meadow and Picksey Mead, Oxford. In M. Jones (ed.), *Archaeology and the Flora of the British Isles*, 76–85. Oxford: Oxford University Committee for Archaeology.
Miles, D. 1986. *Archaeology at Barton Court Farm, Abingdon, Oxon*. London: Council for British Archaeology Research Report, 50.
Miles, D. 1989. Villas and Variety: Aspects of Economy and Society in the Upper Thames Landscape. In K. Brannigan and D. Miles (eds.), *The Economies of Romano-British Villas*, 60–72. Sheffield.

Miles, D. and Palmer, S. (not dated). *Invested in Mother Earth*. Oxford: Oxford Archaeological Unit.

Miles, D. and Palmer, S. 1990. Claydon Pike and Thornhill Farm. *Current Archaeology*, 121, 19–23.

Moffett, L., Robinson, M. A. and Straker, V. 1989. Cereals, Fruit and Nuts: Charred Plant Remains from Neolithic Sites in England and Wales and the Neolithic Economy. In A. Miller, D. Williams and N. Gardener (eds.) *The Beginnings of Agriculture*, 243–61. Oxford: British Archaeological Reports, British Series, 196.

Palmer, N. 1980. A Beaker Burial and Medieval Tenements in the Hamel, Oxford. *Oxoniensia*, 45, 124–225.

Parrington, M. 1978. *The Excavation of an Iron Age settlement, Bronze Age ring-ditches and Roman features at Ashville Trading Estate, Abingdon (Oxfordshire) 1974–1976*. London: Council for British Archaeology Research Report, 28.

Robinson, M. A. 1984. Landscape and Environment of Central Southern Britain in the Iron Age. In B. Cunliffe and D. Miles (eds.), *Aspects of the Iron Age in Central Southern England*, 1–11. Oxford: Oxford University Committee for Archaeology.

Robinson, M. A. and Lambrick, G. H. 1984. Holocene Alluviation and Hydrology in the Upper Thames Basin. *Nature*, 308, 809–14.

Robinson, M. A. and Wilson, R. 1987. A Survey of Environmental Archaeology in the South Midlands. In H. C. M. Keeley (ed.), *Environmental Archaeology: a Regional Review, 2*, 16–100. London: Historic Buildings and Monuments Commission for England Occasional Paper, 1.

Saville, A. 1979. *Excavations at Guiting Power Iron Age Site, Gloucestershire 1974* . Bristol: Committee for Rescue Archaeology in Avon Gloucestershire and Somerset, Occasional Paper 6.

Saville, A. 1983. Excavations at Condicote Henge Monument, Gloucestershire, 1977. *Transactions of the Bristol and Gloucestershire Archaeological Society*, 101, 21–47.

Saville, A. (ed.) 1984a. *Archaeology in Gloucestershire*. Cheltenham.

Saville, A. 1984b. The Iron Age in Gloucestershire: a Review of the Evidence. *In* A. Saville (ed.) 1984a, 140–78.

Saville, A. 1990. *Hazleton North, Gloucestershire 1979–82: The Excavation of a Neolithic Long Cairn of the Cotswold-Severn Group*. London: English Heritage Archaeological Report 13.

Sutton, J. E. G. 1965. Ridge and Furrow in Berkshire and Oxfordshire. *Oxoniensia*, 29/30, 99–115.

Sutton, J. E. G. 1966. Iron Age Hillforts and some other Earthworks in Oxfordshire. *Oxoniensia*, 31, 28–42.

Thomas, R. 1980. A Bronze Age Field System at Northfield Farm? *Oxoniensia*, 45, 310–11.

Thomas, R., Robinson, M., Barrett, J. and Wilson, R. 1986. A Late Bronze Age Settlement Site at Wallingford Oxon. *Archaeological Journal*, 143, 174–200.

Wallis, J. and Lambrick G. H. 1989. Ducklington: Gill Mill Farm. *South Midlands Archaeology*, 19, 49–50.

Williams, A. 1951. Excavations at Beard Mill, Stanton Harcourt, Oxon. *Oxoniensia*, 16, 5–22.

Wymer, J. 1960. Archaeological Notes from Reading Museum. *Berkshire Archaeological Journal*, 58, 52–64.

Young, C. J. 1986. The Upper Thames Valley in the Roman Period. In G. Briggs, J. Cook and T. Rowley (eds.), *The Archaeology of the Oxford Region*. Oxford: Oxford University Department for External Studies.

SECTION 5

Archaeology and alluvial landscapes in southern and eastern Britain: lowland river systems

21

Palaeoliths in alluvium

J. J. Wymer

Summary: *Palaeolithic stone tools and debitage occur in alluvial deposits throughout most of the latter part of the British Quaternary sequence. For the most part they are in secondary contexts and have limited archaeological importance. Rarely, they are found within or beneath bodies of alluvial sediment and can give reliable information on dating, environment and human activities. The manner in which such sites are preserved is discussed and contrasted with the taphonomy of material which has been transported by fluvial agencies from where it was originally discarded. The dangers of misinterpreting the significance of the latter is emphasised, but two examples are given of sites where, although the material is in secondary context, it is still capable of yielding useful archaeological information.*

The first hand-axe ever recorded in Britain was found near King's Cross at the end of the 17th century, and came from a small pit where coarse alluvium (i.e. gravel) was being dug. The deposit itself was virtually ignored in the later, short published report, attention not surprisingly being given to the curiosity of the piece and some elephant teeth found at the same time. Somewhat later, the astute John Frere recognised the significance of the fine alluvial sediment that overlay the palaeoliths he recognised in the old brickyard at Hoxne. He reasoned that the clay, now at the top of a low hill, must have been deposited by water and this could only have happened before the present valley system was formed. This observation was made in 1797 and is the first example of the stratigraphical dating of palaeoliths in alluvium ever published in Britain (Frere 1800). Later observers took little heed of it and were more concerned with the ante or post diluvian aspect of the faunal remains in ancient alluvium and whether humanly-fashioned flint implements were contemporary with them or not. By the later half of the 19th century there was a rational approach to the subject and alluvium was recognised as part of natural fluvial processes, and 'drift gravels' were seen as remnants of past river systems. The presence of palaeoliths in such alluvium attracted much attention (cf. Evans 1872). The realisation that some of the palaeoliths in or under alluvium lay where they had been discarded, as opposed to the obviously derived nature of most of them, was evident to such people as Worthington Smith and Allen Brown in the Thames valley. Work in the Somme valley culminated in the publications of Commont (1914) and Breuil and Koslowski (1932) who demonstrated the complexity of terrace sequences. In Britain, Arkell, Oakley, Shotton, Zeuner and others applied geological science to such deposits. Now, with the advent of Quaternary science and the application and refinement of numerous techniques from many other disciplines, the interpretation of implementiferous alluvium has been revolutionised. Nevertheless, it is the archaeologist who is left with the problem of relating what remains to the human activity during the vast span of time covered by the Palaeolithic period, and this is discussed below. The material found in primary context is considered first, then the material dispersed by various agencies into alluvial deposits.

The number of known Palaeolithic sites in Britain where material lies within or buried by alluvium in the position where it was left or discarded is very few. Each has to be considered separately in order to understand the circumstances which contributed to its preservation. It is obvious that the high energy fluvial agencies that produce coarse alluvium are highly unlikely to leave anything intact on surfaces they scour; stone artifacts and bones are swept away and become constituents of the gravel deposits that finally settled out, although some evidence may remain to indicate that humanly occupied surfaces did once exist in their vicinity (see especially Swanscombe and Furze Platt given as examples below). The only hope of finding material in primary context associated with alluvium is when fine sediments have gently aggraded over river banks or beaches. This can only happen on the contemporary floodplains of any river system, which is very fortunate for Palaeolithic archaeologists as such an environment was attractive to early man for a variety of reasons: the availability of fresh water, relatively easy movement, the attraction of animals for the same reasons, fish and water fowl for the taking, and the availability of flint or other suitable stone for making tools. However, if the conditions were optimum for the accumulation of archaeological sites, they were also the most vulnerable areas for later destruction by the agencies mentioned above.

Few Lower Palaeolithic sites in primary context have been investigated in Britain, and, as yet, there are no reliable methods for determining their chronometric dates. However, some can be related to geological sequences and tentatively assigned to recognised Quaternary stages. These, in turn, can be given estimated dates based on likely correlations with the deep sea oxygen isotope palaeotemperature scale. It seems certain that it cannot be long before refinements and cross-checkings of various methods of absolute dating, such as those based on U/Th, TL or amino-acids, will resolve this problem.

On present evidence the site at Boxgrove, Sussex, is considered to be of a pre-Anglian Stage, i.e. earlier than about 472,000 years, mainly on faunal grounds. Clacton, on the basis of fauna and pollen, is correlated with the Early-temperate zone of the Hoxnian Stage, the Swanscombe Middle Gravels to the end of the same Hoxnian Stage and Hoxne, itself, to the beginning of a cool phase at the end of it. Dates of between about 440,000 to 367,000 have been assessed for the duration of the Hoxnian Stage, but other lines of evidence suggest it was much shorter and, amino acid dating would put Swanscombe and Hoxne into different interglacial episodes. For a fuller discussion of this aspect of Palaeolithic archaeology see Wymer (1988).

The most likely area on a floodplain where Palaeolithic sites in primary context may be preserved is on the edge of an abandoned channel or ox-bow lake, far enough away from the main stream to escape later erosion. Clacton-on-Sea is a good example of such a fortunate situation, and excavated over a sufficient area to show that this was so (Singer *et al* 1973). Figure 21.1 illustrates the distribution of flintwork and bone fragments over part of the occupied area at Clacton, on the edge of a minor channel of a braided river system. The channel has been eroded through London Clay and about a metre of gravel had accumulated on its south bank. It was on the surface of this gravelly beach that the bone and flint litter of people had been discarded. Much remained on the surface, in mint or very fresh condition, but some was found within the body of the gravel. This was interpreted as contemporary with the material on the surface, but trodden into the soft, clayey gravel when it was wet, by humans or animals. Refits demonstrated this, and also the contemporaneity with the initial stages of the fine, calcareous silt which gradually infilled the channel. This presumably explains the decrease in numbers of mint flint artifacts in relation to depth from the surface, and also suggests that some rolled artifacts in the gravel belong to an earlier phase associated with its deposition. Associated mammalian remains such as giant beaver assist with the environmental reconstruction of sites of this nature.

Another, similar Clactonian site in primary context exists within the fine alluvium at Swanscombe, known as the Lower Loam, as excavated by the late John Waechter (Waechter *et al* 1972). Flintwork and faunal remains lie within the alluvium at particular horizons. Refits were made, including several flakes from a large core.

Hoxne is another primary context site, or series of sites to be more precise, of which the levels known as the Lower and Upper Hoxne Acheulian happen to be the two examined in the more recent excavations (Wymer 1985: 148–178; Singer *et al* forthcoming). The Lower Acheulian industry and associated fauna and 'structures' lay near the base but within fine alluvial clay, on the edge of a slack sheet of water. The Upper Acheulian industry occurred at various levels within and on a sandier sediment that is interpreted as floodplain alluvium later covered by a sheet of coarse gravel.

The Palaeolithic site at Boxgrove is very different, for primary context material is within or on marine sands. Fine alluvial sediments were deposited over the sand while people were still active around the old cliff face nearby. The whole sequence is covered by chalky, coarse gravels swept or sludged off the Chalk downs to the north. Beautifully preserved and meticulously excavated (Roberts 1986), the site has demonstrated the exploitation of flint nodules from the old sea cliff and discrete concentrations of knapping debris some distance away resulting from the production of hand-axes. It is also very significant that faunal remains found at a site of this nature are also in primary context, thus the dating evidence they impart cannot be negated by arguments that they have been derived from earlier sediments. In view of Boxgrove being the earliest primary context site in Britain, mainly based on the faunal evidence, this is critical.

High Lodge has a very complex geological history: a great raft of interglacial lacustrine or fluvial sediment had been bodily transported by the ice of the Anglian glaciation. Flintwork contained in this sediment must therefore pre-date the onset of the ice. This is an extreme example of how archaeological material in

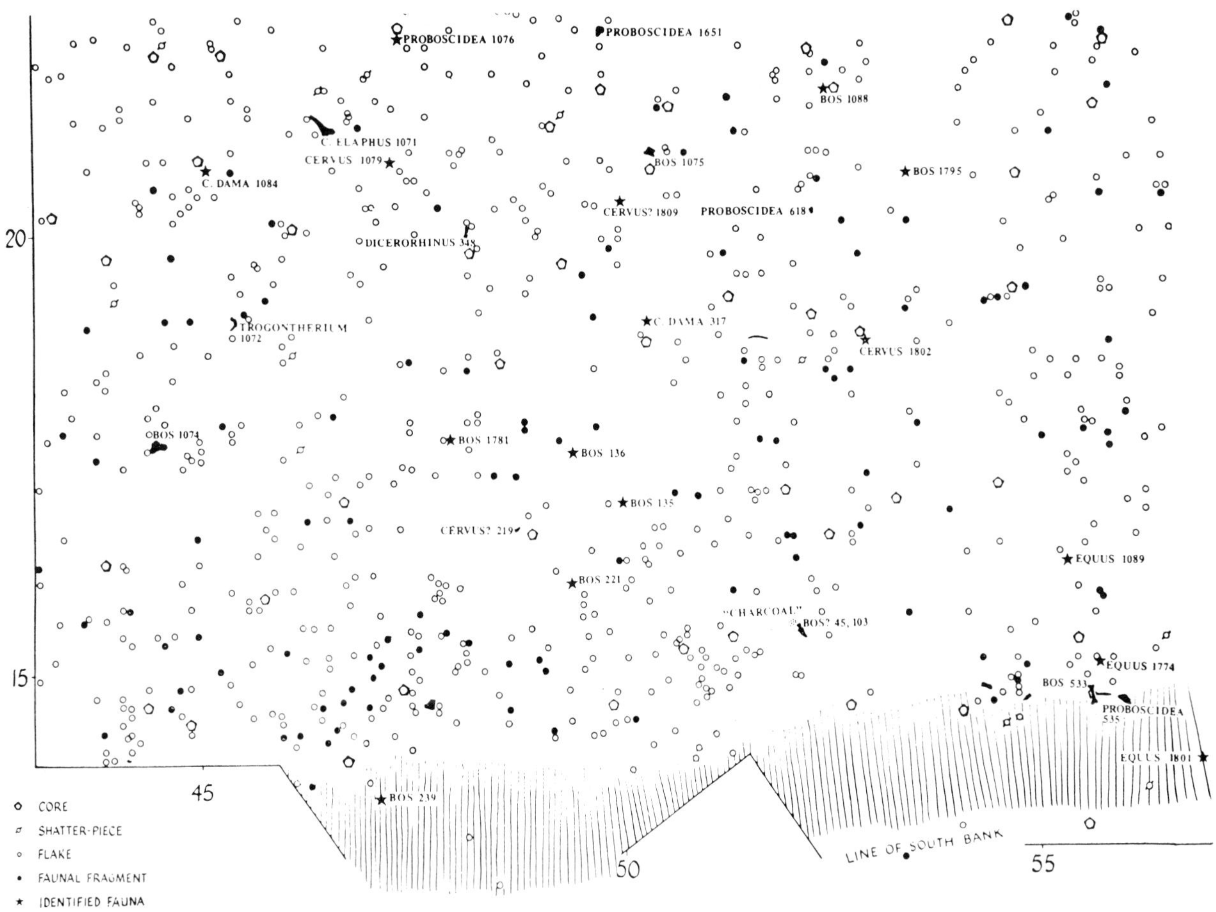

Fig. 21.1. *Part of the distribution pattern of artifacts and faunal remains in primary context at Clacton-on-Sea as excavated in 1970. They lie on and slightly within coarse gravel and are covered by a fine, alluvial marl which, on the evidence of refitted flakes, was being deposited at the time of the occupation. The clay bank of the river channel is shown at the bottom. The numbers beside the frame indicate measurements in metres along the excavation grid.*

Quaternary sediments can be disturbed by geological agencies. Periglacial cryoturbation and solifluction can also result in the intrusion of later material. Thus, glacial periods can be times of much disturbance and re-deposition of existing alluvium and, obviously, the further back in time the greater the number of glacial episodes that have occurred and the chances of such happenings are proportionately more likely.

Barnham, in Suffolk, currently being investigated by the British Museum is a site with a rich Lower Palaeolithic flint industry found in and on soliflucted glacial outwash gravel. The material on this gravel is in primary context. This, in turn, is covered by fine sediment now interpreted as alluvium, although a colluvial element is probably included. The latter contains Acheulian hand-axes and this is therefore one of the rare sites in Britain where two Palaeolithic industries can be separated stratigraphically.

Other sites where palaeoliths occur in primary context within alluvium are Stoke Newington, Crayford, along the Suffolk Stour at Stutton and Harkstead, and possibly at Brundon. Unfortunately, the former two sites are now either dug away or inaccesible beneath urban sprawl, and the others not examined sufficiently to confirm the context. Primary context sites which just happen to be on alluvium but are sealed by sediments produced by such agencies as hill-wash, solifluction or the wind are not directly relevant to this paper. The same applies to the Levalloisian sites at West Drayton and Acton in the Langley Silt Complex (Gibbard 1985), which appear to be buried by colluvial and aeolian deposits rather than fluvial. Pontnewydd is a series of mud flows. The Chiltern brickearth sites of Worthington Smith are more associated with sediments that formed in chalk sink-hollows.

All primary context sites qualify for examination by normal archaeological methods of excavation, the only difference being that the deposits have accumulated through natural agencies instead of human ones. However, when the archaeological evidence has been swept into secondary contexts, the problem is as much geological as archaeological. This will now be discussed.

The dispersal of lithics and bones and whatever else may have been discarded on a river beach or bank by Palaeolithic people would be unlikely to occur except during times when the river was in spate. It is now generally accepted that, as far as lowland Britain is concerned, the sands and gravels remaining from former courses of the major rivers were deposited under cool and cold conditions in braided river systems. This implies a regime of cut and fill and, at some stage, the archaeological and associated material will be subject to these agencies. Such may happen repeatedly and the number of times it does so will obviously affect the dispersal pattern and condition of the artifacts. Palaeolithic sites of this derived nature have rarely been methodically excavated. The opportunity to do so does not often occur, and the horizontal area excavation of alluvial deposits that are implementiferous (usually sands and gravels) and are buried beneath thick, overlying deposits is a time-consuming and expensive matter. The only example in Britain that has been examined on a relatively large scale is the Swanscombe Skull Site, excavated 1955–1960 (Wymer 1964). The actual bed in which the human right parietal was found could be traced by bedding lines, but this was not a level in the true archaeological sense, for eddies and current had left a complex of cross-bedded sands and fine gravels which contained the Palaeolithic artifacts and bones. The only real, stratigraphic markers of consequence were the junction between the Lower Middle and Upper Middle Gravels below and the stoneless current-bedded sands above.

When first discovered, the so-called Skull Level, close to the right parietal fragment, was a fine sand and gravel, bright red with four distinct black bands above it, dipping to the south and conjoining to form one black band which could be followed over a wide area. A coating of dust-like oxide or iron was on some of the flints, which, if not exactly mint, were very sharp. Bones and artifacts were prolific. This was 0.65 m above the junction with the Lower Middle Gravel and apparently comprised one episode of erosion and aggradation. In the absence of total excavation of areas long since eroded away, or more recently dug away commercially, any possibility of tracing the source of the artifacts and bones had gone. There are two likely ways in which this material came to its position of rest:

i) it was washed from a primary context position on a river beach when the river flowed over a body of what now remains of the Lower Middle Gravel, or
ii) it was washed from a beach of sand and gravel that was gradually filling a shallow channel that had already been cut into the Lower Middle Gravel.

The former seems more likely as the concentration of heavy artifacts (i.e. hand-axes) shows at the junction of the Lower and Upper Middle Gravels a similar dispersal pattern to that including the Skull Level, referred to as Layer E (Fig. 21.2). Only the hand-axes are shown on these distribution plans as these were three-dimensionally recorded during the course of the excavation. Flakes were recorded within a grid but not plotted individually. Over 8000 flakes were recorded but only a sample of about 2000 retained and, in retrospect, this is regretted. Refittings could possibly have been made, and probably still could be with the collection now preserved in the British Museum. This might give additional evidence to support the interpretation that most if not all of the flintwork and faunal remains in the gravel and sand of Layer E, including the three fragments of the Swanscombe Skull, came from one discrete activity or occupation area, and was not subjected to any further movement after being dispersed by one flood. It is tempting to see the red and black iron or manganese staining as resulting from the chemical decay of vegetation from a nearby river bank, but other post-depositional factors are just as likely. However, the association is difficult to ignore.

Another prolific site where area excavation with horizontal and vertical recording might have been justified is Furze Platt, Maidenhead. Unfortunately, the only undisturbed areas of the implementiferous gravel is now fossilised beneath the road and buildings of a housing estate. Furze Platt is probably the richest site in the Thames Valley, in terms of the number of artifacts found in relation to the area dug. The majority of the hand-axes occurred in the bottom metre of the gravel overlying the Chalk at Cannoncourt Farm. Little of the material is in a sharp condition like that seen at Swanscombe, and most is rolled to some degree. More than one episode of cut and fill was probably involved. However, the excavation of a small patch of gravel remaining undisturbed on the mainly worked-out floor of the pit revealed one large primary flake, slightly stained but otherwise in mint condition, lying on the Chalk. This implies that palaeoliths had been discarded on a chalk bank and were subsequently swept off during floods which deposited the Lynch Hill Gravel at this point in the valley. A dispersed concentration over a short distance seems likely as a new pit about 300 m to the north in the same gravel of the Lynch Hill Terrace has yielded less than a dozen hand-axes, in spite of frequent observation by R.J. MacRae. The neighbouring Cooper's Pit to the east appears to have been as prolific as the Cannoncourt Farm Pit. The river could only have been flowing south and there were good exposures of the same gravel nearby in that direction some years ago when a housing estate was built, yet no palaeoliths were seen. A localised source for the material, or certainly the great majority of it, seems clearly indicated.

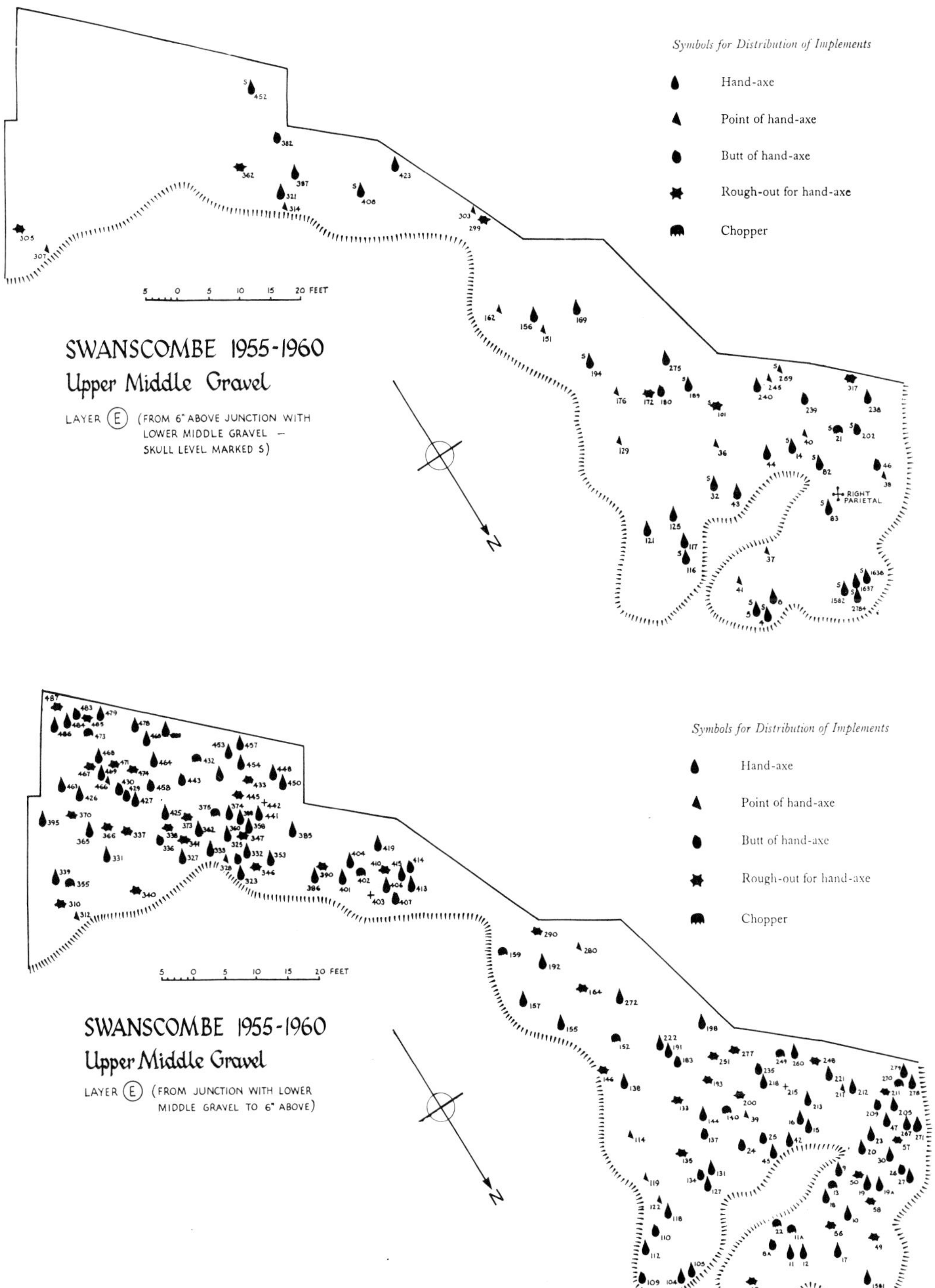

Fig. 21.2. *Distribution patterns of heavy flint artifacts in secondary, derived contexts beneath and within the Upper Middle Gravel at Swanscombe. The lower plan shows the pattern at the junction of the Lower and Upper Middle Gravels and the concentration of material is considered to be partly residual, resulting from erosion of the Lower Middle Gravel. However, there is a similar dispersal pattern on the east side to that on the upper plan. Both originate from the position where the human skull fragments were found (NB: the position of the right parietal is shown but the earlier discoveries of the occipital and left parietal were made beyond the area shown). The material thins out along the direction of the Upper Middle Gravel channel and is thus considered to have been washed off some nearby surface, and progressively swept away as it was moved further into the channel.*

The above two examples serve to indicate that artifacts found under such conditions are almost certainly broadly contemporary with deposition of the alluvium containing them. This may be true wherever fair numbers of palaeoliths in fresh or just slightly rolled conditions are found in one particular gravel. This is not the case with isolated, rolled palaeoliths found in coarse-grade alluvium. They may be very much older than the deposit containing them, washed out of earlier deposits. The same may be true of faunal remains, and great caution is necessary in assessing their age or contemporary environment. Remains of such beasts as straight-tusked elephant may or may not indicate warm conditions. Shells are generally more reliable, but even these may be derived. However, subtle indications may resolve such problems: e.g. opercula of gastropods in position of life, bivalves with their shells still hinged together. Neither would be found in derived mollusc assemblages. Pollen is even more unreliable as an environmental indicator in alluvium as it is inevitably mixed with derived pollen, often from tertiary or older geological periods. Not that the latter is a great problem, for it can be identified and ignored, but that from earlier Pleistocene sediments cannot.

Conversely, it cannot be accepted that rolled and battered palaeoliths in coarse alluvium must be older than fresh ones in the same deposit; some may have been rolled along the river bed for long periods, others just happened to be covered in the initial stages of a cut and fill episode and subjected to no further movement. Thus, fresh and rolled palaeoliths in the same gravel may well be contemporary. Misunderstanding of this has caused much confusion with the interpretation of industrial sequences in the past. However, if palaeoliths do get washed out of terrace gravels into lower ones they are bound to become intensely worn and battered. There are good examples of this in the floodplain gravels of the Thames valley where occasional hand-axes, otherwise apparently restricted to higher terraces, have been found in so rolled a condition that they have become veritable pebbles and nearly unrecognisable as artifacts. With such material, the archaeologists has exactly the same problem as geologists with 'derived fossils'.

As regards palaeoliths in secondary contexts it has to be concluded that great care must be taken in attributing those in coarse alluvium to the date and environment of the deposit containing them, although it has been shown above that there are degrees of probability to be considered. However, the very existence of human flintwork does indicate a human presence in the area at a time no later than the assessed date of the deposit concerned, i.e. a *terminus ante quem*. As greater refinements are made in the geochronometric dating of Quaternary alluvial sequences, so at least our knowledge of the distribution and dating of the Palaeolithic period must improve. Into such a more reliable sequence it will be possible to place the more informative primary context sites, which begin to give some idea of the mode of life at the time.

References

Breuil, H. and Koslowski, L. 1932. Etudes de stratigraphie paléolithique dans le nord de la France, la Belgique et l'Angleterre. *L'Anthropologie*, 42, 27–47.

Commont, V. 1914. *Les hommes contemporains du renne dans la vallée de la Somme*. Amiens.

Evans, J. 1872. *The ancient stone implements, weapons and ornaments of Great Britain*. London: Longmans.

Frere, J. 1800. Account of Flint Weapons discovered at Hoxne in Suffolk, in a letter to the Rev. John Brand, Secretary. *Archaeologia*, 13, 204–5.

Gibbard, P.L. 1985. *The Pleistocene history of the Middle Thames Valley*. Cambridge: University Press.

Roberts, M.B. 1986. Excavation of the Lower Palaeolithic site at Amey's Pit, Boxgrove, West Sussex: a preliminary report. *Proceedings of the Prehistoric Society*, 52, 215–45.

Singer, R., Gladfelter, B.G. and Wymer, J.J. forthcoming. *The Palaeolithic site at Hoxne, Suffolk, England*. Chicago: University Press.

Singer, R., Wymer, J.J., Gladfelter, B.G. and Wolff, R.G. 1973. Excavation of the Clactonian Industry at the Golf Course, Clacton-on-Sea, Essex. *Proceedings of the Prehistoric Society*, 39, 6–74.

Waechter, J.d'A, Hubbard, R.N.L.B. and Conway, B.W. 1972. Swanscombe 1971. *Proceedings of the Royal Anthropological Institute*, 1971, 73–85.

Wymer, J.J. 1964. Excavations at Barnfield Pit, 1955–1960. In C.D. Ovey (ed.), *The Swanscombe Skull: a Survey of Research on a Pleistocene Site*, 19–61. *Occasional Paper of the Royal Anthropological Institute*, 20.

Wymer J.J. 1985. *The Palaeolithic Sites of East Anglia*. Norwich: Geo Abstracts.

Wymer, J.J. 1988. Palaeolithic Archaeology and the British Quaternary Sequence. *Quaternary Science Reviews*, 7, 79–98.

22

A Late Devensian/Early Flandrian site at Three Ways Wharf, Uxbridge: environmental implications.

John S. C. Lewis, Patricia E. J. Wiltshire and Richard Macphail

Summary: *Soil and sediment preservation in the Colne valley permits archaeology, pollen analysis and soil micromorphology to be combined to investigate the pedo/biological environment of two periods of occupation in the Late Devensian and Early Flandrian. Effects of a rise in water-table on the buried occupation soils, and the vegetation history of the area during the Early Mesolithic period, are detailed. A black, charcoal-rich, organic clay, sealing the occupation layer at Three Ways Wharf, is compared with similar deposits in the Colne valley and other tributary valleys of the Thames. The implications of the charcoal within these deposits are explored, along with the effect of rising water-table and organic clay and peat development on Early Mesolithic settlement patterns.*

INTRODUCTION

Fieldwork by Lacaille and others (Lacaille 1961; 1963) established the presence of a number of Mesolithic sites in the Colne valley. The excavation by the Department of Greater London Archaeology, Museum of London, at Three Ways Wharf, Uxbridge, attempted to locate any surviving prehistoric deposits in the bottom of the Colne valley.

The excavation revealed an undisturbed sequence of sediments containing *in situ* flint artefacts and associated fauna and offered the first opportunity for dating and characterising the transition between the Late Devensian and Early Flandrian, a period hitherto poorly understood in this country. Four flint and faunal scatters were defined "A", "B", "C" and "D", with scatters "A" and "C" producing significant quantities of material. These scatters seem to represent distinct phases of occupation: scatter "A" dates to the Late Devensian (Zone III *sensu* Godwin 1940), and scatter "C" to the Early Flandrian. Scatters "B" and "D" are, at present, undated.

A fuller account of the archaeology of Three Ways Wharf has appeared elsewhere (Lewis 1991). This paper will concentrate on palaeoenvironmental evidence and its implications for the Mesolithic period for the Colne valley. It must be stressed, however, that many of the post-excavation analyses are still to be carried out and further findings may modify the views expressed here.

LOCATION

Uxbridge is situated at the north-western periphery of Greater London and adjacent to the river Colne, a major tributary of the Thames (Fig. 22.1). The Lower Colne valley is aligned north-south and cuts through older Thames terrace sediments which rest on Lower Tertiary deposits. North of Denham, the valley cuts through the Upper Chalk (Gibbard 1985: 82). Three Ways Wharf occupies a low-lying site at the north-western edge of Uxbridge in the valley bottom, 75 m from the present course of the river, at approximately 31 m OD.

STRATIGRAPHY

The basal deposits were of fluvial origin and consisted of Colney Street Gravels (Gibbard 1985: 82) overlain by argillaceous sediments (layer 4) deposited by gentle overbank flooding (Fig. 22.2). At the top of the sequence was a grey clay loam (layer 3) containing the *in situ* flint and faunal assemblages (Table 22.2). Overlying and sealing this horizon was a black clay (layer 2) containing large amounts of charcoal and organic material. The sequence appeared to be undisturbed with no major temporal or spatial unconformities (Collcutt, unpublished report). Above the black clay was a thin layer of locally reworked tufa (layer 1). Stratified above layer 1 was a grey/brown clay which produced finds of a possible Late Neolithic/Early Bronze Age date. The rest of the sequence was composed of Medieval and Post-medieval deposits.

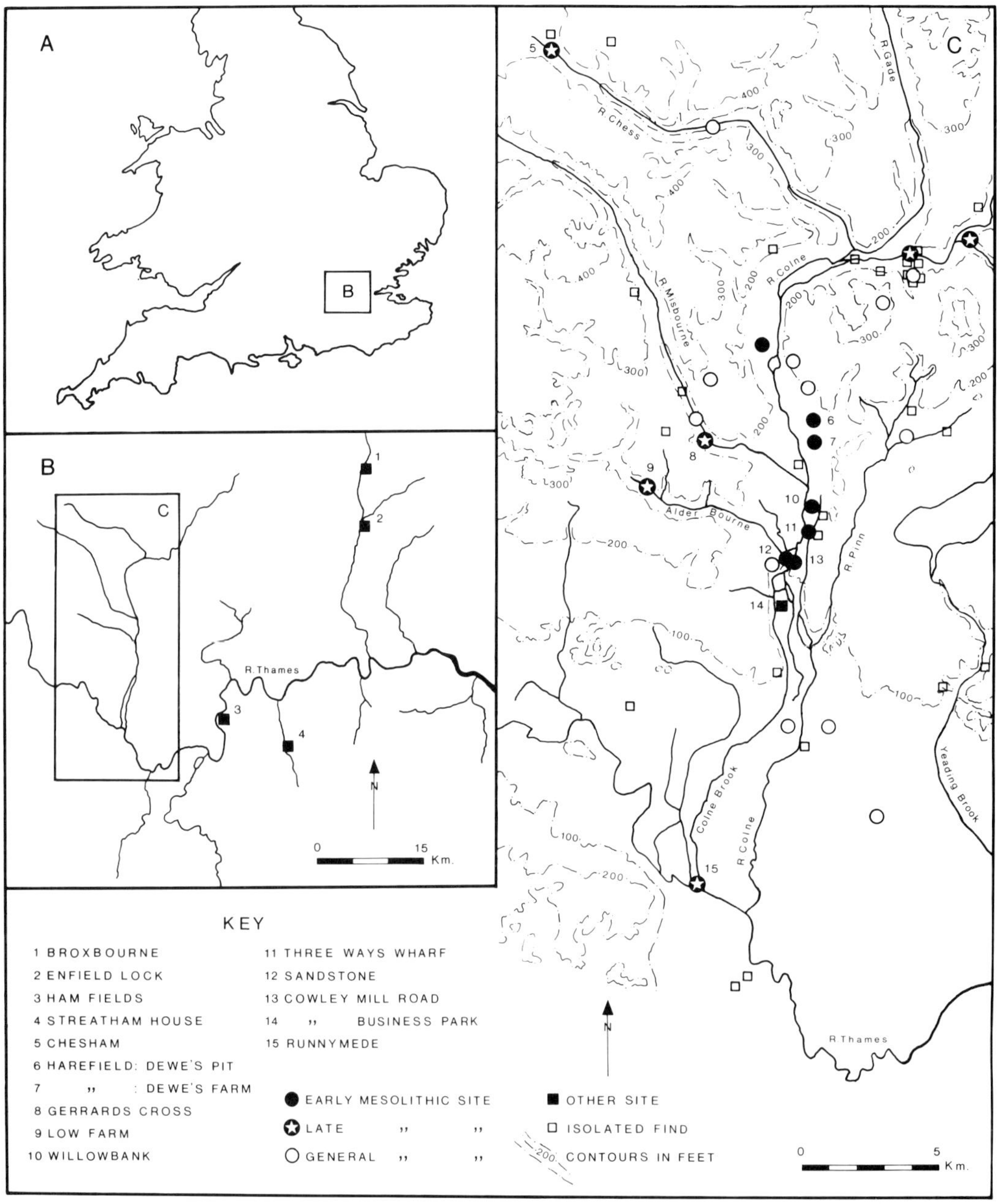

***Fig. 22.1.** Location plan showing Mesolithic sites and finds in the Colne valley and other sites mentioned in the text.*

METHODS

Excavation

An area of ca. 34 m x 24 m was reduced to a depth of ca. 1.8 m (i.e. just above layer 2). All flints larger than 10 mm and bone fragments larger than 50 mm were recorded three-dimensionally. Finds under these sizes were assigned to their respective 0.25 m square and spit. The excavated sediment (in excess of 30 tonnes) was sieved on site or as part of the post-excavation sieving programme.

Soil micromorphology

Three undisturbed soil samples were taken from a monolith located at scatter "A" (Table 22.1) through layers 4, 3 and 2 (black humic clay). Layers 3 and 2 were similarly sampled at scatter "C" by two kubiena boxes. Layers 1 and 2 from the centre of the site were also sampled to investigate the tufa and the base of the overlying later prehistoric level. A further long monolith taken from the exact location of the area which yielded pollen has yet to be studied.

Complementary samples for the bulk analysis of

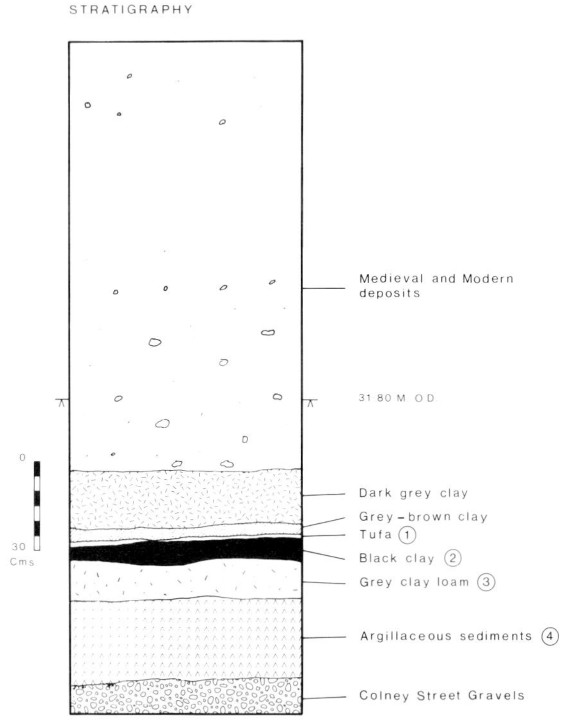

Fig. 22.2. Stratigraphic profile at Three Ways Wharf, Uxbridge.

Sample Number	Depth (cm)	Relative position of sample and corresponding thin section	Archaeological layers	
SCATTER 'A'				
1	97 104	A	Humic clay	(2)
2	107 114	B	Grey clay	(3)
3	117 124	C	Orange clay	(4)
SCATTER 'C'				
4	17 24	D	Humic clay	(2)
5	25 32	E	Grey clay Ferruginous clay	(3)
			Orange clay	(4)
CENTRE OF SITE				
6	12 19	F	Redeposited tufa	(1)
			Humic clay	(2)

Table 22.1. Location of soil micromorphological samples.

	Sample Number	Layer Number	Depth (cm)	Thin Section	Organic Carbon (%)	Calcium Carbonate (%)	Loss on Ignition (%)	Clay (%)	FZ (%)	MZ (%)	CZ (%)	Silt (%)	VFS (%)	FS (%)	MS (%)	CS (%)	VCS (%)	Sand (%)	Texture
Scatter "A:	1	2/3	97-104	A	1.5	0.7	9.6	22	29	14	12	55	7	9	5	<1	<1	23	Clay loam
	2	3/4	107-114	B	1.7	0.77	7.2	27	11	11	19	41	24	4	1	2	1	32	Clay loam
	3	4	117-124	C	8.2	0.7	11.5	25	10	10	19	39	28	4	1	2	1	36	Clay loam
Scatter "C"	4	2	17-24	D	0.08	1.6	15.7	47	11	6	10	27	17	4	2	1	2	26	Clay
	5	3	25-33	E	0.12	0.43	7.9	27	11	12	17	40	20	7	3	1	2	33	Clay loam

Table 22.2. *Soil analytical data.*

grain size, loss on ignition, calcium carbonate and organic carbon (Avery & Bascomb 1974) were also taken (Table 22.2). Undisturbed samples were air-dried, impregnated with polyester resin (Murphy 1986), made into thin sections (Guilloré 1985), described according to Bullock *et al* (1985), and interpreted using the guidelines of Courty *et al* (1989). Full details are to be found in Macphail (1990b).

Soil microfabrics are usually interpreted on a hierarchical basis. At Uxbridge, the earliest features of alluvial sedimentation and pedogenesis, relating to the Upper Palaeolithic and Early Mesolithic archaeology, needed to be differentiated from later features resulting from post-depositional biological activity such as the in-mixing of tufa material. There was also patchy sewage sludge inwash and associated vivianite neoformation from later soil contamination.

Pollen analysis

During the excavation period, a total of six sediment profiles from the site, including areas close to scatters "A" and "C" were analysed. A monolith of 15 x 15 x 50 cm was obtained from each profile, wrapped in polythene, and stored at 3 °C. Preliminary assessment showed that only one monolith of sediments (from immediately adjacent to Scatter "C") was polleniferous and only the black humic clay (layer 2) contained sufficient palynomorphs to make a study feasible; even here, they were exceedingly sparse and generally in a poor state of preservation. Pollen analysis was also carried out on black organic clay obtained from a site situated on the Colne floodplain, 1.4 km south of Three Ways Wharf at Cowley Mill Road, Uxbridge (Fig. 22.1). Here, excavations revealed small *in situ* Early Mesolithic flint scatters (I. Stewart, pers. comm.) sealed beneath a black clay similar to that found at Three Ways Wharf and the nearby site of Sandstone.

In all sections studied at Three Ways Wharf, the sediments underlying and overlying the black clay layer (layers 3 and 1) contained only the occasional grain of Gramineae (grasses) or undifferentiated spores of Filicales (ferns). Microscopic charcoal fragments were abundant in all the layers but were particularly dense within the black layer. Because the upper and lower clay horizons were barren of pollen, ten contiguous samples of 10 mm thickness were prepared from the black polleniferous layer, with samples 1 and 10 being transitional between the black layer and the adjacent clays. Each 10 mm sample consisted of a lateral spit of 20 g of sediment.

The organic, clay matrix was removed using standard treatments for soils (Dimbleby 1985: 155), and the sample suspended in 1.0 ml of glycerol jelly after staining with safranine. Slides were examined with phase-contrast microscopy at x400 magnification, and at x1000 magnification where necessary. The numbers of palynomorphs counted ranged between 638 and 2931 with an average of about 730. Because of the very large concentration of charcoal, it proved impossible to count the fragments by standard techniques (see Clark 1982: 527–31). Palynomorph nomenclature follows that of Moore and Webb (1978: 50–73).

Percentage pollen data were calculated on the basis of total pollen (TP), excluding spores from the sum. Spores were expressed as a percentage of TP and spores. No attempt was made to differentiate between *Corylus* and *Myrica* although 'Coryloid' pollen is assumed to have been *Corylus*. The diagram of selected pollen and spore taxa (Fig. 22.6) was produced using TILIA and TILIA*GRAPH programs (Grimm 1991). The diagram was zoned subjectively and divided into TWW (1) and TWW (2). Minor taxa not shown in Figure 22.6 are listed in Table 22.3.

RESULTS

Archaeology

Four spatially discrete *in situ* lithic and faunal scatters were identified:

Scatter "A" (Late Devensian – Zone III): This area produced ca. 100 fragments of bone and ca. 700 pieces of flintwork bearing some affinity to the "Long Blade" industries of the Late Devensian period (Barton 1989). There was very little post-depositional lateral movement of artefacts. The refitting flints were distributed vertically through approximately 150 mm of sediment, indicating that the scatter represented a limited phase of activity rather than successive occupation horizons (see soil micromorphology).

Taxa	Sample Number									
	1	2	3	4	5	6	7	8	9	10
Salix									0.2	0.4
Sphagnum									0.3	
Cruciferae				0.5				0.2	0.3	0.4
Caryophyllaceae							0.2		0.3	
Bidens type							0.2			
Rosaceae							0.2			
Centaurea nigra type	1.8					0.9				
Rumex undiff.				1.6	0.9					
Plantago lanceolata		2		1						
Ranunculus type				0.5						

Sample 1 = Top of Sequence (0-10 mm)

Table 22.3. *Table of minor pollen taxa.*

The fauna so far identified consisted of a horse mandible and fragments of reindeer bone. One or two pig teeth were also found. Two radio-carbon dates have been obtained from the horse material: 10,270 ± 100 BP (OxA-1778) and 10,010 ± 120 BP (OxA-1902). These dates would place scatter "A" at the end of Zone III.

Scatters "B" and "D" (undated): Scatters "B" and "D" may have originally formed one occupation area but, unfortunately, a now backfilled canal dock has destroyed the relationship between them. The flintwork recovered from both was low in density and undiagnostic.

Scatter "C" (Early Flandrian): The fourth scatter yielded about 7000 pieces of struck flint and about 2000 fragments of bone, some of which had been charred. It would appear that the flint and faunal material had undergone less lateral and vertical movement than the artefacts in scatter "A" had experienced. Preliminary refitting suggests that the scatter represents a limited phase of activity, but further work may suggest otherwise (see soil micromorphology). Burnt flint produced a thermoluminescence date of 8000 ± 800 BP (OxTL 772f).

The lithic assemblage is of Early Mesolithic type whilst the associated faunal remains are dominated by those of red deer. No horse or reindeer material has so far been identified in this scatter.

Interpretation

The occurrence of mammal bone and flint artefacts in discrete scatters, and the presence of both burnt flint and burnt bone among the finds, allows the inference that assemblages "A" and "C" represent the product of hunting and food processing. A topographic study of the site has shown that these scatters were situated on a low bar or "island" within a possibly braided river system (A. Boucher, pers. comm.). The small amount of post-depositional lateral and vertical movement of the artefacts may have been caused by bioturbation.

Dating

At the moment it is difficult to be precise about the length of time which elapsed between the Late Devensian and Early Flandrian occupations (scatters "A" and "C"). It is proposed to obtain a comprehensive series of radiocarbon dates from the faunal remains from scatters "A" and "C", as well as from the black organic layer (2). It is hoped that these will quantify more precisely the chronological difference between Scatters "A" and "C". However, recent work has suggested that due to atmospheric radiocarbon variations around 10,000 BP and 9,500 BP, it may not be possible to achieve this aim (Becker and Kromer: 1991).

Nevertheless, since no artefacts were recovered from layer 2, it would seem that a site which had been sporadically utilised in the Late Devensian /Early Flandrian ceased to be utilised with the development of the black organic layer (2) in the early Boreal (see below).

Soil micromorphology

Scatter "A": The upward fining sequence of alluvium over the gravels comprises some 300 mm of moderately calcareous, coarsely mottled, yellowish brown (10YR5/6) clay loam (Table 22.2, samples 2, 3), containing chalk clasts and mollusc fragments, with some 6 cm of finely mottled, light yellowish-brown (2.5Y6/4), mainly non-calcareous clay loam sediment above (cf. Table 22.2, sample 5, scatter "C"). These are layers 4 and 3 respectively and flint artefacts have been found ranging through layer 3 into the top of layer 4. Colcutt (unpublished report) suggests that the gravel body had produced a dissected topography across an otherwise flat river valley, allowing the low energy deposition of silts (40% silt in the clay loam), this sediment being abandoned by the active stream as it migrated south-eastwards. The sediments (Table 22.1, thin section C) contain evidence of both calcium carbonate depletion (weathered chalk fragments, fine fabric depleted of calcium carbonate), and flushing by calcium carbonate enriched water (micritic pseudomorphic impregnation of roots). These features which are typical of Late Devensian sediments, could relate to the rapid decalcification of sediments under cool and high water flow conditions (Catt 1979). The sediment seems to have been homogenised by biological activity and no microfabrics related to sedimentary structures or cryogenic processes remain. Thus, during the period of Late Devensian occupation, the site was probably damp, but vegetated and the soil biologically active. The upper part of the soil (Table 22.1, thin sections A & B) retains only poor microfabric evidence of later microstructure, which is quite comparable to the soil fabrics extant during the Early Mesolithic activity at scatter "C". In any case, biological activity (over an as yet unknown length of time) is assumed to have continued up to and until increased site wetness caused a valley "swamp" to form during the Early Mesolithic. This conjecture is supported by the "microdebitage sort assessment" (Holman unpublished) which shows the Late Devensian material to be mixed throughout the artefact-bearing zone, whereas the Early Mesolithic material (scatter "C") has been found predominantly in the top two 5 cm thick spits.

Scatter "C": In the field, Layer 3 was divided into the archaeological sub-layers of an upper "grey clay" and a lower "ferruginous clay". Here, at the site of Early Mesolithic activity, evidence of pedological activity in an otherwise iron and clay depleted microfabric, is differentially preserved. This preservation results from strong iron and manganese impregnation of features caused by later hydromorphism (Bouma *et al* 1990), and some artefacts in the "ferruginous clay" had iron concretions on them. The impregnation has preserved a soil (Table 22.1, thin section E and Fig. 22.3) with a fine subangular and channel microstructure that was perforated by medium to coarse roots and well worked by earthworms. This biological activity was present throughout the artefact-bearing zone and suggests a lowering of the water-table at this typical alluvial gley soil (Avery 1980) during the Early Flandrian. This must have had an impact on the present distribution of bone and artefacts, especially those dating to the Late Devensian occupation. Even though near to the stream, this site can be considered to have been fully terrestrial

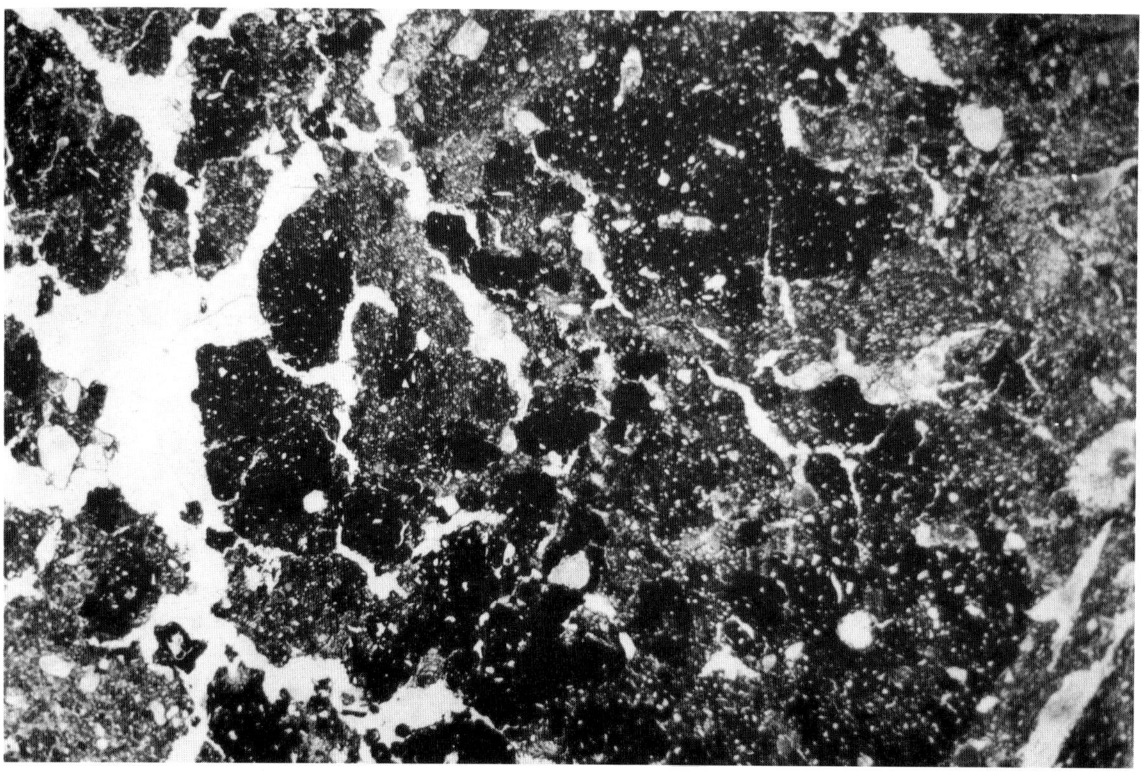

Fig. 22.3. Layer 3, scatter "C", part of thin section E: Biological open soil fabric preserved by iron and manganese impregnation at around 6 cm depth in buried soil. Biological activity in the Early Flandrian soil was responsible for vertically sorting Upper Palaeolithic material, whereas Early Mesolithic material was only partially reworked before a rise in water-table caused hydromorphic conditions and iron and manganese impregnation. Microfiche reader view, longest frame is 1.1 cm.

at this time and may well have had a cover of vegetation which could root successfully in shallowed ripened sediments (Bal 1982). The soil itself contains rare fine charcoal that is believed not to be contamination from the overlying charcoal-rich Layer 2, but may be of Early Mesolithic origin or earlier. In addition, some of the clay coatings and infills affecting the biological porosity may possibly relate to surface soil disturbance brought about by the trampling of humans on the site (Courty *et al* 1989: 124–5, fig. 17.d). In contrast, pale dusty clay inwash into the soil more likely relates to increased wetness on the site.

Increased wetness of the site

Preliminary assessment of the distribution of lithic artefacts in scatter "C" showed that the majority were present in the upper two 5 cm spits of the artefact-bearing zone. Some actually rested on the top of the "grey clay" of layer 3, with their upper surfaces being stained black by the overlying humic clay (layer 2). This may be accounted for by assuming occupation(s), possibly up to the time of the rise in water-table. Only a small portion of the bone and flint was worked down the profile because the increasing wetness curtailed biological activity.

A rise in water-table led to gleying and caused mottling; relic roots were pseudomorphically replaced with iron and manganese. Similar features relating to a rising water-table can be cited from Swanscombe, Kent (Kemp 1985). Soil fabrics in general across the Uxbridge site became depleted of iron and lost their structure (Table 22.1, thin section D and Fig. 22.4). The uppermost 6 cm were most strongly affected by hydromorphic iron and clay depletion and loss of structure on slaking (Bouma *et al* 1990). The deflocculated or dispersed soil caused clay and fine silt to be washed down the profile, and this material probably forms the pale dusty coatings that affect the relic biological structure lower down. Similar effects (slaking, loss of structure, formation of pseudo-layering) on soils caused by increased wetness have been recorded in Neolithic palaeosols in Essex (river Blackwater estuary: Macphail 1990a) and also from soils which had been occupied in the Lower Palaeolithic at Boxgrove, West Sussex (Goldberg and Macphail 1990). Because of these post-depositional transformations, layer 3 (the Mesolithic soil) is now leached and grey in the upper part of the horizon and ferruginous in the lower part. In addition, the uppermost one to two centimetres of layer 3 contain iron-replaced laminae of organic matter (Fig. 22.4), that are considered to relate, not to *in situ* sedimentary features, but to wetting-front phenomena i.e. when organic-rich swamp water infiltrated into the structureless topsoil. Again, similar features are recorded in Essex (Macphail 1990a). All these effects (slaking and loss of structure,

Fig. 22.4. *Layer 3, scatter "C, part of thin section D: Near the surface of the buried soil, a massive/structureless soil is now present as the result of site wetness causing soil structures to collapse through slaking. Soil is also depleted in iron. Note very thin iron replaced organic laminae, which are not sedimentary features, but result from the diffusion of organic matter during the deposition of the humic clay above, as a wetting front effect. Microfiche reader view, longest frame is 1.1 cm.*

and formation of pseudo-layering), and rooting activity of plants, can make it very difficult to identify junctions between buried soils and overlying alluvium.

Charcoal

At Scatter "A", layer 2 contained laminated microscopic charred material (Table 22.2, sample 1). In the area of scatter "C", layer 2 (Table 22.1, thin section D) was essentially a weakly banded, moderately organic clay, with 47% clay and 1.6% organic carbon (Table 22.2, sample 4). It contained much microscopic charcoal as well as charred monocotyledonous epidermis and amorphous organic debris. Indeed, at both scatters "A" and "C", layer 2 contained so much fine charred organic matter that some laminae were almost opaque. In the pollen section the prepared samples showed that although charcoal was exceedingly abundant throughout the layer, it appeared to be concentrated in three particularly dense bands from 10–20 mm, 40–60 mm and 80–90 mm (see Fig. 22.5). However, it must be remembered that the sampling method obscured its true distribution, with the charcoal laminae (revealed by the micromorphology) inevitably being bulked together in each pollen sample.

The size of fragments found in all contexts ranged from <100 μm^2 to >5000 μm^2 so that both microcharcoal and macrocharcoal (sensu MacDonald *et al* 1991: 57–9) were present, although it must be said that the great majority of fragments fell into the microcharcoal size classes and this may be due to the fact that preparation methods would inevitably break up fragile particles.

Pollen

The samples from Cowley Mill Road yielded insufficient palynomorphs for counting but the assemblage of pollen and spore taxa was very similar to that from Three Ways Wharf.

The results from Three Ways Wharf are shown in the pollen diagram (Fig. 22.5) and table of minor taxa (Table 22.3). The patterns observed in the pollen curves strongly suggest that the black humic clay has stratigraphic integrity and it is assumed, therefore, that the pollen record reveals vegetation change in the environs of the site.

When the pollen spectra are examined, it is obvious that species richness was low, with only twenty-two

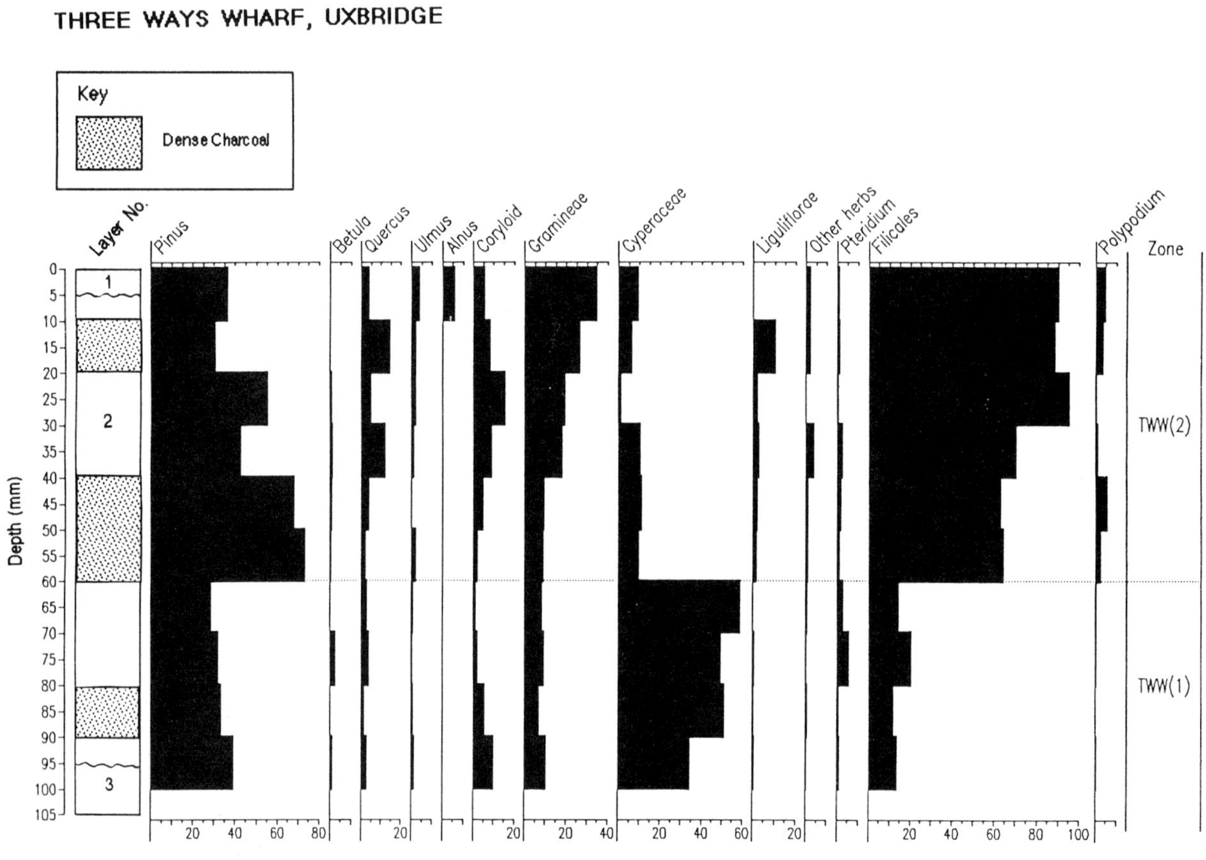

Fig. 22.5. Diagram of selected pollen and spore taxa: Pollen expressed as a percentage of total pollen (TP) excluding spores. Spores expressed as a percentage of total pollen plus spores.

taxa being recorded in the black clay layer. Differential preservation is, therefore, an important factor to be kept in mind during interpretation of the data (Havinga 1964; Havinga 1967; Havinga 1971; Faegri 1971; Elsik 1971; Correia 1971). Much of the observational and experimental evidence for differential susceptibility to decay of pollen and spores is contradictory so that only generalisations can be made. Broadly, the breakdown and disappearance of palynomorphs appears to be related to the oxidation susceptibility and the sporopollenin content of the exine, as well as the redox potential and pH of accreting sediments. The area from which the polleniferous sediment was obtained must, therefore, have offered a different physico-chemical environment from that of the other areas of the site which were studied. It probably remained waterlogged and anaerobic throughout the period during which the black humic clay was deposited whereas elsewhere, the site may have experienced drier, more aerobic conditions, at least periodically.

In spite of the obvious influence of differential preservation, the pollen of supposedly vulnerable taxa such as elm and oak were present in most samples and increased in frequency in the upper part of the sequence along with taxa such as Filicales and *Polypodium* which are considered to be very resistant. It is assumed, therefore, that real vegetational changes are reflected in the pollen diagram rather than simply the effects of differential preservation.

Zone TWW(1) includes the transition between the underlying clay and the black humic layer and, in view of the micromorpholgical evidence and the nature of the pollen spectra, a shallow-water, open environment is suggested. The immediately local vegetation seems to have been dominated by Cyperaceae (sedges) and Gramineae (grasses) possibly indicating sedge/reed swamp, and other herbaceous taxa (see Table 22.3) attest to the open-ness of the site. *Pinus* (pine) was growing in the region along with relatively small numbers of deciduous, temperate taxa such as *Quercus* (oak), *Corylus* (hazel), *Ulmus* (elm) and *Betula* (birch), with Filicales (undifferentiated ferns) possibly being an important component of the ground flora. *Salix* (willow) was also found in the basal two samples. A band of very dense charcoal was found between 80–90 mm and this appears to coincide with a rise in sedges and a fall in other taxa.

The boundary between Zones TWW(1) and TWW(2) is drawn at 60 mm. This coincides with a very dense band of charcoal extending up to 40 mm and also apparently large changes in the pollen spectra. Pine and ferns [excluding bracken but including *Polypodium* (polypody fern)] increase very markedly along with slight rises in elm, hazel, grasses and other herbaceous taxa (although it is difficult to assess the degree of significance of these very small changes in the spectra). Sedges decrease dramatically along with slight declines in bracken and oak. The rest of Zone TWW (2) is characterised by continued low levels of sedges and high levels of ferns. While there is an erratic but sustained decline in pine above the charcoal band, most other taxa show increased values above the charcoal band and a notable appearance was that of *Plantago lanceolata* (ribwort plantain). Another dense charcoal band occurs between 10–20 mm but is not accompanied by very marked changes in the pollen. *Alnus* (alder) was found in only the uppermost sample between 0–5 mm.

Interpretation

Interpretation of the Uxbridge pollen diagram was greatly aided by comparing it with one recently produced by Chambers & Mighall (1991) from Enfield Lock, where the sediments comprised a similar sequence to those at Three Ways Wharf (Bedwin 1991: 162). At Enfield, 'silty clay' was overlain by 50 cm of 'organic mud' which, itself, was overlain by 'shelly marl'. It is thought that the 'organic mud' at Enfield is homologous to the 'black humic clay' (layer 2) in the Uxbridge sequence.

At Enfield Lock, the upper 5 cm of the 'silty clay' and the whole of the 'organic mud' horizon was subjected to detailed pollen analysis, preservation being far superior to that found at Three Ways Wharf or Cowley Mill Road, Uxbridge. A radiocarbon date of 9546 ± 56 BP (UB-3350) was obtained for the basal 2 cm of 'organic mud' at Enfield Lock. This was corroborated by the pollen spectra, so Chambers and Mighall considered that deposition of the organic sediments started in the early Flandrian [pre-Boreal and first part of the Boreal (*sensu* Godwin 1940: 373–4)]. The pollen spectra, and the appearance of alder pollen, strongly suggested a late Boreal date for the sediment just below the marl.

There are differences in the pollen assemblages at Three Ways Wharf and Enfield Lock, for example, birch, willow and grasses were much more abundant, and ferns less abundant at the latter site. These are probably functions of differences in local vegetation, but also of pollen preservation, since the range of taxa found at Uxbridge was depauperate when compared with that of Enfield Lock. However, the main pattern of the pollen curves and the charcoal distribution for both sites are similar.

In the Three Ways Wharf diagram the exaggerated changes in the pollen spectra [particularly at the boundary of Zones TWW(1) and TWW(2)] appear to be associated with the very great increase in charcoal density in the sediment. At Enfield Lock, high charcoal counts were obtained throughout the lower part of the sequence. The rise in pine and Filicales and the drop in sedges, occur at the same level as the very highest charcoal records. These changes are gradual when compared to those at Uxbridge. The very marked variations at Three Ways Wharf may be apparent rather than real and are probably artefacts of sampling. At Enfield the sediments are much thicker, the polleniferous organic muds being 50 cm deep as opposed to the 10 cm (and elsewhere on the site, 4 cm) of black humic silt at Three Ways Wharf. The results from Three Ways Wharf can be readily understood as a crude and telescoped version of those from Enfield Lock.

At both sites, an open, swampy habitat with regional pine woodland prevailed in the early to mid Boreal with frequent and/or intense burning occurring in the area. Pine woodland spread at the expense of open habitat plants and, in the late Boreal, pine declined as the succession of temperate, deciduous woodland proceeded. At both sites there was again evidence of high levels of burning in the uppermost levels of the polleniferous sediments.

Until radiocarbon dates have been obtained, the chronology of the black clay (layer 2) must rely on pollen evidence alone. At Three Ways Wharf, in spite of the consistently low values for birch and relatively low levels for hazel, the pollen spectra point to a Boreal landscape during the period represented by the black humic silt. In Zone TWW (1), the dominance of pine and the low values of deciduous trees imply the early Boreal, possibly Zone V/VIa (sensu Godwin 1940), while in Zone TWW (2), the increasing values for oak, elm and hazel and the appearance of alder suggest a landscape characteristic of the later Boreal, Zones VIb and VIc. The appearance of alder in the uppermost level certainly points to the late Boreal (Devoy 1979: 139; Chambers & Mighall 1991: 14–15). The pollen assemblages from Cowley Mill Road and the nearby site of Sandstone, Iver (Lacaille 1963), also indicate that the black deposit from these sites is of Boreal age.

DISCUSSION

The wider implications of the Three Ways Wharf site have been discussed elsewhere (Lewis 1991: 254). However, in a local context the site can be seen as a Late Devensian and Early Flandrian landscape preserved beneath alluvial deposits.

Micromorphology at Three Ways Wharf has shown that this occupation occurred on mature soils and not on sediments. Lacaille has drawn attention to the stratigraphic sequence of the sites at Harefield, Willowbank, and Sandstone (Lacaille 1961, 1963). At all these sites, the sequence consisted of Early Mesolithic artefacts resting on gravel or alluvial clay and overlain by peat or organic clays. More recently this sequence has been observed at Cowley Mill Road (see above), whilst further downstream, at Cowley Business Park (Fig. 22.1), a similar black organic layer overlying gravels was recorded, though in this case no artefacts were recovered (J. Bennet, pers.comm.). Similar sites have been noted in other tributary valleys of the Thames (Lacaille 1966), though at Thatcham in the Kennet Valley (Wymer 1962), Mesolithic artefacts were already buried beneath algal marl sediments by the time peat developed in Zone VIIa (Churchill 1962: 369). Although no artefacts were recovered, excavations by the Museum of London at Streatham House, Merton, in the Wandle Valley, revealed a black organic layer overlying gravel which gave a radiocarbon date of 9423 ± 72 BP (UB-339) (D. Saxby, pers. comm.).

The well known series of sites at Broxbourne in the Lea Valley were sealed by late Boreal peats (Warren *et al* 1934). More recent excavations at this locality by Jacobi recovered artefacts from beneath a Zone V peat (in Wymer 1977: 129). Lower down the Lea Valley at Enfield Lock a similar sequence of sediments to those at Three Ways Wharf were recorded (as described above), though no artefacts were recovered (Bedwin 1991: 162).

The formation of Early Flandrian peats has often been ascribed to a rise in sea level following the end of the Devensian glaciation (e.g. Lacaille 1966: 4–5; Holgate 1988: 23–4). Devoy (1980: fig 59) has suggested that in the Thames estuary, sea levels rose from -27 m OD to -15 m OD between about 8300 and 8000 BP. Rapid rises such as this must have had an effect on the hydrology of the Thames tributaries and it is difficult not to equate the rising water-table at Three Ways Wharf and subsequent swamp development with a rise in river base level caused by rising sea levels.

The effect of these factors on the Mesolithic economy and settlement patterns is open to debate; we may expect the abandonment of sites in the lower reaches of the valley bottoms and the shift to new sites further upstream in the main valley and its tributaries, or onto the valley sides. On the basis of faunal evidence, Carter (1976: 3) has proposed this model for the Kennet Valley and Holgate (1988: 110) has reviewed the model in the wider context of the Thames Basin.

At Three Ways Wharf, a location which had been utilised over a considerable length of time, seems to have been abandoned with the increased site wetness and formation of the black humic clay (layer 2). In the Lower Colne valley, the majority of sites seem to date to the Early Mesolithic; in contrast, work in the tributaries of the Colne has shown a preponderance of Late Mesolithic sites (Fig. 22.1). For example, the site at Low Farm, Fulmer in the bottom of the Alderbourne Valley produced rod and crescent type microliths (Farley 1978: 608). At Gerrards Cross, Late Mesolithic artefacts were recovered from the lower part of a tufa deposit (Farley 1983: 169), whilst excavations at Chesham in the Chess Valley (Stainton 1989) again produced a Late Mesolithic assemblage, which was possibly redeposited in a soil developed in the Neolithic or Bronze Ages (Farley 1989: 227). This distribution might be seen as supporting the conjectured shift towards higher areas upstream and in the main tributary valleys in the Late Mesolithic. If so, then the presence of charcoal within the organic layers at Three Ways Wharf, Cowley Mill Rd and Enfield Lock may be linked with this change.

The incidence of fire in prehistory, its causes and effects, its correlation with charcoal concentrations in sediments along with observed changes in pollen spectra in pollen diagrams, have long been subjects of debate. The pertinent literature has been cogently reviewed by Bennett et al (1990) and MacDonald *et al* (1991). There are certainly problems in the interpretation of the particle size and abundance of charcoal within sediments in relation to areal extent and frequency of fires in prehistory. But Bennett *et al* (1990: 1 & 6) have argued that the role of fire in British vegetation history cannot be compared with that in North America where large-scale natural burning of woodlands has been commonplace. They interpret the Mesolithic charcoal record as stemming from intensive occupation beside the water's edge rather than a record of natural or man-induced burning of the Boreal forest.

The proximity and extent of the fires are difficult to assess at both Three Ways Wharf and Enfield Lock; whether the charcoal bands were due to local campfires or widespread burning of the pine forest cannot be elucidated. However, the excellent assemblage of artefacts at Uxbridge, including burnt flints and charred bone, attest to a strong Mesolithic presence and the use of fire in the immediate vicinity before the swamp started to form. Furthermore, as already stated, the charred epidermal fragments and amorphous organic material in the black layer indicate very local burning after apparent abandonment of the immediate vicinity. But, the records of charcoal in homologous sediments at such well separated sites as Three Ways Wharf, Cowley Mill Road and Enfield Lock would also suggest that more widespread burning may have occurred at least at some time during the deposition of the black organic layer.

Whilst the theory of a change in settlement pattern as a response to peat formation may be attractive, there are a number of factors which should not be overlooked:

1. The extent of the peat deposits within the Colne valley or those of other Thames tributaries has not been mapped in detail. It is, therefore, unclear if these deposits covered large areas of the valley bottoms or whether they were much more local in extent.
2. The chronological development of these deposits has not been examined and, without a dating programme, it is difficult to reconstruct their developmental sequence.
3. The development of peat deposits may simply have led to a change in the nature of exploitation of a particular area, for example, at Three Ways Wharf, the site may no longer have been used as a kill or processing site but the swamp resources such as reeds, sedges and fowl may have been used. As Clarke (1978: 20) pointed out, swamps and marshes are nutritionally very productive.
4. Little fieldwork has been carried out in the valley bottoms, let alone the surrounding valley sides or the interfluves. The number of sites and finds particularly in the Greater London area is, at present, too small to allow firm conclusions to be drawn.
5. Late Mesolithic implements have been recovered from a palaeo-ground surface at Runnymede at the mouth of the Colne Brook (S. Needham, pers. comm. and this volume), whilst an assemblage recovered from near the Thames at Ham Fields, Surrey, contains several later Mesolithic microliths (Lacaille 1966: 27). These finds and others indicate that Late Mesolithic occupation occurred in the lower reaches of tributary valleys and on the Thames floodplain.

Until these problems are addressed, it is difficult to assess the impact of the development of peats in the valley bottoms on the Early Mesolithic economy.

CONCLUSIONS

The site at Three Ways Wharf was utilised a number of times from the Lateglacial and into the Early Mesolithic, perhaps indicating that the gravel bar or island was a favoured location for specific activities. It is clear from micromorphological studies that these occupations took place on ripened soils and not on sediments.

A rise in water-table led to the abandonment of the site and the formation of a sedge swamp. Palynology records the succession from relatively open conditions, with the local vegetation being dominated by sedge swamp, through Boreal pine-dominated forest to temperate, deciduous woodland. Bands of charcoal have been recorded from organic deposits at a number of sites, though there is no evidence of a relationship between charcoal concentrations and vegetation change. However, the apparent widespread formation of peats and swamps in the bottoms of Thames tributaries may have had an effect on Mesolithic subsistence strategies, which may be reflected in the occurrence of charcoal in these deposits. At the moment, this theory can not be examined fully due to insufficient fieldwork. Nevertheless, the alluvial deposits of the Thames and its tributaries hold great potential, and it is hoped that further multidisciplinary projects such as that carried out at Three Ways Wharf will provide a clearer insight into man's interaction with the environment and landscape in the Lateglacial and Mesolithic periods.

Acknowledgements

We are grateful to Dr Kevin Edwards for drawing the pollen diagram (Fig. 22.5) and to Dr F.M. Chambers and Mr T. Mighall for allowing access to his unpublished report on the sediments from Enfield Lock. Thanks are also extended to Mrs J. Brigham for analyses of bulk samples, the Institut National Agronomique, Paris-Grignon for thin section manufacture, and Dr Marie-Agnès Courty for discussion of the thin sections. We would also like to thank Mr James Rackham for bone identification and all staff of the Museum of London and volunteers of the West London Archaeological Field Group who worked on the site, on the sieving programme and on post-excavation processing and analysis. This work was funded by Trafalgar Brookmount plc, Burton Property Trust Ltd and English Heritage.

References

Avery, B.W. and Bascomb, C.L. 1974. *Soil Survey Laboratory Methods*. Soil Survey Monograph 6, Harpenden.

Avery, B.W. 1980. *Soil classification for England and Wales*. Soil Survey Monograph 14, Harpenden.

Bal, L. 1982. *Zoological Ripening of Soils*. Centre for Agricultural publishing and Documentation, Wageningen, Holland.

Barton, R.N.E. 1989. Long blade technology in southern Britain. In C. Bonsall (ed.), *The Mesolithic in Europe. Proceedings of the Third International Symposium, Edinburgh 1985*, 246–71. Edinburgh: John Donald.

Becker, B. and Kromer, B. 1991. Dendrochronology and radiocarbon calibration of the early Holocene. In R.N.E. Barton, A.J. Roberts, and D.A. Rowe (eds.), *Late Glacial Settlement in north-west Europe*, 22–4. London: Council for British Archaeology Research Report 77.

Bedwin, O. 1991. An early flandrian peat in the Lea Valley; excavations at the former Royal Ordnance Factory, Enfield Lock 1990. *Essex Archaeology and History*, 22, 162–3.

Bennett, K.D., Simonson, W.D. and Peglar, S.M. 1990. Fire and Man in Post-Glacial Woodlands of Eastern England. *Journal of Archaeological Science*, 17, 1–8.

Bouma, J., Fox, C.A. and Miedema, R. 1990. Micromorphology of hydromorphic soils: application for soil genesis and land evaluation. In L. Douglas (ed.), *Developments in Soil Science 19*, 257–78. Amsterdam: Elsevier.

Bullock, P., Fedoroff, N., Jongerius, A., Stoops, G. and Tursina, T. 1985. *Handbook for Soil Thin Section Description*. Wolverhampton: Waine Research Publications.

Carter, H.H. 1976. Fauna of an area of Mesolithic occupation in the Kennet Valley considered in relation to contemporary eating habits. *Berkshire Archaeological Journal*, 68, 1–3.

Catt, J.A. 1979. Soils and Quaternary geology in Britain. *Journal of Soil Science*, 30, 607–647.

Chambers, F.M. and Mighall, T. 1991. Palaeoecological investigations at Enfield Lock: Pollen, pH, Magnetic Susceptibility and Charcoal Analyses of Sediments. Unpublished report, Environmental Research Unit, University of Keele.

Churchill, D.M. 1962. The stratigraphy of the Mesolithic sites III and IV at Thatcham, Berkshire, England. *Proceedings of the Prehistoric Society*, 28, 362–370.

Clarke, D. 1978. *Mesolithic Europe: The Economic Basis*. London: Duckworth.

Clark, R.L. 1982. Point count estimation of charcoal in pollen preparations and thin sections of sediments. *Pollen et Spores*, 24, 523–535.

Collcutt, S.N. 1988. Untitled and unpublished report on the sediments studied in the field at Three Ways Wharf, Uxbridge.

Courty, M.A., Goldberg, P. and Macphail, R.I. 1989. *Soils, Micromorphology and Archaeology*. Cambridge: Cambridge University Press.

Correia, M. 1971. Diagenesis of sporopollenin and other comparable organic substances: application to hydrocarbon research. In J. Brooks, P.R. Grant, M. Muir, P. van Gijzel, and G. Shaw (eds.), *Sporopollenin*, 569–620. London: Academic Press.

Dimbleby, G.W. 1985. *The Palynology of Archaeological Sites*. London: Academic Press.

Devoy, R.J.N. 1979. Flandrian sea level changes and vegetational history of the Lower Thames Estuary. *The Philosophical Transactions of the Royal Society of London*, 285, 355–407.

Devoy, R.J.N. 1980. Post-glacial Environmental Change and Man in the Thames Estuary: A Synopsis. In F.H. Thompson (ed.), *Archaeology and Coastal Change*, 134–148. London: Society of Antiquaries.

Elsik, W.C. 1971. Microbiological degradation of sporopollenin. In J. Brooks, P.R. Grant, M. Muir, P. van Gijzel, and G. Shaw (eds.), *Sporopollenin*, 480–511. London: Academic Press.

Faegri, K. 1971. The preservation of sporopollenin membranes under natural conditions. In J. Brooks, P.R. Grant, M. Muir, P. van Gijzel, and G. Shaw (eds.), *Sporopollenin*, 256–72. London: Academic Press.

Farley, M. 1978. Excavations at Low Farm, Fulmer, Bucks. 1: The Mesolithic Occupation. *Records of Buckinghamshire*, 20(4), 601–616.

Farley, M. 1983. Archaeological notes from Buckinghamshire County Museum. *Records of Buckinghamshire*, 25, 169.

Farley, M. 1989. Archaeological notes from Buckinghamshire County Museum. *Records of Buckinghamshire*, 31, 226–227.

Gibbard, P.L. 1985. *The Pleistocene History of the Middle Thames Valley*. Cambridge: Cambridge University Press.

Godwin, H. 1940. Pollen analysis and forest history of England and Wales. *New Phytologist*, 39, 370–400.

Goldberg, P. and Macphail, R.I. 1990. Micromorphological evidence of Middle Pleistocene landscape and climatic changes from southern England: Westbury-sub-Mendip, Somerset and Boxgrove, West Sussex. In L.A. Douglas (ed.), *Soil Micromorphology*, 441–8. Amsterdam: Elsevier Press.

Grimm, E. 1991. *Tilia and Tilia*Graph*. Illinois State Museum, Springfield, Illinois.

Guilore, P. 1985. *Méthode de Fabrication Méchanique et en Série des Lames Minces*. Paris: Institut National Agronomique.

Havinga, A.J. 1964. Investigation into the differential corrosion susceptibility of pollen and spores. *Pollen et Spores*, 6(2), 621–635.

Havinga, A.J. 1967. Palynology and pollen preservation. *Review of Palaeobotany and Palynology*, 2, 81–98.

Havinga, A.J. 1971. An experimental investigation into the decay of pollen and spores in various soil types. In J. Brooks, P.R. Grant, M. Muir, P. van Gijzel, and G. Shaw (eds.), *Sporopollenin*, 446–79. London: Academic Press.

Holgate, R. 1988. *Neolithic settlement of the Thames Basin*. Oxford: British Archaeological Reports (British Series) 194.

Holman, G.R. 1989. The microdebitage sorting assessment from Three Ways Wharf, Uxbridge. Museum of London: Unpublished report.

Kemp, R.A. 1985. The decalcified Lower Loam at Swanscombe, Kent: a buried Quaternary soil. *Proceedings of the Geological Society*, 96, 343–354.

Lacaille, A.D. 1961. Mesolithic Facies in Middlesex and London. *Transactions of the London and Middlesex Archaeological Society*, 20(3), 101–149.

Lacaille, A.D. 1963. Mesolithic Industries beside Colne Waters in Iver and Denham, Buckinghamshire. *Records of Buckinghamshire*, 17(3), 143–181.

Lacaille, A.D. 1966. Mesolithic Facies in the Transpontine Fringes. *Surrey Archaeological Collections*, 63, 1–43.

Lewis, J.S.C. 1991. Excavation of a Late Glacial and Early Flandrian site at Three Ways Wharf, Uxbridge: Interim report. In R.N.E. Barton, A.J. Roberts, and D.A. Rowe (eds.), *Late Glacial Settlement in north-west Europe*, 246–55. London: Council for British Archaeology Research Report 77.

MacDonald, G.M., Larsen, C.P.S., Szeicz, J.M. and Moser, K.A. 1991. The reconstruction of Boreal forest fire history from lake sediments: a comparison of charcoal, pollen, sedimentological, and geochemical indices. *Quaternary Science Reviews*, 10, 53–71.

Macphail, R.I. 1990a. Soil report on the Hullbridge Project: the sites of Purfleet, the Stumble and others on the Blackwater River. Unpublished Ancient Monuments Laboratory Report 39/90. English Heritage.

Macphail, R.I. 1990b. Soil report on Three Ways Wharf, Oxford Road, Uxbridge, Middlesex. Unpublished Ancient Monuments Laboratory Report 27/91. English Heritage.

Moore, P.D. and Webb, J.A. 1978. *An Illustrated Guide to Pollen Analysis*. London: Hodder & Stoughton.

Murphy, C.P. 1986. Thin Section Preparation of Soils and Sediments. Berkhamsted: AB Academic Publishers.

Rheineck, H.E. and Singh, I.B. 1986. *Depositional Sedimentary Environments*. Berlin: Springer-Verlag.

Stainton, B. 1989. Excavation of an early prehistoric site at Stratford's Yard, Chesham. *Records of Buckinghamshire*, 31, 49–74.

Warren, S.H., Clark, J.G.D., Godwin, H., Godwin, M.E. and Macfadyen, W.A. 1934. An early Mesolithic site at Broxbourne sealed under Boreal peat. *Journal of the Royal Anthropological Institute*, 64, 101–128.

Wiltshire, P.E.J. 1989. Interim report on the microscopic analysis of buried sediments associated with Palaeolithiic and Mesolithic artefacts at Uxbridge, Middlesex. Unpub. Ancient Monuments Laboratory Report 18/89. English Heritage.

Wymer, J.J. 1962. Excavations at the Maglemosian sites at Thatcham, Berkshire, England. *Proceedings of the Prehistoric Society*, 28, 329–361.

Wymer, J.J. 1977. *Gazeteer of Mesolithic Sites in England and Wales*. London: Council for British Archaeology Research Report 20.

23

Holocene alluviation and interstratified settlement evidence in the Thames valley at Runnymede Bridge

Stuart Needham

Summary: *Rescue and research excavations at Runnymede Bridge have provided not only a detailed insight into two prehistoric settlement phases, but also a wealth of topographic and environmental evidence on the development of the local floodplain and its river system. The alluvial history of the site, as understood from deep cuttings, is outlined in terms of alluvial parcels and the stratigraphic positions of human evidence are set out. Sedimentary bodies of varied origins and character are relatively well dated by a combination of cultural debris, radiocarbon, palaeomagnetic and thermoluminesence dating, backed up by pollen evidence. The sequence reconstructed can be seen to represent most parts of the Holocene.*

Attention is focussed on different trends in the type of alluviation and implications for the evolution of the channel system, rates of migration and rates of flooding. An anastomosed system is regarded as likely through much of the Holocene, with a probable trend towards fewer, better defined channels with time. Beaver activity in the catchment is considered along with human interference as potential agent of destabilisation of the ecology and enhanced alluviation. The particular survival/loss balance and alluvial parcel interrelationships found at Runnymede are explained in terms of the alluvial history experienced and the position of the site along the valley. Flood frequency and severity are discussed with particular reference to a high-magnitude event dating to circa 4000 BP. Other evidence for the Lower Thames basin is drawn in to try to differentiate localised phenomena from wider alluvial patterns such as the pre late Boreal rise in water levels, or the Thames IV silting in the lower reaches of the river. This approach is, however, limited at present by a shortage of well-dated sequences in the region.

Footnote: *All calibrated radiocarbon ranges are based on the intercept method using 2-sigma limits as quoted in Pearson and Stuiver 1986 or, for dates prior to 4500 cal BP, using the plots in Pearson et al 1986.*

INTRODUCTION

> ...sensitive interpretation of the local lithostratigraphy allows much more than relative dating. It provides a contextual appreciation, however imprecise, of the wavelengths and amplitudes of environmental variability and of the relative roles of frequent low-magnitude oscillations versus the less common high-magnitude events. (Butzer 1982: 71).

Runnymede Bridge is a site for which these aspirations might perhaps be realised for the Holocene period. The British Museum Research Excavations set out to address the questions of environmental variability and cultural interaction with the local, dynamic environment, starting always with the vital issue of taphonomic process. The primary purpose of this paper is to provide a geomorphological framework for a small patch of alluvial terrain spanning the Holocene period. It therefore concentrates on the physical structure of the deposits, but draws in vital environmental and cultural evidence when these have a bearing on interpretation of formation and transformation. The intention is to demonstrate that, when the survival/destruction balance in an alluvial setting is favourable, archaeological deposits can assist in establishing an unusually refined sequence. This is not purely for the benefit of geomorphological studies, but may also repercuss on a range of cultural explanations – from the endurance of society in the face of adverse conditions, to food production constraints and seasonal mobility, to name but three.

The archaeological site of Runnymede Bridge lies in the middle reaches of the Thames, 40 km west of central London, or nearly 60 river-km from the City. It is situated at the confluence of the the Thames and the Colne Brook, a stream which drains southwards from headwaters on the dip slope of the Chilterns. The river level normally oscillates around 13.00 m OD giving an average downstream river gradient of about 25 cm km^{-1} at the present day. The site itself lies on Staines alluvium, surrounded locally by Shepperton (floodplain) gravel (Gibbard 1985). The alluvial sediments are calcareous with pH values of about 8 and the consequent good preservation of shell and bone. Below the permanent water-table, pollen, macroscopic plant remains and insects are generally well preserved. Much of this material has been studied already and is leading to a detailed picture of the local environment for the duration of the Holocene. The first substantial publication of this diverse evidence is to be found in Needham 1991. Although individual studies will be poorly represented in this paper, the interpretation of the site, its economy and its setting is wholly dependent on a battery of specialists working on biological remains and on the sediments themselves.

Archaeological material at Runnymede Bridge covers a number of periods, but the great mass belongs to two phases of prehistoric settlement, in the Middle Neolithic and the Late Bronze Age. These prehistoric settlement deposits were discovered in the course of planning and building the M25 orbital motorway. During the main rescue campaign in 1978 it quickly became apparent that they were interstratified with informative sedimentary sequences, which were laterally rather variable, and which appeared to relate to both channel fills and overbank alluviation (Needham 1991). After the end of the rescue phase, in 1980, the British Museum ran a campaign of research excavations between 1984 and 1989. One of several stated objectives was to expand knowledge of geomorphological change and particular period topographies; work has had to concentrate on the archaeologically rich site itself and it is recognised that a broader understanding of the deposits would result from more intensive studies in adjacent parts of the valleys, particularly upstream reaches.

THE ALLUVIAL HISTORY OF THE SITE

The main geomorphological developments on the site are reconstructed from a series of trenches, including a number of deep cuttings scattered across the two-hectare site (Fig. 23.1). Inevitably a degree of interpolation and extrapolation is involved in plotting palaeochannels and other boundaries. I shall refer to blocks of sediment layers which are temporally confined and bounded by definable horizons relating to significant hydrological change as *alluvial parcels*. The concept of an alluvial parcel, first outlined in Needham 1989, is defined in an appendix to this paper. Horizons of cultural activity are not in themselves used to define parcels. However, the substantial settlements are centered on stable floodplain surfaces and in these circumstances such horizons will mark the top of a parcel. In detail the associated assemblages may be suspended in a soil profile topping the given parcel, downward migration having accompanied faunal and climatic turbation. In these situations the renewed accretion following occupation implies, by default, changed hydrological conditions or a dramatic hydrological event and is likely therefore to require the definition of a new parcel or sub-parcel. Two parcel attributions of Needham 1991 have been modified in the light of further research: the layers formerly placed in sub-parcel 1B are now attributed to 2B and 2F, whilst previous parcel 4 is now sub-classified as 4A to accommodate a related suite subsequently discovered to the north-west (4B).

The earliest 'alluviation' on the site is, on archaeological grounds, pre Middle Neolithic (sub-parcels 1A,1C-1H); the deposits survive for up to 1.5 m above water-table and have been recognised across a lateral distance of 130 m, thus forming a north-west/south-east aligned spine which escaped erosion by any later river activity. There is variation in the sediments along this axis, although they are predominantly clayey silts above some early gravel (Areas 7/8) and marl (Area 19 especially). The deposits are consistently bedded near horizontally, but it is possible that one or more palaeobanks exist between the recorded exposures. In the absence of such information, it is at present unclear which if any of sub-parcels 1A,1C,1D and 1E are contemporary. However, a combination of pollen and palaeomagnetic evidence may ultimately give some indication of relative dating. At the south-east end of the spine, in Area 8 (sub-parcel 1A), a small cluster of struck flints was embedded in one layer; a burin of probable Mesolithic age was amongst them (Saville 1991: FL3). In Area 19 (sub-parcel 1C) underlying deposits of marl give way to clayey alluvium – thought to be from overbank flooding (awaiting confirmation) – before soil development and Neolithic occupation took place. The sequence observed at the north-west end (Area 13 North – sub-parcel 1E) is different again, and incorporates a Boreal pollen suite at its base (Rob Scaife – pers. comm.), associated with a radiocarbon date of 7790 ± 80 BP (BM-2550).

A 30–metre broad tract of floodplain on the north-east flank of the early spine is occupied by a sequence of channel fills spanning Mesolithic to final Neolithic times, as seen in the long sections of Areas 7/8 (Needham 1991: figs. 10–12) and Areas 22/26/32 (Fig. 23.2). They show a channel which was in general migrating northwards, but this trend was interrupted by phases of incision shown by the river cutting back into earlier bank deposits, for example the two sub-parcel interfaces 1D/1F and 1F/1G. The overall result is a sequence of cut-and-fills, all of which occurred at essentially the same altitude. The latest deposits are likely to be those 13 m

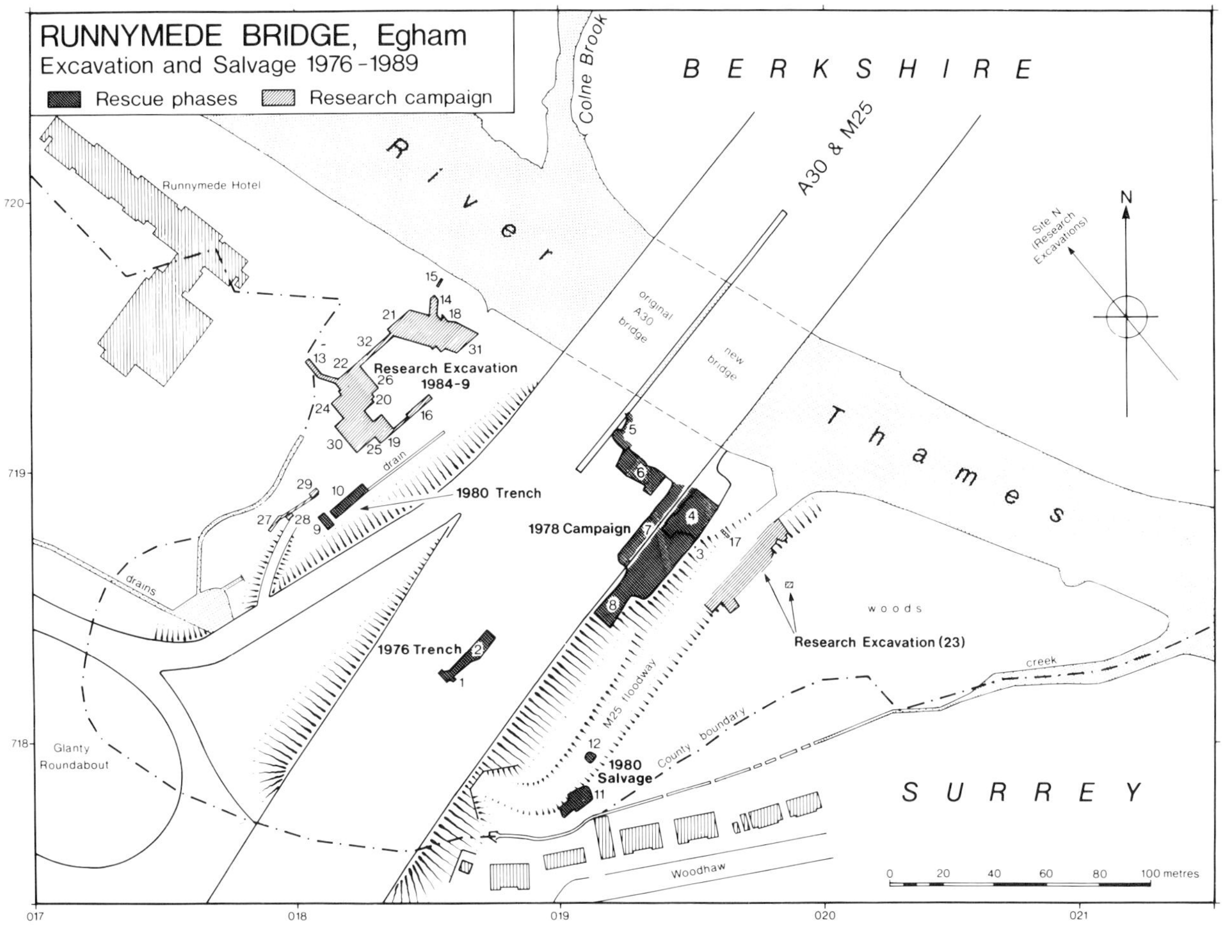

Fig. 23.1. Runnymede Bridge, Berkshire: plan of trenches and deep cuttings.

further north-east (Area 32 East); the base of this sequence has yielded wood recently dated to 5130 ± 50 BP (BM-2657), 5990–5740 cal BP, which cannot be much earlier than the beginning of the Neolithic occupation. The deposits representing this occupation are distributed across the top of this and all the earlier sub-parcels. They contain a large assemblage of Middle Neolithic cultural debris and there are associated structures, stake- and post-built, dating roughly within the period 5800–5300 cal BP (Needham 1991: fig. 132). On the highest part of the dry bank, that is overlying sub-parcels 1C and 1D, the cultural assemblage along with intact hearths are suspended in a developed soil profile, which itself implies an interlude following the effective cessation of aggradation. Time depth is also indicated by a few deposits stratigraphically earlier than the main settlement phase, amongst them some burnt flint clusters, which have yielded mean TL dates of 5580 ± 780 BP and 5750 ± 810 BP (N. Debenham — pers. comm.). A small amount of flintwork at this horizon includes several microliths. In fact, the implied stability was to last for some considerable time afterwards, for there is no evidence of either serious flooding or sediment deposition on the high bank until ca. 4000 cal BP. During the final millennium-and-a-half of this stable period, embracing the Late Neolithic, the site was only episodically visited, resulting in casual losses on the dry bank and the disposal of a group of butchered cattle bones in the flanking marsh in Area 6. The bones are dated to 4270 ± 110 BP (HAR-6136), ca. 5250–4530 cal BP.

Late Mesolithic/Neolithic channel deposits in Areas 6/7/8, perhaps partly equivalent to those already described as sub-parcels 1G and 1H, have been labelled sub-parcel 2A. The degree of temporal overlap is uncertain (Fig. 23.5). They comprise a broad system of lensed deposits in which the main features are the bank cut back into sub-parcel 1A and a bar, perhaps a channel bar, some 20 m distant. This bar contained important dating evidence: a palaeomagnetic date range of 6150–5600 cal BP (Clark and David 1991), and radiocarbon ranges between 5920 and 5330 cal BP for worked, but prostrate timbers near its upper surface (Needham 1991: 350–1). These timbers and associated concentrations of Middle Neolithic rubbish, none being

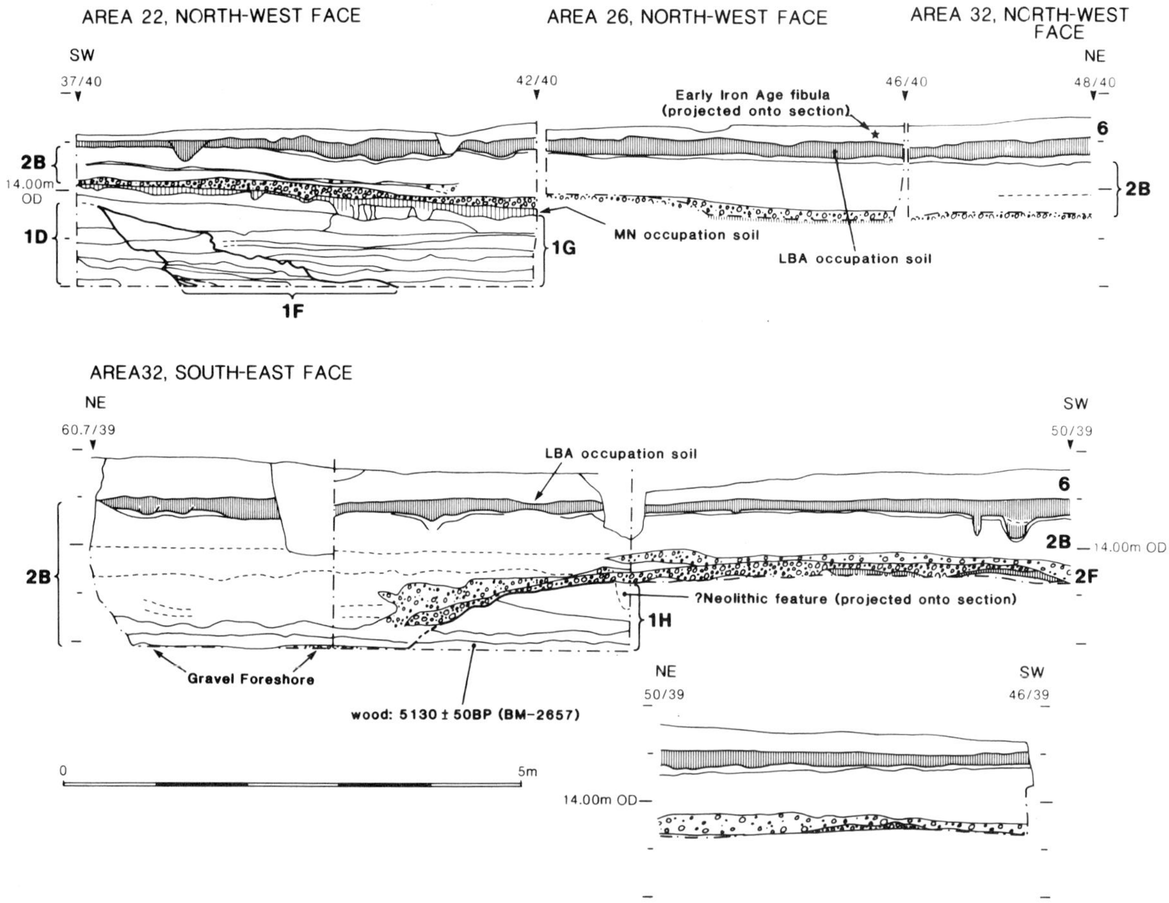

Fig. 23.2. Long sections through Areas 22, 26 and 32.

primary refuse, provide *termini post quem* for the enveloping silts. In contrast, elsewhere in sub-parcel 2A an *in situ* marsh-consolidation platform was recovered. A quantity of domestic debris was associated with brushwood layers through which several piles had been driven; this is now dated to 5600–5325 cal BP on the basis of a new date (4760 ± 50 BP; BM-2773) in addition to two dated piles (Needham 1991: fig. 132). There were localised erosion lines indicating late incision into sub-parcel 2A (sub-parcel 2C), but generally there is a trend in this zone towards quiet backwater conditions and a transition to mud-flat habitats (Evans 1991). The latest date for waterlogged sediments within this sequence comes from a bundle of beaver-gnawed wood from a late recut in Area 14 (sub-parcel 2E): 3760 ± 45 BP (BM-2435 & 2436, weighted mean), ca. 4290–3990 cal BP.

Close to this date, one or more major flood events took place with dramatic effects on the central high bank and on the margins of channels at least. The main result across the high bank was the deposition of a layer of dense gravel of variable thickness (sub-parcel 2F). It was generally 10 cm or more thick, but depth has probably been exaggerated by differential downward migration subsequent to deposition. The cobbles are not large, ranging from a few centimetres at most down to pea-gravel. This effectively blanketed the land surface, but not before some scouring, probably involving stripping of turf, the carving of minor gullies and perhaps some eddying into the tops of softer feature fills. The Middle Neolithic cultural assemblage on the site would by this stage have been well incorporated into the soil profile, but a proportion was nevertheless eroded out, sorted and re-deposited within the incursive gravel matrix. In this process the *in situ* archaeological deposits, those that had escaped the scouring action, now became re-sealed. Close dating of the gravel deposition is not easy, but several sherds of abraded Beaker pottery within its matrix are currently seen as providing a *terminus post quem*. This would place the flood towards 4000 cal BP at the earliest.

It is interesting to see the consequences of this flood action on the flanks of the high bank, at the channel margins. On the north-east side, Area 32 shows the gravel layer tipping down a steep erosion line, at the same time dissipating rather rapidly, doubtless due to initial and subsequent reworking on the channel margin. It is likely that some cutting back of the buried bank immediately preceded this deposition. A pene-contemporaneous gravel deposit, probably a thin layer, survives immediately outside this ancient bank and could be regarded as a foreshore. The deposits sealing this foreshore are comparable to those better exposed in Area 14 and attributed to sub-parcel 2D. The detailed relationship to 2D and radiocarbon-dated 2E is, however, uncertain on the present evidence.

On the south-western flank of the early spine, in Area 8, gravel again spreads down a bank, but this time a large wedge of it was left stranded in the angle of the bank having escaped later erosion by the channel subsequently filled with parcel 3 deposits (Needham 1991: fig. 12).

Looking beyond the banks, the semi dried out northern channel of Neolithic times experienced widespread truncation which, in particular, de-capitated the channel bar of Area 6. The dating evidence available places this truncation after about 2600 cal BC and therefore in the same time bracket as the gravel incursion; it seems likely that they are manifestations of the same serious flooding.

During the succeeding millennium, ca. 4000–3000 cal BP, a new stability seems to have been achieved . In the northern part of the site the backwater areas of the later Neolithic saw renewed slow silting after the flood truncation. This steadily raised the zone out of its mud-flat condition, taking it through a phase of damp grassland, eventually to become drier grassland, probably late within the period. In the process, this sub-parcel 2B silt encroached on the high bank with its gravel capping.

The floodplain aggradation of the 4th millennium BP considerably extended the area of dry bank alongside the earlier silt spine; there were additions on both the north-east, as described, and to a lesser known extent, the south-west. One consequence at Runnymede was an apparent reduction in flooding of the raised floodplain, thereby allowing soil formation to get the better of mineral accretion (Limbrey 1991: 220). These were the immediate preconditions of the Late Bronze Age settlement, which occupied the newly raised banks as well as the still proud central spine. The settlement seems to have been confined by contemporary channel features, which I shall now describe.

In the north a channel was certainly active during the Middle to Late Bronze Age, its southern bank describing a sinuous line alongside the settlement. The westernmost part of the bank (through Areas 21 and 14) appears to have been straight or a little convex. In Area 18 it turned through a marked angle, leading roughly towards the concave line interpolated across Areas 4,5 & 6 (Fig. 23.3). The interpolated bank line does not make a smooth, simple meander. Furthermore, neither the deposits against the preserved bank, nor those under the bank are part of a point bar aggradation; instead silts began to build up as a bar a little away from the maximum channel limit, only later infilling the slough

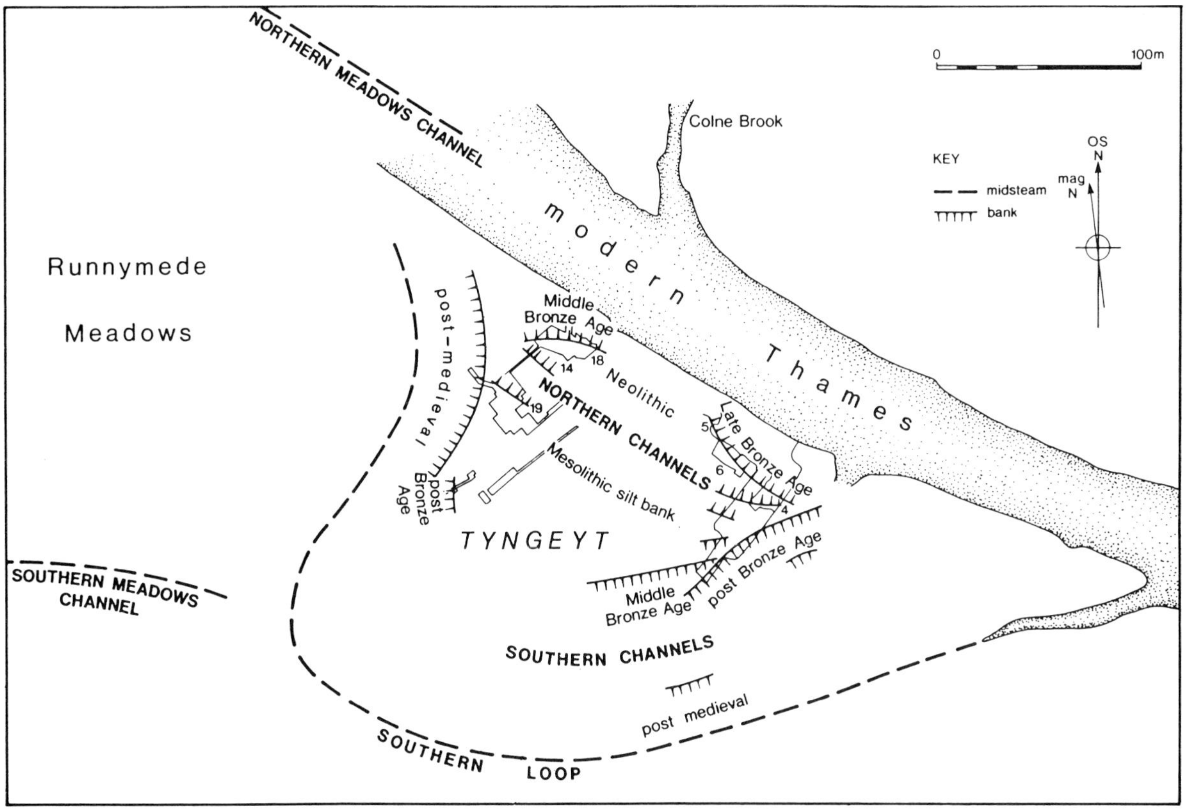

***Fig. 23.3.** Interpolated Holocene channels at Runnymede Bridge.*

in between. It seems possible that the eastern and western reach banks, although linked, had different origins. The concave eastern part could have originally been the bank of a tributary on a similar line to the Colne Brook today, which flowed in an eastward curving sweep before meeting a branch of the Thames. The western sector of parcel 4 (4B) would in this interpretation be associated with a breakthrough channel (neck cutoff) linking a presumed upstream meander of a Thames branch to the lowest reach of the Colne Brook, thereby roughly re-establishing the course of the prior Neolithic channel.

It may be a mistake to try to explain the line of the bank too closely, especially if, as suspected, the river at the time was a meandering-anastomosed system. Some flow changes would be effected by switching, or avulsion into dormant relief channels with consequent rapid and simultaneous erosion on two banks of the new conduit. Such a situation might also give rise to reaches having sudden changes in alignment, differing from neat predictions of sinuosity. Lewis and Lewin (1983) have shown just how variable may be the channel configurations involved in cutoffs.

Whatever the precise origin of this sinuous bank line, from the late 4th millennium cal BP onwards it began to aggrade by means of the aforementioned sand/gravel bar, probably a mid-channel bar (parcel 4), its crest being 6–8 metres from that of the bank. At the western, upstream end (Areas 14,18,21 & 31) aggradation started before a horizon yielding a radiocarbon date of 2900 ± 50 BP (BM-2661), or 3210–2880 cal BP, and the original bank was already largely buried by the time Late Bronze Age occupation began (ca. 2850 cal BP), being recognisable only as a shallow trough between the crests of the bank and bar. The bar in this reach may have forced the supposed cutoff channel northwards, or alternatively it began to choke it, thereby forcing all flow back round the southern end of the site.

Downstream a similar, though less pronounced bar extended for some distance (Areas 6 to 4 at least; it had a slightly greater amplitude than seen in Fig. 23.4 further to the east). Whether these were both parts of one long formation cannot be ascertained because of the intervening modern bridges. The writer finds it difficult to classify these bars. They seem to have much in common with the 'complex of lateral bars' with dead sloughs trapped behind illustrated by Lewin (1981: 47 fig.2.10.e), who states that point and lateral bars vary widely. Channel bars concentrated close to the inside of bends on the river Spey are considered not to be point bars by Ferguson (1981: 115). At Runnymede the eastern bar is the more curious, as it seems to belong to the less familiar patterns of silting against the concave bank which Lewin has collectively dubbed 'counterpoint accretion' (Lewin 1983), unless the channel was narrow and effectively spanned by the recorded sections.

Aggradation of the bar in this eastern reach (Areas 4 and 6) was temporally retarded relative to upstream; at the time occupation began (as marked by layers 24 and 25c in Fig. 23.4), although aggradation was underway, the original river's edge was still aquatic. Bar crest and adjacent trough continued to build up during the early stages of occupation (2850–2750 cal BP; Needham 1991: 60–2, 352). During this period the channel edge changed from slow-flowing water to perennially stagnant, and then became a peat pool. There is

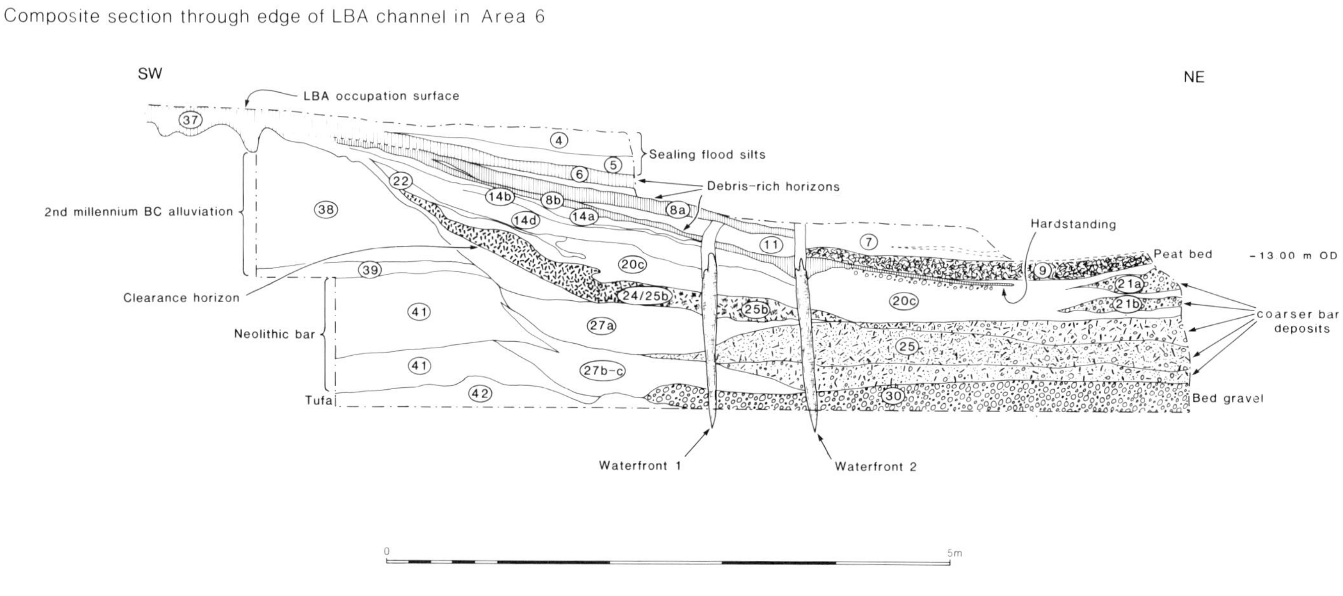

Fig. 23.4. Composite section through sub-parcel 4A deposits, Area 6: edge of river channnel with Late Bronze Age structures. Some important horizons are highlighted.

evidence, however, of continued sedimentary inputs from flowing water nearby – perhaps from a continuing course of the postulated tributary. Some important waterfront structures were erected during this sequence. Most notable were the piled foundations of two phases of what is interpreted as an enclosure stockade. A little later, probably during the lifetime of waterfront 2 and before the peat pool phase, hardstandings for beaching craft were laid on the soft mud of the foreshore (Needham 1991: 63–5, 91).

In addition to the structural remains, large quantities of cultural debris became engulfed in the accumulating silt. Eventually the whole sequence became buried in sterile flood silt, owing to the return of regular overbank flooding. At first this flooding reworked the upper parts of the abandoned Late Bronze Age floor surfaces, middens and collapse deposits – these in combination being manifested as a thick deposit of dark earth spread all over the occupied zone (of about 2 hectares). Presently the new flood silts accumulated on the surface of the occupied island, but the chronology of this 'sealing' stage is uncertain as yet. This is not the place to examine the character of the dark earth, but suffice it to say that it is very rich in artifacts and it contains *in situ* structures such as hearths, oven bases and cobbled areas, in amongst evidence for timber buildings (in the form of post holes; Needham and Sørensen 1988).

River channel activity around the west and south of the site is not directly attested during Late Bronze Age settlement, but might be inferred given evidence for channels shortly before and again afterwards. On the south-eastern edge (Area 8) there is a possible Late Neolithic channel, then a small channel about 8 metres across filled by parcel 3 late in the 4th millennium cal BP, i.e. just a little before LBA occupation. This may have been subsidiary to a main channel alongside to the south which later manifested itself, both here and on the south-western shore (Area 27), as an eroding bank cut back into the profile topped by the *in situ* LBA horizon. Palaeomagnetic dating of the silting against these erosion lines would seem to point to a date range not later than about the Late Iron Age (Clark – pers.comm.); this silting pre-dates overbank alluvium sealing the occupied island. It may be suggested that these shifts represent a phase of incision in response to new hydrological circumstances, and that after some adjustment the channel and steady overbank alluviation achieved a new stability. The channel inferred can probably be regarded as an early stage of the southern loop, a meander which can be documented in Medieval and later times, during which period there seems to have been a steady migration. This was particularly marked in the south-east, where a succession of deposits runs through Areas 8, 23 and 12. This channel was probably carrying little current by Late Medieval times, and the thick upper silts of its fill accumulated during Post-medieval times. The abandonment of the southern loop would only have been possible after the drying out of a suspected feeder coming across the southern meadows upstream (Fig. 23.3), even though the straight course of the Thames may have been concurrent for some considerable time.

DISCUSSION

Although the Runnymede Bridge site only represents a tiny fragment of the Middle Thames valley bottom, the multiple exposures and relatively good dating control might perhaps allow some wider consideration of a number of key issues:

1) The type of river present in the Runnymede reaches during the Holocene. This can be assessed from documented channel changes and the architectural forms preserved in the deposits.
2) The channel changes themselves and consequent effects on (a) the preservation of land forms and integral pre-existing evidence, and (b) subsequent utilisation, through their control of local conditions.
3) The date ranges and possible causes of major alluvial phases locally.
4) Flood frequency and severity.
5) The extent and nature of the effects of human agency.

Many of these issues interrelate and the ensuing discussion does not compartmentalise them.

At Runnymede processes of channel migration and fluctuation and significant overbank alluviation were in train from Mesolithic times, certainly before any possibility of local agriculture (though not before other possible forms of human destabilisation of the environment), and possibly from very early in the Holocene. It would appear that river levels and the associated fluvial deposition at Runnymede Bridge remained within a fairly constant altitudinal range through much of the Holocene, certainly from around the 9th millennium cal BP. This will have been dependent in large measure on the position of the site along the thalweg. The lack of overall vertical incision or aggradation has resulted in the lateral juxtaposition of parts of many sequences (or parcels) where otherwise sequences might have been largely denuded or be out of view well below present day water-table. One possible period for which deposits at Runnymede might lie in an overall lower altitudinal range is the Late Devensian/ Early Holocene, prior to about 9000 cal BP. This was a period of rapid sea-level rise due to glacial retreat; the rise is thought to have had significant effects on the base levels of tributary rivers in the Lower Thames (e.g. Lewis *et al*, this volume). A bed of gravel likely to be Late Devensian underlies the site between 7.0 and

12.75 m OD (Borehole survey for the M25 construction). The top is probably represented by interleaved gravels and silts at the base of the Area 7 section (layers 95–97; Needham 1991: fig. 10). The silts become non-organic above layer 93, between 12.50 and 12.75 m OD, whereas Neolithic and later silts have organic preservation up to between 12.75 and 13.10 m OD. Both the marl in Area 19 and the waterlogged organic horizon in Area 13 suggest that the water-table had risen to its current level by late Boreal times. This was the period of swampy conditions in the Colne valley just to the north (Lewis *et al*, this volume). In the Runnymede exposures, however, this seems to have given way fairly quickly to further sedimentation which sealed quiet-water/swamp deposits.

In the period ca. 8000–4000 cal BP most sedimentation seems to be taking place within cut-and-fill cycles with an overall progression north-east from the central spine. Whether this simply involved a shift of the responsible channel, or simultaneously, a trend to increased confinement of channels, making them narrower, deeper and perhaps fewer, is impossible to determine at present. An alder carr, or ribbon, which stretched along the valley floor at this period, might be expected to have greatly slowed the migration of a non-flashy river system. However, late in this period, by the Middle Neolithic at least, the environmental evidence indicates patches of open ground (Evans 1991): riverside clearings of human and animal origin, in addition perhaps to unstable poorly-vegetated bars and banks of longer standing.

The major alluvial body post-dating 4000 cal BP has rather different characteristics. Cut-and-fill cycles have not been identified and much can be regarded as wide-spread overbank alluviation (sub-parcel 2B). In addition to a transition to a largely forest-cleared landscape, this period also saw a transition to better defined, or more confined, channels. The source of the sediment and the cause of its transportation and deposition are uncertain. Developing agriculture during the previous 2000 years could have created a legacy of readily available sediment 'in storage', poised close to rivers. The transfer of this sediment to the active system and its redeposition could have been triggered by a combination of factors, including such things as increased run off during storms, less impediment to sediment flows into rivers, and the incapacity of confined channels to cope with the increased load. The last could have been exacerbated by the Thames III transgression and its upstream backing up effects (Needham 1985: 133). In this way the balance could be tipped towards a new dominant alluvial regime in the local catchment.

Mention should perhaps be made here of the possibilities of some impact from beaver on the local reaches of the valley. Despite some uncertainty as to the behaviour of beavers in prehistoric Europe (Coles – this volume), there is ample scope for the perturbation of river regimes through bank disturbance, deforestation on the flanking floodplain and slopes, and consequent increases in detritus levels in the water itself. Two well-dated pieces of evidence for beaver activity have been excavated at Runnymede, both being clusters of gnawed wood from the northern part of the site. They have yielded radiocarbon dates of 3760 ± 45 BP (weighted mean of BM-2435 & 2436; Ambers *et al* 1989: 17) and 2900 ± 50 BP (BM-2661), and they roughly bracket alluvial sub-parcel 2B. The waterlogged contexts from which the evidence comes are channel edge fills and at least the earlier of the two seems to be part of a backwater system with little or no flow, so local damming cannot be ruled out. Since there is no evidence that water levels rose during this period, the thick non-organic deposits of sub-parcel 2B are unlikely to have been the result of silting *within* a beaver pond; they could, however, conceivably have resulted from enhanced flooding caused by local damming.

The fact that so many different parcels were both created and fragmentarily preserved at Runnymede suggests that although the river system was frequently mobile, it was to a degree under-fit (e.g. Rose *et al* 1980) and thus did not rework its whole floodplain. Although there were some long periods of migration in a particular branch, there is little sign of pronounced meander development here. Even the apparent oxbow, the *southern loop* of Medieval and later times, may have been in part an artificial product of other channel changes – cutoffs and switches. Through much of prehistory the river is best seen as an anastomosed system, in which both branching and limited channel migration were features (see Friend 1983 for different definitions of anastomosed systems). The evidence for multi-thread channels in the locality includes the palaeochannels immediately surrounding the site, an abandoned channel to the west (Langham's pond, the southern meadows channel of Fig. 23.3) and new evidence for a channel running south through part of Egham (Steve Dyer – pers.comm.). It is possible that there was a trend through time towards less branching, although, as has been seen, this would not necessarily rule out the reactivation of relict channels as a result of avulsions. Ferguson has suggested that lowland clay rivers such as the Thames are characterised by an absence of channel migration and by underfit meander patterns (1981: 119). This is clearly only partly right. He also saw eyots in the Thames as ancient channel bars which are no longer active bedforms. The fine-alluvium central spine at Runnymede Bridge, however, which was probably an island at times during the Holocene, suggests that a broader range of formations should be considered.

The history of channel changes as understood at the present is summarised graphically in Fig. 23.5. In this diagram channel activity to the north and south of the central early spine is split into west and east reaches, thereby creating four short reaches bordering the site. From each alluvial parcel recorded it is possible to suggest a chronology for a sequence from channel flow, through channel fill, to overbank flooding in that particular locus. Inevitably there are gaps in the record owing to the vagaries of survival and exploration; however, the picture is already complex and there are clearly some shifts in the major river activity. Most activity before 4000 cal BP is witnessed in the northern sector, although there may have been a switch to the southern a little before the great flood. Late in the fourth millennium cal BP, there was probably simultaneous activity in both zones, but thereafter the weight of evidence swings markedly to the southern channel. Resumption of the northern course for active flow is only assured for the relatively recent past when the southern one was certainly going out of commission. It is possible that these perceived changes literally represent a changing balance of the discharge of the respective channels, but it is perhaps equally possible that the pattern apparent is a product of the 'field of view' dictated by the excavations, and that the distinct channels were alternately weaving in and out of that restricted frame.

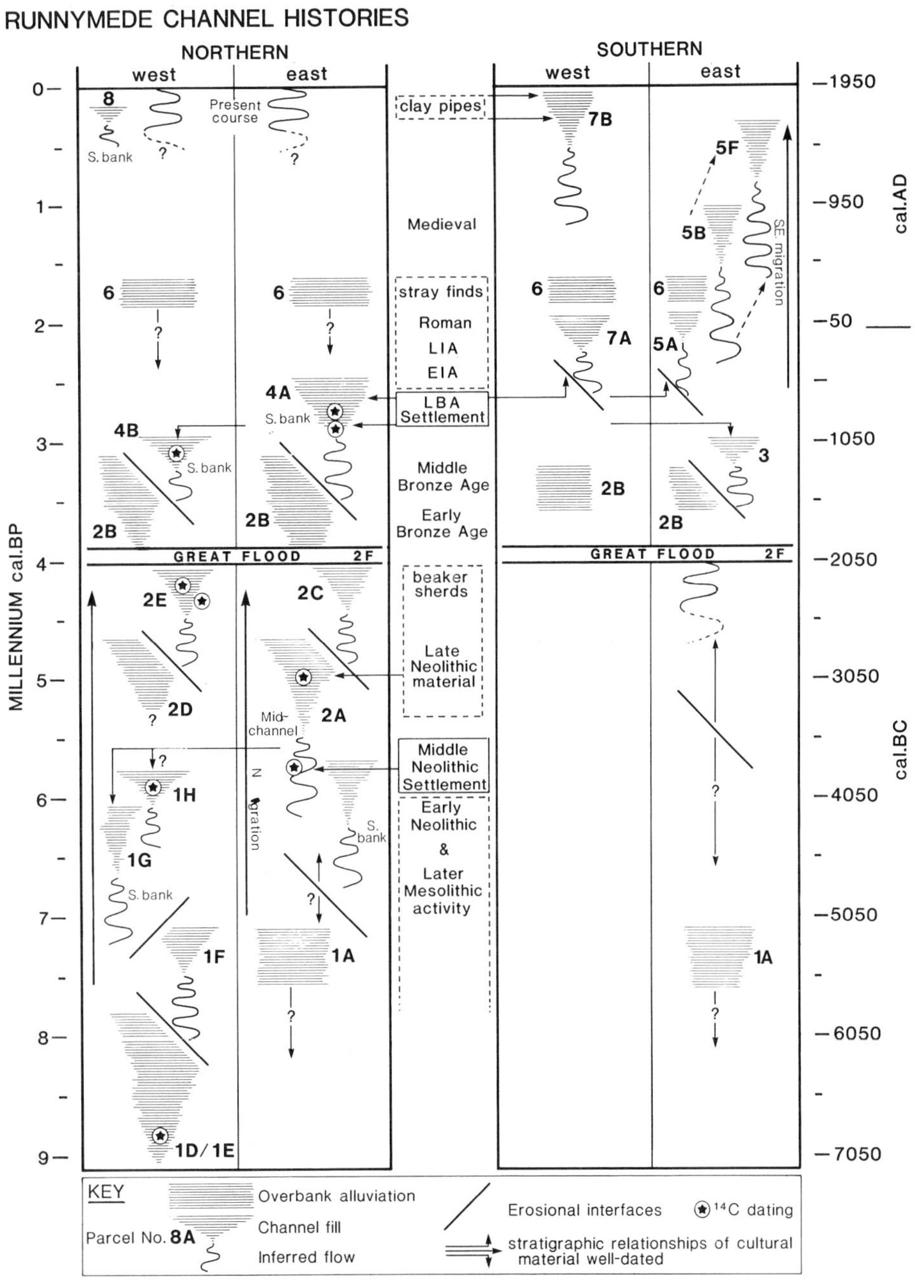

Fig. 23.5. Summary of Runnymede Bridge channel histories.

The survival of the early silt spine is not so much the result of any intrinsic properties resisting erosion, but was more probably due to the nature of the river system and the early establishment of two main channels to north and south. Even when one or other of these channels was largely silted up it would seem that hydrological adjustments, whether dramatic or not, tended to seek them out and re-establish flow on a similar course. This phenomenon may have been encouraged during and after Late Bronze Age times by the peripheral stockade of piles (only proven in one reach). Even when reduced to stumps largely buried in the channel edge, they may have slowed the erosion taking place under standard flow conditions and thereby slowed encroachment on the already ancient landform of the early spine. There is also some evidence for channel margin structures during the Middle Neolithic (Needham 1991: 26–9, 372), but their overall nature and extent are unknown. Whatever the exact combination of factors responsible for the enduring spine, its existence was evidently paramount for the siting of two major settlements, and doubtless also had a bearing on other uses of the site, such as the modern springhead for a bridge.

Humans could have been responsible for other effects on topographic changes; looking at a rather small scale for example, disturbance at the bank caused slumping or accelerated erosion at the beginning of LBA occupation. Furthermore, timber structures built out into the water could have deliberately constricted passage along the rivers. This in turn would have led to changes in flow conditions and scour/aggradation patterns, e.g. 'overdeepening' of the channel (Lewin 1981: 44).

One fluvial phase at Runnymede Bridge which deserves detailed consideration is the massive flood dated around 2000 cal BC. This appears to have been a rather irregular event in terms of the local Holocene sequence and is of interest for questions of the frequency and magnitude of high-magnitude floods. Given the long time-spans of fluvial activity documented for the site, it can confidently be suggested that this flood (or set of floods in quick succession) has a recurrence interval much greater than a hundred years. Indeed it is the only tangible evidence for a flood of such high energy during the whole of the Holocene. While there may have been others which have not left their mark locally, it would seem that we are truly dealing with a very exceptional event. There may be indication of an abnormally severe flood in a sequence recorded at Eden Walk, Kingston (Field *et al* forthcoming), but it is not yet clear that it would have been closely synchronous with Runnymede sub-parcel 2F.

One should perhaps also investigate possible more local causes. It may be of relevance, for instance, that renewed forest clearance is indicated in the pollen record from Area 6 SS1 during the Late Neolithic (Greig 1991: fig. 114 — depths 140–150 cm). This is not particularly closely dated but should pre-date the gravel incursion. If woodland was being attacked on a regional scale over a short period of time, the sudden increase in run-off may well have destabilised the hydrology and severely affected the flood risk.

Costa has stated the importance of Holocene stratigraphic records for assessing frequency of high-magnitude floods (Costa 1978). Rather than his specific methods for flood frequency analysis, it is the overall comprehension of hydrological history at Runnymede which helps to give an idea of flood frequency. The strange 'inversion' seen in the relatively coarse gravel thrown up onto deep fine-grained alluvial deposits in a long fines-dominated sequence recalls Costa's comment that "The most compelling evidence for the rare occurrence of those floods was coarse flood deposits on low terraces composed of relatively fine-grained material."(*ibid*: 626). Such inversions are of course seen more frequently in the stratigraphic sequences of upland Britain (e.g. Ferguson and Werritty 1983: 189 fig.8; see also Tipping, this volume). Serious floods do not necessarily have a lasting effect on river regimes and the effects are often transitory. Gupta (1983) discusses the conditions that can lead to floods having a 'stable imprint' on channel form, one of which is a relative coarseness of the flood-deposited sediment.

Lesser magnitude floods are more difficult to document individually although their aggregate effect is sometimes massive. Clearly the deposition of fines up to one metre or more above standard flow river levels was commonplace at Runnymede during certain phases of the Holocene. More powerful floods may perhaps be indicated by reworking of the gravel capping left by the major flood over the central spine. Reworking is seen in the form of strings of pebbles running off the main gravel bed and over some intervening silt (e.g. Area 32 — Fig. 23.2), or as gullies through the gravel bed, some probably due to later scouring. In both processes the movement of the heavy particles may have been limited and localised, and there is no evidence that the floods in question approached the force of their predecessor.

In due course it will be of value to compare the phases of deep alluviation and those of hydrological adjustments to similar evidence elsewhere in the catchment. At present there are rather few exposures which are both well studied and well dated. Downstream at Chertsey Abbey Meads, work by Phil Jones (Surrey Archaeological Unit — pers.comm.) has established a succession of fluvial deposits, incorporating early shorelines and associated human activity. Here there was some channel infill during the Early or Mid Holocene, but, unlike at Runnymede , there appears to have been no alluviation during the 4th millennium cal BP, although there is a later body of alluvium.

Further downstream at Kingston, the main bodies of alluvium revealed in the sections recorded at Eden Walk (mentioned above) seem to date to the Iron Age and later. These overlie the probably truncated top of an organic-rich backwater accumulation spanning the Neolithic to Middle Bronze Age (Penn *et al* 1984; Field *et al.* forthcoming). Again there is little specific correlation with hydrological conditions at Runnymede; the late alluvium in fact would appear to correlate well with a phase of alluviation (Thames IV) in central London described on various sites by Merriman (this volume).

Care must obviously be exercised in interpreting the presently limited evidence for Holocene alluviation along the Lower Thames. The two sites mentioned probably do not present as comprehensive a section of the valley floor as has been amassed at Runnymede and much more work is needed to see the spatial extent of particular fluvial regimes.

Acknowledgements

Thanks go to all specialists involved in the Runnymede project and especially to Rob Scaife and Tony Clark for useful discussions on as yet unpublished results. Phil Jones, Steve Dyer and Dale Serjeantson kindly provided information on other sites in the catchment. The line drawings are the work of Stephen Crummy.

References

Ambers, J., Matthews, K. and Bowman, S. 1989. British Museum natural radiocarbon measurements XXI. *Radiocarbon*, 31, 15–32.

Butzer, K. 1982. *Archaeology as human ecology: method and theory for a contextual approach.* Cambridge University Press.

Clark, A.J. and David, A.E.U. 1991. Palaeomagnetic dating. *In* S.P. Needham 1991, 354.

Collinson, J.D. and Lewin, J. (eds.) 1983. *Modern and ancient fluvial systems.* International Association of Sedimentologists, Special Publication no.6. Oxford: Blackwell.

Costa, J.E. 1978. Holocene stratigraphy in flood frequency analysis. *Water Resources Research*, 14, 626–32.

Evans, J.G. 1991. Syntheses of the environmental evidence. *In* S.P. Needham 1991, 363–8.

Ferguson, R.I. 1981. Channel form and channel changes. In J. Lewin (ed.), *British Rivers*, 90–125. London: Allen and Unwin.

Ferguson, R.I. and Werritty, A. 1983. Bar development and channel changes in the gravelly River Feshie, Scotland. *In* J.D. Collinson and J. Lewin (eds.) 1983, 181–94.

Field, D., Penn, J., Serjeantson, D. and Shipley, M.P. forthcoming. Eden Walk II, Kingston: environmental reconstruction and prehistoric finds. *Surrey Archaeological Collections.*

Friend, P.F. 1983. Towards the field classification of alluvial architecture or sequence. *In* J.D. Collinson and J. Lewin (eds.) 1983, 345–54.

Gibbard, P.L. 1985. *The Pleistocene history of the Middle Thames Valley.* Cambridge University Press.

Greig, J. 1991. The botanical remains. *In* S.P. Needham 1991, 234–61.

Gupta, A. 1983. High-magnitude floods and stream channel response. *In* J.D. Collinson and J. Lewin (eds.) 1983, 219–27.

Lewin, J. 1981. Contemporary erosion and sedimentation. In J. Lewin (ed.), *British Rivers*, 34–58. London: Allen and Unwin.

Lewin, J. 1983. Changes of channel patterns and floodplains. In K.J. Gregory (ed.), *Background to Palaeohydrology: a perspective*, 303–20. Chichester: Wiley.

Lewis, G.W. and Lewin, J. 1983. Alluvial cutoffs in Wales and the borderlands. *In* J.D. Collinson and J. Lewin (eds.) 1983, 145–54.

Limbrey, S. 1991. The sediments in Area 6. *In* S.P. Needham 1991, 218–22.

Needham, S.P. 1985. Neolithic and Bronze Age settlement on the buried floodplains of Runnymede. *Oxford Journal of Archaeology*, 4, 125–37.

Needham, S.P. 1989. River valleys as wetlands: the archaeological prospects. In J.M. Coles and B.J. Coles (eds.), *The archaeology of rural wetlands in England*, 29–34. Exeter: Wetland Archaeological Research Project, University of Exeter.

Needham, S.P. 1991. *Excavation and salvage at Runnymede Bridge, 1978: the Late Bronze Age waterfront site.* London: British Museum Press.

Needham, S.P. and Sørensen, M-L.S. 1988. Runnymede refuse tip: a consideration of midden deposits and their formation. In J.C. Barrett and I.A. Kinnes (eds.), *The archaeology of context in the Neolithic and Bronze Age: recent trends*, 113–26. Sheffield: Department of Archaeology, University of Sheffield.

Pearson, G.W., Pilcher, J.R., Baillie, M.G.L., Corbett, D.M. and Qua, F. 1986. High-precision ^{14}C measurement of Irish oaks to show natural ^{14}C variations from AD 1840–5210 BC. *Radiocarbon*, 28, 911–34.

Pearson, G.W. and Stuiver, M. 1986. High-precision calibration of the radiocarbon time scale, 500–2500 BC. *Radiocarbon*, 28, 839–62.

Penn, J., Field, D. and Serjeantson, D. 1984. Evidence of Neolithic occupation in Kingston: excavations at Eden Walk, 1965. *Surrey Archaeological Collections*, 75, 207–24.

Rose, J., Turner, C., Coope, G.R. and Bryan, M.D. 1980. Channel changes in a lowland river catchment over the last 13,000 years. In R.A. Cullingford, D.A. Davidson and J. Lewin (eds.), *Timescales in geomorphology*, 159–75. Chichester: Wiley.

Saville, A. 1991. The flintwork: Mesolithic, Neolithic and Bronze Age. *In* S.P. Needham 1991, 125–9.

APPENDIX

Some guidelines for the definition of Alluvial Parcels

When dealing with long and/or complex sequences of fluvial deposits it may be useful to define and label major accretion units to facilitate description and discussion of detailed sequence, process and interaction with cultural evidence. This has been attempted for the part of the floodplain at Runnymede Bridge investigated in the context of prehistoric settlement and a series of *alluvial* parcels has been defined according to the following guidelines.

Alluvial parcels are blocks of sediment which may contain both horizontal and inclined layer members. They are intended to be interpretative units, interpretation being based on the available knowledge of: a) dated horizons; b) sedimentary characteristics; c) changes in hydrological conditions, particularly those creating major stratigraphic interfaces. Parcels will consequently be subject to re-evaluation, especially if new exposures or associated information comes available for study.

An alluvial parcel should have a temporal span which is largely exclusive of those for proximate parcels. When seen in juxtaposition the interface between two parcels will tend to be one of two types:

i) an oblique erosion line (e.g. an ancient river bank) against which one set of deposits is broadly later than that which it abuts (although some temporal overlap is possible depending on detailed depositional circumstances);
ii) a more horizontal surface believed to represent a significant standstill, or indeed a truncation surface.

Definition is non-genetic; it is not intended that a given parcel encompass a limited range of sediment types (although this may be the case) and by implication the immediate environment is likely to change within the span of a parcel. Variations may include steady gradational changes along a particular deposit, dependent on topographic differences. Identification of lateral gradation will depend on the length of exposures available; there is obviously a danger that it may pass unrecognised when two or more short sections are well separated.

Deposits within a given parcel will, then, often reflect changing circumstances, but overall these should be much less drastic than those implied at parcel interfaces which refer to important changes in the hydrological regime. Apart from the environmental variability described above, parcels might also embrace one or more minor reversals seen as localised incisions or standstills where these are not seen to interrupt significantly a particular pattern of aggradation. Such minor reversals might for example be due to expected channel adjustments in the wake of variations in flow. If there were any suggestion of a lengthy hiatus, or of a marked change in the character of accretion, then a parcel boundary should probably be defined instead.

In practice one of the biggest problems in defining parcels is likely to arise from uncertainties in the correlation of separated exposures. This is a particular problem where lateral trends of the sort mentioned above come into play. Where the dating evidence suggests a broad coevality and the sedimentary record suggests compatible hydrological factors, then two sequences may be attributed to the same parcel, but to different sub-parcels to account for their variant sedimentary sequences (e.g. parcel 1). On the other hand, if two units known to be broadly contemporary belong to discrete river channel fills, they are treated as entirely separate parcels (e.g. parcels 3 and 4).

When thought useful to do so, sub-parcels are also used to characterise any internal morphology caused by either minor reversals or trends in immediate environmental conditions (as for example 2A versus 2B).

The use of alluvial parcels at Runnymede is distinct from *stratigraphic units* and *phases*. These are employed for interpretation of different, more detailed aspects of the sequences.

24

Predicting the unexpected: prehistoric sites recently discovered under alluvium in central London

Nick Merriman

Summary: *Apart from river finds, central London is not an area principally known for its prehistoric remains, although it lies in an area of the Middle Thames valley which must have played a key role in the development of prehistoric society. However, a number of recent excavations have begun to alter this view. Four sites are discussed, and it is concluded that the area potentially contains important archaeological information well-preserved beneath alluvium, especially at the edges of former gravel islands. Most discoveries have, however, been made by accident in the course of work aimed at the elucidation of other questions. It is argued that a more coherent strategy for the recovery of archaeology buried beneath alluvium should be formulated, using a combination of borehole information and trial trenching.*

INTRODUCTION

Work over the last twenty years in the Upper and Lower Thames valley, particularly in Oxfordshire and Essex, has demonstrated the intensity and diversity of prehistoric settlement. One area that has received relatively little investigation is that part of the valley now covered by the urban sprawl of London. It is likely to have witnessed the same intensity of occupation as the better investigated areas to the east and west, and study of the area could provide valuable data on many research questions concerned with the development of later prehistoric society in southern England, particularly as it lies in many cases in an area of apparent transition between traditions in the Upper and Lower valleys. Investigation of the London area has been hindered by the fact that the majority of the urban build-up occurred before archaeology had developed as a discipline. Thus, up until quite recently prehistoric material from central London almost exclusively comprised complete artifacts dredged from the Thames. A small scatter of chance finds of different periods made on land were hardly firm indication that the seeming importance of the river was matched by a corresponding intensity of settlement on its banks.

Partly this lack of evidence was explained by lack of exploration, but it was also partly a result of the expectation that no prehistoric material was recoverable, and consequently little effort was expended on trying to recover it. In particular, the old chestnut that the floodplain was not occupied in prehistoric times was a great hindrance to the investigation of the area. A familiar vicious circle was easily reached whereby lack of evidence from the Sites and Monuments Record (SMR) meant that certain areas were not explored, with the result that nothing was found and consequently the SMR remained blank. In addition, even when areas of prehistoric potential were explored, time, resources, the depth of the overlying deposits and the inclinations of the excavators conspired to ensure that little opportunity was taken to examine this potential.

Over the last ten years or so, this vicious circle has gradually been broken through a mutually reinforcing combination of chance discoveries and enhanced appreciation of the potential for the recovery of extremely well-preserved evidence under alluvium. As a result, a number of important sites of prehistoric, Roman and Medieval date have now been investigated in central London, most of them with very high quality preservation of material and environmental evidence. For the prehistoric period at least, this evidence, together with that recovered from 'dry' sites, suggests that the central London area was indeed as densely occupied as the better-documented areas to the east and west. In this paper I wish to outline the preliminary findings from four of these sites and suggest that it is now time to evolve a strategy for the investigation of alluvial sites in urban areas in order to make better use of our scarce resources and to extract the maximum information from them with the greatest efficiency. My discussion will be confined to sites of prehistoric date, and will be relatively brief, as little environmental or dating work has yet been carried out on them.

BREAKING THE CIRCLE

As in many similar situations, prehistoric material has begun to be recovered under alluvium in central London partly due to changed perceptions amongst archaeologists of the potential of the floodplain, caused by chance finds in similar topographic situations elsewhere. The key site, of course, is Runnymede Bridge (Needham 1991), which for the first time has demonstrated beyond any shadow of a doubt that the floodplain was permanently occupied at various stages in prehistory, and has raised the possibility that similar island sites at the confluences of tributaries with the Thames may have existed. In central London we know from documentary sources that a number of these former eyots exist, for example at Westminster, Southwark and Bermondsey (Nunn 1983), and the Runnymede discoveries drew much closer attention to them as possible foci of prehistoric and later settlement. Other floodplain discoveries such as those at Farmoor (Lambrick and Robinson 1979) and the barrows at King's Weir in Oxfordshire (Bowler and Robinson 1980) added weight to the possibility of recovering intact prehistoric surfaces underneath alluvial deposits formerly dismissed as 'natural' and not investigated.

RICHMOND TERRACE

This latter issue came up especially clearly in a site in central London itself, at Richmond Terrace, Whitehall (Fig. 24.1) (Andrews and Merriman 1986). In 1980 development work on the site prompted archaeological investigations which revealed Post-medieval buildings overlying a blue riverine clay. Unfortunately excavations

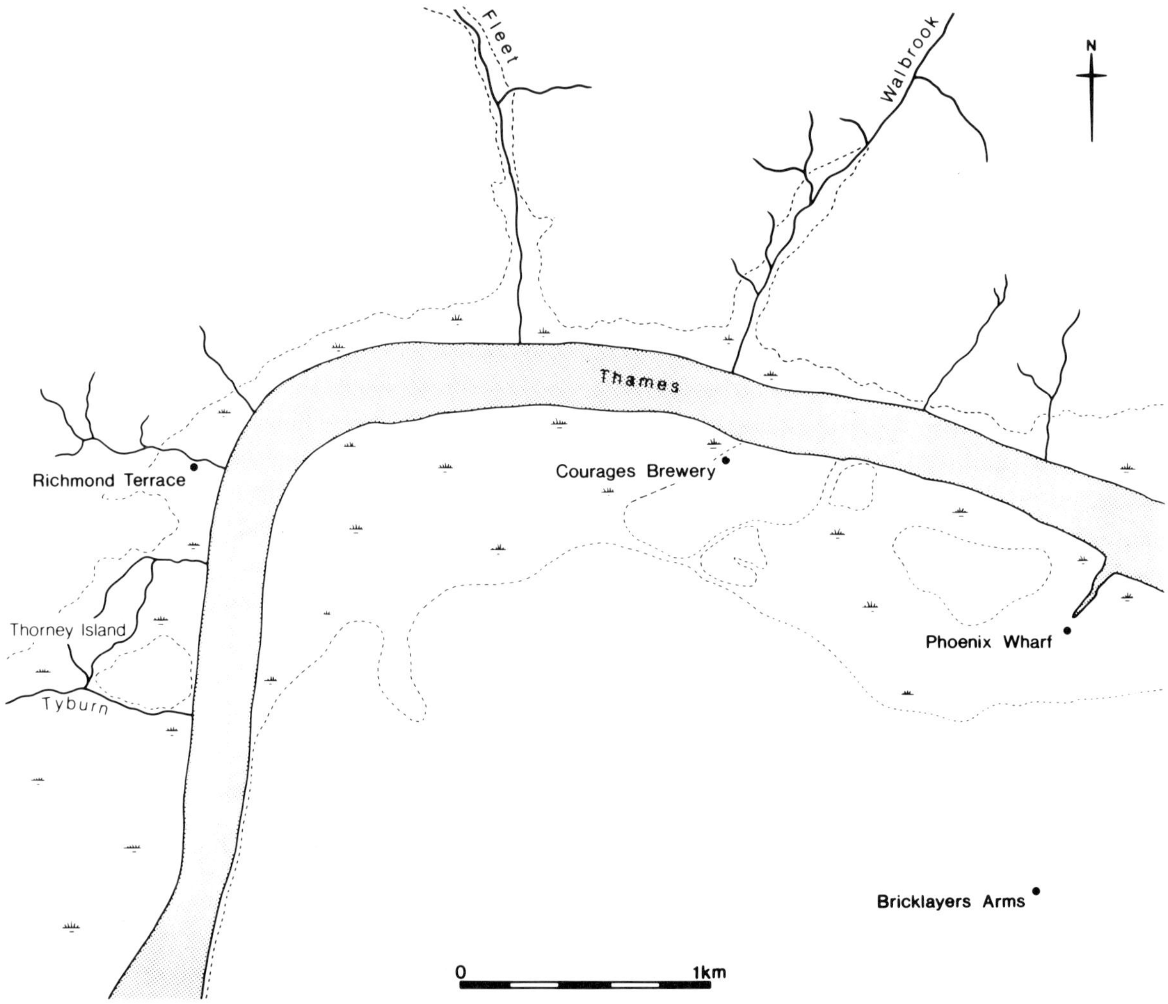

Fig. 24.1. Prehistoric sites in central London recently discovered under alluvium.

stopped at this 'natural' deposit, on the assumption that this deposit was of such great age that no human activity could lie beneath it. In the course of the subsequent watching brief during construction work, a base-plate and upright post of alder were recorded, sealed by up to a metre of this clay. The wood yielded a single radiocarbon date of 2540 ± 70 BP (Har-6393), which calibrates using Stuiver and Pearson (1986) to 830–414 cal BC (intercept, 1–sigma).

As a consequence of this lesson, and also because of the increased scale of archaeological work in the area resulting from the boom in construction in the mid-1980s, greater attention began to be paid to the possibility of recovering prehistoric evidence in central London and indeed elsewhere, for example at Three Ways Wharf in Uxbridge, where a Late Upper Palaeolithic and Early Mesolithic site was found under alluvium in the Colne Valley (Lewis, Wiltshire and Macphail, this volume). In addition, the construction boom meant that other previously uninvestigated areas began to receive archaeological attention, many of them containing large tracts of alluvium. As a result of this greater vigilance and increased pace of work three sites were excavated in 1987 and 1988 which yielded, by London standards, impressive preservation of prehistoric surfaces and deposits.

BRICKLAYERS ARMS

The first site in central London explored primarily for its potential to produce material and environmental evidence for the prehistoric (as well as later) periods was Bricklayers Arms in Bermondsey (Fig. 24.1) (Jones 1988). The area covered by the site was a former railway marshalling yard which was being redeveloped for housing. Borehole data showed that there were peat layers on the site sealed by waterlain clays at an accessible depth, and that consequently there was a good possibility of recovering environmental data and, perhaps, cultural evidence preserved beneath these layers. The trial trenches put in showed a sequence of glacial gravels, succeeded by waterlain clays. Ostracod analysis of the latter by Eric Robinson (1988) showed them to have been laid down under freshwater lake conditions, and pollen analysis by Nick Branch suggests that the phase represents initially open-water conditions with a plant community dominated by grass, sedge and bracken, followed by the formation of an alder carr (Branch 1989). To the north of the site, where the gravels rise up, these clays disappear, showing the land has remained dry through much of prehistory. The date of this clay is not certain as yet, but Branch's pollen work strongly suggests it may represent a local variant of Devoy's Thames III deposition (Devoy 1979), although we should of course be wary of equating processes occurring around Tilbury with those in central London. Within the clay, in the section of one trench were found two flint axes, presumably dropped or thrown into the water.

Right on the uppermost layers of this clay, in another trench, a collection of timber was found at the junction of the clay and the higher gravel on the north of the site. Some of this is deliberately shaped, and much of it on a similar alignment. Initially thought to be driftwood, its alignments, deliberate shaping, and the presence of some flint flakes amongst it, suggest that it may have been laid deliberately, possibly to consolidate marshy ground. Samples of alder and willow from this timber have yielded radiocarbon dates of, respectively, 3260 ± 60 BP (GU-2900) and 3300 ± 60 BP (GU-2901). Above this clay is a peat deposit with heights (+0.34 to -0.14 m OD top; 0.0 to -0.68 m OD bottom) within the range of variation encountered for the so-called Tilbury IV regression peats elsewhere in Southwark (Tyers 1988, table 2). The pollen assemblage here is dominated by alder, oak, hazel and lime, with a decline in lime towards the end of the zone, suggested by Branch to be due to deliberate clearance for agriculture (Branch 1989). In one trench this peat seals a fallen oak bearing tool marks which are consistent with the use of an axe.

Another series of waterlain clays above this peat layer represents another transgression phase. This time ostracod analysis (E. Robinson 1988) shows the water to have been saline, demonstrating that it was now the tidal Thames that was covering the area, and that we are probably witnessing the first occurrence of a branching river system around Bermondsey in the Postglacial period. Pollen analysis shows a substantial decline in arboreal pollen, possibly to be equated with increased clearance, against a background of climatic deterioration leading to what is likely to be a local variant of Devoy's Thames IV transgression (Devoy 1979).

COURAGE'S BREWERY

It is beginning to look as if this sequence (working upwards) of gravels, flood clays, peat, and further flooding, is one that can be identified right across Southwark and Bermondsey. The Courage's Brewery site (Fig. 24.1) is one that is best known for its spectacularly preserved Roman wooden warehouse (Mills and Whittaker 1991: 160–61). Upon excavation the warehouse was shown to cut through a sequence (from top to bottom) of clay, peat and humic waterlain silt. When the layers were removed by hand, it was confirmed that the upper fine flood clays sealed a peat layer whose height (bottom average 0.0 m OD, top average 0.20 m OD) is again within the range of levels for known Tilbury IV deposits in Southwark (ca. 1500–1000 cal BC). This has yet to be confirmed by radiocarbon dating. Flint flakes and cores were found lying on top of this peat, evidently deposited when it was a fairly dry surface, but no further finds were encountered through its 20 cm thickness. Below there was a 20 cm thick layer of humic waterlain silts, in the lower part of which were burnt flints, flint flakes, and a leaf-shaped arrowhead. The excavator believes this lower material to be derived from an underlying gravel surface by means of flood scouring (John Dillon, pers. comm.). This gravel surface was at first thought to be artificial, but then recognised as a natural foreshore of the Thames system itself. Cut into this were a number of hollows with fragments of burnt clay in their fill. Again, these features are as yet undated, but the position of the overlying layer, and its contained finds (which give a *terminus post quem* of the Neolithic period), suggest that layer may represent the Thames III transgression. Current dating of that transgression would suggest that the foreshore is sealed some time before ca. 2000 BC.

The discoveries at Courage's Brewery, then, have revealed an environmental sequence covering at least the Neolithic and the Bronze Age, with accompanying archaeological material. The discovery of early foreshore deposits with evidence of human activity on them demonstrates the potential that exists for uncovering glimpses of prehistoric landscapes preserved under alluvium even in such unpromising areas as central London.

PHOENIX WHARF

Even more revealing of such potential is another site in Bermondsey, that of Phoenix Wharf on the Jamaica Road, opposite St Saviour's Dock, an inlet into the Thames (Fig. 24.1). This site, it seems, lies on the edge of an island, probably the northern fringe of Bermondsey island. Here the excavation was aimed initially at answering questions concerned with the course of Medieval rivers in the area but, as frequently happens, it also uncovered more unusual evidence of a far older date. Here once more a sequence was encountered of flood deposits sealing a peat, which in turn sealed a prehistoric land surface, this time with definite evidence of an occupation sequence (Bowsher 1991).

The earliest feature on the site was a truncated straight-sided pit, at least 1.85 m x 0.95 m in plan and surviving to 0.50 m in depth, filled with burnt flint, charcoal, cinders, patches of clay, and a charred bovine radius. At the eastern edge of the pit was an irregular spread of stake holes, again filled with burnt material, which also lay above and around the area of the pit. A small east-west channel was found just to the south of the pit and the two features might have been associated. This phase has been interpreted by the excavator as the remains of a cooking pit, associated burnt mound and retaining fence line — one of the very few such structures known from the Thames valley. Charcoal from the basal fill of this pit has been dated to 3310 ± 40 BP (BM-2766), which calibrates using Stuiver & Pearson (1986) to 1645–1590 or 1575–1525 cal BC (intercept, 1–sigma).

This mound was, it seems, scattered by the next phase of activity on the site, which was ploughing or, more accurately, ground preparation with an ard. This process had left criss-cross ard marks across the site, including across the fill of the square pit and the east-west channel. Crescentic marks were also found, possibly made by a hoe or spade. Above these marks, and presumably in contemporary use, was a grey silty ploughsoil ca. 15 cm thick, with scatters of burnt flint, densest around the pit, probably representing the ploughed-out burnt mound, and a variety of flintwork and abraded pottery. The only diagnostic flint piece was a plano-convex flint knife, which might point to an Early Bronze Age date, although the pottery hints at a longer time span.

The sand below the ploughsoil has what appear to be cockchafer flight holes in it (James Rackham, pers. comm.), which means they presumably derive from the surface of the ploughsoil, either before it was ploughed, during fallow periods, or after abandonment. This indicates a dry grassland at some stage, and thus a low river level.

Above the transgression layer a regression indicated by a peat at around 0.0 m OD to +0.50 m OD is encountered. This is undated, but its height and stratigraphical position suggest it too may be equated with the similar peats in Southwark radiocarbon dated to the Middle — Late Bronze Age Tilbury IV horizon (Tyers 1988). This was subsequently covered by the silts of successive river channels.

CONCLUSIONS

These discoveries, together with others such as the Roman channel revetments from Guy's Hospital in Southwark (Mills and Whittaker 1991: 160), various boats from the Thames and its tributaries (Lawrence 1929, McGrail 1978), and other sites noted elsewhere in this volume, have clearly demonstrated that even within the densely built-up central London area there exist deep deposits of alluvium preserving high quality archaeological information. It is also clear that some of the most productive areas will be those that lie at the junction between the higher ground on the edge of an island and the adjacent peat and alluvium preserving evidence of human activity that took place when the river levels were lower. Excavations at Flag Fen near Peterborough have also demonstrated the potential for recovering timber structures in former marshes (Pryor *et al* 1986) and those at Farmoor show the possibility of settlement on low-lying ground close to a river channel (Lambrick and Robinson 1979).

So far, though, recovery of this information has been extremely haphazard and dependent more or less on chance. While it will always be difficult to predict accurately the location of deeply buried sites, what is needed is a strategy to make our recovery of them less haphazard, and to provide us with the highest quality information with the most efficient use of resources.

Clearly, the crucial factor in predicting the likely survival of prehistoric environmental and cultural material is knowledge of topography, which will enable us to pinpoint areas where peats and clays meet the edges of former gravel islands, and where buried channels and former lakes and marshes lie. In broad terms basic topographic guidelines can be followed, such as the known presence of islands or streams. However, in very many areas of central London we do not know where the exact edges of islands or buried channels lie, and of course it is highly likely that other undiscovered islands exist in areas of alluvium. In urban areas, of course, the prospection techniques noted in other papers in this volume are inapplicable, so we have to have recourse to other methods. As a first step, we must narrow down the options as to where we investigate. We can do this by using a potentially invaluable, but greatly underused resource, which are the thousands of borehole records available for the area. Because of the amount of development work that has been undertaken, most of central London has dense borehole coverage, much of it stored centrally with the London Residuary Body or with borough planning departments. Although they will only distinguish between, say, gravel and 'made ground', they offer great potential for providing a baseline for predicting the location of productive sites.

A good example of the potential that exists in this area is the work of a firm of consultant engineers providing geotechnical services for the London Docklands Development Corporation. One very useful product of their work has been a contour map of the Thames gravel in the Docklands area. By extracting spot heights from the borehole records for the area and computerising the information onto a mapping system, they have been able to produce a 1:10,000 scale map of the top of the gravel in the area (Fig. 24.2). On the basis of this map, then, a number of areas in Docklands which may have been relatively high gravel islands, and thus foci for prehistoric and later occupation, can be highlighted for further investigation. Similarly, areas with very deep alluvium which will not have received any Postglacial occupation, can be pinpointed and ruled out of further investigation (other than that of an environmental nature). When the opportunity becomes available to investigate some of them more thoroughly, individual borehole records can be consulted as part of the evaluation process, to indicate more accurately the nature of deposits and the likelihood of encountering alluvial deposits sealing former landsurfaces. Once an area of potential has been pinpointed, trial trenching and auguring then seem to be the only options to narrow down even further the area for investigation.

This sort of work could be usefully continued to cover the whole of the central London area, and indeed other urban locations where sites may lie buried under alluvium. The thousands of borehole records are a great, virtually untapped resource that, if used, could lead to a much better use of our limited archaeological resources by removing the need for trial trenching in areas that are likely to be unproductive. A long-term borehole project could begin by reviewing and storing centrally all existing borehole information for the area, with the aim of producing a map of subsurface gravel contours for the floodplain. When this is achieved, a means would exist for the more prudent allocation of resources in the future. The increasing stress laid on evaluation work before planning permission is granted (DoE 1990) might allow borehole surveys to be undertaken as part of this process. When new borehole work is undertaken, more careful monitoring would enable more detailed recording of subsurface deposits. A parallel project to review finds from alluvial areas would also help to pinpoint possible areas of occupation. Alluvial deposits are also likely to preserve high quality environmental information, so that borehole work is likely to form a vital part of the evaluation process for floodplain deposits even if direct evidence of occupation is not revealed.

The value of such work is twofold. First, on a regional basis, it will enable us to pinpoint areas where high quality data on the prehistory of the Greater London area may be recovered. A rapid assessment of floodplain deposits could provide some badly needed information on changes to the prehistoric environment of the area. The basic framework of dating and material culture sequences still remains to be constructed, and the possibility of deposits on the floodplain that would begin to provide this is a clear argument for concentrating efforts in this area. On a wider canvas, as noted at the beginning of the paper, the position of the area between the middle reaches of the Thames and the estuary, where sequences of occupation are demonstrably different, means that it may hold some clues to the understanding of variations in human settlement in the valley, and to hydrology and alluviation of the valley itself. In particular, the Greater London area should be able to shed light on the apparent shift some time in the Bronze Age from a core area centred on the upper Thames to one centred on the middle to lower reaches. With careful management of the opportunities presented, it is clear that central

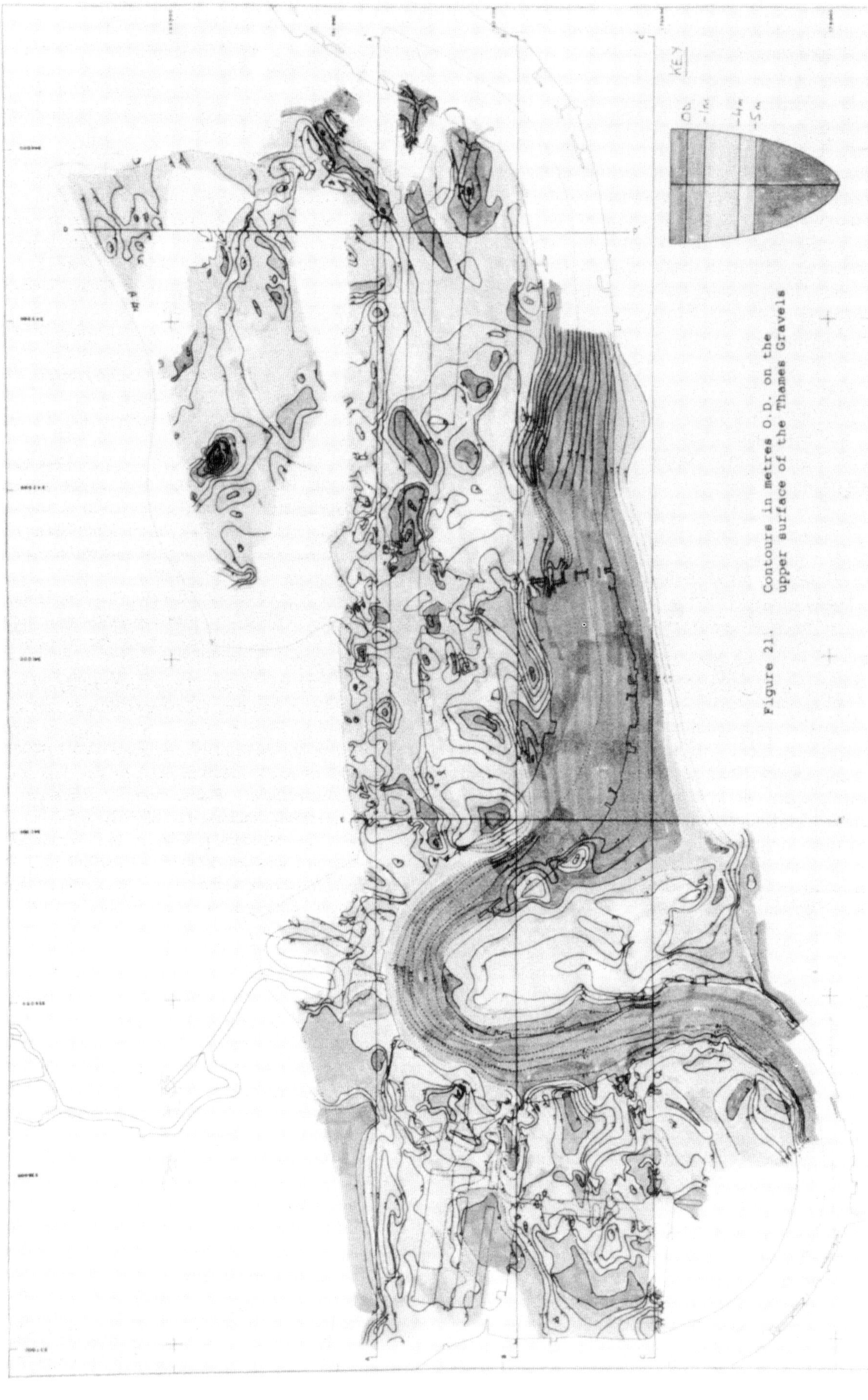

***Fig. 24.2.** Contour plot of the gravel/made ground interface in the London Docklands Development area (by kind permission of A F Howland Associates, Norwich).*

London will in the future reveal some important archaeological evidence sealed under alluvium: what has been presented here is only a glimpse of the potential that exists.

Acknowledgements

This paper is the synthesis of fruitful collaboration with colleagues in the Greater London Environmental Archaeology Unit and the Department of Greater London Archaeology of the Museum of London. In particular, I would like to thank Julian Bowsher, John Dillon and Helen Jones for allowing me to use information from Phoenix Wharf, Courage's Brewery and Bricklayers Arms prior to publication, and James Rackham and Ian Tyers for helpful comments on earlier drafts of the paper. David Bentley kindly provided Figure 24.1. Figure 24.2 was supplied by A F Howland Associates of Norwich, to whom special thanks are due. All errors and inconsistencies are my own responsibility.

References

Andrews, D. and Merriman, N. 1986. A prehistoric timber structure at Richmond Terrace, Whitehall. *Transactions of the London and Middlesex Archaeological Society*, 37, 17–21.

Bowler, D. and Robinson, M. 1980. Three round barrows at King's Weir, Wytham, Oxon. *Oxoniensia*, 45, 1–8.

Bowsher, J. 1991. A burnt mound at Phoenix Wharf, south-east London: a preliminary report. In M.A. Hodder and L.H. Barfield (eds.), *Burnt Mounds and Hot Stone Technology, Papers from the Second International Burnt Mound Conference, Sandwell, 12–14 October 1990.* Sandwell Metropolitan Borough Council, 11–19.

Branch, N. 1989. Pollen Analysis of the Sediment and Peat Columns from Bricklayers Arms, Southwark, London. Archive Report, Royal Holloway and Bedford New College, Egham.

Devoy, R.J.N. 1979. Flandrian sea level changes and vegetational history of the lower Thames estuary. *Philosophical Transactions of the Royal Society of London*, 285. Series B, 355–407.

Department of the Environment. 1990. *Planning Policy Guidance 16: Archaeology and Planning*. London: Department of the Environment.

Jones, H. 1988. Excavations at Bricklayers Arms Railway Depot Site (Rolls Road, London SE1). Archive Report, Department of Greater London Archaeology, Museum of London.

Lambrick, G. and Robinson, M. 1979. *Iron Age and Roman riverside settlements at Farmoor, Oxfordshire*. London: Council for British Archaeology, Research Report 32.

Lawrence, G.F. 1929. Antiquities from the Middle Thames. *Archaeological Journal*, 86, 69–98.

McGrail, S. 1978. *Log Boats of England and Wales*. Oxford: British Archaeological Reports (British Series) 51.

Merriman, N.J. 1987. A Prehistory for Central London? *The London Archaeologist*, 5 (12), 318–26.

Mills, J. and Whittaker, K. 1991. Southwark. *Current Archaeology*, 124: 155–162.

Needham, S.P. 1991. *Excavation and Salvage at Runnymede Bridge, 1978: the Late Bronze Age waterfront site.* London: British Museum Press.

Nunn, P.D. 1983. The development of the River Thames in Central London. *Transactions of the British Institute of Geographers*, new series, 8, 187–213.

Pryor, F., French, C., and Taylor, M. 1986. Flag Fen, Fengate, Peterborough I: discovery, reconnaissance and initial excavation (1982–5). *Proceedings of the Prehistoric Society*, 52, 1–24.

Robinson, E. 1988. An ostracod fauna from the Bricklayers Arms site. Archive Report, University College London.

Stuiver, M. and Pearson, G.W. 1986. High-precision calibration of the radiocarbon time scale, AD 1950–500 BC. In M. Stuiver and R.S. Kra (eds.), *Proceedings of the Twelfth International Radiocarbon Conference – Trondheim Norway, Radiocarbon*, 28 (2B), 805–838.

Tyers, I. 1988. The prehistoric peat layers. In P. Hinton (ed.) *Excavations in Southwark 1973–76 and Lambeth 1973–79.* London and Middlesex Archaeological Society and Surrey Archaeological Society Joint Publication No 3, 5–12.

Concluding remarks

25

Archaeology under alluvium: human agency and environmental process. Some concluding thoughts.

Martin Bell

Summary: *Some themes which ran through the conference are critically considered with particular reference to the integration of evidence from archaeology and the earth sciences. The difficulties of discriminating between the effects of human activity, climate and other natural factors are discussed and the need for greater chronological precision is highlighted. It is argued that there is a need for greater focus on how past communities perceived and responded to environmental hazards. Recent developments in evaluating the archaeological potential of river valley sites are noted as is the need for an integrated approach to archaeological and natural environmental conservation.*

ARCHAEOLOGICAL FORMATION PROCESSES

Exploration of the common ground between archaeologists and earth scientists is much to be welcomed in Britain where geoarchaeological research has not been given the same emphasis that it has in the United States or in the Netherlands (e.g. Louwe Kooijmans 1991). Archaeologists have become increasingly interested in river valleys because sites and associated palaeoenvironmental evidence are well preserved. Interpretation of the archaeology is inevitably bound up with investigation of the formation processes of the sedimentary context and indeed the distinctive archaeological site formation processes within a river valley (Macklin and Needham, chapter 2).

In focusing on areas where a high quality archaeological record is preserved we must acknowledge that those areas may be unrepresentative of the landscape as a whole. This is not always the case: there are some dry-land sites which have subsequently been inundated and waterlogged, examples being the Neolithic and Bronze Age surfaces buried by increased flooding and alluviation in the Thames Valley during the first millennium BC (Robinson, chapter 19) or the dry-land Neolithic soils and settlements preserved by coastal sedimentation in Essex (Murphy and Wilkinson 1991). Such profound changes highlight the dangers of looking at river valleys, wetlands or coastal habitats as divorced from the neighbouring dry-land resources with which their way of life would originally have been closely integrated (Louwe Kooijmans 1991). Going beyond the individual site, to perception of the wider landscape, is a particular challenge for the future. In attempting to do this valley sites offer some positive advantages. They receive and store sediments, biota and artifacts from a defined catchment and may thus provide information about the wider environment (Macklin and Needham, chapter 2). On the negative side, however, there are problems of interpretation in view of taphonomic factors, the seriousness of which is determined partly by catchment size.

INTEGRATION AND CONTRASTS

The conference heard a succession of papers about some remarkable sites where archaeological and geomorphological data sets had been masterfully integrated. Not so much emphasis was accorded to contrasts and disagreements between data sets which might also have been instructive. Only when these critical issues are highlighted can they become the focus of future research programmes. Scaife and Burrin's research (chapter 8) in Wealden river valleys illustrates this. Their evidence, both for anthropogenic impact in some valleys as early as the Boreal and for significant Bronze Age and Iron Age human activity, represents an interesting mismatch with the traditional archaeological picture which suggests only limited prehistoric activity in that area. There are, however, clear indications that this reflects a lack of archaeological investigation which is, in part, the result of traditional preconceptions of the Weald as uncleared wildwood until Medieval assart.

Detailed surveys are now beginning to reveal much more evidence of prehistoric activity (Tebbutt 1974; Gardiner 1990).

PEOPLE AND CLIMATE

One controversial topic emerged as a sort of sub-theme for the conference. It underpinned many of the papers but curiously was only occasionally directly confronted. This was the issue of the relative importance of anthropogenic and natural factors as causes of changes in river regime. A clear dichotomy emerged between those who, implicitly or explicitly, favoured either climate or human activity. Sometimes the impression was left that this was based more on individual preference than on specific aspects of the data under discussion. There also emerged a political dimension to such preferences. Current concerns about global warming and the environmental crisis mean that palaeoenvironmental studies are at last beginning to enter the realm of major political debate. The Natural Environmental Research Council (N.E.R.C.) has embarked on a programme of research concerned with long-term climatic change. This is a welcome and important development but may lead to a situation in which researchers applying for funding might be tempted to highlight those aspects of the evidence pertinent to the role of climatic factors and give rather less emphasis to human activity. There is a very real need for research designed to disentangle these two factors but that objective cannot be helped by separate funding bodies for work on the natural environment (N.E.R.C.), archaeological science (Science and Engineering Research Council) and indeed, with regard to planning and hazard perception and response (see below), social sciences (Economic and Social Research Council).

A range of climatic parameters such as precipitation (quantity and distribution), evaporation, frozen ground etc affect river regimes. The relationship is mostly clearly defined and easily studied in relation to episodic flooding events. A surprising point which emerged in discussion was the very limited number of detailed published studies of individual British floods. Some speakers (e.g. Dinn and Roseff, chapter 13) mentioned recent events as analogues for the conditions under which their sediments were laid down. Discrimination of individual events in the palaeoenvironmental record and comparative investigations of sediments and biota from past and present floods would be profitable.

Studies which emphasise the role of anthropogenic factors focus on the effect of people in clearing woodland and creating agricultural landscapes, which modern analogues show can have dramatic hydrological effects, leading to substantially increased soil erosion and thus sediment supply. The relationship between past and present soil erosion formed the subject of a one day meeting at the Institute of Archaeology, London on 15 May 1991 which in a sense complemented the present conference by focusing paticularly on erosion processes on the scale of individual fields. That conference is also being published (Bell and Boardman 1992) and like the present volume it aims at the integration of perspectives from archaeology and the earth sciences. Most erosion in Britain today occurs during heavy rainfall events, which, however, only have an effect under specific land-use regimes, particularly those, such as autumn sowing, which create bare soil conditions at vulnerable times of the year (Boardman 1990a). The proportion of the soil which is eroded from agricultural land and ends up in streams and rivers is seldom quantified. Much of that sediment will be sorted and redeposited but some, especially the finer fractions, will be flushed out to sea or redeposited as estuarine sediments. The contribution of soil erosion to stream sediments is further discussed by Brown (1992). It is also illustrated by the sequence at Everley Water Meadow below Hambledon Hill, Dorset, where prehistoric to Romano-British palaeochannel deposits had interleaved layers of chalk granules eroded, probably by rilling, from 'Celtic' fields upslope (Bell 1992). Evans (1990) has attempted to quantify Holocene soil erosion on the basis of the volume of Soil Survey mapped colluvial *and* valley sediments.

Although clearance, combined with agricultural activity, is the most widespread way in which people have affected river regimes, individual conference papers highlighted the significance of more local factors including the possible effects of fish weirs, mills, and other river structures (Salisbury, chapter 14; Clay, chapter 15; Needham, chapter 23). Attention was also directed to the possible effects of the establishment and management of water meadows on alluvial stratigraphy and hydrological conditions as emphasised at the conference by Susan Limbrey (chapter 6), although the date at which this practice originated requires further clarification.

OTHER AGENTS OF CHANGE

In attempting to distinguish the role of people and climate there is the danger that we will ignore the ecological context of these processes and give insufficient consideration to other causes of environmental change. People are not the only faunal agents bringing about purposeful and large-scale environmental changes. The sometimes widespread effects of beavers were highlighted again by Coles (chapter 9) who showed that their effects may be difficult to distinguish from human activity. It is likely that woodland openings, especially fertile beaver lawns left by silted or breached ponds, would have been opportunistically exploited by people. We also need to give greater consideration to the possible effects of other faunal agents such as aurochsen and red deer, especially in ecotonal situations such as river valleys, coastal margins and lake edges where animals may have congregated and thus have helped to create, or maintain, more open habitats. Other natural environmental changes which also need to be considered include sea-level and coastal change in the lower reaches of rivers.

EPISODIC CHANGE

A number of recent studies have identified periods of especially high erosion and/or deposition separated by longer periods of relative stability (e.g. Tipping, chapter 11; Dinn and Roseff, chapter 13). This is true both of studies which favour climatic causation (Starkel 1985; Macklin and Lewin, forthcoming) and those which emphasise the role of human activity (Richter 1986; van Andel *et al* 1990). Wendland and Bryson (1974) even claimed a sequence of globally synchronous Holocene environmental changes in a paper which is, however, marked by an entirely uncritical use of dates and a highly questionable apparent assumption that the major determinant of cultural change is climatic. Papers that make more discriminating and careful use of the dating evidence (e.g. Macklin, Passmore and Rumsby, chapter 12; Macklin and Lewin, forthcoming) have also identified discrete episodes of environmental change some of which correspond to the claimed globally synchronous environmental discontinuities which Wendland and Bryon identified. This leads to Macklin and Lewin's conclusion (Macklin and Lewin, forthcoming) that the processes of river alluviation are climatically driven but culturally blurred. A converse proposition would be that cyclical and stochastic changes continue throughout time but in geographical areas where the fluctuations are not extreme they may only register geomorphologically under specific land-use regimes.

REGIONAL CONTRASTS

Despite hints of synchronous changes over large areas, recent work has also highlighted regional contrasts in the extent to which climatic change produced clear geomorphological signals. Several contributors to the conference noted that work in North America has tended to emphasise the role of climate (Knox 1984). This is also the case with studies in Eastern Europe (Starkel 1985) and Alpine lakes (Joos 1982). The impression is that continental interiors produce clearer palaeoclimatic signals than maritime areas. Rainfall is probably the key factor; in continental areas with low rainfall small changes in precipitation can affect vegetation cover and the boundaries between steppe/prairie and woodland (Webb *et al* 1984) and this produces a detectable signal in the sequence of riverine sediments (Knox 1985). By contrast, in the river sediment sequences of the eastern woodland of North America, the inferred climatic signal does not register to the same extent (Knox 1984). Furthermore, in comparing the European and American evidence we should note that less than 7% of North American pollen diagrams show evidence of anthropogenic impact before European contact (McAndrews 1988). After the advent of European agriculture, widespread palynological changes occur, together with changes of river regime (Butzer 1977) and greatly increased sediment input to lakes (Davis 1976). In Britain, where almost all areas show evidence of at least some anthropogenic impact in pollen diagrams from soon after 5000 BP, it will inevitably be much more difficult to disentangle climatic and anthropogenic effects. The conference drew attention to contrasts between the lowland zone of south and east Britain where, as a generalisation, there is evidence for earlier alluviation than in the west and north, where the main alluviation episodes mostly post-date 3000 BP (Macklin and Needham, chapter 2). This may be explicable in terms of the dates at which widespread clearance occurs. Many parts of the lowland zone show significant human impact on woodland in the Neolithic or Bronze Age whereas many sites in the west and north were only extensively cleared in the Iron Age. For instance, northern England and the Central Valley of Scotland show evidence of major clearance in the two hundred years on either side of the Roman conquest (Turner 1979).

CHRONOLOGY

Our ability to identify the causes of changes in river regime depends very much on the precision of our chronology (Macklin and Needham, chapter 2), in particular on whether changes are coeval regionally and to what extent they coincide with local or regional land-use and social changes. For this reason it is important that we are very clear and frank about the basis of our chronology and precise about its margins of error. In Britain most of the existing dates rest on artifactual evidence, which is often rather imprecise, or more commonly on radiocarbon. With the latter the attendant problems of standard deviations, calibration issues, possible reworking and the dating of heartwood, etc are likely to mean that some existing dates are only reliable to within a few hundred years. In investigations of the chronology of environmental change it might be a useful and salutary exercise to point-score the reliability of individual dates rather as Mead and Meltzer (1985) have done in investigations of late Pleistocene megafaunal extinctions.

The greatest potential for investigating relationships between environmental and social change lies with dendrochronology, which was not considered in detail at the conference. The potential of this technique has been highlighted by the linking of tree-ring sequences across Europe (Pilcher *et al* 1984). Its potential is admittedly limited to those areas where sufficient waterlogged wood of the right size and tree type is preserved. These are, however, the very situations where, because of either aridity or flooding, etc., human activity is particularly subject to the vicissitudes of environmental change. Highly successful investigations have taken place in the American South West (Dean 1988) where dendrochronology provides a framework for study of the relationship between climatic change, river regime and the settlement histories of pueblos (Gumerman 1988). The potential for addressing such issues in a British context is illustrated by the recent dating of the Sweet Track, Somerset to winter/spring 3807–6 BC (Hillam *et al* 1990). The trackway itself represents evidence of increased wetness which corresponds to a period between 4000–3800 cal BC. At this time there is other proxy evidence for wetter conditions in the form of floodplain oaks in the river Trent and the drowning of coastal submerged forests (Morgan *et al* 1987), which is likely to have occurred during episodes of increased storminess. Current experimental research on the application of luminescence dating and palaeomagnetism is described by Bailiff (chapter 3) and Clark (chapter 4) respectively. These techniques are important in facilitating the dating of inorganic sediment sequences but in the immediate future they are not likely to offer the chronological precision to confront most issues of causation.

HAZARD PERCEPTION AND RESPONSE

Frequently during the conference the tacit assumption was made that a particular change in river regime produced a human response. Such reasoning can easily become uncritically deterministic. The underlying assumption sometimes is that a particular set of environmental parameters will give rise to only one human response. That ignores the remarkable human capacity for planned independent long-term action. Hazards, such as floods, erosion, storms, etc. may be perceived and responded to in very different ways depending on the social context in which they occur. Hazards are after all essentially defined in relation to a subject — that which they affect. The extent to which hazard is socially defined is emphasised by the role of coping mechanisms, that is social practices such as reciprocity, sharing, storage, mobility, diversification etc, which help to buffer societies against environmental change. Watts (1983) emphasised that traditional coping mechanisms of the nineteenth century and earlier in northern Nigeria were effective in reducing the effects of episodic drought. However, under the monetary economy of colonial administration even small-scale rainfall changes had much more dramatic social repercussions. In other words the perceived hazard has increased because social mechanisms are less able to cope with the effects of rainfall variation.

Research on present-day flood hazard (White 1974; Smith and Tobin 1979) cannot provide a neat set of formulae for the interpretation of past events. It can, however, help to focus our thoughts on the range of possible human reactions to flooding. Examples are the Lynmouth flood of 1952 (Kidson 1953), the Louth flood of 1926 (Latter 1932) and the Till flood of 1841 (Cross 1967). Each tells a remarkable story of human adaptability and charitable response by communities beyond the affected area, which again warns us against interpretations based on simplistic determinism.

Archaeological approaches to these issues could clearly benefit greatly from research on environmental perception (Saarinen *et al* 1984), behavourism (Boal and Livingstone 1989) and hazard research (Burton *et al* 1978; Hewitt 1983). Recently archaeologists have begun to address issues of hazard and coping strategy, albeit from a palaeoeconomic point of view (Halstead and O'Shea 1989). The inference of hazard from the archaeological record is manifestly not a straightforward matter. We can formulate theoretical frameworks informed by present-day hazards research, the ethnohistorical record and social theory. Ultimately, however, these theories need to be tested against the evidence for a behavioural response attested in the archaeological record.

In considering these problems, the timescale of human decision-making emerges as an important factor. Ways of life and coping strategies can be seen as adapted to environmental change within certain perceived limits. Periodically those limits will be exceeded as a result of stochastic events and secular trends. Human perception of the likelihood of change will be related to the duration and perceived recurrence interval of known past changes. It is logical to suggest that the timescale of this knowledge will vary from one society to another. It will depend on the length of the individual and community memory and thus on mechanisms for the transmission of knowledge from one generation and geographical area to the next. An ethnohistorical example illustrating the duration of memory of geomorphological change is the oral legend of a volcanic eruption in Papua New Guinea between 1630 and 1670; this event was perceived in very different ways, positive as well as negative, by the various communities which it affected (Blong 1982). Another example is the memory of the native American site of Ozette which was engulfed by a mud slide around 1750 AD (Coles and Coles 1989; Gleeson and Grosso 1976). Such situations where we have both detailed geomorphological evidence of environmental change and an ethnohistorical record of its human perception are particularly important in providing opportunities to compare and contrast palaeoenvironmental and social perspectives.

The extent to which it is possible to address issues of hazard, response and their timescales, varies according to the quality of the archaeological and palaeoenvironmental records and the precision of their chronology. On these counts river valleys, along with other wetlands, emerge as key foci for future research.

RIVER VALLEYS AS A CULTURAL RESOURCE

Many of the papers, particularly Lambrick (chapter 20) noted that river valleys represent an archaeological and palaeoenvironmental resource under severe pressure from quarrying and other activities. Management of that resource by planning authorities and others requires an adequate level of knowledge. In some sections of river valleys long archaeological sequences are preserved by sedimentation whilst in others the archaeology has been destroyed by erosion. Lewin (1983) has shown that some Welsh rivers have reworked virtually their entire floodplain in some stretches over the last century. Where adequate knowledge of riverine geomorphology is available it is becoming possible to distinguish between sediments of a catchment where most of the early evidence may have been destroyed and others with the potential for preservation (Lewin, chapter 10). It must at the same time be acknowledged that even fairly recent deposits, which are unimportant in terms of traditional archaeology, can nonetheless provide vital palaeoenvironmental records. Adequate knowledge depends partly on the information amassed by regular observation, which may even produce a detailed picture of river changes in such built-up areas as London (Merriman, chapter 24) and Lyon, as Jean-Paul Bravard described at the conference. Particularly impressive is the range of evidence from the Trent valley (Salisbury, chapter 14). Even in this well-studied area it is noteworthy that sites as late as the Medieval period may be so deeply buried as to be archaeologically invisible until they are in the process of destruction (Clay, chapter 15). Where it is necessary to evaluate the archaeological potential of large areas of river valleys, current developments in airborne remote sensing techniques may have much to offer, as described at the conference by Jenny Allsop and D. Greenbaum.

There is also a pressing need to refine smaller-scale methodologies for dealing with river valleys and other areas of deeply buried archaeology. Dinn and Roseff (chapter 13) demonstrate how detailed research and assessment of one site can provide the basis for development of more widely applicable evaluation techniques. Environmental assessments are becoming an increasingly important part of archaeological work in response to recent legislation complying with an EEC directive. Rick Turner discussed these developments at the conference on the basis particularly of recent experience in the Severn Estuary (Severn Estuary Levels Research Committee 1991; 1992). He emphasised the importance of integrating assessment work within the broad discipline of archaeology by publishing accounts of the methodology and results of assessments. Equally important is the integration of individual assessments within more broadly based academic research programmes for the regions or types of site concerned.

As archaeologists have shifted focus away from the isolated site and towards landscapes and people/environment relationships, so we have precipitated something of an identity crisis with regard to what we mean by archaeological sites. Clearly this has legislative and planning implications. Pleas were made at the conference for river sediments to be regarded as part of the archaeological record (Macklin and Needham, chapter 2). Most British river valley sequences will contain at least some information about past human activity, whether or not they contain significant individual sites and artifacts. Perhaps the importance of such sequences should be evaluated, not just in archaeological terms, but as part of a wider resource relevant both to the natural landscape and to interactions involving people. There is common ground here with the recently reorganised bodies responsible for nature conservation (Lambrick 1985) and its more active cultivation would be very much in the spirit of the interdisciplinary theme so successfully developed by the conference.

References

Andel, T.H. van, Zangger, E. and Demitrack, A. 1990. Landuse and soil erosion in prehistoric Greece, *Journal of Field Archaeology*, 17, 379–396.

Bell, M. 1992. The prehistory of soil erosion. In M. Bell and J. Boardman (eds.), *Past and Present Soil erosion*, 21–35. Oxford: Oxbow Books.

Bell, M. and Boardman, J. 1992. *Past and Present Soil Erosion*. Oxford: Oxbow Books.

Blong, R.J. 1982. *The Time of Darkness*. Canberra: Australia National University Press.

Boal, F.W. and Livingstone, D.N. (eds.) 1989. *The Behavioural Environment: essays in reflection, application and re-evaluation*. London: Routledge.

Boardman, J. 1990a. Soil erosion on the South Downs: a review. In J. Boardman, I.D.L. Foster and J.A. Dearing (eds.), *Soil Erosion from Agricultural Land*, 87–105. Chichester: John Wiley.

Boardman, J. 1990b. *Soil Erosion in Britain: costs, attitudes and policies*. Falmer: Educational network for environment and development, Sussex University.

Brown, A.G. 1992. Slope erosion and colluviation at the floodplain edge. In M. Bell and J. Boardman (eds.), *Past and Present Soil Erosion*, 77–87. Oxford: Oxbow Books.

Burton, I., Kates, R.W. and White, G.F. 1978. *The Environment as Hazard*. New York: Oxford University Press.

Butzer, K.W. 1977. *Geomorphology of the Lower Illinois Valley as a spatial-temporal context for the Koster Archaic site*. Reports of Investigations of the Illinois State Museum, no 34.

Coles, B. and Coles, J. 1989. *People of the Wetlands*. London: Thames and Hudson.

Cross, D. 1967. The great Till flood of 1841. *Weather*, 22, 430–433.

Davis, M.B. 1976. Erosion rates and land use history in southern Michigan. *Environmental Conservation*, 3, 139–48.

Dean, J.S. 1988. Dendrochronology and palaeoenvironmental reconstruction on the Colorado Plateaus. In G.J. Gumerman (ed.), *The Anasazi in a Changing Environment*, 119–167. Cambridge: University Press.

Evans, R. 1990. Soil erosion: its impact on the English and Welsh landscape since woodland clearance. In J. Boardman, I.D.L.Foster and J.A. Dearing (eds.), *Soil Erosion on Agricultural Land*, 231–254. Chichester: Wiley.

Gardiner, M. 1990. The archaeology of the Weald: a survey and a review. *Sussex Archaeological Collections*, 128, 33–53.

Gleeson, P. and Grosso, G, 1976. Ozette site. In D.R. Croes (ed.), *The Excavation of Water Saturated Archaeological Sites (Wet Sites) on the north-west coast of North America*. Ottawa: National Museum of Man, Mercury Series, Archaeological Survey of Canada Paper 50.

Gumerman, G.J. (ed.) 1988. *The Anasazi in a Changing Environment*. Cambridge: University Press.

Halstead, P. and O'Shea, J. (eds.) 1989. *Bad Year Economics: cultural responses to risk and uncertainty*. Cambridge: University Press.

Hewitt, K. (ed.) 1983. *Interpretations of Calamity*. Boston: Allen and Unwin.

Hillam, J., Groves, C.M., Brown, D.M., Baillie, M.G.L., Coles J.M. and B.J. 1990. Dendrochronology of the English Neolithic. *Antiquity*, 64, 210–220.

Joos, M. 1982. Swiss Midland lakes and climatic changes. In A.F. Harding (ed.), *Climatic Change in Later Prehistory*, 44–51. Edinburgh: University press.

Kidson, C. 1953. The Exmoor storm and the Lynmouth floods. *Geography*, 38, 1–9.

Knox, J.C. 1984. Response of river systems to Holocene climates. In H.E. Wright Jnr. (ed.), *Late Quaternary Environments of the United States*, 2, The Holocene, 26–41. London: Longman.

Knox, J.C. 1985. Response of floods to Holocene climatic change in the Upper Mississippi Valley. *Quaternary Research*, 23, 287–300.

Lambrick, G. 1985. *Archaeology and Nature Conservation*. Oxford: University Department for External Studies.

Latter, P.R. 1932. The Louth Flood. *The Lincolnshire Magazine*, 1 (2), 73–77.

Lewin, J. 1983. Changes of channel patterns and floodplains. In K.J. Gregory (ed.), *Background to Palaeohydrology*, 303–319. Chichester: Wiley.

Louwe Kooijmans, L.P. 1991. Wetland exploitation and upland relations of prehistoric communities in the Netherlands. *East Anglian Archaeology*, 50.

Macklin, M. and Lewin, J. Forthcoming. Holocene river alluviation in Britain. In, I. Douglas and J. Hagedorn (eds.), *Geomorphology and Geoecology, Fluvial Geomorphology*. Zeitschrift für Geomorphologie (supplement) 85.

McAndrews, J.H. 1988. Human disturbance of North American forests and grasslands: the fossil pollen record. In B. Huntley and T. Webb (eds.), *Vegetation History*, 674–697. Dordrecht: Kluwer Academic Publishers.

Mead, J.I. and Meltzer, D.J. (eds.) 1985. *Environments and extinctions: Man in late-glacial North America*. Orono: Centre for the Study of Early Man.

Morgan, R.A., Litton, C.D. and Salisbury, C.R. 1987. Trackways and tree trunks — dating Neolithic oaks in the British Isles. *Tree Ring Bulletin*, 47, 61–69.

Murphy, P. and Wilkinson, T. 1991. Survey and excavation on the tidal foreshore zone. In J.M. Coles and D.M. Goodburn (eds.), *Wet Site Excavation and Survey*, 10–15. Exeter: Wetland Archaeology Research Project.

Pilcher, J.R., Baillie, M.G.L., Schmidt, B. and Becker, B. 1984. A 7272-year tree-ring record for western Europe. *Nature*, 312, 150–152.

Richter, G. 1986. Investigations of soil erosion in Central Europe. In C.P. Burnham and J.I. Pitman (eds.), *Soil Erosion*, 14–27. London: South East England Soils Discussion Group 3.

Saarinen, T.F., Seamon, D. and Sell, J.L. 1984. *Environmental Perception and Behaviour: an inventory and prospect*. Chicago: University Department of Geography, Research Paper 209.

S.E.L.R.C. 1991. *Severn Estuary Levels Research Committee 1991*. Lampeter: St. David's University College.

S.E.L.R.C. 1992. *Severn Estuary Levels Research Committee 1992*. Lampeter: St. David's University College.

Smith, K. and Tobin, G.A. 1979. *Human Adjustment to the Flood Hazard*. London: Longman.

Starkel, L. 1985. Lateglacial and postglacial history of river valleys in Europe as a reflection of climatic changes. *Zeitschrift für Gletscherkunde und Glazialgeologie*, 21, 159–64.

Tebbutt, C.F. 1974. The prehistoric occupation of the Ashdown Forest area of the Weald. *Sussex Archaeological Collections*, 112, 34–43.

Turner, J. 1979. The Environment of North East England during Roman times as shown by pollen analysis. *Journal of Archaeological Science*, 6, 285–290.

Watts, M. 1983. On the poverty of theory: natural hazards research in context. In K. Hewitt (ed.), *Interpretations of Calamity*, 231–62. Boston: Allen and Unwin.

Webb. T.III, Cushing, E.J. and Wright, H.E. Jnr. 1984. Holocene changes in the vegetation of the Midwest. In H.E. Wright Jnr. (ed.), *Late Quaternary Environments of the United States*, 2, The Holocene, 142–165. London: Longman.

White, G.F. 1974. *Natural Hazards: Local, national, global*. New York: Oxford University Press.

Wendland, W.M. and Bryson, R.A. 1974. Dating climatic episodes of the Holocene. *Quaternary Research*, 4, 9–24.

Index